International Symposium of Cavitation and Multiphase Flow

(ISCM 2014)

IOP Conference Series: Materials Science and Engineering
Volume 72

Beijing, China
18 - 21 October 2014

ISBN: 978-1-63439-883-1
ISSN: 1757-8981

Printed from e-media with permission by:

Curran Associates, Inc.
57 Morehouse Lane
Red Hook, NY 12571

Some format issues inherent in the e-media version may also appear in this print version.

Printed by Curran Associates, Inc. (2015)

Institute of Physics
Dirac House, Temple Back
Bristol BS1 6BE UK

Phone: 44 1 17 929 7481
Fax: 44 1 17 920 0979

techtracking@iop.org

Additional copies of this publication are available from:

Curran Associates, Inc.
57 Morehouse Lane
Red Hook, NY 12571 USA
Phone: 845-758-0400
Fax: 845-758-2634
Email: curran@proceedings.com
Web: www.proceedings.com

Preface

Forum 1 - Invited Lectures

Forum 2 - Hydrodynamic Cavitation

Forum 3 - Pump Cavitation

Forum 4 - Interdisciplinary Research on Cavitation and Multiphase Flow

Forum 5 - Acoustic Cavitation

Forum 6 - Super Cavitation

IOP Conf. Series: Materials Science and Engineering **72** (2015) 011001 doi:10.1088/1757-899X/72/1/011001

Foreword

The International Symposium on Cavitation and Multiphase Flow (ISCM 2014) was held in Beijing, China during 18th-21th October, 2014, which was jointly organized by Tsinghua University, Beijing, China and Jiangsu University, Zhenjiang, China. The co-organizer was the State Key Laboratory of Hydroscience and Engineering, Beijing, China.

Cavitation and multiphase flow is one of paramount topics of fluid mechanics with many engineering applications covering a broad range of topics, e.g., hydraulic machinery, biomedical engineering, chemical and process industry. In order to improve the performances of engineering facilities (e.g. hydraulic turbines) and to accelerate the development of techniques for medical treatment of serious diseases (e.g. tumors), it is essential to improve our understanding of cavitation and Multiphase Flow. For example, the present development towards the advanced hydrodynamic systems (e.g. space engine, propeller, hydraulic machinery system) often requires that the systems run under cavitating conditions and the risk of cavitation erosion needs to be controlled.

The purpose of the ISCM 2014 was to discuss the state-of-the-art cavitation and multiphase flow research and their up-to-date applications, and to foster discussion and exchange of knowledge, and to provide an opportunity for the researchers, engineers and graduate students to report their latest outputs in these fields. Furthermore, the participants were also encouraged to present their work in progress with short lead time and discuss the encountered problems.

ISCM 2014 covers all aspects of cavitation and Multiphase Flow, e.g., both fundamental and applied research with a focus on physical insights, numerical modelling and applications in engineering. Some specific topics are:

Cavitating and Multiphase Flow in hydroturbines, pumps, propellers etc
Numerical simulation techniques
Cavitation and multiphase flow erosion and anti-erosion techniques
Measurement techniques for cavitation and multiphase flow detection
Fluid-structure interaction induced by cavitation and multiphase flow
Multi-scale modelling of cavitating flows and Multiphase Flow
Cavitation nuclei: theory and experiments
Supercavitation and its applications
Synergetic effects of cavitation and silt-laden erosion
Shock waves and microjets generated by cavitation
Nonlinear oscillations of gas and vapour bubbles
Fundamentals of physics of acoustic cavitation
Sonochemistry and sonoluminescence

IOP Conf. Series: Materials Science and Engineering **72** (2015) 011001 doi:10.1088/1757-899X/72/1/011001

Biomedical applications of cavitation effects
Ultrasonic cavitation for molten metal treatment
Cavitation for enhanced heat transfer

The ISCM 2014 brought together 95 scientists, researchers and graduate students from 11 countries, affiliated with universities, technology centers and industrial firms to debate topics related to advanced technologies for cavitation and Multiphase Flow, which would enhance the sustainable development of cavitation and Multiphase Flow in interdisciplinary sciences and technology. The technical committee selected 54 technical papers on the following topics: (i) Hydrodynamic Cavitation, (ii) Super Cavitation, (iii) Pump Cavitation, (iv) Acoustic Cavitation, (v) Interdisciplinary Research of Cavitation and Multi-Phase Flows, and 13 invited plenary and invited forum lectures, which were presented at the symposium, to be included in the proceedings.

All the papers of ISCM 2014, which are published in this Volume of IOP Conference Series: Materials Science and Engineering, had been peer reviewed through processes administered by the editors of the ISCM 2014, those are Yulin WU, Shouqi YUAN, Zhengwei WANG, Shuhong LIU, Xingqi LUO, Fujun WANG and Guoyu WANG. The papers published in this Volume include 54 technical papers and 3 full length texts of the invited lectures.

We sincerely hope that the International Symposium on Cavitation and Multiphase Flow is a significant step forward in the world wide efforts to address the present challenges in the modern science and technology.

Professor Yulin WU
Chairman of the Local Organizing Committee
International Symposium on Cavitation and Multiphase Flow (ISCM 2014)
October, 2014

IOP Conf. Series: Materials Science and Engineering **72** (2015) 011002 doi:10.1088/1757-899X/72/1/011002

International Symposium of Cavitation and Multiphase Flow (ISCM 2014)

Tsinghua University, Beijing, China

October 18th – 21th, 2014

Qinghua Garden, 100084, Beijing, China

International Advisory Committee

Principals:
Xiaoxing PENG (China Ship Scientific Research Center, China)
Wei SHYY (Hong Kong University of Science and Technology, China)
Members:
Yu AN (Tsinghua University, China)
Jacques André ASTOLFI (Institut de Recherche de l'Ecole Navale, France)
Gerard BOIS (Laboratoire de Mécanique de Lille-UMR CNRS 8107, France)
Sönke BRODERSEN (KSB AG, Germany)
Xavier ESCALER (Politech. Univ. of Catalunya, Spain)
Dmitry ESKIN (Brunel University, UK)
Jianren FAN (Zhejiang University, China)
Jean-Pierre FRANC (LEGI, France)
Eric GONCALVES (LEGI, France)
Fengqing HAN (South China University of Sci. & Tech., China)
D-H HELLMANN (KSB AG, Germany)
Chenguang HUANG (Institute of Mechanics, Chin. Acad. Sci., China)
Toshiaki KANEMOTO (Kyushu Institute of Technology, Japan)
Smaine KOUIDRI (Université Pierre et Marie Curie (Paris 6), France)
Peter LEE (University of Manchester, UK)
Shengcai LI (University of Warwick, UK)

IOP Conf. Series: Materials Science and Engineering **72** (2015) 011002 doi:10.1088/1757-899X/72/1/011002

International Technical Committee

Guoyi PENG (Nihon University, Japan)
Ehsan ROOHI (Ferdowsi University of Mashhad, Iran)
Xueming SHAO (Zhejiang University, China)
Stephen SHAW (Xi'an Jiaotong-Liverpool University, China)
Qinghua SHI (Dongfang Electric Machinery Co., Ltd, China)
William STRAKA (Pennsylvania State University, USA)
Tom van TERWISGA (Netherlands & Delft Technical Univ. Netherland)
Qianxi WANG (University of Birmingham, UK)
Satoshi WATANABE (Kyushu University, Japan)
Xianzhu WEI (Harbin Electric Machinery Co., Ltd, China)
Haizhen XIAN (North China Electric Power University, China)
Jinliang XU (North China Electric Power University, China)
Baoshan ZHU (Tsinghua University, China)

Organizing Committee

Chairmen: Yulin WU (Tsinghua University)
Shouqi YUAN (Jiangsu University)
Executive Chairman: Zhengwei WANG (Tsinghua University)
Co-Chairmen:
Yu AN (Tsinghua University)
Huashu DOU (Zhejiang Sci-Tech University)
Jianzhong LIN (Zhejiang University)
Shuhong LIU (Tsinghua University)
Houlin LIU (Jiangsu University)
Xingqi LUO (Xi'an University of Technology)
Fujun WANG (China Agriculture University)
Guoyu WANG (Beijing Institute of Technology)
Members:
Shuliang CAO (Tsinghua University)
Jianjun FENG (Xi'an University of Technology)
Rennian LI (Lanzhou University of Technology)
Fengchen LI (Harbin Institute of Technology)
Chao LIU (Yangzhou University)
Li LU (China Inst. of Water Resources and Hydropower Research)
Xuelin TANG (China Agriculture University)
Zhongdong QIAN (Wuhan University)
Daqing QIN (Harbin Electric Machinery Co., Ltd)
Dezhong WANG (Shanghai Jiaotong University)
Dazhuan WU (Zhejiang University)
Minguan YANG (Jiangsu University)
Yongzhi ZHAO (Dongfang Electric Machinery Co., Ltd)
Zuchao ZHU (Zhejiang Sci-Tech University)
Yuning ZHANG (North China Electric Power University)

IOP Conf. Series: Materials Science and Engineering **72** (2015) 011002 doi:10.1088/1757-899X/72/1/011002

Forum: Acoustic Cavitation
Chairmen:Yuning ZHANG, Yu AN, Weizhong CHEN
Forum: Hydrodynamic Cavitation
Chairmen: Guoyu WANG, Xianwu LUO
Forum: Corrosion, Erosion and Abrasion
Chairmen: Xingxing HUANG, Lingjiu ZHOU
Forum: Super Cavitation
Chairman: Huashu DOU
Forum: Cavitation Enhanced Heat and Mass Transfer
Chairmen: Yuning ZHANG, Haizhen XIAN
General Secretaries
Xianwu LUO (Tsinghua University, luoxw@mail.tsinghua.edu.cn)
Hong LI (Jiangsu University, hli@ujs.edu.cn)
Secretariat Members:
Bin JI and Qi WU
Dept. of Thermal Eng., Tsinghua University, Beijing 100084, China
E-mail: iscm@tsinghua.edu.cn
Phone: +86-10-62782251, +86-10-62794735
Fax: +86-10-62782251, +86-10-62794735

IOP Conf. Series: Materials Science and Engineering **72** (2015) 011003 doi:10.1088/1757-899X/72/1/011003

International Symposium on Cavitation and Multiphase Flow (ISCM 2014) – Photographs

IOP Conf. Series: Materials Science and Engineering **72** (2015) 011003 doi:10.1088/1757-899X/72/1/011003

IOP Conf. Series: Materials Science and Engineering **72** (2015) 011003 doi:10.1088/1757-899X/72/1/011003

IOP Conf. Series: Materials Science and Engineering 72 (2015) 011003 doi:10.1088/1757-899X/72/1/011003

ISCM 2014
International Symposium of Cavitation and Multiphase Flow 2014
Opening Ceremony

ISCM 2014
International Symposium of Cavitation and Multiphase Flow 2014
Opening Ceremony

Open
INTERNATIONAL SYMPOSIUM OF CAVITATION
AND MULTIPHASE FLOW 2014

Cavitation reg
Secondary flows
INTERNATIONAL SYMPOSIUM OF CAVITATION
AND MULTIPHASE FLOW 2014

IOP Conf. Series: Materials Science and Engineering **72** (2015) 011003 doi:10.1088/1757-899X/72/1/011003

IOP Conf. Series: Materials Science and Engineering **72** (2015) 011003 doi:10.1088/1757-899X/72/1/011003

IOP Conf. Series: Materials Science and Engineering **72** (2015) 011003 doi:10.1088/1757-899X/72/1/011003

IOP Conf. Series: Materials Science and Engineering **72** (2015) 011003 doi:10.1088/1757-899X/72/1/011003

IOP Conf. Series: Materials Science and Engineering **72** (2015) 011003 doi:10.1088/1757-899X/72/1/011003

IOP Conf. Series: Materials Science and Engineering **72** (2015) 011003 doi:10.1088/1757-899X/72/1/011003

The Simulation

IOP Conf. Series: Materials Science and Engineering **72** (2015) 011003 doi:10.1088/1757-899X/72/1/011003

4.Result
ISCM 2014
INTERNATIONAL SYMPOSIUM OF CAVITATION
AND MULTIPHASE FLOW 2014
CAVITATION AND MULTIPHASE FLOW

IOP Conf. Series: Materials Science and Engineering **72** (2015) 011003 doi:10.1088/1757-899X/72/1/011003

IOP Conf. Series: Materials Science and Engineering **72** (2015) 011003 doi:10.1088/1757-899X/72/1/011003

INTERNATIONAL SYMPOSIUM OF CAVITATION
AND MULTIPHASE FLOW 2014

INTERNATIONAL SYMPOSIUM OF CAVITATION
AND MULTIPHASE FLOW 2014

IOP Conf. Series: Materials Science and Engineering **72** (2015) 011003 doi:10.1088/1757-899X/72/1/011003

ISCM 2014 Beijing
International Symposium of Cavitation and Multiphase Flow
Beijing, China, 2014
Bubble jet impact on a rigid wall of different stand-off parameters
Li Shuai
College of Shipbuilding Engineering
Harbin Engineering University
Harbin 150001, China
INTERNATIONAL SYMPOSIUM OF CAVITATION AND MULTIPHASE FLOW 2014

IOP Conf. Series: Materials Science and Engineering **72** (2015) 011003 doi:10.1088/1757-899X/72/1/011003

清華大学
Numerical Simulation of
Unsteady Flow Characteristics
for Cavitation around a 3-D Hydrofoil
Institute of Fluid Machinery and Fluid Engineering, Tsinghua University
INTERNATIONAL SYMPOSIUM OF CAVITATION
AND MULTIPHASE FLOW 2014

IOP Conf. Series: Materials Science and Engineering **72** (2015) 011003 doi:10.1088/1757-899X/72/1/011003

1. Introduction
Blade inlet angles
Hub diameters
Incidence angle
passage throat area
Blade angle distribution
Inlet incidence angle and relative velocity
Previous Research:
Actually, the blade profile affects directly on the development of low pressure zone in suction surface and the motion of cavities region.
Yongyao

IOP Conf. Series: Materials Science and Engineering **72** (2015) 011003 doi:10.1088/1757-899X/72/1/011003

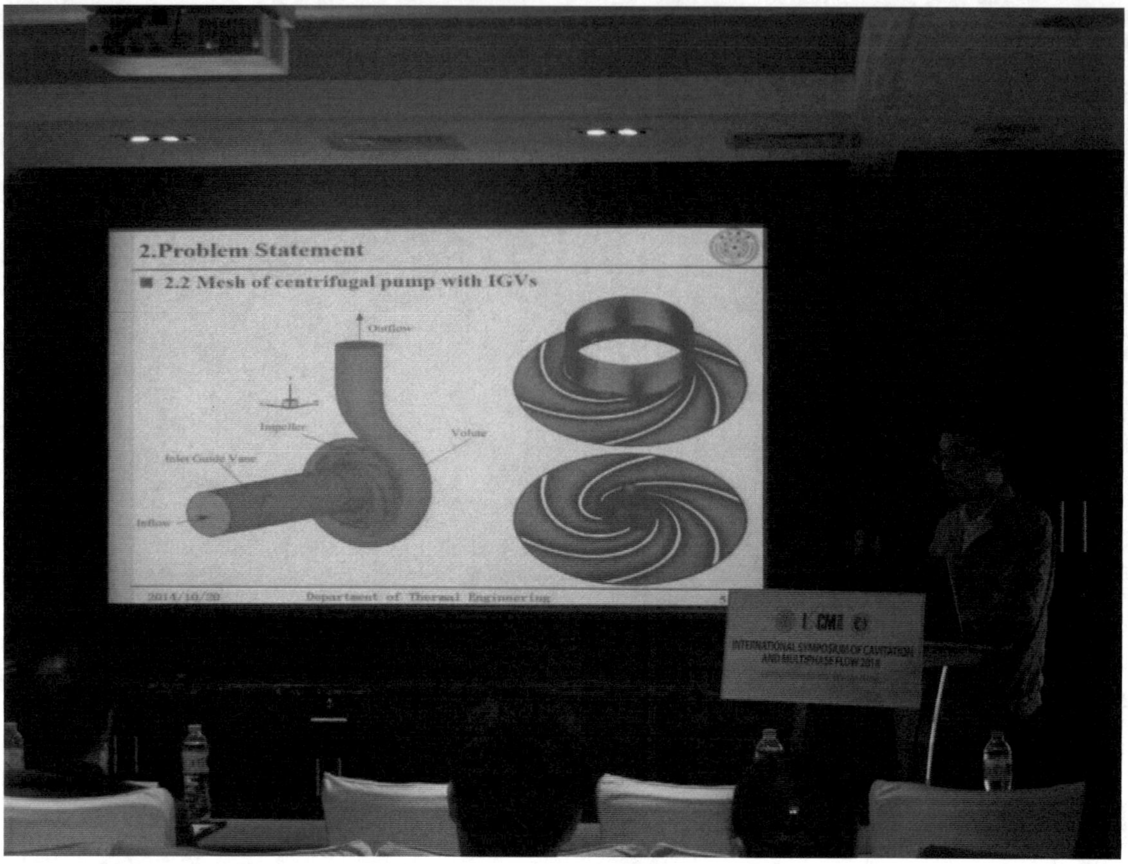

IOP Conf. Series: Materials Science and Engineering **72** (2015) 001004 doi:10.1088/1757-899X/72/0/001004

Peer review statement

All papers published in this volume of *IOP Conference Series: Materials Science and Engineering* have been peer reviewed through processes administered by the proceedings Editors. Reviews were conducted by expert referees to the professional and scientific standards expected of a proceedings journal published by IOP Publishing.

The reviewers who selected and reviewed the papers included in the Proceedings of International Symposium of Cavitation and Multiphase Flow (ISCM 2014) are:

Yulin WU (Tsinghua University)

Shouqi YUAN (Jiangsu University)

Zhengwei WANG (Tsinghua University)

Huashu DOU (Zhejiang Sci-Tech University)

Shuhong LIU (Tsinghua University)

Xingqi LUO (Xi'an University of Technology)

Fujun WANG (China Agriculture University)

Guoyu WANG (Beijing Institute of Technology)

Leqin WANG (Zhejiang University)

Dazhuan WU (Zhejiang University)

Bin JI (Tsinghua University)

Yuning ZHANG (North China Electric Power University)

IOP Conf. Series: Materials Science and Engineering **72** (2015) 012001 doi:10.1088/1757-899X/72/1/012001

Two-dimensional direct numerical simulation of bubble cloud cavitation by front-tracking method

G Peng[1], G Tryggvason[2] and S Shimizu[1]
[1] Department of Mechanical Engineering, College of Engineering, Nihon University, Koriyama, Fukushima 963-8642, JP
[2] Department of Aerospace and Mechanical Engineering, University of Notre Dame, Notre Dame, South Bend, IN 46556-5637, USA

E-mail: peng@mech.ce.nihon-u.ac.jp

Abstract. Unsteady bubble cloud cavitation phenomenon caused by negative pressure pulse has been treated numerically by applying a front tracking method. The behaviour of bubble cloud expanding and contracting is evaluated by tracking the motion of all bubble interfaces. Numerical investigation demonstrates that: (1) In the collapsing of bubble cloud micro liquid jets toward the inner bubbles are formed while the outer layer bubbles contract extremely, and then a high impact pressure is released when the inner central bubble contacts to its minimum. (2) The oscillation of bubble cloud depends upon the void fraction greatly. In the case of high void fraction, the frequency of cloud oscillation is lower than that of individual bubble and the decay of the oscillation becomes much slowly also.

1. Introduction

Cavitation is an important phenomenon often observed in variety of hydraulic systems such as nozzles and hydraulic machineries, where small vapour bubbles are formed when the surrounding liquid pressure drops sufficiently low in some region of relative high velocity [1]. Corresponding to the variation of surrounding pressure these bubbles expand and contract in a very short time and the impact pressure released at bubble collapsing may reach to several thousand atmospheric pressures [2]. Experimental investigations has been demonstrated that when a cloud of bubbles collapse coherently it results in much greater material damage and much strong noise compared to cumulative effect resulted by collapsing of the individual bubbles made up the cloud [3]. That is to say, impact pressures released at collapsing of bubble cloud are much higher than that released at collapsing of one single bubble. In order to utilize the high cavitation impact actively high-speed water jets have been developed and widely applied [4, 5]. Recently, active utilizations of cavitation extended to various fields such as ultrasound medical applications and drag delivery for cancer therapy [6]. However, the precise physical phenomena involved in cloud cavitation have not yet been completely identified. For clarifying the mechanism of cavitation phenomena it is important to understand the collapse and interaction of multiple bubbles. Generally, two approaches are developed to treat the dynamics of bubble cloud. One deals with the macroscopic motions of bubbles by using the continuum two-fluid method, in which the motions of liquid and bubbles are treated separately by interactively solving two sets of governing equations derived for liquid and gas phases [7, 8]. Usually, the volume change of bubbles is described by the Rayleigh-Plesset equation or similar ones and the exchanges of mass, momentum, and energy are estimated explicitly as transfer terms. However, the transfer terms depends upon the estimation of the interaction between the liquid and bubbles, which is difficult since there is not a general physical model describing the interaction of liquid and bubbles [9]. The other one is to analyze the microscopic motions of individual bubble directly by tracking the motion of sharp bubble surfaces [10]. For the high requirement of computation resource for solving Navier–Stokes equations, the direct numerical simulations of cavitation bubbles are very limited. Popinet & Zaleski [11] studied the effects of viscosity on creation of re-entrant jets during collapse of a cavitation bubble near a rigid boundary. Dabiri et al. [12] studied the deformation of a cavitation bubble in shear flows.

Under this background, this work makes a try to treat bubble cloud cavitation phenomenon. Focused on the mutual interaction of bubbles cavitation of a bubble cloud caused by negative pressure pulse is simulated numerically by applying the front tracking method. By tracking the motion of all bubble interfaces the behavior of bubble cloud expanding and contracting is simulated and the coherent structure of bubble oscillation is then analyzed.

2. Method of DNS simulation and computational setup of bubble cloud

Cavitation phenomena of bubble cloud are concerned in consideration of mutual interactions among multi bubbles. So, the behaviour of a bubble cloud responding to the action of a negative pressure pulse simulated directly by applying the front tracking method developed by Tryggvason et al. [13-15]. The density field is expressed uniformly by using a Heaviside function defined along fluid interface, which is calculated by tracking the marker points on the front. The working fluids are treated as viscous ones and the surface tension acted on bubble interfaces is counted. In order to simplify the complex cavitation phenomena the following assumptions are adopted. (1) The compressibility of liquid phase is negligible compared to that of the gas phase including in bubbles. (2) Bubbles are consisted of non-condensable gas, which obeys the perfect gas law. (3) Pressure inside a bubble is uniformly distributed. (4) Coalescence and fragmentation of bubbles do not happen in the present process. With above assumptions one set of flow governing equations is written for both liquid and gas phases involved, and the phase boundary is treated as an imbedded interface by adding the appropriate source terms to the conservation laws. These source terms are in the form of delta-functions localized at the interface and are selected in such a way to satisfy the correct matching conditions at the phase boundary. For both of the incompressible liquid and compressible gas phases, the momentum conservation equation is given as follows:

$$\frac{\partial(\rho \boldsymbol{u})}{\partial t} + \nabla\cdot(\rho \boldsymbol{u}\boldsymbol{u}) = -\nabla p + \nabla\cdot\left[\mu(\nabla\boldsymbol{u} + \nabla^T\boldsymbol{u})\right] + \rho\boldsymbol{F}_b + \boldsymbol{F}_\sigma \tag{1}$$

where

$$\boldsymbol{F}_\sigma = \sigma\int_A \kappa_f \boldsymbol{n}_f \delta(\boldsymbol{x}-\boldsymbol{x}_f)dA_f \tag{1.1}$$

Here, $\boldsymbol{u}$ is the velocity, p is the pressure, and ρ and μ are the discontinuous density and viscosity fields, respectively. $\boldsymbol{F}_b$ denotes the body force and $\boldsymbol{F}_\sigma$ does the surface tension acted on the interface. The subscript f denotes the front. δ is a high-order delta-function constructed by repeated multiplication of one-dimensional delta functions. κ is the mean curvature of the front. $\boldsymbol{n}$ is a unit vector normal to the front. Formally, the integral is over the entire front segment, thereby adding the delta-functions together to create a force that is concentrated at the interface, but smooth along the front. $\boldsymbol{x}$ is the point at which the equation is evaluated and $\boldsymbol{x}_f$ is the position of the front segment.

Concerning the mass conservation, for the continuous incompressible liquid phase the following equation is supplemented.

$$\nabla\cdot\boldsymbol{u} = 0 \tag{2}$$

When combined with the above momentum equation these equation leads to an elliptic equation for the liquid pressure, which is linked to the bubble inner pressure across the interface. Therefore mass conservation of the gas phase contained in bubbles needs to be concerned to determine the pressure inside a bubble. For the gas inside a bubble it density varies greatly with surround liquid pressure, and thus the following equation is adopted.

$$\nabla \cdot \boldsymbol{u} = -\frac{1}{\rho}\frac{D\rho}{Dt} \tag{3}$$

For the convenience of numerical treatment the gas included in bubbles is assumed to be non-condensable one that conforms to the adiabatic law, and the equation of state is written as $p / p_0 = (\rho / \rho_0)^\gamma$ by neglecting the variation of temperature. Then, the density variation is related to pressure variation and the mass conservation for the gas phase is rewritten to the following form.

$$\nabla u = -\frac{1}{\gamma p}\frac{Dp}{Dt} \tag{4}$$

where γ is the ratio of specific heats taken to be 1.4, and D/Dt represent the Lagrangian derivative.

As shown in Fig. 1, an arbitrarily arranged bubble cloud in an open rectangular channel of liquid is concerned. In order to induce the bubble cloud to cavitation a negative pressure pulse given by the following sine waveform is acted on the bubble cloud.

$$p_{in} = \begin{cases} p_{L0}[1+\cos(2\pi t / T_0)]/2, & (t \le T_0) \\ 0 & (t > T_0) \end{cases} \tag{5}$$

where p_{L0} denote the initial liquid pressure around the cloud and T_0 the duration time of the pressure pulse. Then, the boundaries in the horizontal and vertical directions are set to be a liquid boundary of given pressure. The boundaries in the span-wise direction are thought to be periodic ones. For convenience of computation above equations are solved in non-dimensional way and all variables are normalized as follows.

$$x^* = \frac{x}{R_0},\ t^* = \frac{t}{R_0}\left(\frac{p_0}{\rho_0}\right)^{1/2},\ \boldsymbol{u}^* = \boldsymbol{u}\left(\frac{\rho_0}{p_0}\right)^{1/2},\ \rho^* = \frac{\rho}{\rho_0},\ p^* = \frac{p}{p_0} \tag{6}$$

As an important physical parameter of bubbly mixture the void fraction of gas phase is often used to express the amount of gas per unit volume [16]. Usually, it is defined by the total volume of bubbles in a given domain. However, differing to a uniform bubbly flow a bubble cloud occupies only its local area rather the whole domain. Geometrically, the size of a bubble cloud is determined by the envelope surface of all the outer layer bubbles, and then the void fraction of a bubble cloud is defined as follows.

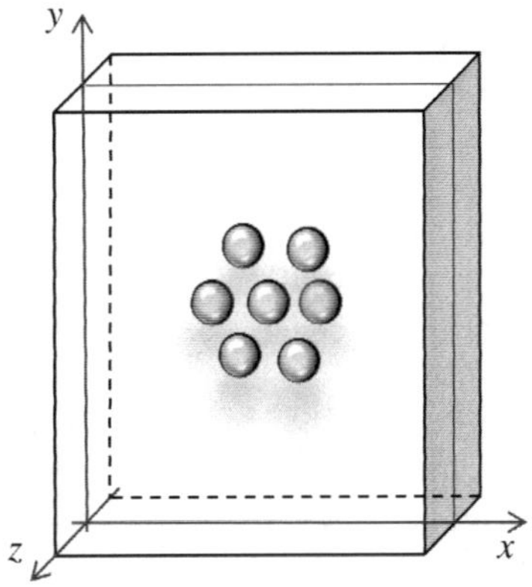

Figure 1. Sketch of computational domain

$$\alpha = \frac{3\oint I(\boldsymbol{x})dV}{4\pi R_c^3} \tag{7}$$

where $I(\boldsymbol{x})$ is an indicator function whose value is given to be 1.0 in the interior of all the bubbles and 0 elsewhere, and $\boldsymbol{x}$ denotes the coordinates of small volume dV taken within the cloud. R_c denotes the equivalent radius of a bubble cloud and α does the void fraction of the cloud. For convenience the superscript * denoting dimensionless parameters are omitted hereafter.

3. Results and discussions

Figure 2 shows, as an example, the cavitating behaviour of a bubble cloud composed of seven same bubbles at a sequence of time. At the initial state shown in Fig. 2 (a) the void fraction and the radius of the bubble cloud are evaluated to be that $\alpha_0 = 0.51$ and $R_{C0} = 6.5$, where R_{C0} is also normalized by the initial bubble radius R_0. Figure 2 (b) shows the state of bubble cloud expanding, where the bubble surface moving velocity demonstrated by vectors is relatively small and the cloud is approximately reaching to its biggest size. Figure 2 (c) shows a moment that the outer layer bubbles begin to contract although the inner bubble is still in expanding. Figure 2 (d) shows the state that the inner bubble begins to contract also under the action of outer bubbles. Figure 2 (e) shows the collapsing of outer layer bubbles where re-entry jets towards to the inner bubble are formed although the inner bubble is still in contracting. Figure 2 (f) shows the moment that the inner bubble gets into collapsing (contracting to the smallest one) where a strong impulse pressure high as 35 times of the initial pressure is demonstrated. From the figure we may see that strong re-entry jets are generated in the outer layer bubbles while they are closing to be broken.

Figure 2 (g) to (i) show the variation of bubble shapes in the process of rebounding where the inner bubble expanding nearly uniformly in all directions but the outer layer bubbles expand in the bilateral directions fast and they present to be horse-like shape instantaneously. The results demonstrate that the outer layer bubbles shields the inner bubble and get into collapsing earlier due to the effect of outside liquid pressure. As expected, re-entry jet occurs at the outer layer bubble surfaces and it works to promote subsequent collapsing of the inner bubble.

Figure 3 shows the variation of radii of all bubbles included in the cloud. The thick solid line presents the radius R_{b1} of the central bubble and the thin solid lines do the radii of outer layer bubbles numbered as 2 to 7. According to the result we know that the central bubble expands continually while the outer layer bubbles begin to contract. But the collapsing velocity of the inner bubble is much fast than that the outer layer bubbles. So, the central bubble contracts to the minimum size closing to breaking before the outer layer ones get into collapsing at the moment denoted by line A in Fig. 3 (b) and then rebound quickly. At the moment denoted by line B the inner bubble begins to contract again for the shielding effect of outer layer bubbles and then gets into its minimum state closing to breaking again at the moment denoted by line C. Therefor two large impact pressures are generated before and after the out layer bubbles contacts to minimum ones as shown by the thick dashed line in Fig. 3 (b). The result demonstrates that the behaviour of bubble cloud cavitation is much more complex for the interaction of bubbles in different positions compared to the case of single bubble. The above result agrees qualitatively with the experimental results reported by N. Bremond et al. [17] although the arrangement of bubbles and the intensity of negative pressure pulse are different [18].

Corresponding to expanding and contracting of individual bubbles in a bubble cloud, the size and the void fraction of bubble cloud vary temporally. Figure 4 shows, as example, the oscillations of three different bubble clouds composed of the same bubbles distributed differently, where Case A denotes the bubble cloud shown in Fig. 2 and Case B and Case C do bubble clouds whose initial radius are increase to $R_{C0} = 13.1$ and $R_{C0} = 23.5$ respectively by increasing the distance among bubbles. Figure 3 (a) shows the variation of bubble cloud size and Fig. 3 (b) does the variation of void fraction responding to the given pressure pulse. According to Fig. 3 (a), we understand that the bubble cloud of Case A expands and contacts periodically at frequency lower than that of single bubble except a

IOP Conf. Series: Materials Science and Engineering **72** (2015) 012001 doi:10.1088/1757-899X/72/1/012001

disorder near t=38. This high frequency oscillation should be actually the effect of interaction between the outer layer bubbles and the inner bubble. Comparing the results of Case A, B and C we known that the frequency of bubble cloud oscillation becomes higher while the void fraction decreases corresponding to the increase of the distance among bubbles. In the Case C the oscillation of bubble cloud radius closes to that of an individual bubble since the interaction of surrounding bubbles becomes very weak. Comparing Fig. 3 (a) and (b) we understand that the void fraction of bubble cloud oscillates in a good periodic manner. The damping of bubble cloud oscillation becomes quicker with the decrease of initial void fraction since the resonance of bubbles become weaker gradually.

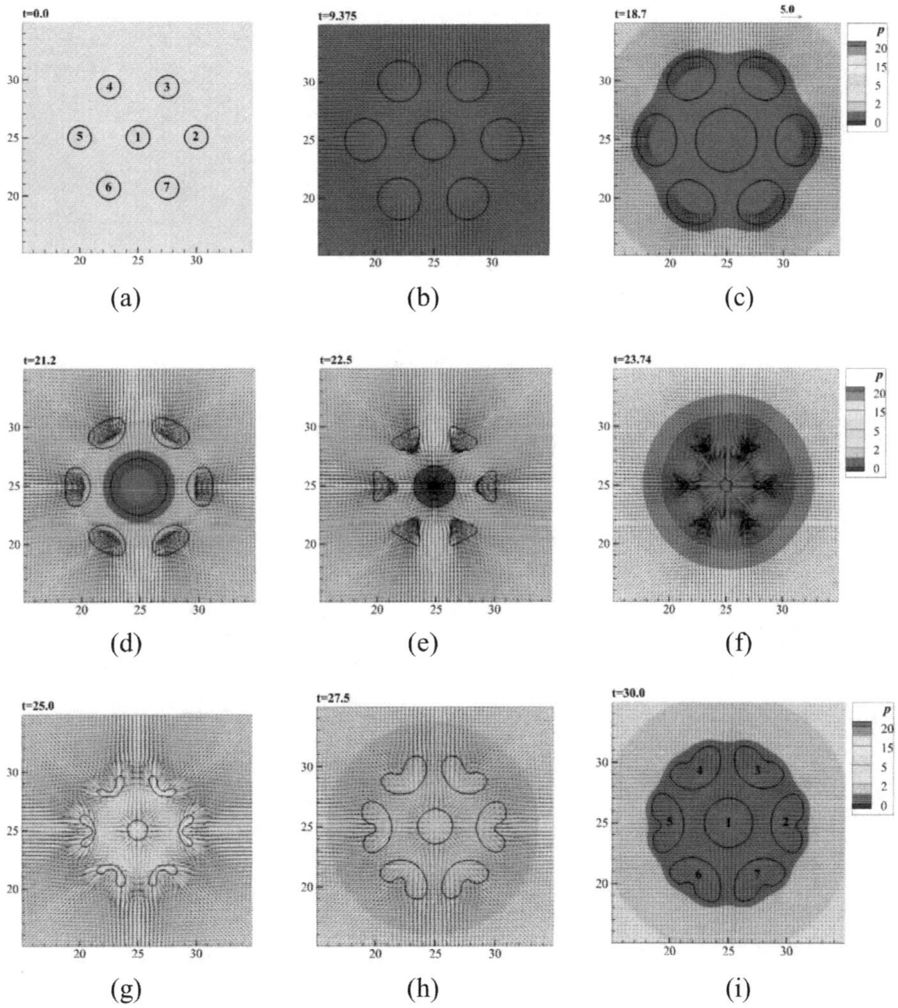

Figure 2. Expansion and collapse of a bubble cloud (RC0=6.5, α0= 0.182) responding to a negative pressure pulse: (a) to (i) flow distributions and the shape of bubbles at a sequence of times.

4. Conclusions

Direct numerical simulation of bubble cloud cavitation were performed by applying the front tracking method and the response of a arbitrarily distributed bubble cloud to a negative sinusoidal pressure pulse has been investigated numerically. The behaviour of bubble cloud expanding and collapsing reproduced by numerical simulation qualitatively agrees with the experimental ones reported by N. Bremond et al. [17] and the capability of present method to treat critical bubble cloud cavitation has been demonstrated.

Numerical investigation demonstrates that: (1) In the collapsing of bubble cloud micro liquid jets toward the inner bubbles are formed while the outer layer bubbles contract extremely, and then a high impact pressure is released when the inner central bubble contacts to its minimum. (2) The oscillation of bubble cloud depends upon the void fraction greatly. In the case of high void fraction, the frequency of cloud oscillation is lower than that of individual bubble and the decay of the oscillation becomes much slowly also.

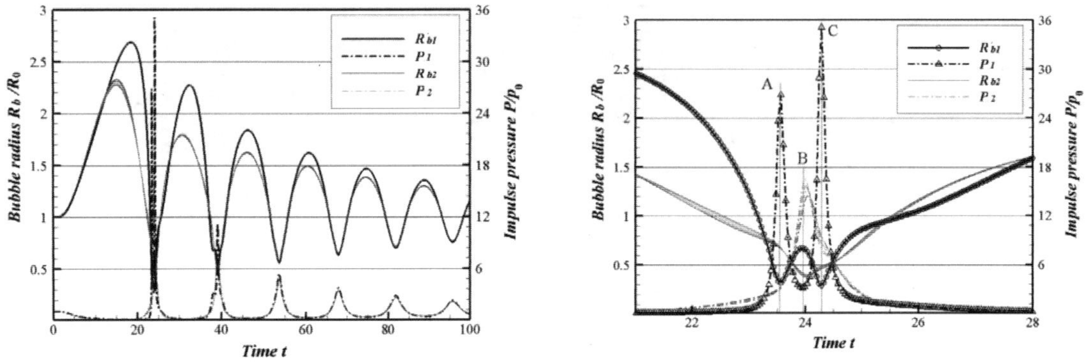

(a) Oscillation of bubble radius (b) Interaction of bubbles at cloud collapsing

Figure 3. Oscillation of bubble radii and impulse pressures generated at bubble collapsing

References

[1] Brennen C E 1995 *Cavitation and bubble dynamics* (New York: Oxford University Press).
[2] Reisman G E, Wang Y-C and Brennen C E 1998 *J. Fluid Mech.* **355** 255-83
[3] Reisman G E and Brennen C E 1996 Pressure Pulses Generated by Cloud Cavitation: Proc. 1996 Fluids Engineering Division Conference (New York: ASME FED-236) 319-28
[4] Shimizu S and Peng G 2009 *Water jetting technology for LOHAS* (Tokyo: Int. Academic Printing Co. Ltd.)
[5] Soyama H, Kusaka T and Saka M 2001 *J. materials science letters* **20** 1263-5
[6] Matsumoto Y, Allen J S, Yoshizawa S, Ikeda T and Kaneko Y 2005 *Experimental Thermal and Fluid Science* **29**(3) 255–65
[7] Saurel R and Lemetayer O 2001 *J. Fluid Mech.* **431** 239-71
[8] Tamura Y and Matsumoto Y 2009 *J. Hydrodynamics (B)* **21**(1) 41-6
[9] Matsumoto Y and Yoshizawa S 2004 *Int. J. Numerical Methods in Fluids*, **47**(6) 591–601
[10] Yu P-W, Ceccio S L and Tryggvason G 1995 *Phys. Fluids* **7** 2608
[11] Popinet S and Zaleski S 2002 *J. Fluid Mech.* **464** 137–63
[12] Dabiri S, Sirignano W A and Joseph D D 2010 *Phys. Fluids* **22** 042102
[13] Unverdi S and Tryggvason G 1992 *J. Comput. Phys.* **100** 25-37
[14] Tryggvason G, et al. 2001 *J. Comput. Phys.* **169** 708-59
[15] Tryggvason G, Scardovelli R and Zaleski S 2011 *Direct Numerical Simulations of Gas-Liquid Multiphase Flows* (Cambridge, UK: Cambridge University Press)
[16] Fuster D and Colonius T 2011 *J. Fluid Mech.* **688** 352-89
[17] Bremond N, Arora M, Ohl C-D and Lohse D 2006 *Phys. Rev. Lett.* **96** 224501 1-4
[18] Peng G and Shimizu S 2013 *J. Hydrodynamics* **25**(4) 502-9

Acknowledgments
This work was partly supported by JSPS, Grant-in-Aid for Scientific Research (C) (N0. 26420124).

Enhancement of heat and mass transfer by cavitation

Y N Zhang[1,3], Y N Zhang[2], X Z Du[1] and H Z Xian[1]
[1]Key Laboratory of Condition Monitoring and Control for Power Plant Equipment, School of Energy, Power and Mechanical Engineering, North China Electric Power University, Beijing 102206, China
[2]School of Engineering, University of Warwick, Coventry CV4 7AL, UK

Email: zhangyn02@gmail.com

Abstract. In this paper, a brief summary of effects of cavitation on the heat and mass transfer are given. The fundamental studies of cavitation bubbles, including its nonlinearity, rectified heat and mass diffusion, are initially introduced. Then selected topics of cavitation enhanced heat and mass transfer were discussed in details including whales stranding caused by active sonar activity, pool boiling heat transfer, oscillating heat pipe and high intensity focused ultrasound treatment.

1. Introduction
Cavitation has attracted much attention in the past decades for its unique characteristics, e.g. strong collapse force, generation of high temperature field and oscillating fluid flow around bubbles. Under irradiation of acoustic waves, the bubbles will also oscillate, termed as "acoustic cavitation" [1-6]. Comparing with the traditional hydrodynamic cavitation, acoustic cavitation can be more easily controlled by choosing appropriate acoustic parameters. Based on those principles, new mechanisms and techniques are being investigated intensively.

In this paper, we briefly reviewed of the physical oscillations of cavitation bubbles and several selected topics associated with cavitation enhanced heat and mass transfer. In Sec. 2, a wide range of physical effects relating with cavitation bubbles are introduced; in Sec. 3, recent progress on cavitation enhanced heat and mass transfer are discussed in details with four examples including whale stranding induced by active sonar, pool boiling heat transfer, oscillating heat pipe and high intensity focused ultrasound treatment.

2. Physical oscillations of cavitation bubbles
The basic equation of bubble motion can be described by various kind of models. A Rayleigh-Plesset models or its variants have been developed mainly by Plesset [7], Noltingk and Neppiras [8], Poritsky [9]. A more advanced model was developed by Gilmore [10] by incorporating the effects of sound irradiation for oscillating bubbles. Keller and Miksis [11] further considered the retarded time in the equations. Prosperetti and Lezzi [12,13] explored the connections between various models and built a more sophiscated model for bubble motion. This equation of bubble motion coupled with heat and mass transfer equation around bubbles are the main equations to be solved for bubble dynamics.

[3] Corresponding author.

2.1. Nonlinear response

Figure 1 shows the nonlinear response of bubbles to the acoustic excitation. In Figure 1, one can observe five types of nonlinear phenomena:

a. Main resonance (marked as "1/1" in Figure 1);
b. Harmonic resonance (marked as "2/1", "3/1" etc in Figure 1);
c. Subharmonic resonance (marked as "1/2" in Figure 1);
d. Ultraharmonic resonance (marked as "3/2", "5/2" etc in Figure 1).
e. Jump phenomenon caused by nonlinear feature of bubble motion equation (marked as dotted line near each resonance).

For more details about definitions of each resonance and jump phenomenon, readers are referred to the classic paper by Lauterborn [14].

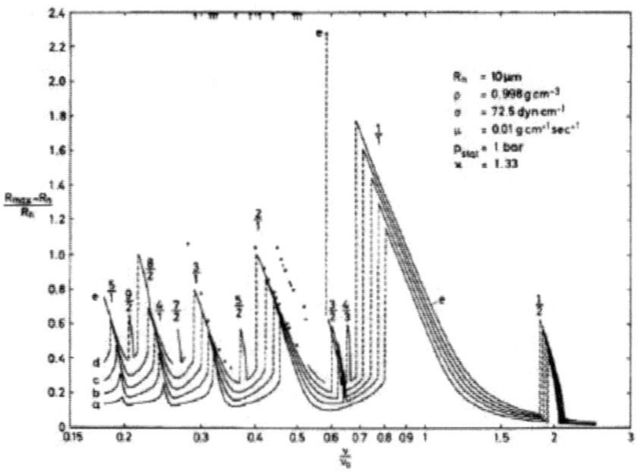

Figure 1. Frequency response curve of bubble with radius $R_0 = 10\mu m$ under irradiation of acoustic field with different pressure amplitudes (a) 40 kPa; (b) 50 kPa; (c) 60 kPa; (d) 70 kPa; (e) 80 kPa. This figure was adapted from Figure 9 of Lauterborn and Kurz [5].

Other nonlinear response includes bifurcation and even chaos as shown in Figure 2. For review of this topic, readers are referred to Lauterborn and Parlitz [15].

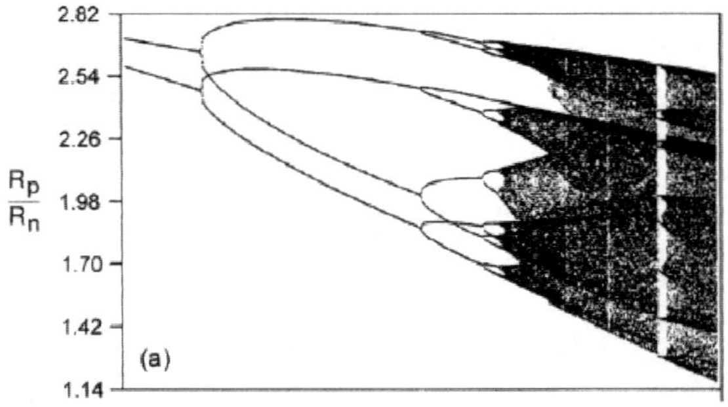

Figure 2. Frequency bifurcation diagram and to chaotic oscillations. This figure was adapted from Figure 15 of Lauterborn and Kurz [5].

2.2. Rectified mass diffusion

During bubble oscillations, its surface area and gas concentration inside bubbles change. At the same time, the bubble can generate a (radical) flow field around bubbles, which will influence the mass transfer through convection terms. As a result, a rectified effect exist during bubble oscillations, leading to a small amount of gas diffused into or out of bubbles. After a long duration (e.g. thousands of cycles), this rectified effect will be prominent and changes the bubble radius effectively. This phenomenon is called "rectified (mass) diffusion". This effect has been studied both experimentally and numerically by numerous researchers [16-22].

Figure 3 shows the variations of bubble radius versus time obtained both by experiments and analytical solution. One can find that there exists a threshold value of pressure amplitude corresponding to the balance of the gas diffused into and out of the bubbles. For higher pressure

IOP Conf. Series: Materials Science and Engineering **72** (2015) 012002 doi:10.1088/1757-899X/72/1/012002

amplitude, the bubble will grow gradually; for lower pressure amplitude, the bubble will dissolve. "SW" in Figure 3 refers to the generation of surface wave on the bubbles. A much clear experimental observation was given by Lee et al. [17] as shown in frame 7 of Figure 4.

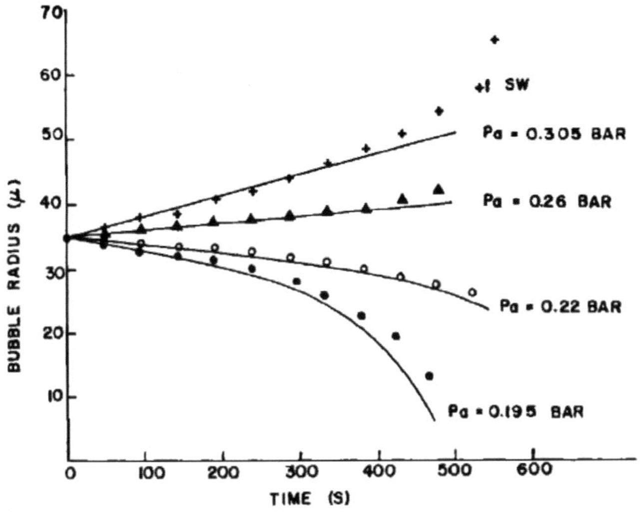

Figure 3. Variations of bubble radius versus time through rectified mass diffusion effects. This figure was adapted from Figure 4 of Crum [16].

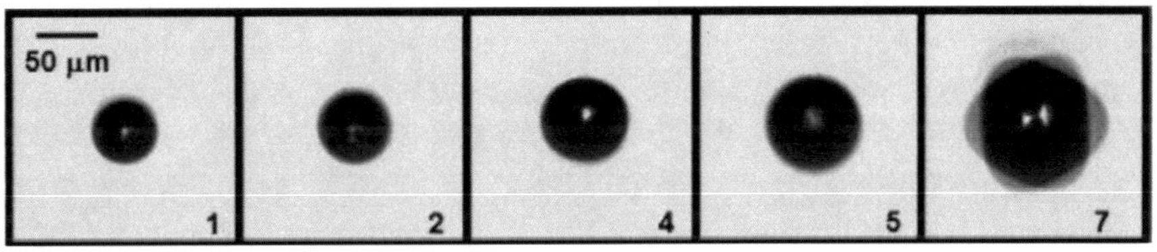

Figure 4. Variations of bubble radius versus time through rectified mass diffusion effects with the presence of surfactants. The marked numbers refer to the number of frames. Inter-frame time duration is 30s. This figure was adapted from Figure 1 of Lee et al. [17].

For analytical solution, Zhang and Li [18] developed generalized equations of rectified mass diffusion by incorporating effects of surface tension, non-uniform pressure inside bubbles and liquid compressibility. Zhang and Li [19] further extended the modelling of rectified mass diffusion into the viscoelastic materials (e.g. human tissues).

2.3. Rectified heat diffusion

Similar like the rectified effects of gas bubbles shown in Sec 2.2, there also exists rectified heat diffusion of vapor bubbles. The physical process of rectified heat diffusion is quite similar as introduced in Sec 2.2 but it is much more intense as the values of thermal conductivity exceeds those of mass diffusivity by two orders. For review of this topic, readers are referred to Prosperetti and Hao [23].

Figure 5 shows an example of the computation of the process of rectified heat diffusion of a vapor bubble excited by acoustic field. The mean radius shows a quick growth up to the resonance value, which is followed by a much limited growth. Gumerov [24] pointed out that the bubble growth could reach a limited value after the order of hundreds of thousands of cycles. Hao and Prosperetti [25] found that resonance effects only play an important role on the dynamics of vapor bubbles for a few cycles at most and the phase of sound amplitude could influence the behaviour of certain small bubbles (e.g. their collapse or growth).

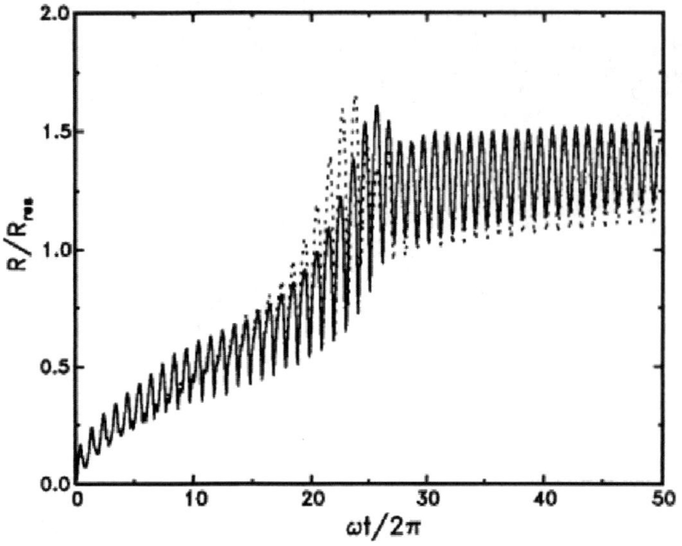

Figure 5. Variations of bubble radius versus time through rectified heat diffusion effects in saturated water with 101.3 KPa. The sound amplitude is 30.39 kPa and the frequency is 1 kHz. The bubble radius is normalized using linear resonance bubble radius. The solid line is obtained by the complete model while the dotted line is the result of an approximated model. This figure was adapted from Figure 6 of Prosperetti and Hao [23].

2.4. Other effects

Due to the limited length of present paper, other important effects are not included in our discussions, e.g. wave scattering effects [26,27], damping mechanisms [29-32], bubble instability [33,34] and wave propagation in bubbly mediums [35,36]. For example, the wave speed in bubbly mediums is a strong function of both frequencies and bubble sizes as shown in Figures 6 and 7. When the bubble are excited by acoustic excitation with multiple frequencies, the problem will be much complex, leading to a lot of new phenomena [37].

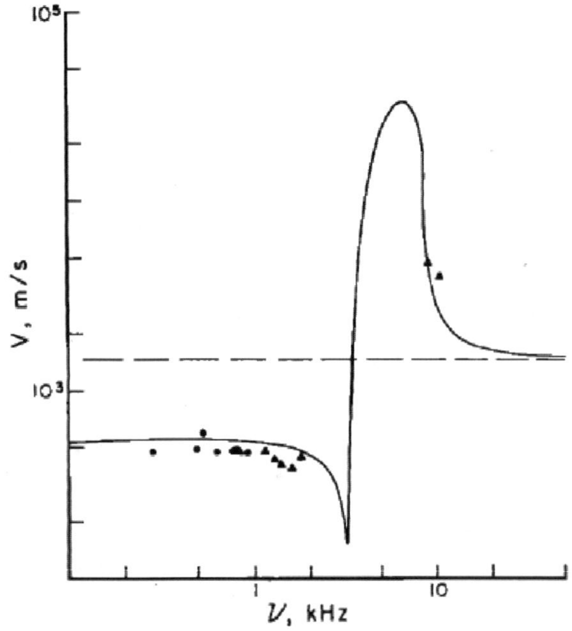

Figure 6. A typical example of speed of acoustic wave during its propagation in bubbly mediums. This figure was adapted from Figure 8 of Commander and Prosperetti [36].

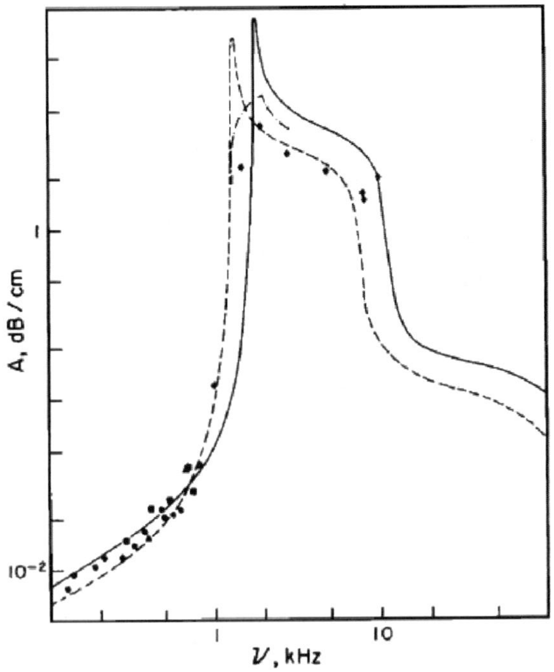

Figure 7. A typical example of attenuation coefficient of acoustic wave during its propagation in bubbly mediums. This figure was adapted from Figure 3 of Commander and Prosperetti [36].

3. Cavitation enhanced heat and mass transfer

In this section, several topics including whale stranding by active sonar activity, pool boiling heat transfer, oscillating heat pipe and high intensity focused ultrasound treatment are selected as demonstrating examples for cavitation enhanced heat and mass transfer.

3.1. Whales stranding by active sonar activity

Nowadays, during the marine military training, active sonar is being frequently used to increase the detection range and the accuracy. When those active sonar passing through certain marine mammals (e.g. beaked whales and dolphins), a serious damage will be caused by a rapid process of bubble activity as shown in Figure 8.

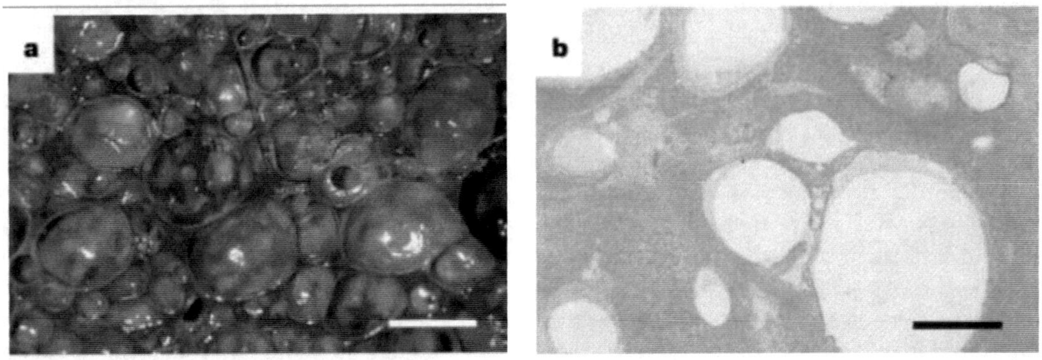

Figure 8. Gas-filled cavities in a common stranded dolphin. This figure was adapted from Figure 1 of Jepson et al. [38].

The reason for generations of those damage is possibly due to the rectified mass diffusion process introduced in Sec. 2.2. Although under normal conditions, the process of the rectified mass diffusion of bubbles is slow, it is rather rapid under the oversaturation condition. Crum et al. [39] has proved that in ex vitro supersaturated tissues, extensive bubble can be produced under irradiation of short pulses of low frequency sound as those used during active sonar activity (referring to Figure 9). For detailed reviews of this topic, readers are referred to Cox et al. [40].

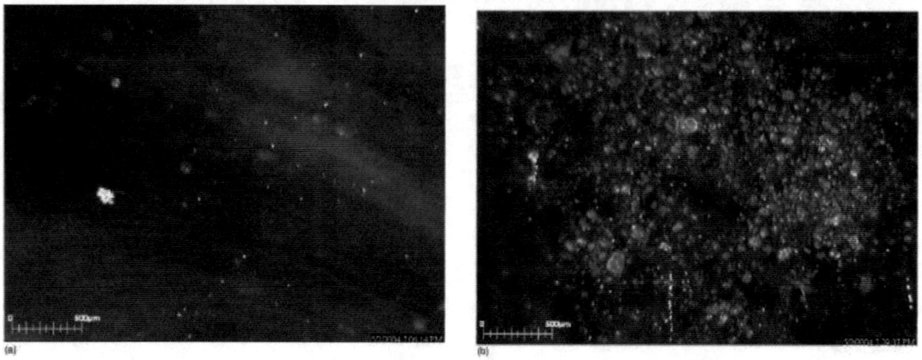

Figure 9. Supersaturated tissue with and without ultrasound (b and a respectively). This figure was adapted from Figure 3 of Crum et al. [39].

3.2. Pool boiling heat transfer

With cavitation generated during pool boiling, it has been observed that the pool boiling heat transfer can significantly promoted, e.g. the improvement of critical heat flux [41-45] and increment of heat transfer coefficient [46]. Various kinds of mechanism for cavitation enhancement of pool boiling heat transfer have been proposed e.g. bubble size reduction and frequency of bubble formation increment [47, 48]. However, the cavitation enhanced pool boiling heat transfer is a quite complex phenomenon because the influential parameters are numerous (e.g. frequency and amplitude of applied acoustic field, heat flux, surface roughness). Hence, in the literature, both negative and positive influences of cavitation on boiling heat transfer have been all reported. Wong and Chon [49] showed that below a critical sound pressure the cavitation effects on heat transfer are negligible. If the pressure is above this threshold, natural convection is significantly enhanced while there is still a negligible effect on nucleate boiling. Kim et al. [50] also found that the enhancement is severely reduced at the initial stage of boiling. Furthermore, the enhancement effects are also influence by the types of pool boiling. Park and Bergles [51] found that for saturated boiling the heat flux decreases with the acoustic fields applied while for subcooled boiling, an improvement with the order of 5–10% is observed.

Figure 10 shows an example of the variations of heat transfer coefficient versus the acoustic pressure amplitude of the applied ultrasonic fields [52]. Krishnan et al. [52] divided the response into three regimes depending on the acoustic pressure amplitudes. In zone I, a slight deterioration effect of pool boiling is observed. After passing a critical pressure amplitude (e.g. about 1.05 bar in Fig.10), it enters into zone II, in which the heat transfer coefficient increases remarkably with the acoustic pressure amplitude. When the acoustic pressure amplitude is larger than 1.7 bar, the enhancement of heat transfer coefficient is limited and enters into saturation zone (zone III). The above behaviour is strongly influenced by acoustic frequency. For example, for frequency 140 kHz, only zones I and II are observed. For frequency 185 kHz, no significant change of heat transfer coefficient is observed over a wide range of values of acoustic pressure amplitude. Krishnan et al. [52] explained the phenomenon based on the hypothesis of liquid-vapor-surface interaction through the process of nucleation and its size density.

Douglas et al. [45] proposed that the contact line interaction with the acoustic field is a possible mechanism for the enhancement of boiling pool heat transfer. Figure 11 shows the capillary wave generation during the detachment of the bubbles with the presence of the acoustic waves. Hence the bubble remove from the surface could be facilitated. Another mechanism for the enhancement of bubble removal is the Bjerknes forces generated with the presence of the acoustic field through the radiation of the adjacent bubble nearby. According to the rough estimation, comparing with the primary Bjerknes force and buoyance force, the secondary Bjerknes forces plays an important role during bubble detachment process.

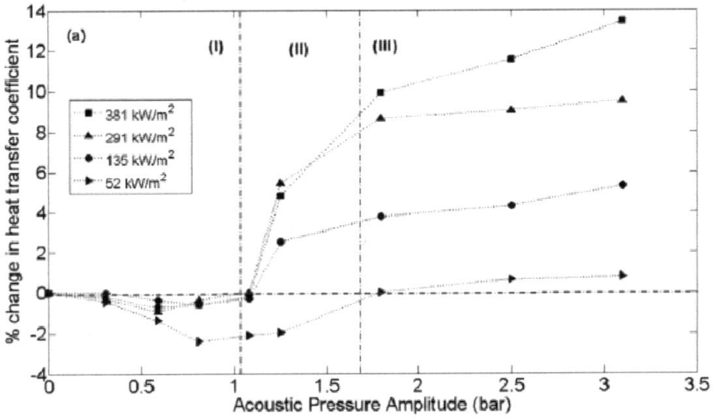

Figure 10. Change in heat transfer coefficient versus acoustic pressure amplitude for different heat fluxes. Ultrasonic frequency is 47 kHz. This figure was adapted from Fig.7a of Krishnan et al. [52].

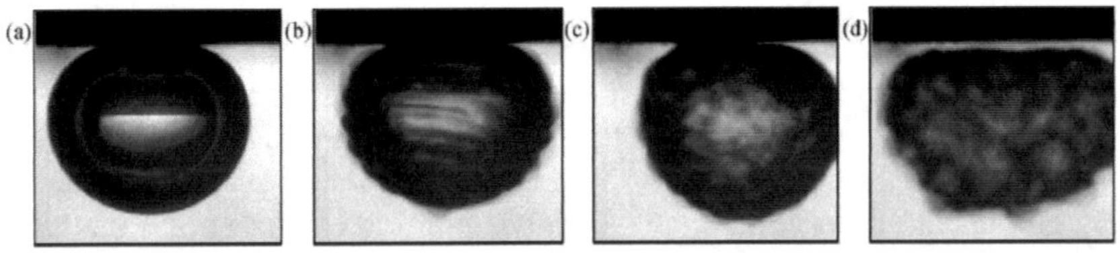

Figure 11. Detachment process of an individual air bubble with the presence of acoustic field. This figure was adapted from Fig.3 of Krishnan et al. [52].

3.3. Oscillating heat pipe

Oscillating heat pipe (OHP) is typically a meandering tube with capillary dimension, which turns in many times between condensation sections and evaporation sections [53-55]. When OHP operated, a self-sustaining oscillatory flow will be generated inside OHP, during the cycles of which the heat is transferred mainly as latent heat. In the past decade, the features of OHP has been intensively studied both experimentally [56-58] and numerically [59-61].

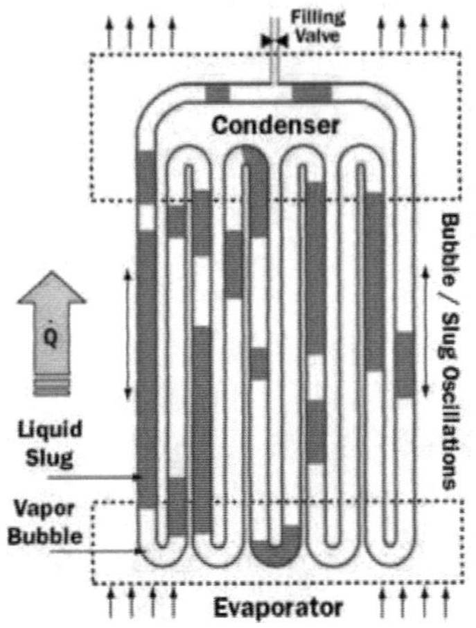

Figure 12. An example of oscillating heat pipes. This figure was adapted from Fig.1 of Khandekar [54].

Recently, the effects of acoustic field on the performance of OHP have attracted much attentions [62, 63]. Generally, the input thermal power for the start-up process of the OHP forming an oscillating motion is relative high. Zhao et al. [62] found that with the presence of an acoustic field with a very small power, the start-up power of can be significantly reduced while the effective thermal conductivity of the OHP can increase (referring to Figure 13). Zhao et al. [63] further found that the presence of acoustic fields can also affect the oscillatory motion inside OHP, enhance the heat transfer performance in a low power input region and reduce the thermal resistance (referring to Figure 14).

3.4. High intensity focused ultrasound treatment

High intensity focused ultrasound (HIFU) treatment is rapidly developing modality for the treatment of the deep-seated solid tumors. HIFU generates a high intensity focused ultrasonic beams on the target, leading to the significant rise of the local temperature for killing the tumors. For a comprehensive review, readers are referred Coussios and Roy [64].

Figure 15 shows the temperature rise increment versus pressure amplitude of HIFU. The measurement in Figure 15 was obtained in a tissue-mimicking phantom consisting of a suspension of graphite particles in an agar gel using a fine wire thermocouple positioned near the focal region [65,66]. Initially, for pressure amplitude less than 1.1 MPa, the temperature rise linearly with the pressure amplitude of HIFU. In this region, the prediction ("orange line" in Figure 15) correlates with the experimental measurement well. Then, a large increment of temperature is shown. This is caused by the generation of harmonics in the beam as the energy could be converted from low to high frequencies due to the nonlinearity of the pressure waves. Because the absorption of HIFU energy in the tissue increases with the increase of the frequency [67,68], energy deposition could be enhanced, leading to a sharp increase of temperature. Further increase of the pressure amplitude do not increase the temperature prominently due to saturation effects. For larger pressure amplitude, near-boiling temperature can be attained within a short exposure, leading to appearance of the boiling bubbles. Those boiling bubbles could scatter the HIFU waves and attenuate the waves strongly (called "shielding effects"). Additionally, those bubbles could produce unpredictable and uncontrollable tadpole-type lesions [69,70].

4. Conclusion

Cavitation process many unique features. For example, heat and mass can diffused into or out of bubbles during the oscillations of bubbles, termed as "rectified diffusion". Cavitation enhanced heat or mass transfer not only can promote traditional mechanical and chemical engineering devices (e.g. pool boiling heat transfer, oscillating heat pipe and chemical reactor) but also play an important role during biomedical treatment (e.g. high intensity focused ultrasound treatment). It also further deepens our understanding towards many interesting phenomena (e.g. whales stranding during military training). There are still many problems to be explored, leading to a promising direction for future research.

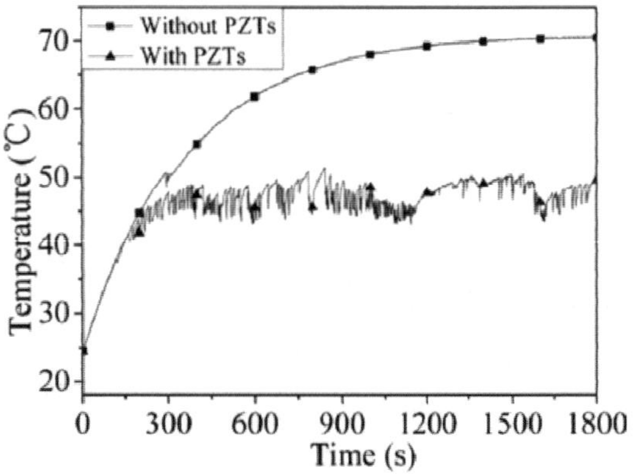

Figure 13. An example of the effects of acoustic field on the oscillating temperature in the oscillating heat pipes. "With PZTs" refers to the presence of acoustic field; "Without PZTs" refers to the absence of acoustic field. This figure was adapted from Fig.4 of Zhao et al. [62].

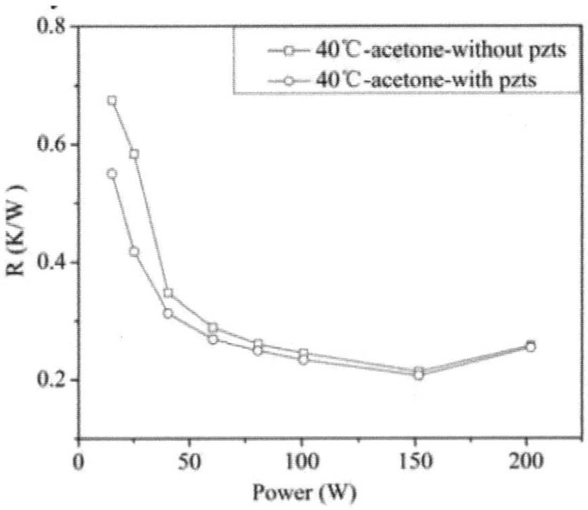

Figure 14. Thermal resistance of oscillating heat pipe with and without acoustic field. This figure was adapted from Fig.3 of Zhao et al. [63].

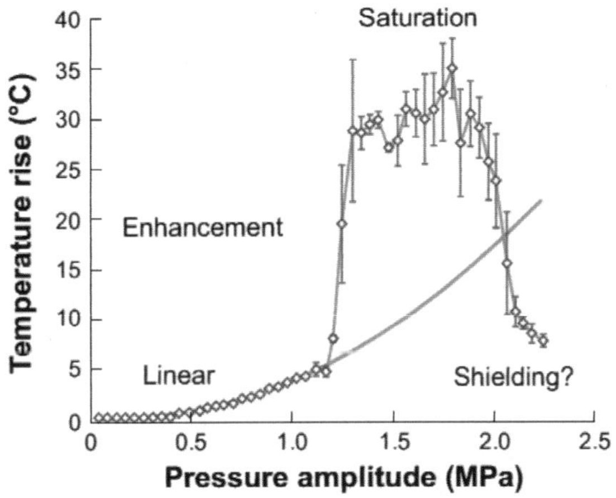

Figure 15. Experimentally measurement (blue line) and numerical predictions (orange line) of temperature rise at the HIFU target. This figure was adapted from Fig.2 of Coussios and Roy [64].

Acknowledgement
The authors acknowledge the financial support from "the Fundamental Research Funds for the Central Universities" (Project No.: 2014ZD09).

References
[1] Plesset M and Prosperetti A 1977 *Annu. Rev. Fluid mech.* **9** 145
[2] Prosperetti A 1984a *Ultrasonics* **22** 69
[3] Prosperetti A 1984b *Ultrasonics* **22** 115
[4] Feng Z and Leal L 1997 *Annu. Rev. Fluid mech.* **29** 201
[5] Lauterborn W and Kurz T 2010 *Rep. Prog. Phy.* **73** 106501
[6] Ashokkumar M 2011 *Ultrason. Sonochem.* **18** 864
[7] Plesset M S 1949 *J. Appl. Mech.* **16** 277–82
[8] Noltingk B E and Neppiras E A 1950 *Proc. Phys. Soc. Lond. B* **63** 674–85
[9] Poritsky H 1952 Proc. First US National Congress for Applied Mechanics (New York) ed E Sternberg (New York: ASME) pp 813–21
[10] Gilmore F R 1952 Report No 26-4
[11] Keller J B and Miksis M 1980 *J. Acoust. Soc. Am.* **68** 628–33
[12] Prosperetti A and Lezzi A 1986 *J. Fluid Mech.* **168** 457–78
[13] Lezzi A and Prosperetti A 1987 *J. Fluid Mech.* **185** 289–321

[14] Lauterborn W 1976 *J. Acoust. Soc. Am.* **59** 283
[15] Lauterborn W and Parlitz U 1988 *J. Acoust. Soc. Am.* **84** 1975
[16] Crum L A 1980 *J. Acoust. Soc. Am.* **68** 203
[17] Lee J, Kentish S and Ashokkumar M 2005 *J. Phys. Chem. B* **109** 14595
[18] Zhang Y and Li S C 2014 *ASME J. Heat Transf.* **136** 042001
[19] Zhang Y and Li S C 2013 *International Journal of Heat and Mass Transfer* **69** 106
[20] Fyrillas M M and Szeri A J 1994 *J. Fluid Mech.* **277** 381
[21] Leong T, Collis J, Manasseh R, Ooi A, Novell A, Bouakaz A, Ashokkumar M, and Kentish S 2011 *J. Phys. Chem. C* **115** 24310.
[22] Webb I R, Arora M, Roy R A, Payne S J and Coussios C C 2010 *J. Fluid Mech.* **663** 209
[23] Prosperetti A and Hao Y 2002 *Annals of the New York Academy of Sciences* **974** 328
[24] Gumerov N A 2000 *Phys. Fluids* **12** 71
[25] Hao Y and Prosperetti A 1999 *Physics of Fluids* **11** 2008
[26] Medwin H 1977 *Ultrasonics* **15** 7
[27] Zhang Y 2013 *ASME J. Fluids Eng.* **135** 091301
[28] Devin C 1959 *J. Acoust. Soc. Am.* **31** 1654
[29] Zhang Y 2013 *International Communications in Heat and Mass Transfer* **43** 1
[30] Zhang Y and Li S C 2012 *Journal of Hydrodynamics* **24** 760
[31] Zhang Y and Li S C 2010 *J. Acoust. Soc. Am.* **128** EL306
[32] Zhang Y and Li S C 2014 *International Communications in Heat and Mass Transfer* **53** 43
[33] Blake J R, Keen G S, Tong R P and Wilson M 1999 *Philosophical Transactions of the Royal Society of London. Series A: Mathematical, Physical and Engineering Sciences* **357** 251.
[34] Zhang Y, Du X, Xian H and Wu Y 2014 *Ultrasonic Sonochemistry*, in press. DOI: 10.1016/j.ultsonch.2014.07.021
[35] van Wijngaarden L V 1972 *Annual Review of Fluid Mechanics* **4** 369
[36] Commander K W and Prosperetti A 1989 *The Journal of the Acoustical Society of America* **85** 732
[37] Zhang Y 2012 *International Communications in Heat and Mass Transfer* **39** 1496
[38] Jepson P D, Arbelo M, Deaville R, Patterson I A P, Castro P, Baker J R, Degollada E, Ross H M, Herráez P, Pocknell A M, Rodríguez F, Howie F E, Espinosa A, Reid R J, Jaber J R, Martin V, Cunningham A A and Fernández A 2003 *Nature* **425** 575
[39] Crum L A, Bailey M R, Guan J, Hilmo P R, Kargl S G, Matula T J and Sapozhnikov O A 2005 *ARLO* **6** 214
[40] Cox T M, Ragen T J, Read A J, Vos E, Baird R W, Balcomb K, Barlow J, Caldwell J, Cranford T, Crum L A, d'Amico A, d'Spain G, Fernandez A, Finneran J, Gentry R, Gerth W, Gulland F, Hildebrand J, Houser D, Hullar T, Jepson P D, Ketten D, MacLeod C D, Miller P, Moore S, Mountain D, Palka D, Ponganis P, Rommel S, Rowles T, Taylor B, Tyack P, Wartzok D, Gisiner R, Mead J, Llowry L and Benner L 2006 *J. Cetacean Res. Manage.* **7** 177
[41] Isakoff S E 1956 Heat Transfer and Fluid Mechanics Institute, Stanford University Press, Stanford, CA, p. 15
[42] Markels S M, Durfee R L and Richardson R 1960 AEC Report No. NYO-9500.
[43] Iida Y and Tsutsui K 1992 *Exp. Therm. Fluid Sci.* **5** 108
[44] Heffington S and Glezer A 2004 *Therminic. Sophia Antipolis*, Cote d'Azur
[45] Douglas Z, Boziuk T R, Smith M K and Glezer A 2012 *Phys. Fluids* **24** 052105
[46] Zhou D W, Liu D Y, Hu X G and Ma C F 2002 *Exp. Therm. Fluid Sci.* **26** 931
[47] Ornatskii A P and Shcherbakov V K 1959 *Teploenergetika* (Moscow, Russ. Fed.), **6** pp. 84–85.
[48] Romie F E and Aronson C A 1961 Experimental Investigation of the Effects of Ultrasonic Vibration on Burnout Heat Flux With Boiling Water, Final Summary Report
[49] Wong S W and Chon W Y 1969 *AIChE J.* **15** 281
[50] Kim H Y, Kim Y G and Kang B H 2004 *Int. J. Heat Mass Transfer* **47** 2831
[51] Park K A and Bergles A E 1988 *Int. J. Heat Mass Transfer* **31** 664

[52] Krishnan S, Das S K and Chatterjee D 2014 *Journal of Heat Transfer* **136** 031501

[53] Kutz M. 2006 Pulsating Heat Pipes. *Mechanical Engineers; Handbook, Energy and Power*, Third Edition, Book 4, Chapter 9

[54] Khandekar S and Groll M 2004 *International Journal of Thermal Sciences* **43** 13

[55] Zhang Y and Faghri A 2008 *Heat Transfer Engineering* **29** 20

[56] Ma H B, Wilson C, Borgmeyer B, Park K, Yu Q, Choi S U S and Tirumala M 2006 *Applied Physics Letters* **88** 143116

[57] Ma H B, Wilson C, Yu Q, Park K, Choi U S and Tirumala M 2006 *Journal of Heat Transfer* **128** 1213

[58] Xian H, Xu W, Zhang Y, Du X and Yang Y 2014 *International Journal of Heat and Mass Transfer* **79** 332

[59] Khandekar S, Schneider M and Groll M 2002 Mathematical modeling of pulsating heat pipes: state of the art and future challenges. Heat and Mass Transfer, SK Saha, SP Venkateshen, BVSSS Prasad, and SS Sadhal, eds., Tata McGraw-Hill Publishing Company, New Delhi, India, 856-862.

[60] Qu W and Ma H B 2007 *International Journal of Heat and Mass Transfer* **50** 2309

[61] Shafii M B, Faghri A and Zhang Y 2001 *Journal of Heat Transfer* **123** 1159

[62] Zhao N, Zhao D and Ma H B 2013 *Journal of Heat Transfer* **135** 074503

[63] Zhao N, Fu B, Zhao D and Ma H 2013 Experimental Investigation of Ultrasonic Effect on an Acetone Oscillating Heat Pipe. In *ASME 2013 4th Int. Conf. on Micro/Nanoscale Heat and Mass Transfer* (pp. V001T12A004-V001T12A004). American Society of Mechanical Engineers.

[64] Coussios C and Roy R 2008 Annu. Rev. Fluid mech. 40 395

[65] Holt R G and Roy R A 2001 *Ultrasound Med. Biol.* **27** 1399

[66] Holt R G and Roy R A 2005 Bubble dynamics in therapeutic ultrasound. In *Bubble and Particle Dynamics in Acoustic Fields: Modern Trends and Applications*, ed. A Dionikov, pp. 108–229. Kerala, India: Transw. Res. Netw.

[67] Goss S A, Frizzell L A and Dunn F 1980 *J. Acoust. Soc. Am.* **67** 1041

[68] Goss S A, Johnston R L and Dunn F 1978 *J. Acoust. Soc. Am.* **64** 423

[69] Chen W S, Lafon C, Matula T J, Vaezy S and Crum L A 2003 *Acoust. Res. Lett. Online* **4** 41

[70] Khokhlova V A, Bailey M R, Reed J A, Cunitz B W, Kaczkowski P J and Crum L A 2006 *J. Acoust. Soc. Am.* **119** 1834

Bubble dynamics under acoustic excitation with multiple frequencies

Y N Zhang[1, 2], Y N Zhang[1] and S C Li[2]
[1]Key Laboratory of Condition Monitoring and Control for Power Plant Equipment, School of Energy, Power and Mechanical Engineering, North China Electric Power University, Beijing 102206, China
[2]School of Engineering, University of Warwick, Coventry CV4 7AL, UK

Email: yuning.zhang@warwick.ac.uk

Abstract. Because of its magnificent mechanical and chemical effects, acoustic cavitation plays an important role in a broad range of biomedical, chemical and mechanical engineering problems. Particularly, irradiation of the multiple frequency acoustic wave could enhance the effects of cavitation. The advantages of employment of multi-frequency ultrasonic field include decreasing the cavitation thresholds, promoting cavitation nuclei generation, increasing the mass transfer and improving energy efficiency. Therefore, multi-frequency ultrasonic systems are employed in a variety of applications, e.g., to enhance the intensity of sonoluminenscence, to increase efficiency of sonochemical reaction, to improve the accuracy of ultrasound imaging and the efficiency of tissue ablation. Compared to single-frequency systems, a lot of new features of bubble dynamics exist in multi-frequency systems, such as special properties of oscillating bubbles, unique resonances in the bubble response curves, and unusual chaotic behaviours. In present paper, the underlying mechanisms of the cavitation effects under multi-frequency acoustical excitation are also briefly introduced.

1. Introduction

Under radiation of acoustic waves, the bubbles will oscillate, termed as "acoustic cavitation" [1,2]. Acoustic cavitation has attracted much attention for many years because of its unique physical complexity [3], chemical applications [4] and biomedical significance [5]. Particularly, a considerable body of work has been produced treating the cavitation under multiple-frequency acoustic wave irradiation, i.e., two or more acoustic waves with the same or different frequencies acting on cavitation bubbles simultaneously (as shown in Figure 1).

In the literature, many positive effects of multi-frequency excitation have been reported. For example, the use of multi-frequency ultrasound field could lower the cavitation thresholds [6], generate more new cavitation nuclei, increase the active cavitational volume [8] and improve energy efficiency [9]. Hence, the multi-frequency systems are employed to enhance the intensity of sonoluminescence [10-17], to increase efficiency of sonochemical reaction [18-22], to improve the accuracy of ultrasound imaging [23-25] and the efficiency of tissue ablation [26,27].

The high energy focusing during bubble collapsing under acoustical wave could induce light emission, termed "sonoluminescence (SL)" [28]. Multi-bubble sonoluminescence (MBSL) and Single-bubble sonoluminescence (SBSL) were found by Marinesco and Trillat [29] and Gaitan et al. [30] respectively. For resent reviews on sonoluminescence, readers are referred to refs. [28,31,32]. Plenty of works have proved that the use of multi frequency acoustical excitation is an efficient way to boost sonoluminescence. Holzfuss et al. [11] employed driving acoustic wave with a combination of the first and the second harmonics to increase the light emission during bubble oscillations up to 300% higher than the system with only fundamental frequency excitation. The relative phase and pressure amplitudes plays an important role on the intensity of sonoluminenscence. Krefting et al. [13] investigated the influential parameters on the dual-frequency excited SBSL more systematically. By experimental measurements and numerical simulations, the region of light emission is mapped into the three-dimensional parameter space spanned by the two driving pressure amplitudes and their relative phase. The addition of the second frequency was able to amplify SBSL in particular parameter zone. Ciuti et al. [6,17] studied the factor of the enhancement of MBSL in ultrasound field with highly different frequencies. In most of the cases, the addition of low frequency into the high frequency field led to stronger light emission, except the intensities of both frequencies were much higher than the corresponding cavitation thresholds. For other resent results of sonoluminescence under multi frequency excitations, readers are referred to refs. [7,10,14,16,18].

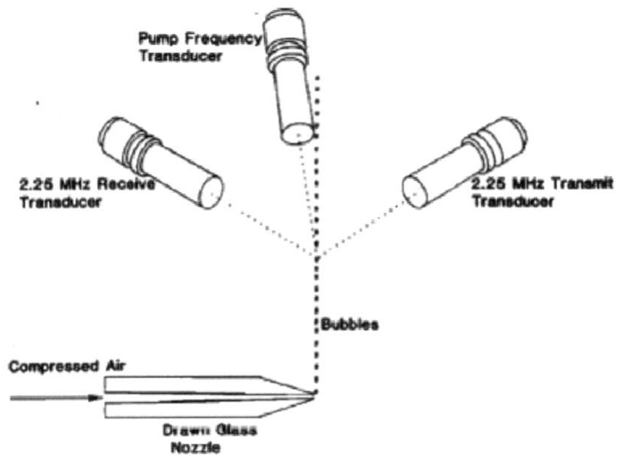

Figure 1. (Figure 3 of ref. [53]) Schematic of the multi-frequency system. Three ultrasonic transducers focus on air bubbles simultaneously. Two of them are transducers with the same frequencies (2.25 MHz) and the other one has a different frequency.

The high temperature and pressure created by bubble collapse can also promote chemical activities within or near the bubbles, which termed as "sonochemistry" [33]. The mechanisms and applications of sonochemistry were reviewed recently in a number of works [34-38]. The efficiency of sonochemical reactions could be enhanced by multi-frequency ultrasound systems. Feng et al. [21] employed both dual-frequency and three-frequency systems by combining ultrasonic transducers with frequencies of the kHz order and the MHz order. The release of iodine, the change of electroconductivity, and the fluorescence formation during sonication were all enhanced by multi-frequency sound source. In past decade, the application of multi-frequency systems was studied widely by researchers. Researchers also made efforts on optimizing the cavitation effects in sonochemical reactors [9,22,39,40]. Moholkar et al. [41,42] revealed that the mode and spatial distribution of the cavitation could be controlled by adjusting the parameters of the dual-frequency field. For instance, the phase difference between the two ultrasound waves had greater influence on production of radicals than the frequency ratio. And it is possible to overcome the directional sensitivity of the cavitation events and erosion of the sonicator surface by inducing the additional ultrasound waves. Kanthale et al. [18] investigated the effect of intensity and frequency of dual-frequency system numerically and compared the results with experimental results in literature. Their work indicated that there exists optimum value of the ultrasound intensity and frequencies.

Acoustic cavitation has wide application in the field of biomedicine. It can enhance the accuracy of ultrasound imaging with the help of the microbubbles [24,43], promote drug and gene transfer into tissue and cells [44,45], treat tumors and ablate tissues [46-47]. Recently, researchers tried to benefit these practices by induce multi-frequency systems. He et al. [26] introduced a high intensity focused ultrasound (HIFU) device which could work both at single-frequency mode and dual-frequency mode. The experimental results showed that the dual-frequency HIFU induced larger tissue lesion than the conventional one under single-frequency mode, which meant that dual-frequency could improve the efficiency of tumor ablation. Guo et al. [27] placed tissues under the focus of HIFUs under single-frequency, dual-frequency and tri-frequency modes respectively. The multi-frequency mode could yield higher temperature during tissue ablation and higher raise rate of the temperature. The numerical simulation agreed well with the experimental results.

In present paper, the recent development of the bubble dynamics under multiple frequencies are reviewed together with the physical mechanisms of cavitation effects. The whole paper is organized as follows: in Sec.2, the oscillation of bubbles and its response to the multi-frequency acoustic waves are discussed together with its nonlinear features (e.g. chaos); in Sec.3, physical mechanisms relating with promotion of cavitation effects under multiple-frequency waves are briefly introduced.

2. Fundamentals of bubble dynamics under multiple frequencies

In the physical, chemical and biomedical effects of cavitation and corresponding applications mentioned above, oscillations of bubbles play an essential role. Therefore, bubble dynamics is one of the paramount problems in the cavitation dynamics. In this section, the basic features of bubble oscillations under multi-frequency excitations are introduced.

2.1. Oscillations of bubbles

For solving the radial oscillations of gas bubbles in liquids, various models have been developed [1,3,48-50]. By solving the bubble motion equations, the variation of bubble radius with time could be obtained as shown in Figures 2 and 3 (solutions from the Keller-Miskis model [51]). For single-frequency sinusoidal acoustic excitation (as shown in Figure 2), the gas bubble shows steady oscillation with the amplitude and frequency unchanged while for dual-frequency acoustic excitation (shown in Figure 3), the oscillation of gas bubble is much more complicated with the frequency and amplitude oscillating.

2.2. Response of bubbles

The bubble oscillators can be described by response curves, i.e., the amplitude of the steady state oscillation plotted versus the driving frequency. Lauterborn [52] gave a thorough investigation of the basic properties of nonlinear oscillations of gas bubbles in liquids. The response curves of a bubble to a single-frequency field were calculated and displayed in Figure 4. The expression n/m (Here, m and n are two integers) above the peaks represents the order of the resonance. Cases with m =1 and n =2, 3… correspond to harmonics; cases with m =2, 3… and n =1 correspond to subharmonics; cases with m =2, 3… and n =2, 3… correspond to ultraharmonics. For the definitions of these resonances, readers are referred to Page 284 of ref. [52]. However, the response curves of bubbles under multi-frequency excitations have not been studied yet.

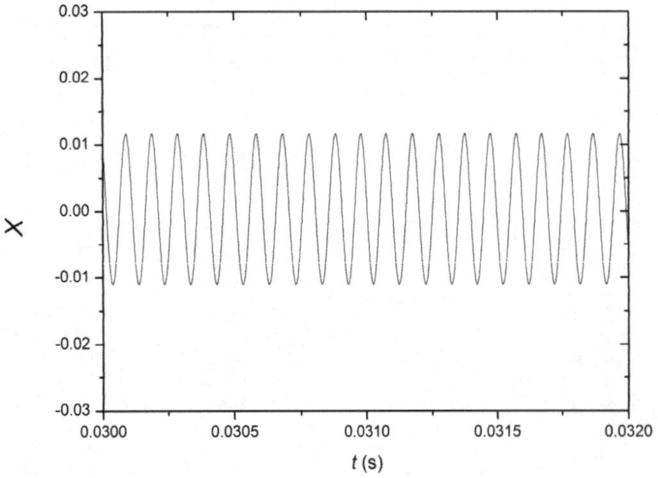

Figure 2 Instantaneous bubble radius excited by the single-frequency acoustic excitation against time. The equilibrium bubble radius: $R_0 = 10\mu m$. The angular frequency of sound wave: $\omega_s = 0.03\omega_0$. The pressure amplitude of sound wave: $P_A = 0.05P_0$. $X = (R - R_0)/R_0$. ω_0 is the natural frequency of the bubble. P_0 is the ambient pressure.

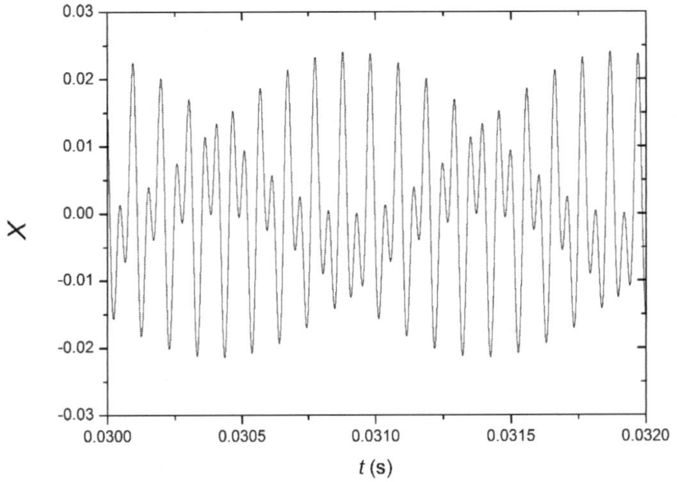

Figure 3 Instantaneous bubble radius excited by the dual-frequency acoustic excitation against time. The equilibrium bubble radius: $R_0 = 10\mu m$. The angular frequency of sound wave one: $\omega_1 = 0.03\omega_0$; sound wave two: $\omega_2 = 0.057\omega_0$. The pressure amplitude of sound wave: $P_{A1} = P_{A2} = 0.05P_0$. $X = (R - R_0)/R_0$. ω_0 is the natural frequency of the bubble. P_0 is the ambient pressure.

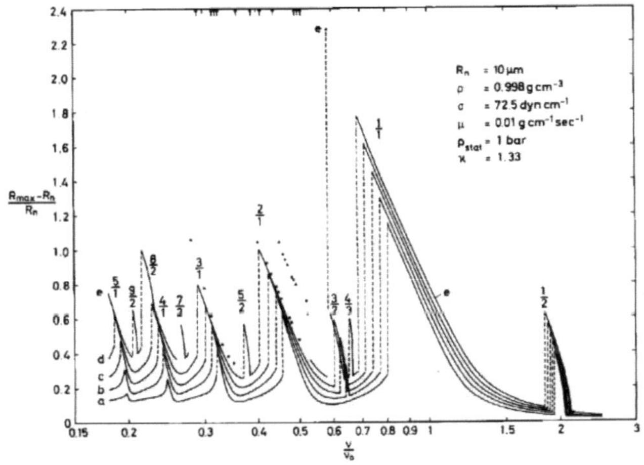

Figure 4 (Fig 3 of ref. [52]) Frequency response curves for a bubble in water with a radius at rest of $R_n = 10\mu m$ for different sound pressure amplitudes P_A of (a) 0.4, (b) 0.5, (c) 0.6, (d) 0.7, and (e) 0.8 bar. ν is the frequency of the driving sound field. ν_0 is the nature frequency of the bubble oscillation. $R_{\max}$ is the maximum radius of the bubble during its steady-state oscillation. The numbers marked above the peaks are the orders of the resonances, represented as n/m.

Via the Fourier transform, the "time domain" diagram (e.g., Figures 2 and 3) could transform to a "frequency domain" diagram, i.e., the power spectrum. Figure 5 is a typical power spectrum of bubble oscillation under single-frequency excitation. The corresponding frequencies of the "lines" are the driving frequency ν (main resonance), its integer multiples (harmonics), $\nu/2$ (subharmonic), $3\nu/2$ and $5\nu/2$ (ultraharmonics). Compared to the single frequency system, the echo of bubble oscillators in a dual-frequency acoustical field with frequencies f_1 and f_2 would contain the bands at $f_1 \pm f_2$ (Figure 6 shows the schematic spectrum of the bubble echo) besides the typical bands in single-frequency systems [53,54]. Furthermore, the echoes at $f_1 \pm f_2$ are much "sharper" than those at the main resonant frequency, which lead to more accurate bubble size measurement [53-55].

Nevertheless, in multi-frequency systems, the scattered echo contains more bands than the difference and sum frequencies of driving frequencies. Figure 7 [56] is an experimental spectra of scattered signal of bubbles excited by a dual-frequency field (with frequencies f_1 and f_2). In this figure, there are bands corresponding to main resonances (marked by ■), their harmonics (marked by □), subharmarnics (marked by ▲) and sum and difference (marked by △). Besides these, there are other peaks (marked by ◆), of which corresponding frequencies are $2f_1 - f_2$ and $2f_2 - f_1$ and their magnitudes are at the same order with the harmonics. All of these suggest the unique properties of bubble oscillations under multi-frequency excitations, which are worthy to investigate quantitatively and systematically.

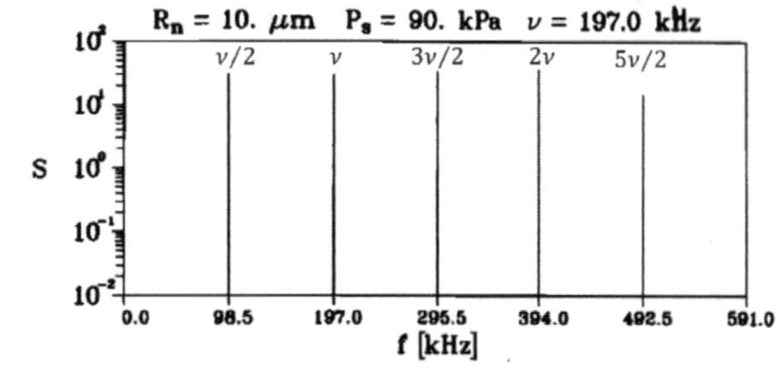

Figure 5. (Figure 13 of ref. [57]) Power spectrum of bubble oscillator. Equilibrium bubble radius $R_n = 10\mu m$. sound pressure amplitude P_S : 90kPa. Driving frequency ν : 197kHz

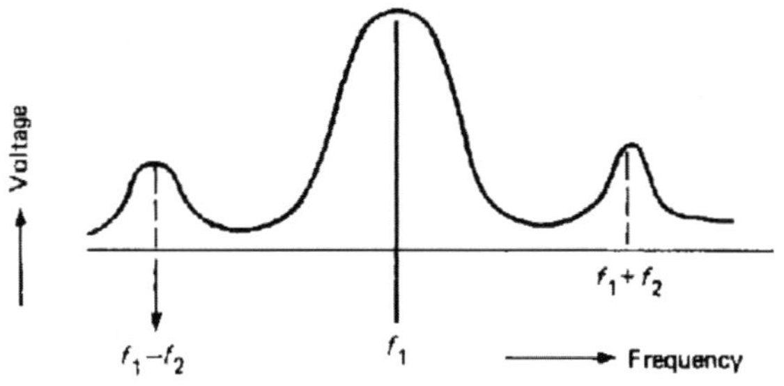

Figure 6. (Figure 1 of ref. [54]) The schematic spectrum of the bubble echo under dual-frequency excitation with frequency f_1 and f_2 .

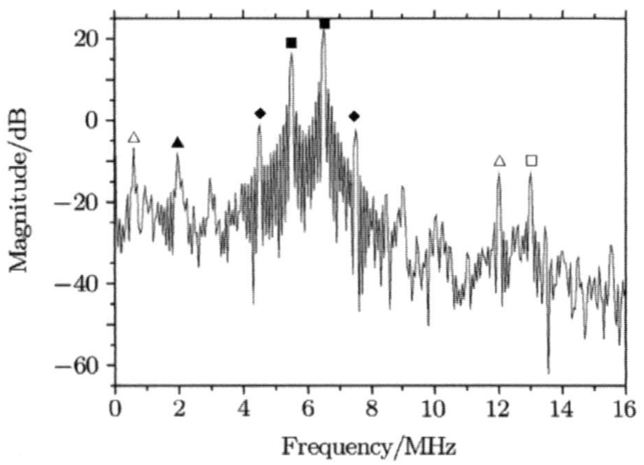

Figure 7. (Figure 3 of ref. [56]) Experimental spectra of the scattered signal from bubbles excited by a dual-frequency ultrasound field. The frequencies of the two sound waves are $f_1 = 5.5\text{MHz}$, $f_2 = 6.5\text{MHz}$, respectively. Sound pressure amplitude 0.5 MPa.

2.3. Chaotic oscillations

For a bubble oscillator forced by sound wave with period *T*, the period of its radial oscillation is equal to *T* according to the linear theory (as the oscillator shown in Figure 2). The phase space diagram [58] of the bubble oscillating with the driving frequency would be like Figure 8 (a), which is also named "limit cycle". When the control parameter of the system, μ, becomes μ_1, the period doubles to *2T* [Figure 8 (b)], as the bubble oscillating at the subharmonic resonance. If the parameter further changes to μ_2, the period reaches *4T* [Figure 8 (c)]. Then after infinitely successive doubling, there is no certain period for bubble oscillating [Figure 8 (d)], i.e., it turns into *chaotic oscillation*. The changes of the state of the bubble at particular values of the parameter are called *bifurcations*. If the limit sets (named "attractors") of the bubble oscillator are plotted against the control parameter, a bifurcation diagram can be obtained. Figure 9 (a) illustrates the bifurcation diagram of the normalized bubble radius driven by single-frequency sound wave. In the range of the parameters of the figure, the bubble oscillates chaotically.

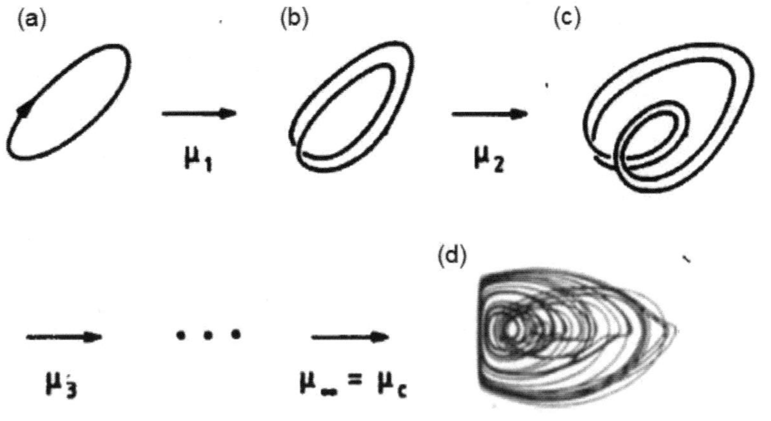

Figure 8. (Figure 8 of ref. [57]) Demonstration of route to chaos via an infinite cascade of period-doubling bifurcations. The trajectories are the phase space diagrams of bubble oscillation. μ_x indicates one of the parameters of the system.

In the chaos, a system can obey certain physical laws so that its future behaviour is deterministic while the system could exhibit unpredictable behaviour owing to the sensitivity to initial conditions [59]. It widely exists in dynamical systems. Lauterborn and his colleagues [57,60-62], have introduced research methods of chaos and deepen the understanding of the bifurcation structure and strange attractors of acoustical bubble oscillation. Because of its unpredictability, chaotic oscillation should be avoided in the practices of sonochemistry, ultrasound enhanced drug delivery and surgery. The addition of the second forcing term is one method to control chaos, which has been successively applied in pendulum system [63], Duffing oscillator [64,65] and electric circuit [66]. Behnia et al. [67] found that the addition of the second sound wave, i.e., the usage of the dual-frequency system, could also reduce the chaotic oscillations of the bubbles to the regular ones (as the comparison of the bifurcation diagrams between single- and dual- frequency shown in Figure 9). They discussed the influence of the frequency of the second wave and the phase difference between two waves. And numerical simulations were performed on the progress of HIFU tumor ablation.

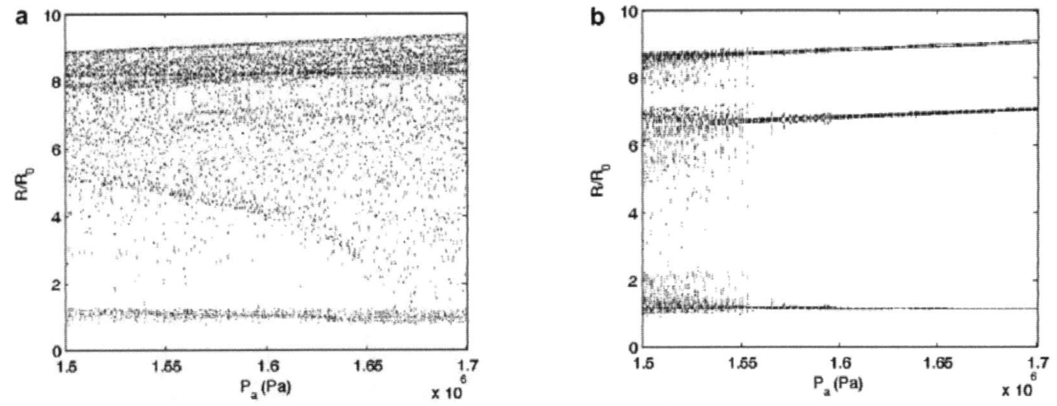

Figure 9. (Figure 1 of ref. [67]) Bifurcation diagrams of the normalized bubble radius with the initial radius 10μm. a) Single-frequency excitation, driving frequency 200kHz. b) Dual-frequency excitation, driving frequencies 200kHz and 500kHz.

3. Physcial mechanisms

The underlying mechanisms of the effects of multi-frequency acoustical excitations are still not clear. Moreover, for different cavitation systems, e.g., single-bubble cavitation and multi-bubble cavitation, the mechanisms of multi-frequency actions are also different. In this section, some of the mechanisms are introduced briefly.

3.1. Dissociation hypothesis

Ketterling and Apfel [68] explained the multi-frequency sonoluminescence using phase space diagrams based on the dissociation hypothesis (DH) initially proposed by Lohse and Hilgenfeldt [69]. Lohse et al. [69,70] proposed that because of the high temperature during SBSL, the nitrogen and oxygen in air bubbles may dissociate to free radicals •*O* and •*N*, which will compose water solvable chemicals (e.g., NO_3^- and NH_4^+) in subsequent reactions. Meanwhile, the inert gas can accumulate in bubbles under strongly acoustically driven [70,71]. Therefore, the mass transfer through chemical reaction and rectified diffusion should be considered in SBSL. Multi-frequency acoustical excitation could enhance these effects.

Based on Ketterling and Apfel [68], a phase diagram was constructed based on calculation of the equation of bubble motion, diffusive equilibrium and the Mach criterion (assuming the ratio of the bubble wall velocity and speed of gas is larger than one), which separated the response of the bubbles into four regions, i.e., stable SL, unstable SL, stable non-SL and unstable non-SL. Compared to the experimental data of Holzfuss et al. [11], an excellent quantitative agreement was found.

3.2. Nucleation

For multi bubble cavitation, Cuiti et al. [6,17] proposed that in an ultrasound composed by two highly different frequencies, the large amount of the new nuclei generated by the added low frequency (LF) is the main factor of the enhancement of the light intensity of MBSL.

According to Figure 10, at the moment of switching the LF field off the SL intensity did not fall down immediately but, on the contrary, it increased firstly and then decreased smoothly. This phenomenon suggested that the fragments induced by bubble collapse became new nuclei with smaller radii than the initial equilibrium radii which were more likely to collapse in the high frequency (HF) field. This mechanism could also explain easily the “after effect” of the LF field on the intensity of sonoluminescence generated by the HF field in the case of switch-on the LF and HF fields sequentially (as shown in Figure 11). New nuclei generated by LF could live for such a sufficient long time that a higher sonoluminescence intensity occurred in the HF field.

Feng et al. [21] investigated the cavitation yield from the aspect of sonochemisitry. They also pointed out that the production of new bubbles by the LF field is one possible mechanism of the enhancement of sonochemical efficiency.

3.3. Suppression of forming the stable clusters of cavitation bubbles

During multi-bubble cavitation, the bubbles in clusters are close to each other and interact strongly by shock waves and Bjerknes forces [72]. Therefore, bubbles deform under these interactions in the early stage of collapse. The nonspherical collapse is less efficient than spherical collapse in terms of the energy concentration, which is considered to be one of the reasons decreasing the intensity of MBSL by bubble cluster [73]. The added low frequency sound wave can impede the formation of bubble cluster by the shock waves and liquid microjets produced by collapse of large LF bubbles. So that, the total efficiency of the energy concentrated by cavitation bubbles may raise in the multi-frequency systems.

3.4. Periodical decrease in the total quasi-static pressure in the LF field

If the frequency of the LF field is substantially (10 times or more) lower than that of the HF field, then the LF field is quasi-static in relation to the HF field. The quasi-static pressure during the negative pressure amplitude half-period of the LF field lowers the cavitation threshold of the HF field by increasing bubble size and increasing the number of bubbles driven by that field. In the compression half-period, the increase of the LF-field quasi-static pressure may cause an increase of the efficiency of the HF bubbles collapses [74-76].

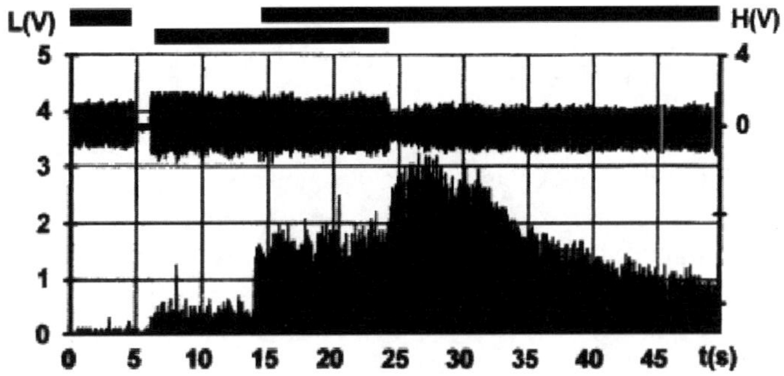

Figure 10. (Fig 2 of ref. [17]) Time history of the hydrophone output H (upper record) and of the photomultiplier L (lower record, which indicates the intensity of light emission). High frequency (HF) field parameters: pulse period 100 ms; pulse duration 2 ms; driving voltage $U_{HF} = 55V$. Low frequency (LF) field parameters: $U_{LF} = 750V$. The intensity of the ultrasound: $I_{HF} > I_{HF,th}$, $I_{LF} \ll I_{LF,th}$ (the subscript *th* represents the threshold of cavitation).

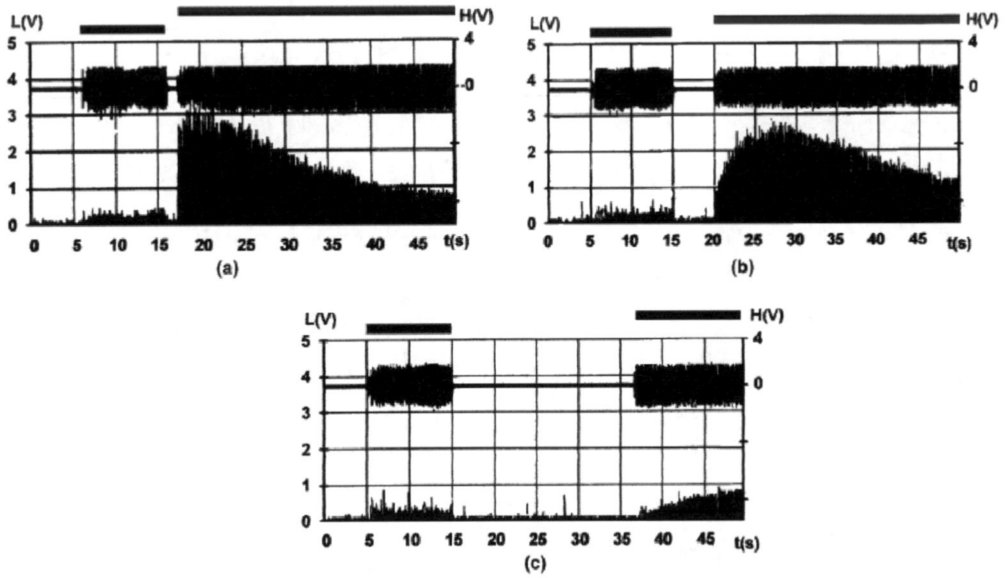

Figure 11. (Figure 3 of ref. [17]) Time history of the hydrophone output H (upper record) and of the photomultiplier L (lower record) for different time intervals Δt between low frequency (LF) field off and high frequency (HF) field on: $\Delta t \approx 2s$ (a), 5s (b), 22.5s (c). Other parameters are the same as in Figure 10.

4. Conclusion

With the increasing interests on the effects of acoustical cavitation, the multi-frequency acoustical systems also attract more and more attentions in the fields of sonochemistry, ultrasound imaging, and ultrasound enhanced biomedicine *etc*. The ability of multi-frequency systems on boosting the sonoluminescence, improving sonochemical efficiency, and promoting tumor treatment has been proved both theoretically and experimentally. And researchers have also done lots of efforts on the optimization of the multi-frequency systems.

However, compared to the wide applications of the multi-frequency acoustical systems, the mechanisms of the multi-frequency induced effects are not clear and the investigations of the fundamental issues of bubble dynamics under multi-frequency excitations are lack. For instance, the studies of Behnia et al. [67] were performed in a very narrow range of parameter zone (the amplitude of sound pressure was from 1.55 MPa to 1.7 Mpa). Actually, multi-frequency excitation may lower the critical pressure amplitude for chaotic oscillations [77], which means the addition of the second sound wave may also induce chaos. Furthermore, as far as we know, many basic problems of bubble oscillations, e.g., acoustical scattering cross section and Bjernes forces, have not been studied in the multi-frequency systems. Therefore, for a deeper understanding of bubble behaviour under multi-frequency excitation and better applications of multi-frequency systems in the fields mentioned before, a systematic investigation of the bubble dynamics under multi-frequency excitation is necessary.

Acknowledgement

The authors acknowledge the financial support from China Scholarship Council, the Ph.D. tuition fees from the University of Warwick and the EPSRC WIMRC grant (RESCM 9219), "the Fundamental Research Funds for the Central Universities" (Project No.: 2014ZD09).

References

[1] Plesset M and Prosperetti A 1977 *Annu. Rev. Fluid mech.* **9** 145
[2] Brenner M, Lohse D and Dupont T 1995 *Phys. Rev. Lett.* **75** 954
[3] Lauterborn W and Kurz T 2010 *Rep. Prog. Phy.* **73** 106501
[4] Ashokkumar M 2011 *Ultrason. Sonochem.* **18** 864
[5] Coussios C and Roy R 2008 *Annu. Rev. Fluid mech.* **40** 395
[6] Ciuti P, Dezhkunov N, Francescutto A, Kulak A, and Iernetti G 2000 *Ultrason. Sonochem.* **7** 213
[7] Chen W, Chen X, Lu M, Miao G, and Wei R 2002 *J. Acoust. Soc. Am.* **111** 2632
[8] Servant G, Laborde J, Hita A, Caltagirone J and Gerard A 2003 *Ultrason. Sonochem.* **10** 347
[9] Sivakumar M, Tatake P and Pandit A 2002 *Chem. Eng. J.* **85** 327
[10] Hargreaves K and Matula T 2000 *J. Acoust. Soc. Am.* **107** 1774
[11] Holzfuss J, Rüggeberg M and Mettin R 1998 *Phys. Rev. Lett.* **81** 1961
[12] Kanthale P, Brotchie A, Ashokkumar M and Grieser F 2008 *Ultrason. Sonochem.* **15** 629
[13] Krefting D, Mettin R and Lauterborn W 2002 *J. Acoust. Soc. Am.* **112** 1918
[14] Brotchie A, Grieser F and Ashokkumar M 2008 *J. Phys. Chem. C.* **112** 10247
[15] Dezhkunov N 2003 *J. Eng. Phy. Therm phy.* **76** 142
[16] Brotchie A, Ashokkumar M and Grieser F 2007 *J. Phys. Chem. C.* **111** 3066
[17] Ciuti P, Dezhkunov N, Francescutto A, Calligaris F and Sturman F 2003 *Ultrason. Sonochem.* **10** 337
[18] Kanthale P, Gogate P and Pandit A 2007 *Chem. Eng. J.* **127** 71
[19] Brotchie A, Grieser F and Ashokkumar M 2010 *J. Phys. Chem. B.* **114** 11010
[20] Servant G, Laborde J, Hita A, Caltagirone J and Gerard A 2003 *Ultrason. Sonochem.* **10** 347
[21] Feng R, Zhao Y, Zhu C and Mason T 2002 *Ultrason. Sonochem.* **9** 231
[22] Tatake P and Pandit A 2002 *Chem. Eng. Sci.* **57** 4987
[23] Zheng H, Mukdadi O, Kim H, Hertzberg J and Shandas R 2005 *Ultrasound Med. Bio.* **31** 99
[24] Barati A, Mokhtari-Dizaji M, Mozdarani H, Bathaie Z and Hassan Z 2007 *Ultrason. Sonochem.*

14 783
[25] Wyczalkowski M and Andrew J 2003 *J. Acoust. Soc. Am.* **113** 3073
[26] He P, Xia R, Duan S, Shou W and Qian D 2006 *Ultrason. Sonochem.* **13** 339
[27] Guo S, Jing Y and Jiang X 2013 *IEEE Trans. Ultrason. Ferroelectr. Frequency. Control.* **60**
[28] Brenner M, Hilgenfeldt S and Lohse D 2002 *Rev. Mod. Phy.* **74** 425
[29] Marinesco N and Trillat J 1993 *Proc. R. Acad. Sci.* **196** 858
[30] Gaitan D, Crum L, Church C and Roy R 1992 *J. Acoust. Soc. Am.* **91** 3166
[31] Suslick K and Flannigan D. 2008 *Annu. Rev. Phys. Chem.* **59** 659
[32] Putterman S and Weninger K 2000 *Annu. Rev. Fluid mech.* **32** 445
[33] Storey B and Szeri A 2000 *Proc. Royal Soc. London. Series A. Math. Phy. Eng. Sci.* **456** 1685
[34] Henglein A 1987 *Ultrasonics* **25** 6
[35] Adewuyi Y 2001 *Indus. Eng. Chem. Res.* **40** 4681
[36] Einhorn C, Einhorn J and Luche J 1989 *Synthesis* **1989** 787
[37] Mason T 1999 *Philos. T. R. Soc. B.* **357** 355
[38] Nikitenko S, Venault L, Pflieger R, Chave T, Bisel I and Moisy P 2010 *Ultrason. Sonochem.* 17 1033
[39] Yasuda K, Torii T, Yasui K, Iida Y, Tuziuti T, Nakamura M and Asakura Y 2007 *Ultrason. Sonochem.* **14** 699
[40] Brotchie A, Mettin R, Grieser F and Ashokkumar M 2009 *Phys. Chem. Chem. Phys.* **11** 10029
[41] Moholkar V, Rekveld S and Warmoeskerken M 2000 *Ultrasonics* **38** 666
[42] Moholkar V amd Vijayanand S 2009 *Chem. Eng. Sci.* **64** 5255
[43] Wu J, Pepe J and Dewitt W III 2003 *Ultrasound Med. Bio.* **29** 555
[44] Song Y, Hahn T, Thompson I, Mason T, Preston G, Li G, Paniwnyk L and Huang W 2007 *Nucleic Acids Res.* **35** e129
[45] Hernot S, and Klibanov A 2008 *Adv. Drug. Delivery Rev.* **60** 1153
[46] Xu Z, Ludomirsky A, Eun L, Hall T, Tran B, Fowlkes J and Cain C 2004 *IEEE Trans. Ultrason. Ferroelectr. Frequency. Control.* **51** 726
[47] Maxwell A, Wang T, Cain C, Fowlkes J, Sapozhnikov O, Bailey M and Xu Z 2011 *J. Acoust. Soc. Am.* **130** 1888
[48] Prosperetti A 1984a *Ultrasonics* **22** 69
[49] Prosperetti A 1984b *Ultrasonics* **22** 115
[50] Feng Z and Leal L 1997 *Annu. Rev. Fluid mech.* **29** 201
[51] Keller J and Miksis M 1980 *J. Acoust. Soc. Am.* **68** 628
[52] Lauterborn W 1976 *J. Acoust. Soc. Am.* **59** 283
[53] Newhouse V and Shankar P 1984 *J. Acoust. Soc. Am.* **75** 1473
[54] Shankar P, Chapelon J and Newhouse V 1986 *Ultrasonics* **24** 333
[55] Phelps A and Leighton T 1994 *Bubble Dynamics and Interface Phenomena* (Netherland: Springer) p 475
[56] Ma Q, Qiu Y, Huang B, Zhang D and Gong X 2010 *Chinese Phy. B.* **19** 094302.
[57] Lauterborn W and Parlitz U 1988 *J. Acoust. Soc. Am.* **84** 1975
[58] Jordan D and Smith P 2007 *Nonlinear ordinary differential equations: an introduction for scientists and engineers* (New York)
[59] Ott E 2002 *Chaos in dynamical systems* (Cambridge university press)
[60] Parlitz U., Englisch V, Scheffczyk C and Lauterborn W 1990 *J. Acoust. Soc. Am.* **88** 1061
[61] Lauterborn W, Holzfuss J and Billo A 1994 *Proc. Ultrasonics Symposium (IEEE)* Vol. 2
[62] Lauterborn W, Thomas K, Mettin R, Philipp K, Kroeninger D and Schanz D 2008 *Arch. Acoust.* **33** 609
[63] Braiman Y and Goldhirsch I 1991 *Phy. Rev. Lett.* **66** 2545
[64] Chacón R and Díaz Bejarano J 1993 *Phy. Rev. Lett.* **71** 3103
[65] Jing Z and Wang R 2005 *Chaos Solitons Fractals* **23** 399
[66] Tamura T, Inaba N and Miyamichi J 1999 *Phy. Rev. Lett.* **83** 3824

[67] Behnia S, Sojahrood A, Soltanpoor W and Jahanbakhsh O 2009 *Ultrason. Sonochem.* **16** 502
[68] Ketterling J and Apfel R 2000 *J. Acoust. Soc. Am.* **107** 819
[69] Lohse D and Hilgenfeldt S 1997 *J. Chem. Phys.* **107** 6986
[70] Lohse D, Brenner M, Dupont T, Hilgenfeldt S and Johnston B 1997 *Phy. Rev. Lett.* **78** 1359
[71] Hilgenfeldt S, Lohse D and Brenner M 1996 *Phy. Fluids. (1994-present)* **8** 2808
[72] Leighton T 1994 *The acoustic bubble* (Academic press)
[73] Evans A 1996 *Phy. Rev. E.* **54** 5004
[74] Carpendo L, Ciuti P, Francescutto A, Iernetti G and Johri G 1987 *Acoust. Lett.* **11** 178
[75] Wolfrum B, Kurz T, Lindau O and Lauterborn W 2001 *Phys. Rev. E.* **64** 046306
[76] Iernettia G, Ciutia P, Dezhkunovb N, Realic M, Francescutto A and Johrie G 1997 *Ultrason. Sonochem.* **4** 263
[77] Szeri A and Leal L 1991 *Phy. Fluids A. Fluid Dyn. (1989-1993)* **3** 551

IOP Conf. Series: Materials Science and Engineering **72** (2015) 022004 doi:10.1088/1757-899X/72/2/022004

Observations and measurements in cloud cavitating flows

G H Chen, G Y Wang, C L Hu, B Huang and M D Zhang
Department of Thermal Energy Engineering, Beijing Institute of Technology, Beijing 100081, China

E-mail: chenguanghao@bit.edu.cn

Abstract. The main purpose of this study is to shed light on the cloud cavitating flow and associated characteristic of pressure fluctuation near wall. A simultaneous sampling technique is used to synchronize the observations of cavitation instantaneous behaviour and the measurements of pressure signals near wall in a convergent-divergent channel. The results show that, a typical quasi-periodical sheet/cloud cavitation can be categorized into three stages: (1) the growth of attached cavity; (2) the shedding of the attached cavity; (3) the development and collapse of the detached cavities. At the stage one, the magnitudes of pressure fluctuation under the attached cavity are limited. However, they become significant in the closure region of attached cavity, especially, when attached cavity reaches its maximum length. At the stage two, the attached cavity begins to shed small detached cavity, leading to the generation of small local pressure fluctuations with higher frequency. At the stage three, a large detached cavity is gradually formed in the rear of the channel. When it collapses rapidly in the downstream, pressure pulses with the magnitudes of the order of several atmospheres are detected. The propagation speeds of pressure pulses in different region are found to be related with the bubble density in the flow field.

1. Introduction

In liquid flows, cavitation generally occurs if the pressure in low-pressure regions of high-speed fluid machinery drops below the vapor pressure, and consequently the negative pressures are relieved by means of forming gas filled or gas and vapor filled cavities. The very destructive effects which are observed to occur when such cavitation form and collapse in a flow have been documented by Soyama and Kato[1].

By decreasing the cavitation numbers, four regimes can be identified: inception cavitation, sheet cavitation, cloud cavitation, and super cavitation[2]. Sheet/cloud cavitation usually causes the severe oscillating phenomena known as cavitation instabilities, which can result in severe pressure fluctuations, noise, vibration and erosion. A lot of researchers have analyzed experimentally the sheet/cloud cavitating flow in cavitation tunnels[3-7]. Venturi-type sections and two-dimensional foil sections are usually used to study unsteady sheet/cloud cavitation mechanism. Kubota et al.[3] have investigated the unsteady structure of cloud cavitation. The velocity field can be characterized by large-scale and small-scale structures. The overall cloud cavitation structure is associated with the large-scale velocity, while the small-scale velocity is noticeable around cavity boundaries. Furthermore, the structure of cloud cavitation moves at a velocity smaller than the mean free stream velocity and has a concentrated vorticity region at its centre. Stutz and Reboud[8] have experimentally studied the two-phase cavitating flow structure. Both quasi-steady sheet cavitation and large-scale, unsteady cavitation in convergent-divergent nozzles have been examined. The mean values of the void fraction, flow velocity and bubble size inside the cavities have been measured. The occurrence of a re-entrant jet under the cavity, that flows upstream and results periodically in cavity break off, is demonstrated for example by Stutz et al.[8] with double optical probe measurements and by Pham et al.[9] with electrical impedance probes. More recently, the pressure fluctuation in sheet/cloud cavitation has been investigated around a hydrofoil by Leroux et al.[6], shock waves produced by cloud cavity collapse is believed to influence the frequency of sheet/cloud cavitation. Reisman and Brennen et al.[10] discuss the pressure pulses of large amplitude and short duration during the rapid collapse of cloud cavitation that are very similar to pressure pulses in the present study, several types of propagating structures, the so-called bubbly shock waves, have been observed. The largest impulsive and radiated noise has been attributed to the coherent collapse of a well-defined separate cloud in regions of high pressure. A brief review of these recent works indicates that we still have inadequate understanding on the correlation between pressure fluctuation signal variation and unsteady cavitation behaviours, since it's difficult to measure the unsteady pressure fluctuation and moreover synchronize cavitation behaviours with pressure fluctuation.

The present study focuses on the unsteady cavitating flows in a convergent-divergent channel by experiment method. A high-speed camera is synchronized with four piezo-electric pressure transducers by a simultaneous sampling technique to study the correlation between unsteady cavitation behaviours and unsteady wall-pressure fluctuation. Corresponding post-processing method is used to analyze the power spectral density and magnitude of pressure impulse. Both quasi-periodic pressure fluctuations and pressure pulses are analyzed together with the corresponding sheet/cloud cavitation structures.

2. Approach and set-up

The experiments are carried out in a closed-loop cavitation tunnel[11].

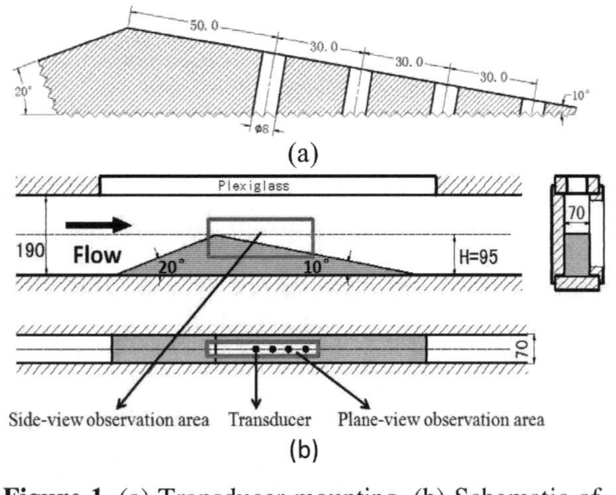

Figure 1. (a) Transducer mounting. (b) Schematic of test section and test model. Units in millimeter.

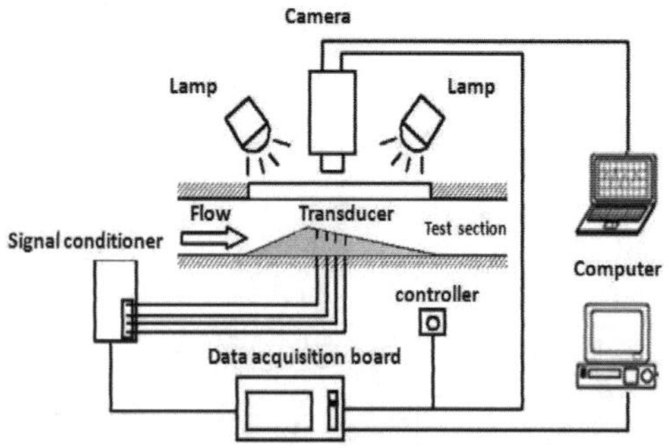

Figure 2. Layout of simultaneous sampling system.

Experiments are conducted in a rectangular test section of 70mm in width and 190mm in height as shown in Figure1(b). There are two windows, one on the top and one on the side, which are made of perspex for optical access. A triangle model is installed on the bottom wall to form a convergent-divergent channel, the convergent and divergent angle are 20 and 10 degree respectively, contraction ratio is 0.5. A separated-vortex type of cavitation generated from the leading edge of the divergent part can be observed from both top and side wall of the channel. The pressure fluctuations induced by unsteady cavity behaviours are measured by piezo-electric transducers installed flush with the wall surface as illustrated in Figure 1(a).

The Reynolds number and Cavitaiton number are defined as below:

$$Re= U_t H / \nu \quad (1)$$

$$\sigma=(P_\infty - P_v) / (0.5\rho_l U_t^2) \quad (2)$$

where, U_t, P_∞, P_v, ρ_l are, respectively, the average velocity at the throat of convergent-divergent test section, the static pressure, the vapor pressure, and the water density. The Reynolds number is fixed at 1.14×10^6.

The experimental conditions are maintained to within 1% uncertainly on the angle of convergent-divergent section, and 2% uncertainty on both the flow velocity and the upstream pressure. Together, the Cavitation number can be controlled to within 5% uncertainty.

The cavitation phenomena are documented by a high-speed digital camera(HG-LE,by Redlake), up to a rate of 105 frames per second(fps). In order to maintain desirable spatial resolutions, much lower recording speed is adopted. Specially, depending on the focus of the investigation, 3000 fps is used in the present study.

The unsteady wall-pressure fluctuations generated by the sheet/cloud cavitation behaviours are measured by 4 piezo-electric pressure transducers(sensibility 7.3mV/kPa), and the sampling rate is 1.024MHz.

Figure 2 shows the layout of simultaneous sampling system, a high-speed visualization and unsteady pressure fluctuation measurement set ups are combined by a controller. When the controller is triggered, the cavitation images and unsteady wall-pressure will be captured simultaneously, but at a different sampling rate. Even though the sampling rate of high-speed camera is far less than that of piezo-electric transducers, each cavity image has its own corresponding pressure fluctuation value, and the uncertainty of simultaneous sampling is negligible compared with the unsteady sheet/cloud cavitation development period.

3. Experiment results and discussion

3.1. Quasi-periodic behaviours of sheet/cloud cavitation

About one and a half cycles of sheet/cloud cavitation images captured from plane-view are shown in Figure 3. A typical quasi-periodic sheet/cloud cavitation development process can be divided into three main stages: (1) the process of growth of attached cavity, (2) the process of shedding of attached cavity, (3) the process of development and collapse of detached cavities, which are marked by line1, 2, 3, correspondingly.

In the stage one, as shown in Figure 3, the cavity image at T_1+0ms refers to the origin of a cycle. During $t\in$[T_1+0ms, T_1+8ms], the small vortex cavities coalesce with each other to form an attached cavity behind the throat. As the time develops, the attached cavity gradually grows and covers the wall surface. It reaches its maximum length at about T_1+32ms, while the four transducers are covered by attached cavity. In the stage two, the attached cavity begins to shed at the rear part and some small detached cavities are observed, finally, it's broken up totally at about T_1+48ms. In the stage three, a new cycle of growth begins when vortex cavities reappear near the throat and immediately form a new attached cavity. The small detached cavities in the rear of attached cavity gradually grows into a large detached cavity, and the growth of the large detached cavity is basically in phase with that of the newly formed attached cavity until it reaches its maximum size at about T_1+72ms. Finally, the large detached cavity collapses rapidly when it's convected to the downstream. It should be noted that, between T_1+0ms and T_1+28ms, small detached cavities also show a similar process of development and collapse in the rear of attached cavity.

3.2. Unsteady wall-pressure fluctuation

3.2.1. Pressure fluctuations associated with the unsteady sheet/cloud cavitation behaviours. As shown in Figure 4, about one and a half cycles of pressure fluctuation measured simultaneously with the sheet/cloud cavitation images in Figure 3 are presented. As marked by the number 1, 2, 3 on the top of Figure 4, they correspond separately to the three stages of sheet/cloud cavitation: (1) the process of growth of attached cavity, (2) shedding of attached cavity, (3) development and collapse of detached cavity as indicated in Figure 3(a).

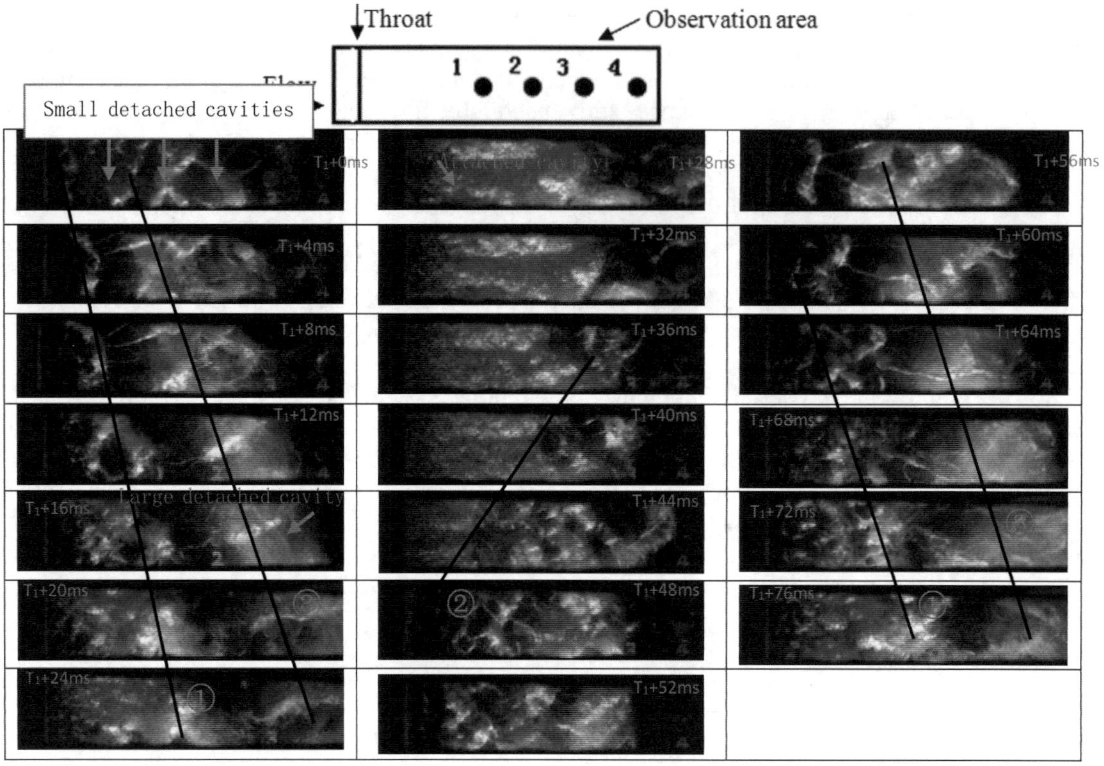

Figure 3. Plane view observation of sheet/cloud cavitation during about one and a half cycles. $\sigma = 0.75$

The locations of transducers in streamwise direction are indicated by the black points as shown in the schematic in Figure 3(a). After T_1+0ms, an attached cavity appears near the throat, as the attached cavity grows, it covers transducer 1# and 2# in succession. Accordingly, transducer 1# and 2# experience, one after another, a decrease in pressure fluctuation as indicated by the declining arrow as shown in Figure 4. When attached cavity obtains its maximum length at about T_1+30ms, severe cavity fluctuation is observed in the rear of attached cavity, accordingly, strong pressure fluctuations are captured on transducer 3# and 4# as shown in the rectangle frame labeled by c. This characteristic is consistent with that by Leroux et al.[6].

At about T_1+30ms, the attached cavity begins to shed and produce some small detached cavities. Due to the variation of these small detached cavities, small local pressure fluctuations are detected as indicated by the black vertical arrows.

After T_1+48ms, a large detached cavity is formed in the rear of a newly generated attached cavity, as it moves downstream gradually, a decrease, followed by an increase in pressure fluctuation is detected as shown by the lines labeled by d and e. (see also, for instance, label a and b between T_1+0ms and T_1+30ms). Such kind of correlation between detached cavity and pressure wave propagation agrees well with previous numerical results[12-14].

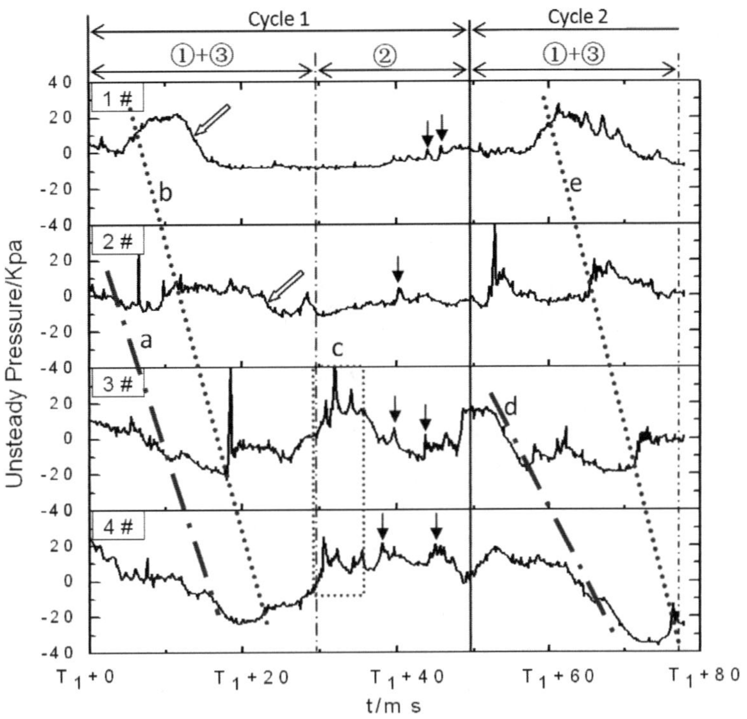

Figure 4. Instantaneous pressure signals during quasi-periodic cavity development process, $\sigma = 0.75$

3.2.2. Pressure pulses associated with the collapse of detached cavity. Figure 5 shows the cavitation structures in the process of rapid and coherent collapse of large detached cavity associated with the larger pressure pulses. The large detached cavity is also amplified for better observation, and the collapse region is indicated by the ellipse frame. As can be seen in the rectangle frame region, the large detached cavity collapses rapidly near transducer 3#. As the time develops, the collapse region becomes larger and larger. It should be noted that the collapse of large detached cavity does not involve large change in the overall dimensions of the cloud, rather, collapse involves large changes in the bubble density within the cloud. As shown in Figure 6, the pressure pulse with the magnitude of the order of several atmospheres are detected on transducer 3# which is located in the collapse region, three other transducers also detect the same pressure pulse as indicated by the black arrows with a discernible time delay. The significant difference of magnitudes of pressure pulse detected on the 4 transducers may due to the energy dissipation in the propagation process. By examining the time intervals separating the arrival of pressure pulses at the neighbouring transducers, it is possible to obtain the propagation speeds of the pulses. As presented by the bar graphs in Figure 7, the propagation speeds represent the rates at which the pressure pulses are propagated away from the detached cavity collapse region. In all cases, the propagation speeds are smaller than the sonic speed in the pure liquid. In the region between transducer 1# and 2#, the bubble density is high inside the attached cavity(as can be seen in Figure 5), and the corresponding propagation speed is about 240m/s. The bubble density is much lower in the region between transducer 2# and 3#, and the corresponding propagation speed is about 600m/s. In the region where large detached cavity collapses, the propagation speed is about 410m/s. The difference of propagation speed in different region may due to the fact that the sonic speed is decreased with the increase of bubble density in the flow field.

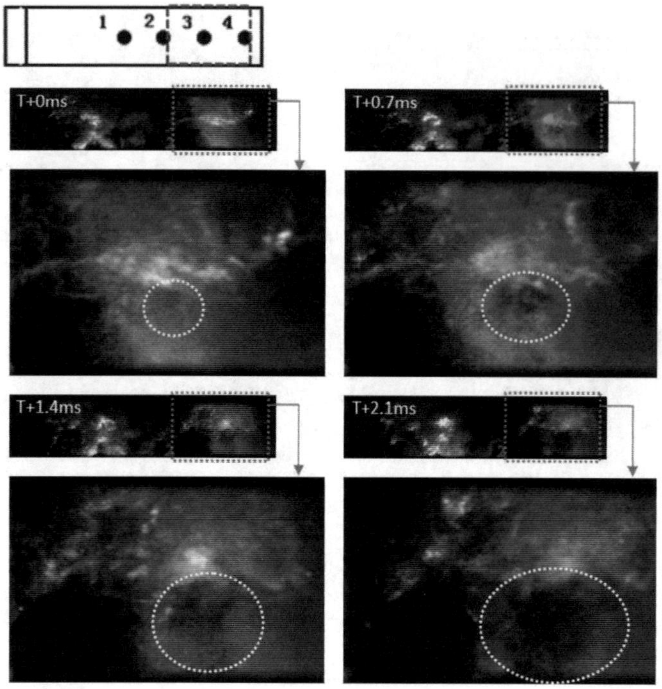

Figure 5. Cavitation structure in the process of rapid and coherent collapse of large detached cavity.

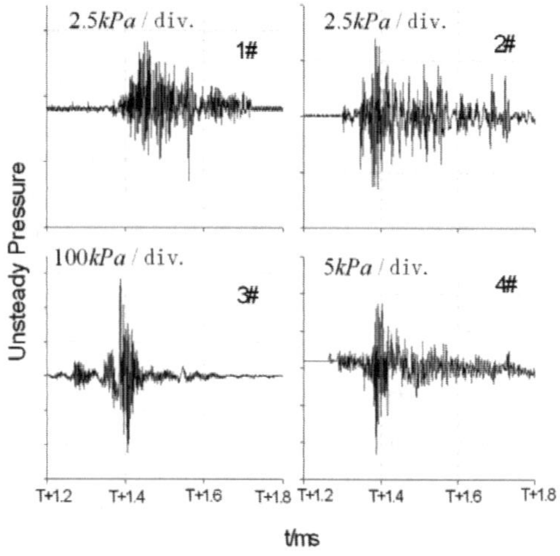

Figure 6. Time evolution of pressure pulse.

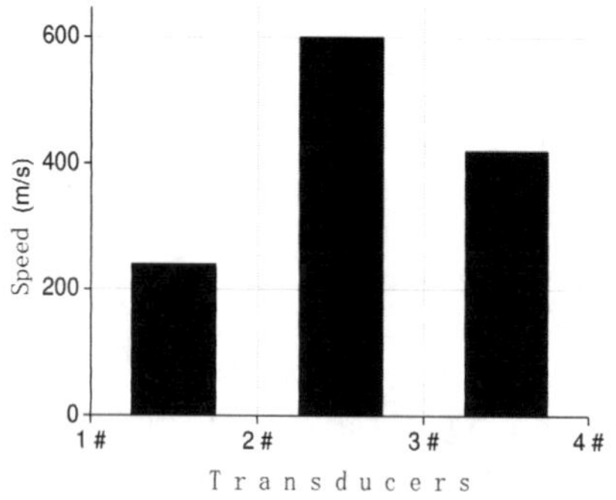

Figure 7. Propagation speed of pressure pulse.

4. Conclusion

The unsteady cavitating flows and corresponding wall-pressure fluctuation in a convergent-divergent channel is studied by a simultaneous sampling system. The following is a summary of the main findings:

(1) A typical quasi-periodic sheet/cavitation development process is characterized by three stages: (1) the process of growth of attached cavity, (2) the shedding of attached cavity, (3) the development and collapse of detached cavities.

(2) In the stage one, the magnitude of pressure fluctuations under the attached cavity are small,

however, it's large in the closure region of attached cavity, especially, when attached cavity obtains its maximum length. In the stage two, the attached cavity begins to shed and produce some small detached cavities. Due to the variation of these small detached cavities, small local pressure fluctuations with higher frequency are detected. In the stage three, a large detached cavity is formed in the rear of the channel, as it moves downstream gradually, a decrease, followed by an increase in pressure fluctuation is detected.

(3) When the detached cavity collapses in the downstream, and pressure pulses with the magnitude of the order of several atmospheres are detected. The difference in the propagation speed of pressure pulse in different cavitation region is found to be related with the bubble density in the flow field.

References

[1] Soyama H, Kato H and Oba R 1992 Cavitation observations of severely erosive vortex cavitation arising in a centrifugal pump *3rd I. Mech. E. Int. Conf. on Cavitation (Cambridge, UK, 9-11 December 1992)*
[2] Wang G Y, Senocak I and Shyy W 2001 *Prog. Aerosp. Sci.* **37**:551–81
[3] Kubota A, Kato H, Yamaguchi H and Maeda M 1989 *J. Fluids Eng.* **111**: 204-10
[4] Foeth E J, van Terwisga T and van Doorne C 2008 *J. Fluids Eng.-Trans. ASME* **130**, 071303
[5] Luo X, Ji B, Zhang Y and Xu H Y 2012 *Chin. Phys. Lett.* **29**(1): 016401
[6] Leroux J B, Astolfi J A and Billard Y 2004 *J. Fluids Eng.* **126**: 94-101
[7] Coutier-Delgosha O, Stutz B, Vabre A and Legoupil S 2007 *J. Fluid Mech.* **578**:171-222
[8] Stutz B and Reboud J L 1997 *Phys. Fluids.* **9**(12):3678–86
[9] Pham T M, Larrarte F and Fruman D H 1999 *J. Fluids Eng.* **121**: 289–96
[10] Reisman G E, Wang C and Brennen C E 1998 *J. Fluids Mech.* **355**: 255–83
[11] Huang B, Young Y L and Wang G Y 2013 *J. Fluids Eng.* **135**(7): 071301
[12] Luo X, Ji B and Peng X X 2012 *J. Fluids Eng.* **134**(4):041202
[13] Ji B, Luo X, Arndt R and Wu Y 2014 *Ocean Eng.* **87**: 64-77
[14] Ji B, Luo X, Arndt R, Peng X and Wu Y 2015 *Int. J. Multiphase Flow* **68**: 121-34

IOP Conf. Series: Materials Science and Engineering **72** (2015) 022005 doi:10.1088/1757-899X/72/2/022005

Numerical analysis of the interactions of sheet cavitation and cloud cavitation around a hydrofoil

D Z Pan, D S Zhang[1], H Y Wang, W D Shi and L Shi
Research Center of Fluid Machinery Engineering and Technology, Jiangsu University, Zhenjiang 212013, China

Email: zds@ujs.edu.cn

Abstract: Partial cavitation and cloud cavitation on NACA66(MOD) was studied based on PANS turbulence model combined with the Zwart cavitation model. The results agree well with the data from the experimental results. The existence of a kind of unsteady characteristics of cloud cavitation is analysed in details. The reasons for the inhibition mechanism of the sheet cavitation are discussed. Pressure shock wave occurs near the tailing edge of the hydrofoil when the collapse of cloud is so large leading to great influence on the cavity closed to the leading edge. The evolution of the pressure peak and its propagation toward the leading edge were investigated through the study of the time domain signal of the monitor points along the streamline of flow. Finally, the propagation of the pressure wave on the suction side was further investigated through analysing the spatial-time history of wall pressure.

1. Introduction

In the past years, many research about cavitation were carried out using experiment and simulation [1-4], which show that the turbulence model is crucial determining the accuracy of simulation prediction. The traditional RANS model over predicts the turbulence viscosity of the cavity rear part. In order to solve the problem, Coutier D et al [5] proposed an eddy viscosity correction for RNG k-ε model, and obtained good simulation predicting results. With the development of the computer technology, large eddy simulation (LES) began to applied in the unsteady cavitation simulation, but it is difficult to obtain a grid-independent solution [6]. Girimaji et al [7] proposed a partially-averaged Navier-Stokes(PANS) model, to achieve a seamlessly transition from RANS to direct numerical simulation(DNS) of the Navier-Stokes equations. Ji et al. [8-9] used the PANS model in the cavitation simulation for the twist hydrofoil and obtained an agreement results.

In the past years, the question in all cavitation researches concerns the origin of the flow instability. Leroux et al. [10] found the relation of the length of the cavity and the foil chord exists in the experiment. Later, Leroux [1] found the inhibition phenomenon of residual cavity. Jung H S et al. [11] employed the Shin cavitation model [12] to further investigate the problem.

In the present paper, the cavitation field around the two-dimensional NACA66(MOD) hydrofoil were investigated, trying to further explore the interactions of sheet cavitation and cloud cavitation.

[1] Corresponding author.

2. Governing equations and numerical method

2.1. Turbulence model

The PANS model can transit smoothly from RANS to DNS through changing the two control parameters (f_k and f_ε), which are defined as follow:

$$f_k = k_u / k, \quad f_\varepsilon = \varepsilon_u / \varepsilon \tag{1}$$

Where k and ε are total turbulence kinetic energy and turbulence eddy dissipation respectively. The subscript u refers to the unresolved quantities. Based on the standard k-ε model, Girimaji [7] shows the governing equations of the PANS model as follow:

$$\frac{\partial(\rho k_u)}{\partial t} + \frac{\partial(\rho u_j k_u)}{\partial x_j} = \frac{\partial}{\partial x_j}[(\mu + \frac{\mu_t}{\sigma_{ku}})\frac{\partial k_u}{\partial x_j}] + P_{ku} - \rho\varepsilon_u \tag{2}$$

$$\frac{\partial(\rho\varepsilon_u)}{\partial t} + \frac{\partial(\rho u_j \varepsilon_u)}{\partial x_j} = \frac{\partial}{\partial x_j}[(\mu + \frac{\mu_t}{\sigma_{\varepsilon u}})\frac{\partial \varepsilon_u}{\partial x_j}] + C_{\varepsilon 1} P_{ku}\frac{\varepsilon_u}{k_u} - C^*_{\varepsilon 2}\rho\frac{\varepsilon_u^2}{k_u} \tag{3}$$

Compared to the standard k-ε model, PANS model modified some coefficients as follow:

$$\sigma_{ku} = \sigma_k f_k^2 / f_\varepsilon, \quad \sigma_{\varepsilon u} = \sigma_\varepsilon f_k^2 / f_\varepsilon \tag{4}$$

$$C^*_{\varepsilon 2} = C_{\varepsilon 1} + f_k / f_\varepsilon (C_{\varepsilon 2} - C_{\varepsilon 1}) \tag{5}$$

Where σ_k=1.0, σ_ε=1.3, $C_{\varepsilon 1}$=1.44 and $C_{\varepsilon 2}$=1.92.

2.2. Cavitation model

Cavitation model is a mathematical model using describing the mass transfer between the liquid and vapor, which is defined as

$$\frac{\partial \rho_v \alpha_v}{\partial t} + \frac{\partial(\rho_v \alpha_v u_j)}{\partial x_j} = m \tag{6}$$

The source terms for the mass transfer rate are given by

$$m = \begin{cases} -F_e \dfrac{3 r_{nuc}(1-\alpha_v)\rho_v}{R_B}\sqrt{\dfrac{2}{3}\dfrac{p_v - p}{\rho_l}} & (p < p_v) \\ F_c \dfrac{3\alpha_v \rho_v}{R_B}\sqrt{\dfrac{2}{3}\dfrac{p - p_v}{\rho_l}} & (p > p_v) \end{cases} \tag{7}$$

Where F_e=50, F_c=0.01, r_{nuc}=5×10^{-4}, R_B=10^{-6} based on the work by Zwart et al [13].

2.3. Boundary conditions and grid

The simulation used the NACA66(MOD) hydrofoil with a chord length c=150 mm. Figure 1 shows the classical boundary conditions for hydrofoil: imposed inlet velocity and fixed outlet pressure. A no-slip wall is used for upper and lower walls and hydrofoil surface. The total grid elements are 89172 and the y+ values are within 0~10.

According to the experiment [12], the angle of attack and the cavitation number are 6°, 0.99 respectively. Velocity U_{in}=5.33 m/s is fixed at the inlet, corresponding to Reynolds number 8×10^5. The outlet pressure is determined by the cavitation number. In the unsteady simulation, the time step Δt was set to 10e-3s.

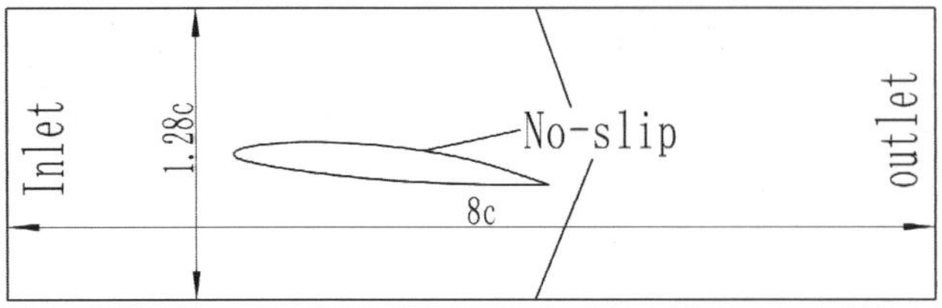

Figure 1. Computational domain and boundary conditions

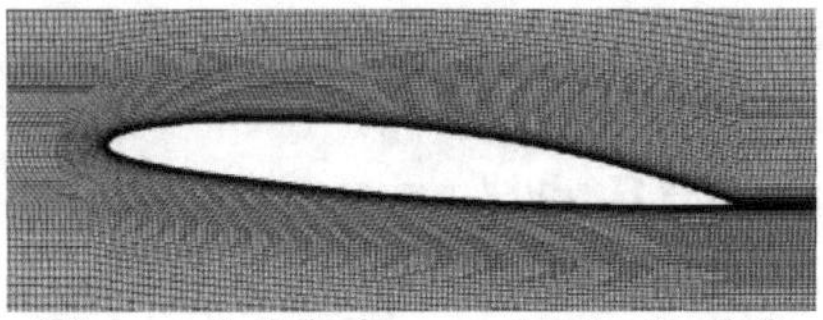

Figure 2. Mesh near the hydrofoil

3. Results and discussion

3.1. One-cycle cavitation oscillation

According to the arrangement of the Leroux's experiment, the cavitation field around the NACA66(MOD) hydrofoil has been investigated. Employed of the homogeneous flow model, the volume fraction contour in the picture represents the distribution of the bubbles in the flow field. The two extremes of the rainbow chart, the red and blue correspond to pure vapor and water. The comparison between the experiment and simulation of the cavitation cavity, named dynamics 1 by Leroux, shows in the follow pictures.

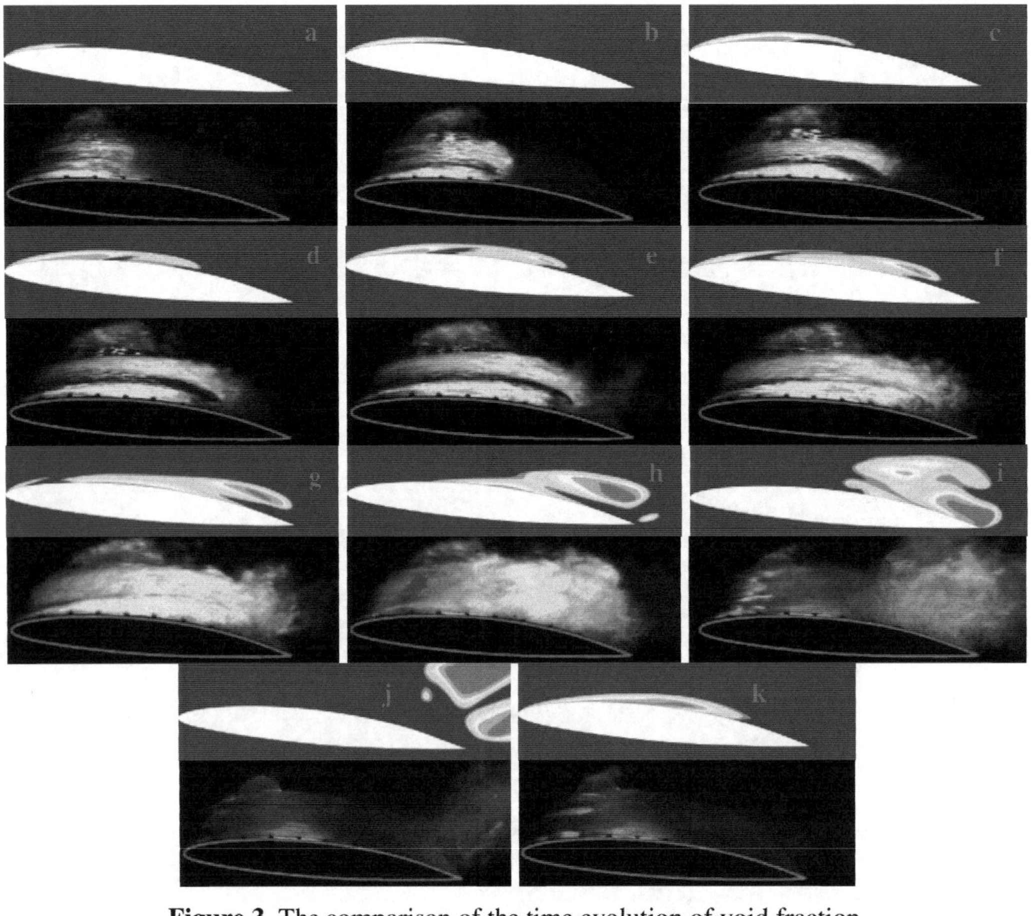

Figure 3. The comparison of the time evolution of void fraction.

Based on the above results, cloud cavitation of dynamics 1 within each cycle can be divided into the following steps:(a) cavitation first appeared at the leading edge in the form of the attached cavity and gradually expand along the hydrofoil surface toward the tail edge;(b) With the attached cavity continued to advance toward the tail edge direction, the length of the cavity unceasing increases and the adverse pressure gradient at the rear part of the cavity gradually increases.(c) The shear layer closed to rear part of the cavity clusters began to become unstable and thus the re-entrant jet initial forms under the adverse pressure gradient; (d)With the re-entrant jet becoming stronger, the cavity were broken up to two parts and the front part of the broken cavity fast shrink toward the leading edge and the rear part formed the cloud bubble clusters entrained by vortex;(e)Along with the cloud cavity moving to the downstream, the front portion of the residual attach cavity continues to grow and begin a new cycle.

As seen in the figure i and j, the residual sheet cavity closed to the leading edge is suppressed and then disappears after the cloud cavity forms at the trailing edge of the hydrofoil. What does suppress the development of the sheet cavity at the front part of the hydrofoil?

3.2. The interactions between the sheet cavitation and cloud cavitation

In order to eliminate the effects of the model size, a dimensionless parameter $(p-\overline{p})/\rho_0 u_0^2$ was selected. Where $\overline{p}$ is the averaged pressure. Figure 4 shows some pressure contour near the instant of the figure i, j within the figure3. From the current numerical calculation, it can be seen that the local pressure peak is appeared during the process of the formation and shedding of the cloud cavity. These local pressure peaks gradually spread toward the surroundings through the pressure wave. Figure 4 shows the local pressure peaks and the spreading pressure wave induced by it suppress the development of the sheet cavity at the front part of the hydrofoil. The possible reason is that the area of the passage is changed due to the formation of the cloud cavity. And the lager cloud cavity has a blocking effect on the flow field. Then the pressure peaks form at both sides of the cloud cavity. And the pressure peak at the front part of the cloud cavity and the enough stronger of the pressure wave induced by it suppress the development of the sheet cavity at the front part of the hydrofoil. The inhibition on the sheet cavity become weaken with the decreasing of the stronger of the pressure peak and its moving to downstream due to the cloud cavity shedding toward the downstream. Subsequently, the sheet cavity regenerates.

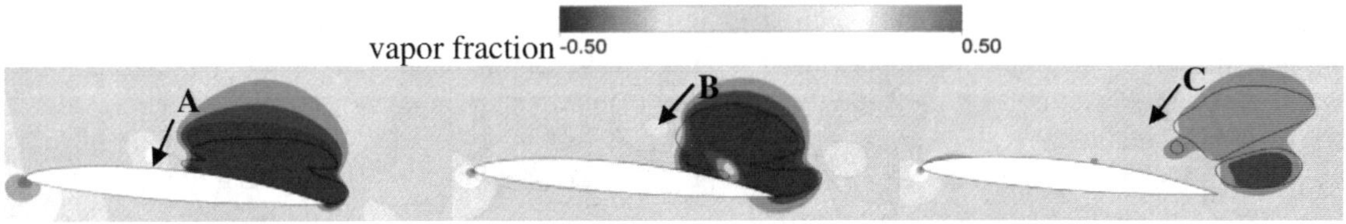

Figure 4. Local pressure peaks and its propagating to surroundings.

Figure 5 shows the simulation results of the pressure pulse within the two cycles at the three monitor points, namely *x/c=0.3, 0.5, 0.7*. As seen in the figure, the pressure fluctuates in the range from the high value (corresponds to the pure liquid) to low value (corresponds to the appearance of the bubbles). Sequence 1 and 2 respectively correspond to the development of the cavity and the shedding of the cloud cavity. The pressure drop indicated by the black arrows indicates the appearance of the bubbles here. It can clearly see that after the collapse of the cloud cavity, the pressure wave propagate from the trailing edge to leading edge, followed by the pressure increases progressively, and then pressure spikes occur. With the receding of the stronger pressure wave induced by the collapse of the cloud cavity, the pressure on the suction side is repeatedly fluctuate, as seen in the region 2 of the picture. Subsequently, the pressure falls down again and the cavitation reappears, and the pressure on the suction side reduces from the leading edge to trailing edge, and the attached sheet cavity also gradually extends as the order. With the attached sheet cavity along the surface expands toward the trailing edge, the pressure gradient at the trailing part of the cavity gradually increases and the stronger of the re-entrant jet gradually becomes larger. The cavity breaks up after the re-entrant jet becoming enough stronger. Figure 5 the region marked A circled by green dash lines indicates the influences of the breakup of the cavity on the pressure monitor curves. It can conclude that the more close to the leading edge, the bigger influences the breakup of the cavity impacts on and the longer the duration time.

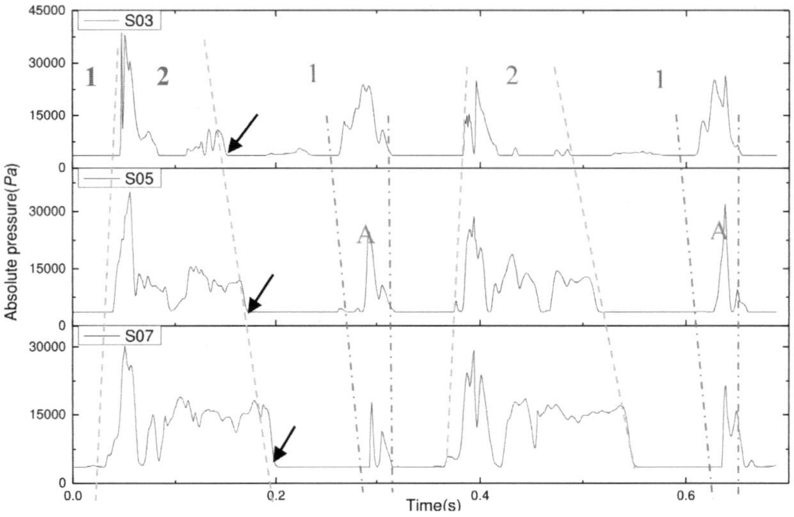

Figure 5. Instantaneous pressure signals during cavity growth/destabilization.

In order to further investigate the propagation of the pressure wave on the suction side, the spatial-time history of wall pressure is made, about the instant i, j of the figure 3. As seen in the picture, the wall pressure is divided into two distinct parts. At the front part the pressure is maintained at a high level, thereby suppress the development of the sheet cavity near the front part of the hydrofoil. However, as seen in the figure, the first part of the distinct parts has been divided into two half part by pressure dimples. Overall, the pressure of the suction side gradually decreases over time. Judging from the fragments of the local time, the pressure of the first part repeatedly changes in the cycle of the fall-rise-fall. The pressure dimples dividing the first part into two parts become more clearly. The latter part, where the cloud cavity occupies, a small pressure peak rises first over time and then spreads toward the downstream, and then rises again and pulls the pressure of the zone originally occupied by the cloud cavity and propagates toward upstream.

4. Conclusions

The present work used simulation to reproduce the inhibition phenomenon of sheet cavity near the front part of the hydrofoil found and named dynamics 1 in the Leroux's experiments. Through analysing the pressure contour of the inhibition moment of the development of the sheet cavity, the local pressure peaks induced during the process of the formation of the cloud cavity and its shedding is found. The local pressure peak and the pressure wave induced by it play an important role in suppressing the development of the sheet cavity near the front part of the hydrofoil. Then the time domain signal of the pressure recorded by some monitor points at the suction side of the hydrofoil indicates the collapse of the cloud cavity results in the emergence of the pressure spike. The breakup of the cavity induced by the re-entrant jet has a different influence on the various parts of the hydrofoil. And the more close to the leading edge of the hydrofoil, the bigger of the influences, and the longer of the duration time. Through the spatial-time history of wall pressure in the suction side, the propagation of the pressure wave on the suction side was further investigated and the phenomenon of the pressure fluctuation distinctly divided into two parts is found. The higher pressure level of the first part of the pressure curves suppresses the development of the sheet cavity near the front part of the hydrofoil.

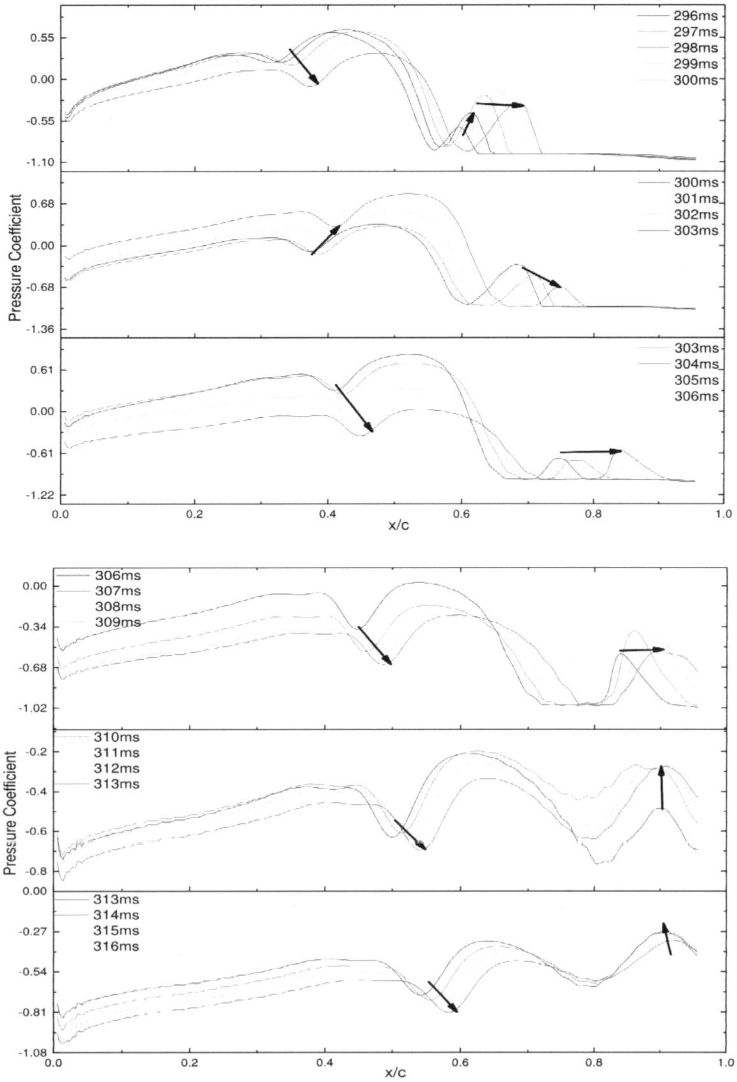

Figure 6. Spatial-time history of wall pressure in the suction side of the hydrofoil during the formation of the cloud cavity and its shedding.

Acknowledgments

This work was supported by National Natural Science Foundation of China (Grant No.51109093), Graduate Students research and Innovation Program of Jiangsu Province (CXLX13_659), Natural Science Foundation of the Higher Education Institutions of Jiangsu Province (13KJD470001) and PAPD.

References

[1] Leroux J B, Coutier-Delgosha O and Astolfi J A 2005 *Physics of Fluids* **17**(5) 052101
[2] Wang G Y, Senocak I, Shyy W, Ikohagi T and Cao S 2001 *Progress in Aerospace Sci.* **37** 551-81
[3] Coutier-Delgosha O, Reboud J L and Delannoy Y 2003 *Int. J. for numerical methods in fluids* **42**(5) 527-48
[4] Ducoin A, Huang B and Yong Y L 2012 Int. J. of Rotating Machinery **2012** Article ID 215678
[5] Coutier-Delgosha O, Fortes-Patella R and Reboud J L 2003 *J. Fluid Eng.* **125** 38-45

[6] Moin P 2002 *Int. J. of Heat and Fluid Flow* **23** 710-20
[7] Girimaji S S 2006 *J. of Applied Mechanics* **73** 413-21
[8] Ji B, Luo X W, Arndt R E A and Wu Y L 2014 *Ocean Engineering* **87** 64-77
[9] Ji B, Luo X W, Wu Y L, Peng X X and Duan Y L 2013 *Int. J. of Multiphase Flow* **51** 33-43
[10] Leroux J B, Astolfi J A and Billard J Y 2004 *J. Fluid Eng.* **126**(1): 94-101
[11] Seo J and Lele S 2009 Numerical investigation of cloud cavitation and cavitation noise on a hydrofoil section *Proc. of the 7th Int. Symp. On Cavitation (Ann Arbor, Michigan, USA, 17-22 Aug 2009)*
[12] Shin B R, Iwata Y and Ikohagi T 2003 Computational Mechanics **30**(5-6) 388-95
[13] Zwart P J, Gerber A G and Belamri T 2004 A Two-Phase Flow Model for Predicting Cavitation Dynamics *5th International Conference on Multiphase Flow (Yokohama, Japan, 30 May-4 June 2004)*

IOP Conf. Series: Materials Science and Engineering **72** (2015) 022006 doi:10.1088/1757-899X/72/2/022006

Applications of a curvature correction turbulent model for computations of unsteady cavitating flows

Y Zhao, G Y Wang, B Huang and C L Hu
Beijing Institute of Technology, Beijing 100081, China

Email: zhaoyu2011@bit.edu.cn

Abstract. A Curvature Correction model (CCM) based on the original k-ε model is proposed to simulate unsteady cavitating flows. The objective of this study is to validate the CCM model and further investigate the unsteady vortex behaviors of cavitating flows around a Clark-Y hydrofoil. Compared with the original k-ε model, predicted results are improved in terms of the cavity detachment and hydrofoil fluctuations. Results show that streamline curvature correction of CCM model overcomes the over-predictions of turbulence kinetic energy and eddy viscosity in cavitating vertical region with the original k-ε model, which leads to better simulation abilities for the unsteady cavitating flow computations. Based on computations, it is proved that the vortex structure is significantly modified by the transient cavitation, especially with respect to the cavity shedding behaviors. Complex vortex interactions and corresponding cavity shedding process near hydrofoil trailing edge lead to various load frequencies.

1. Introduction

Cavitation is a dynamic phase-change phenomenon that occurs in liquids when the static pressure drops below the vapor pressure of liquid[1] [2]. It is well known that the unsteady cavitation in turbo-machinery and marine control surfaces will lead to problems such as material damage, vibration, noise and reduced efficiency[3] [4]. Actually, cavitation physical mechanisms are not well understood due to the complex, unsteady flow structures associated with turbulence and phase change.

Cavitating flows are generally relatively high Reynolds number flows and hence the turbulence modeling plays an important role in the capture of unsteady behaviors. Conventional eddy-viscosity RANS turbulence models, such as the k-ε model, have been widely used in many turbulent flows of practical interest. However, many researchers [5] [6] have indicated that high eddy viscosity of the original Launder–Spalding version of the k-ε model may dampen the vortex shedding behavior and excessively restrain the cavitation instabilities. To improve the predicted results for unsteady cavitating flows with conventional eddy viscosity models, many approaches have already been developed. A widely used approach is the density correction model (DCM) [7], which takes into account the compressibility characteristics of cavitating flows. Better predictions of unsteadiness of cavitating flows with DCM model have been proved [8] [9] [10]. Similarly, the present paper is devoted to the refinement of conventional eddy-viscosity RANS model considering the streamline curvature effect in strong vertical cavitating flows.

It's well known that instabilities of cavities often result in the formation of large-scale vortex structures[11] and various researches have proved the strong correlation between cavitation and vortex structures. Gopalan and Katz[12] observed that the collapse of the vapor structure is a primary mechanism of vorticity production leading to the generation of hairpin vortices in the downstream region. Lyer and Ceccio[13] [14] found the streamwise velocity fluctuations increased with cavitation, which is confirmed that the collapse of the vapor cavities is a source of vorticity generation. An inherent shortcoming of the conventional eddy-viscosity RANS turbulence models is their property of material frame indifference[15]. In other words, these models do not contain streamline curvature production terms, and hence are not capable of capturing effects of streamline curvature in strong vertical cavitating flows. An effective approach for resolving this issue is proposed by Spalart and Shur[16], which has been introduced to many conventional eddy-viscosity RANS turbulence models, including the Spalart–Allmaras (S-A) model[17] and the Shear Stress Transport (SST) model[18]. With revised models, better predictions of streamline curvature effect in turbulent flows with conventional eddy-viscosity RANS models have been documented including 2D flow in a channel with U-duct, 3D foil tip vortex flow and so on.

In the present paper, the streamline curvature correction is introduced to the *k-ε* model to achieve the curvature correction model (CCM) for simulations of sheet/cloud cavitating flows around a Clark-Y hydrofoil. Validations of CCM model are obtained comparing with experimental data as well as the baseline model, and hence advantages of CCM model are highlighted. Based on the numerical results, unsteadiness of vortex structures and induced lift fluctuations in cloud cavitating flows are discussed.

2. Numerical model in cavitation computation

2.1. Favre-averaged continuity and momentum equations

The set of governing equations comprises the conservative form of the incompressible Favre-averaged Navier-Stokes equations, coupling with classical cavitation model and turbulence closure. The mass continuity, momentum equations are given below.

$$\frac{\partial \rho_m}{\partial t}+\frac{\partial(\rho_m u_j)}{\partial x_j}=0 \tag{1}$$

$$\frac{\partial(\rho_m u_i)}{\partial t}+\frac{\partial(\rho_m u_i u_j)}{\partial x_j}=-\frac{\partial p}{\partial x_i}+\frac{\partial}{\partial x_j}[(\mu_L+\mu_T)(\frac{\partial u_i}{\partial x_j}+\frac{\partial u_j}{\partial x_i}-\frac{2}{3}\frac{\partial u_k}{\partial x_k}\delta_{ij})] \tag{2}$$

The mixture property φ_m, can be expressed as

$$\varphi_m=\varphi_l\alpha_l+\varphi_v(1-\alpha_l) \tag{3}$$

where ρ_m is the mixture density, *u* is the velocity, *p* is the pressure, μ_L and μ_T are the laminar and turbulent viscosity, subscripts *i*, *j*, *k* are the directions of the axes, subscripts *l* and *v* present liquid and vapor, and φ can be density, viscosity and so on.

2.2. Curvature correction turbulent model(CCM)

2.2.1. Baseline model. The original *k-ε* turbulence model has become the workhorse of practical engineering flow calculations since it was proposed by Launder and Spalding. The *k-ε* model shows two partial differential equations for the transport of the turbulence kinetic energy *k* and dissipation rate *ε*:

$$\frac{\partial(\rho_m k)}{\partial t}+\frac{\partial(\rho_m k u_j)}{\partial x_j}=P_k-\rho_m\varepsilon+\frac{\partial}{\partial x_j}\left[\left(\mu_m+\frac{\mu_T}{\sigma_k}\right)\frac{\partial k}{\partial x_j}\right] \tag{4}$$

$$\frac{\partial(\rho_m\varepsilon)}{\partial t}+\frac{\partial(\rho_m\varepsilon u_j)}{\partial x_j}=C_{\varepsilon 1}\frac{\varepsilon}{k}P_k-C_{\varepsilon 2}\rho_m\frac{\varepsilon^2}{k}+\frac{\partial}{\partial x_j}\left[\left(\mu_m+\frac{\mu_T}{\sigma_\varepsilon}\right)\frac{\partial\varepsilon}{\partial x_j}\right] \tag{5}$$

where the production term of turbulence kinetic energy (P_k) and the Reynolds stress tensor (τ_{ij}) are defined as: $P_k=\tau_{ij}\frac{\partial u_i}{\partial x_j}$ and $\tau_{ij}=-\rho_m u_i' u_j'=\frac{2}{3}\rho_m k\delta_{ij}-\mu_T\left(\frac{\partial u_i}{\partial x_j}+\frac{\partial u_j}{\partial x_i}\right)$, with $C_{\varepsilon 1}=1.44$, $C_{\varepsilon 2}=1.92$, $\sigma_\varepsilon=1.3$, and $\sigma_k=1.0$.The turbulent eddy viscosity is defined as:

$$\mu_T=\frac{C_\mu\rho_m k^2}{\varepsilon},\quad C_\mu=0.09 \tag{6}$$

2.2.2. Curvature correction method. Shur[16] suggested a modification to describe the streamline curvature effect in strong vertical flows. Correction function $f_{Rotation}$ is defined as:

$$f_{\text{Rotation}}=(1+C_{r1})\frac{2r^*}{1+r^*}\left[1-C_{r3}\tan^{-1}(C_{r2}\tilde{r})\right]-C_{r1} \tag{7}$$

$\tilde{r}$ and r^* here are defined as

$$\tilde{r}=2\omega_{ik}S_{jk}\left(\frac{dS_{ij}}{dt}\right)\frac{1}{\omega D^3},\quad r^*=\frac{S}{\omega} \tag{8}$$

The strain rate and vorticity tensor are defined, respectively, using Einstein summation convention as

$$S_{ij}=\frac{1}{2}\left(\frac{\partial u_i}{\partial x_j}+\frac{\partial u_j}{\partial x_i}\right),\quad \omega_{ij}=\frac{1}{2}\left(\frac{\partial u_i}{\partial x_j}-\frac{\partial u_j}{\partial x_i}\right) \tag{9}$$

where $S=\sqrt{2S_{ij}S_{ij}}$, $\omega=\sqrt{2\omega_{ij}\omega_{ij}}$, $D=\sqrt{\max\left(S^2,0.09\omega^2\right)}$. The empirical constants C_{r1}, C_{r2} and C_{r3} are set to 1.0, 2.0 and 1.0, respectively. The curvature correction function f_r is used as a multiplier of the production term P_k in Eq.(4).

$$P_k \rightarrow P_k \cdot f_r \tag{10}$$

where $f_r=\max\left(0,1+C_{scale}\left(f_{r1}-1\right)\right)$ and $f_{r1}=\max\left(\min\left(f_{Rotation},1.25\right),0\right)$.

The original curvature correction function is limited in the range from 0 up to 1.25. The lower limit is introduced for numerical stability reasons, whereas the upper limit is needed to avoid over-generation of the eddy viscosity. The scaling coefficient has been set with different values in this research to govern the effect of the curvature correction needed for cavitating flow.

2.3 Computational setup

Computational set-up is shown in figure 1. The computational domain and boundary conditions are given according to the experimental set-up[9]. The Clark-Y hydrofoil with the chord length of c=0.07m is located at the center of the test section with the angle of attack of 8°. The two important dimensionless parameters are the Reynolds number Re and the cavitation number σ, which is defined based on the outlet pressure p, the saturated vapor pressure p_v, and the inlet velocity U_∞.

$$Re=\frac{U_\infty c}{\nu},\quad \sigma=\frac{p-p_v}{\rho U_\infty^2/2} \tag{11}$$

Computations are performed for cloud cavitation condition (σ=0.8). Constant velocity is imposed at the inlet, U_{∞}=10m/s, with the corresponding Reynolds number of $Re=7\times10^5$. The vapor pressure of water at 25 °C is p_v=3169 Pa. Details of the grid solutions are shown in Huang's previous work[9]. Basically, the numerical grids used here are enough to the wall function requirement.

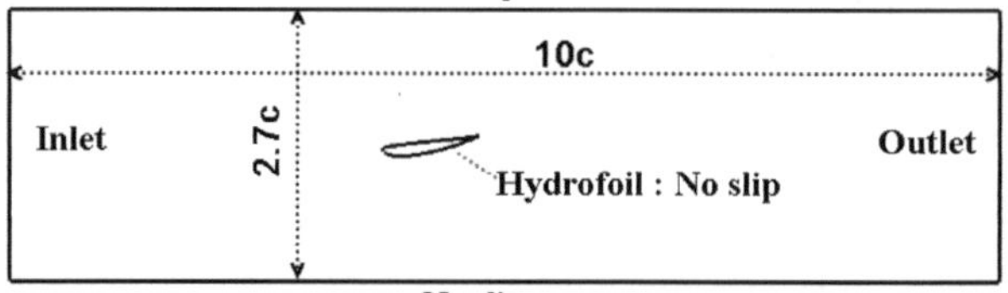

Figure1. Boundary conditions for Clark-Y hydrofoil

3. Results and discussions

3.1. Time-averaged cavity visualization and flow structures

The time-averaged flow structure and cavity shape with different models are shown in figure 2. The cavitation structures consist of two parts, which are attached and detached cavity respectively. The attached cavity is located in the leading edge of the hydrofoil, while the detached cavity is near the trailing edge. In the figure, attached cavity can be predicted with both the original k-ε and CCM models. The detached cavity cannot be captured with the original k-ε models, while it can be well predicted by the CCM models with appropriate scaling coefficient (C_{Scale}=10, 20).

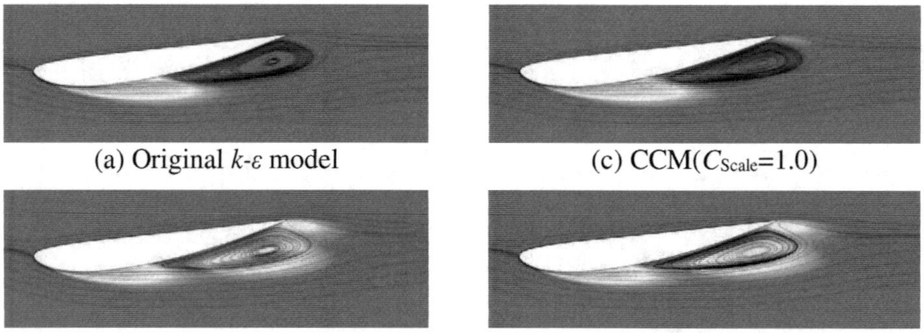

(a) Original k-ε model (c) CCM(C_{Scale}=1.0)

(b) CCM(C_{Scale}=10) (d) CCM(C_{Scale}=20)

Figure 2. Time-averaged vapor volume fraction contour and representative streamlines

Previous researches [9] [18] indicate that the detached cavity is formed due to the re-entrant jet and overlaps with the vertical zone near the trailing edge. The formation and significance of the re-entrant jet with CCM model (C_{Scale}=10) are first highlighted. In Figure3(a), a vertical zone consists of the re-entrant jet in the upper part and incoming flow from upstream in the lower part and the front of the re-entrant jet will determine the cavity end. The vertical zone will grow in size while the re-entrant jet pushing the attached cavity toward upstream will also become stronger, and then the cavity is detached in Figure3(b) with a low density region near the center of the vertical zone.

It is known that the original k-ε turbulence model over-predicts the turbulent viscosity, causing the re-entrant jet in cavitating flows to lose momentum. Thus, the re-entrant jet is not able to cut across the cavity sheet, which significantly modifies the predictions of detached cavity. To assess the CCM models, time averaged turbulence eddy viscosity contours are shown in Figure4. In the figures, significant decrease of time averaged turbulence eddy viscosity in vertical zone with CCM models can be observed. It is due to the decrease of predicted time averaged turbulence kinetic energy discussed

IOP Conf. Series: Materials Science and Engineering **72** (2015) 022006 doi:10.1088/1757-899X/72/2/022006

before. As over-predictions of eddy viscosity with the original k-ε model can be well solved, better predictions of detached cavity can be achieved with CCM model.

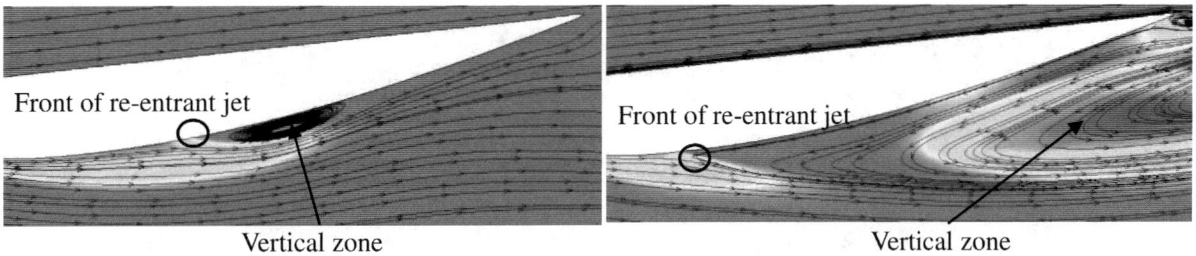

(a) Formation of the re-entrant jet (b) Highlight of the detached cavity

Figure 3. The generation of re-entrant jet and detached cavity (by CCM model(C_{Scale}=10) with streamline)

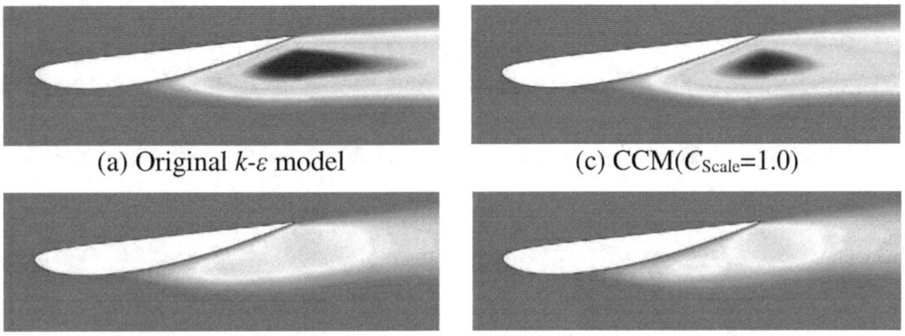

(a) Original k-ε model (c) CCM(C_{Scale}=1.0)

(b) CCM(C_{Scale}=10) (d) CCM(C_{Scale}=20)

Figure 4. Time averaged turbulence eddy viscosity contours

3.2. Instantaneous Cavity Shapes and flow structures

Comparisons of the temporal evolutions of the computational and experimentally observed cavity structures are shown respectively in Figure5. The results demonstrate that the CCM model (C_{Scale}=10) is capable of capturing the time-dependent cavity performances, in accordance with the experimental observations. The vertical zone growing (t_0~t_0+28.6% *Cycle*) and hence detached cavity formatting process shown in Figure2 can be captured. Moreover, the detached cavity will be dissipated when it travels toward downstream (t_0+55.5%~t_0+74.5% *Cycle*). Then, the re-entrant jet and the vertical zone will become weaker, and meanwhile the attached cavity will grow up again to form the next cycle. On the contrary, the original k-ε model fails to capture the time-dependent cavity performances.

The curvature correction function f_r, and hence revised turbulence kinetic energy predicted with the CCM model (C_{Scale}=10) are shown in Figure6. Predicted turbulence kinetic energy contours at representive times with the original k-ε model are also shown as a comparison. From the figures, it is shown that as the attached cavity expands towards hydrofoil trailing edge (t_0~t_0+28.6% *Cycle*), large value of curvature correction function can be observed inside of the vertical zone. Curvature correction during this process results in the decrease of predicted turbulence kinetic energy. And hence turbulence eddy viscosity decreases, which keeps predicted attached cavity stretching until covering the whole hydrofoil suction side. As the attached cavity breaks up and partially sheds downstream (t_0+55.5%~t_0+74.5% *Cycle*), curvature correction function is remarkable in the vertical zone. Decrease of predicted turbulence kinetic energy hence turbulence eddy viscosity can be seen during this process, which leads to better capture of the unsteadiness of cavitation.

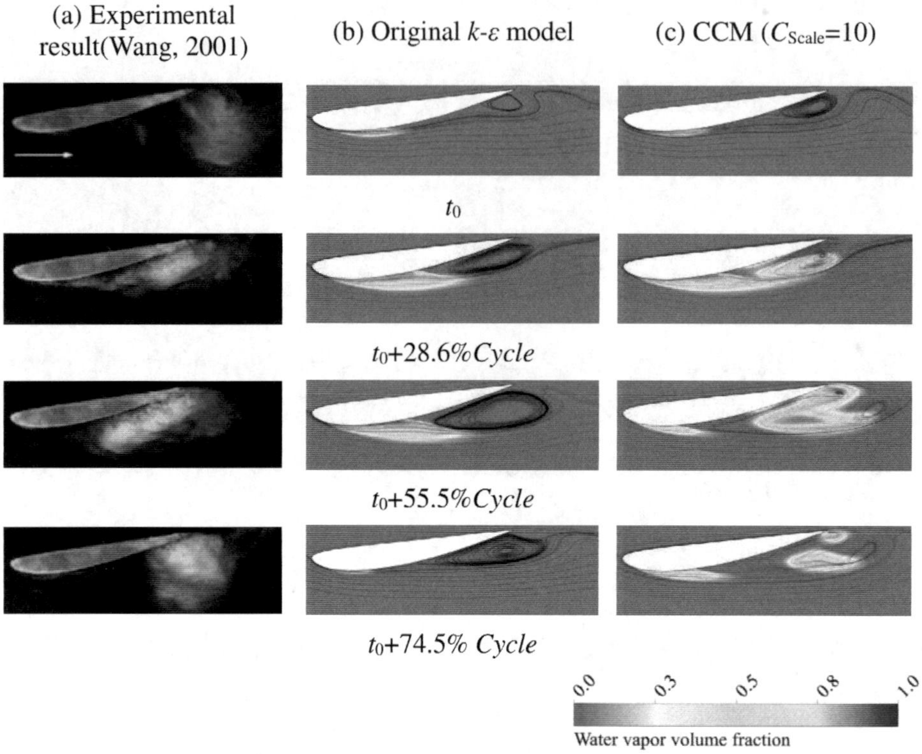

Figure 5. Instantaneous vapor fraction contours and streamlines

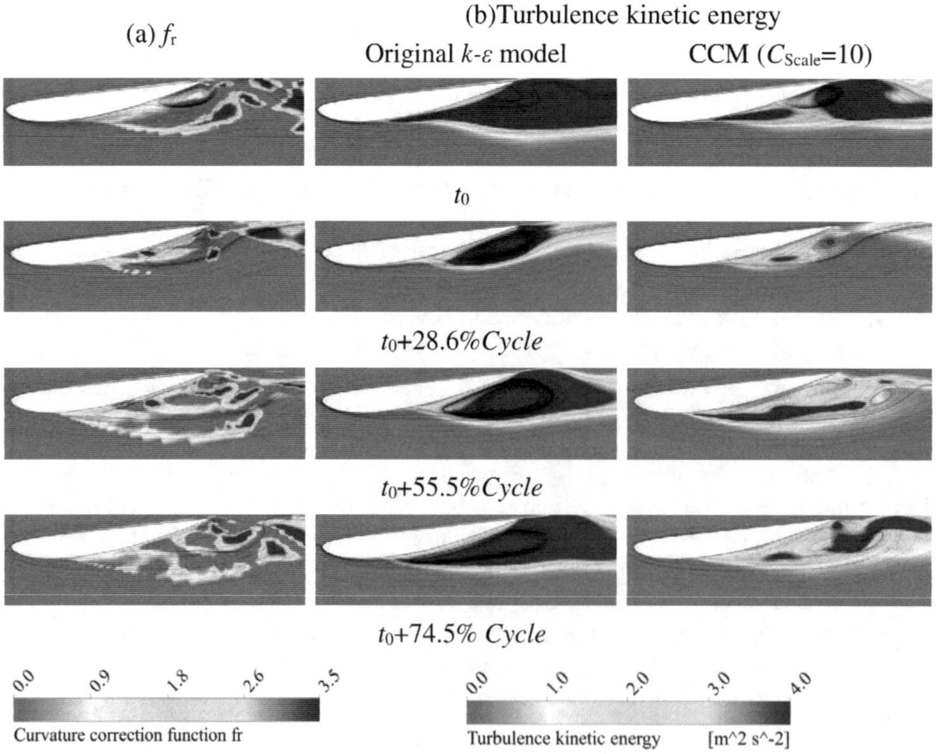

Figure 6. Time evolutions of turbulence kinetic energy contours

In order to compare the space-time evolution of the cavity shapes predicted with different schemes, the vapor volume fractions of the cavity are plotted as a function of the time and space. Comparisons between the shedding processes predicted with different models are presented in Figure 7. The x- axis is dimensionless time, T_{ref}, defined as:

$$T_{ref} = c/U_{\infty} \tag{12}$$

Where c is the chord length and U_{∞} is the free stream velocity. The y-axis is dimensionless position at the foil suction side. L/L_{ref}=0 corresponds to the leading edge, and L/L_{ref}=1.0 corresponds to the trailing edge.

In Figure7, it is found that time evolutions of cavity volumes with two models are both periodic, and the predicted periods are almost the same. However, the cavity volume distributions along the foil suction side are significantly different. The maximum length of cavities predicted with the original k-ε model is just nearly half of the hydrofoil chord, and no shedding cavity is observed. Predicted attached cavity with the CCM model (C_{Scale}=10) can develop until covering the whole hydrofoil suction side. Cloud cavity shedding behavior can be captured. At the same time, partial cavity shedding process near the trailing edge of hydrofoil can also be observed.

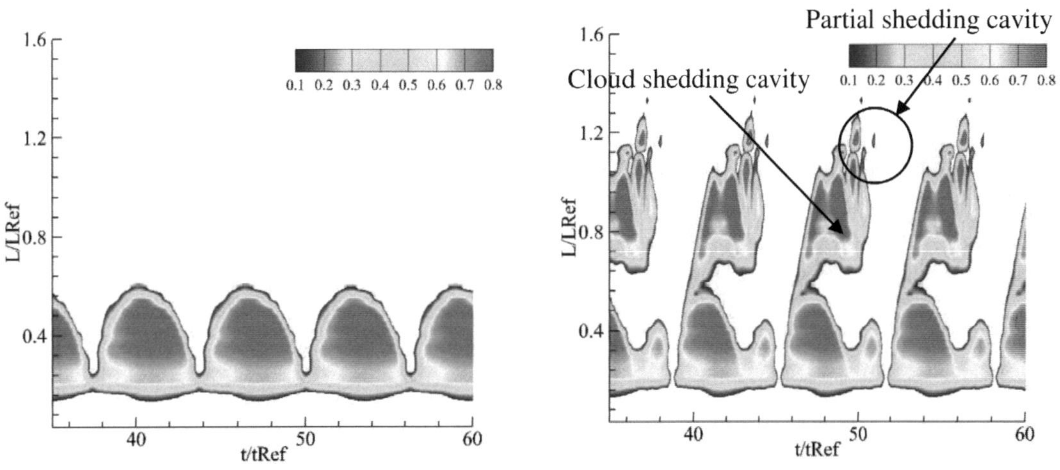

(a) Original k-ε model (b) CCM (C_{Scale}=10)

Figure 7. Time evolutions of the water vapor fraction in various sections

4. Conclusions

In this paper, a curvature correction model (CCM) is proposed, which takes into account the strong streamline curvature effect in cavitating vortex structures and modifies the production term of turbulence kinetic energy. Details of the modifications are discussed. The CCM models with different scaling coefficients are assessed by experimental data as well as those with the original k-ε model, interactions between cavitation and vortex structures are also discussed to get an insight of improved predictions with the CCM model. Compared with the original k-ε model, predicted results are improved, including the cavity detachment and hence more hydrofoil load fluctuations. Streamline curvature correction of CCM model overcomes the over-predictions of turbulence kinetic energy and eddy viscosity in cavitating vertical region with the original k-ε model, which leads to better simulation abilities for the unsteady cavitating flow computations.

Acknowledgment

This work was supported by the National Natural Science Foundation of China (Grant No. 11172040) and the state Key Program of National Natural Science Foundation of China (Grant No. 51239005).

References

[1] Knapp R T, Daily J W and Hammitt F G 1970 (New York : McGrawHill)
[2] Brennen C E 1995 *Oxford Engineering & Sciences Series 44* (Oxford: Oxford University Press)
[3] Arndt R E A 1981 *Annu. Rev. Fluid Mech.* **13**(1) 273-326
[4] Joseph D D 1995 *Phys. Rev. E.* **51** 1649–50.
[5] Coutier-Delgosha O, Fortes-Patella R, Reboud J L 2003 *J. Fluid Eng.* **125**(1) 38-45.
[6] Ji B, Luo X, Arndt R and Wu Y 2014 *Ocean Eng.* **87** 64-77.
[7] Zhou L J and Wang Z W 2008 *J. Fluid Eng.* **130**(1) 1-7.
[8] Li X, Wang G, Yu Z and Shyy W 2008 *Acta Mechanica Sinica* **25**(2) 161-72
[9] Huang B, Young Y L, Wang G Y and Shyy Wei 2013 *J. Fluid Eng.* **135**(7) 071301-16
[10] Arndt R E, Song C C and Kjeldsen M *Proceedings of 23rd symposium on naval hydrodynamics. (Val de Reuil, France, Sep 2000)*
[11] Gopalan S and Katz J 2000 *Phys. Fluids* **12** 895-911
[12] Lyer C and Ceccio S 2002 *Phys. Fluids* **14** 3414-30
[13] Laberteaux K R, Ceccio S L and Mastrocola V J 1998 *Exp. Fluids* **24** 489-98
[14] Rumsey C L, Gatski T B and Morrison J H 1999 *AIAA paper* 99-0157
[15] Spalart P R and Shur M L 1997 *Aerospace Sci. and Tech.* **1**(5) 297-302
[16] Shur M L, Strelets M K, Travin A K and Spalart P R 2000 *AIAA Journal* **38**(5) 784-92
[17] Smirnov P E and Menter F R 2008 *ASME Paper* GT 2008-50480
[18] Wang G Y, Senocak I and Shyy W 2001 *Progress in Aerospace Sci.* **37**(6) 551-81

IOP Conf. Series: Materials Science and Engineering **72** (2015) 022007 doi:10.1088/1757-899X/72/2/022007

Partially Averaged Navier-Stokes method based on *k-ω* model for simulating unsteady cavitating flows

C L Hu, G Y Wang and Z Y Wang
Beijing Institute of Technology, Beijing 100081, China

E-mail:qhclq@163.com

Abstract. The turbulence closure is significant to unsteady cavitating flow computations as the flow is frequently time-dependent accompanied with multiple scales of vortex. A turbulence bridging model named as PANS (Partially-averaged Navier-Stokes) purported for any filter-width is developed recently. The model filter width is controlled through two parameters: the unresolved-to-total ratios of kinetic energy f_k and dissipation rate f_ω. In the present paper, the PANS method based on k-ω model is used to simulate unsteady cavitating flows over a Clark-y hydrofoil. The main objective of this work is to present the characteristics of PANS k-ω model and evaluate it depending on experimental data. The PANS k-ω model is implemented with various filter parameters (f_k=0.2~1, $f_\omega=1/f_k$). The comparisons with the experimental data show that with the decrease of the filter parameter f_k, the PANS model can reasonably predict the time evolution process of cavity shapes and lift force fluctuating in time. As the PANS model with smaller f_k can overcome the over-prediction of turbulent kinetic energy with original k-ω model, the time-averaged eddy viscosity at the rear of attached cavity decreases and more levels of physical turbulent fluctuations are resolved. What's more, it is found that the value of ω in the free stream significantly affects the numerical results such as time-averaged cavity and fluctuations of the lift coefficient. With decreasing f_k, the sensitivity of ω-equation on free stream becomes much weaker.

1. Introduction

Cavitation typically occurs when the static pressure decreases below its vapor pressure.[1] The turbulent cavitating flows strongly affect the performances of blades, the propeller, the pump, et al. in engineering applications. Due to the complex, unsteady flow structures associated with turbulence and cavitation dynamics, there are significant computational issues in regard to stability, efficiency, and robustness of the numerical algorithm for turbulent unsteady cavitating flows.[2]

The turbulence model has a significant effect on the simulation of unsteady cavitating flows. The Reynolds-averaged Navier–Stokes (RANS) model fails to predict the vortex shedding behavior and excessively restrain the cavitation instabilities due to the high eddy viscosity.[3] Recently, several hybrid modeling approaches to improve the traditional RANS models have been proposed and developed, such as, detached eddy simulation (DES)[4], a filter-based model (FBM)[5], PANS k-ε model [6] et al. These hybrid models are applied to simulating the cavitating flows and obtain expected results, see ref.[7][8][9][10].

Inspired by their work, the present paper is devoted to investigate the recent PANS *k-ω* model based on the unsteady cavitating flow over a Clark-y hydrofoil and evaluate it by comparisons with experimental data. We mainly focus on the influence of the filter with f_k on the simulation and the sensitivity of the freestream *ω*.

IOP Conf. Series: Materials Science and Engineering **72** (2015) 022007 doi:10.1088/1757-899X/72/2/022007

2. Theoretical formulation

2.1 Favre-averaged continuity and momentum equations

The set of governing equations comprises the conservative form of the incompressible Favre-averaged Navier-Stokes equations, coupling with cavitation model (Zwart cavitation model[11]) and turbulence closure. The mass continuity, momentum equations are given below.

$$\frac{\partial \rho_m}{\partial t} + \frac{\partial(\rho_m u_j)}{\partial x_j} = 0 \quad (1)$$

$$\frac{\partial(\rho_m u_i)}{\partial t} + \frac{\partial(\rho_m u_i u_j)}{\partial x_j} = -\frac{\partial p}{\partial x_i} + \frac{\partial}{\partial x_j}[(\mu_L + \mu_T)(\frac{\partial u_i}{\partial x_j} + \frac{\partial u_j}{\partial x_i} - \frac{2}{3}\frac{\partial u_k}{\partial x_k}\delta_{ij})] \quad (2)$$

The mixture property, ϕ_m, can be expressed as

$$\phi_m = \phi_l \alpha_l + \phi_v (1-\alpha_l) \quad (3)$$

where α_l is theliquid volume fraction, ϕ can be density, viscosity, and so on.

2.2 Turbulence Model

The PANS model is proposed by Lakshmipathy and Girimaji[12] based on the standard k-ω RANS model. The filter width in PANS k–ω model is controlled by specifying the ratios of unresolved-to-total kinetic energy, f_k and unresolved-to-total turbulent frequency, f_ω. The two parameters are defined as:

$$f_k = \frac{k_u}{k}, \quad f_\omega = \frac{\omega_u}{\omega} \quad (4)$$

The k and ω are the total turbulent kinetic energy and turbulent frequency, respectively. Throughout, subscript u indicates PANS unresolved statistics. The smaller the f_k , the finer is the filter. The value of f_k can vary from 0 to 1: f_k =1 represents RANS and f_k =0 indicates DNS. Here, the Reynolds number considered in the flow fields is considerably high, so f_ω=1/ f_k. The two-equation PANS k–ω model can be summarized as

$$\frac{\partial k_u}{\partial t} + U_j \frac{\partial k_u}{\partial x_j} = \frac{\partial}{\partial x_j}\left[\left(\nu + \frac{\nu_u}{\sigma_{ku}}\right)\frac{\partial k_u}{\partial x_j}\right] + P_u - \beta^* k_u \omega_u \quad (5)$$

$$\frac{\partial \omega_u}{\partial t} + U_j \frac{\partial \omega_u}{\partial x_j} = \frac{\partial}{\partial x_j}\left[\left(\nu + \frac{\nu_u}{\sigma_{\omega u}}\right)\frac{\partial \omega_u}{\partial x_j}\right] + \alpha P_u \frac{\omega_u}{k_u} - \beta' \omega_u^2 \quad (6)$$

$$\beta' = \alpha\beta^* - \frac{\alpha\beta^*}{f_\omega} + \frac{\beta}{f_\omega} \quad (7)$$

$$\sigma_{ku} = \sigma_k \frac{f_k}{f_\omega}, \sigma_{\omega u} = \sigma_\omega \frac{f_k^2}{f_\omega} \quad (8)$$

$$\frac{\nu_u}{\nu} = \frac{k_u/\omega_u}{k/\omega} = \frac{f_k}{f_\omega} \quad (9)$$

The values for various coefficients are:

$$\beta^* = 0.09, \quad \alpha = 5/9, \quad \beta = 0.075, \quad \sigma_k = 2.0, \quad \sigma_\omega = 2.0 \quad (10)$$

3. Results and discussions

Figure 1 shows the computational domain and boundary conditions which are given according to the experimental set-up[13]. The Clark-y hydrofoil with the chord length of 0.07m is located with the angle of attack of 8°. The two important dimensionless parameters, the Reynolds number Re and the cavitation number σ, are defined as

$$Re = \frac{U_\infty c}{\nu} \quad , \quad \sigma = \frac{p - p_v}{\rho U_\infty^2 / 2} \tag{11}$$

where U_∞ the inlet velocity, p is the outlet pressure, p_v is the saturated vapor pressure , c is the hydrofoil chord, ν is the kinetic viscosity and ρ is the water density. Computations are performed for cloud cavitation condition (σ=0.8). Here, U_∞=10m/s is imposed at the inlet, with the corresponding Reynolds number of Re =7×10^5. The saturated pressure of water at 25 °C is p_v=3169 Pa.

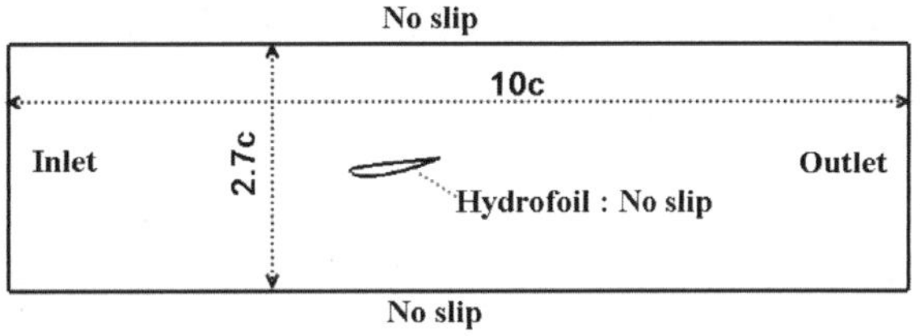

Figure 1. Boundary conditions for Clark-Y hydrofoil, c is the hydrofoil chord

Figure 2 shows the contours of time-averaged water vapor volume fraction and eddy viscosity predicted with different f_k respectively. Wang[14]found that the cloud cavity structure consists of two parts, the attached front portion and the detached rear region. From figure 2, the PANS with higher f_k ,such as f_k=1.0 ~0.6, underestimates the detached part and obtains the whole cavity of smaller scale, compared with that of lower f_k. As is reported that the poor prediction of cavity shedding may be due to over-prediction of the turbulent viscosity in the rear part of the cavity. [15] The differences of the cavity shape at the rear region predicted by PANS with different f_k imply that the eddy viscosity distributions are handled differently. In figure 2, the eddy viscosity at the rear of hydrofoil decreases with smaller f_k and a dramatic decrease is conducted by the PANS with f_k =0.2. Mathematically, to some extent, the eddy viscosity in PANS is proportional to the square of f_k value, as is shown in Equation (9).

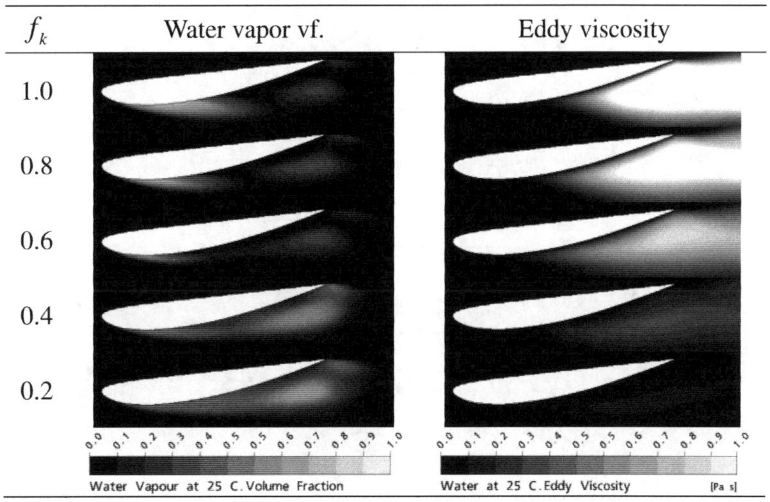

Figure 2. Time-averaged water vapor volume fraction and eddy viscosity contours

IOP Conf. Series: Materials Science and Engineering **72** (2015) 022007 doi:10.1088/1757-899X/72/2/022007

To further investigate the PANS model, figure 3 shows the time-evolutions of the computational and experimentally observed cavity structures. It demonstrates that the PANS model with various f_k performs differently in simulating the cloud cavitating flow. Compared with the experimental results, it is found that the PANS with f_k=0.2 is capable of capturing the cavity inception at the time of t_0, growth toward the trailing edge at t_0+38%T, attached cavity break up at t_0+58%T and finally large scale cavity shedding at t_0+84%T, in accordance with the qualitative features observed experimentally. In contrast, the cases of f_k=1.0 and f_k=0.6 present the failure of predicting the attached cavity in larger size and the shedding process of the detached cavity. This is probably because the higher eddy viscosity restrains the cavity growing and induces unexpected unsteady behaviors.

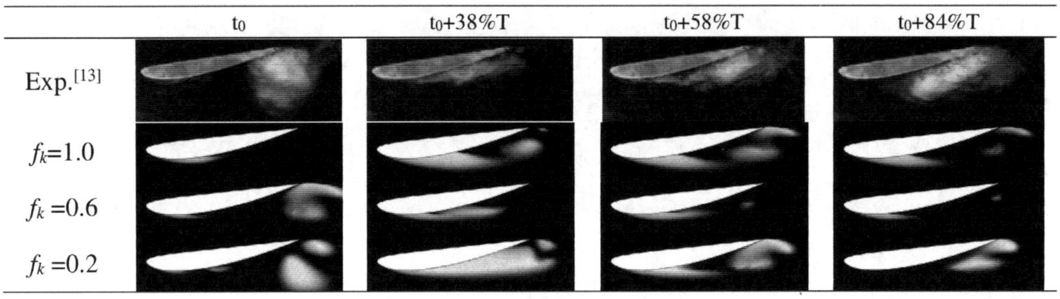

Figure 3. Instantaneous contours of water vapor volume fraction

The influence of f_k in PANS on the unsteady characters of cloud cavitating flows can be demonstrated by the time-dependent lift coefficient as shown in figure 4. Obviously, although the lift signals predicted by the PANS with different f_k are fluctuating periodically in time, yet there are substantial differences in the fluctuation detail. The PANS with higher f_k, such as f_k =1.0 and f_k =0.8, can predict a relatively steady lift curve with little fluctuations. With decreasing f_k, the PANS can resolve more scales in the field and the lift coefficient curve presents more fluctuations in time which is in good agreement with the experiment.

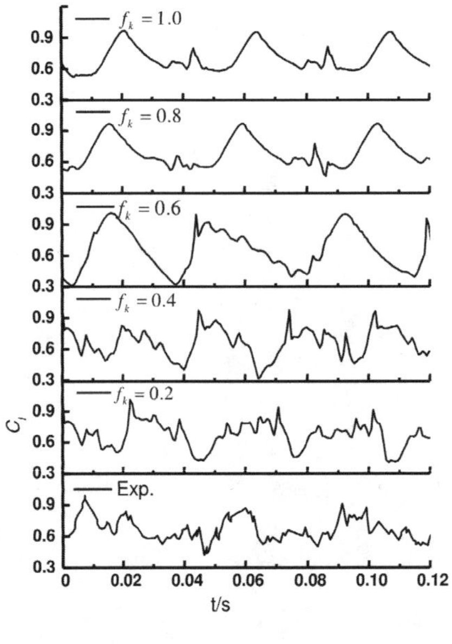

Figure 4. Time-evolutions of hydrofoil lift coefficient

Figure 5 gives the time-averaged u-velocity component profiles tracked along y-direction at different specified positions in the chordwise. It is found that the trend of velocity curve predicted by any PANS model is similar to the experimental result at any position. From the figure 5(a)~(c), the time-averaged attached cavity becomes thicker, and accordingly the boundary layer also expands along the y-direction. Although the differences between predictions and experimental data are more substantial at the rear region, such as at x/c=0.8, x/c=1.0 and x/c=1.2, still the agreement is reasonable considering the difficulties in experimental measurement. To draw a comparison between two PANS methods, the results of PANS k-ε model[16]are also shown in figure 5. In contrast, the distributions of the velocity in the boundary layer obtained by PANS k-ε model with f_k=0.2 is closer to the experimental results, while the PANS k-ω model with f_k=0.2 predicts velocity profiles even further away from the experiments. The reason for the poor performance of the k-ω model may lie in its freestream dependency.

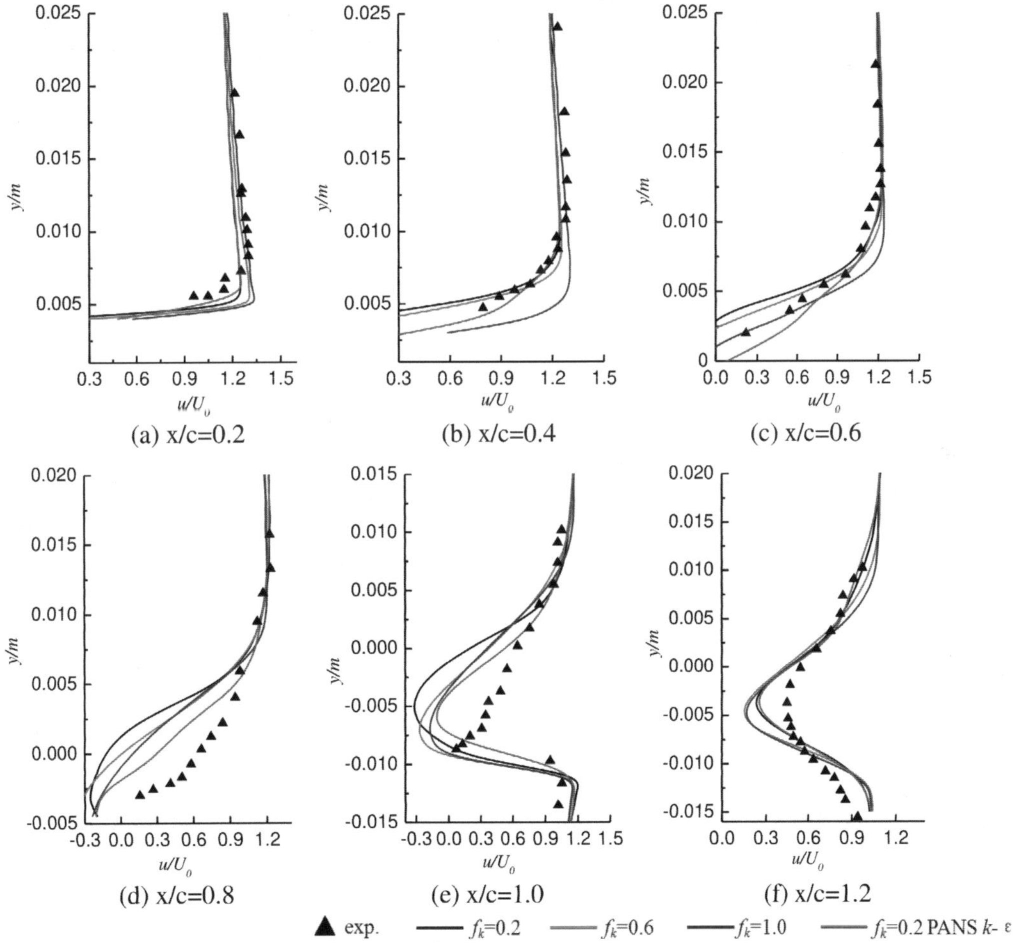

Figure 5. Time-averaged x-direction velocity at different locations

In order to investigate the sensitivity of the freestream ω to the PANS k-ω model, figure 6 and figure 7 give the contours of time-averaged water vapor volume fraction and the time-dependent lift coefficient curve predicted by PANS with different f_k and ω, respectively. From figure 6, it is found that the size of the time-averaged cavity increases with decreasing the value of f_k for the both cases of ω=1 and ω=60. When keeping the f_k to be constant, the contour of time-averaged water vapor volume fraction presents different distributions with different values of ω. Moreover, in figure 7, the lift coefficient signals fluctuate in the different forms under condition of different values of ω but the same f_k. In contrast, the difference caused by ω becomes smaller gradually with decreasing f_k. Just as seen in Fig7(c), when f_k=0.2, the lift coefficient curves predicted by the PANS models with different ω almost overlap each other.

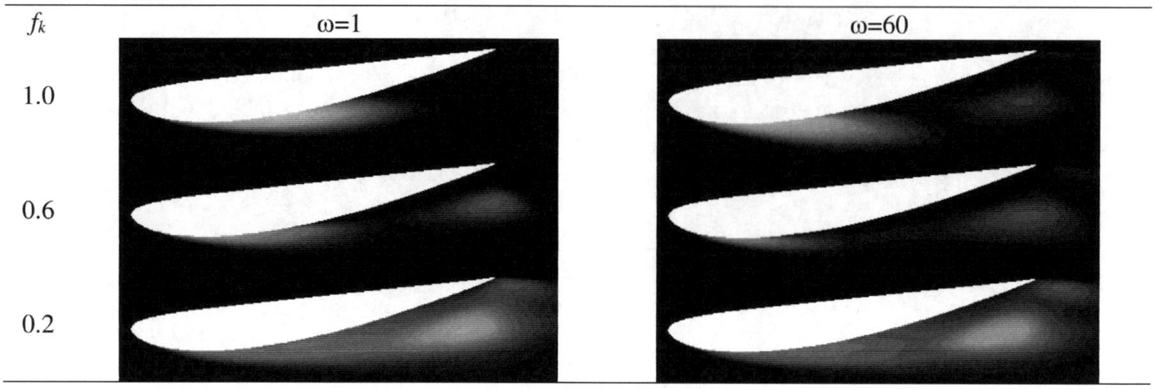

Figure 6. Time-averaged water vapor volume fraction

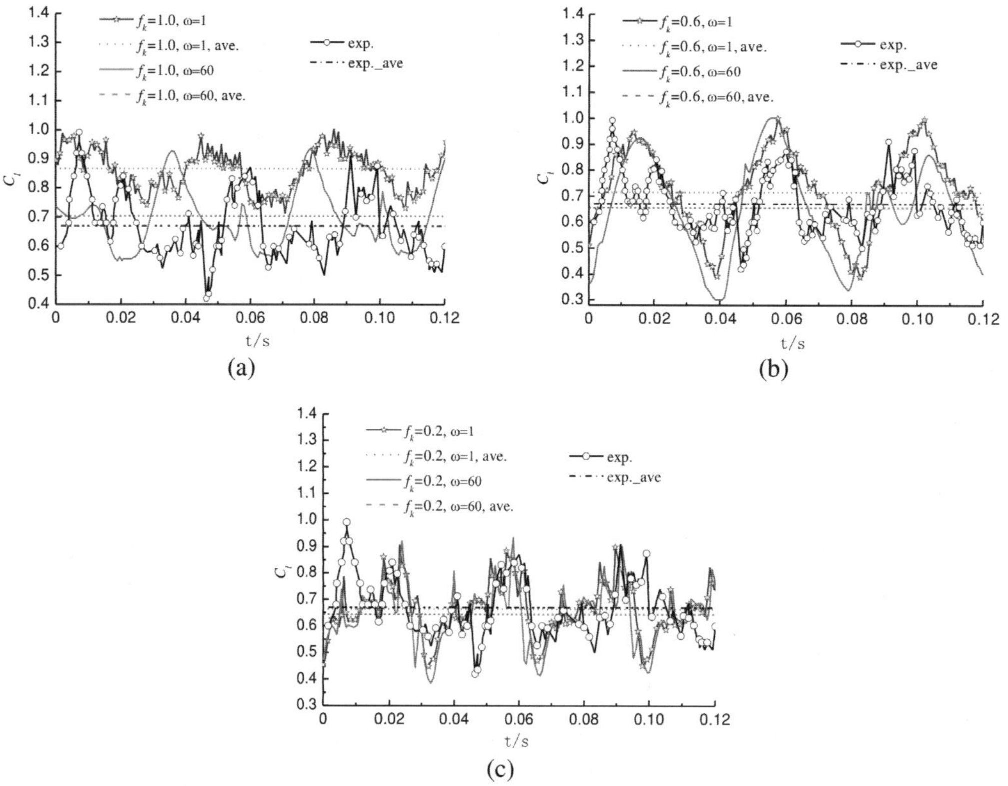

Figure 7. Time-evolutions of hydrofoil lift coefficient

4. Conclusions

In this paper, the PANS k-ω model is used for simulating the unsteady cavitating flow over a Clark-y hydrofoil. Based on the experimental data, the PANS model is evaluated. The important conclusions are the followings:

1) Compared with the experimental data, the PANS model can reasonably predict the time evolution process of cavity shapes and lift coefficient fluctuating with decreasing the filter parameter f_k. What's more, as the PANS model with smaller f_k can overcome the over-prediction of the time-averaged eddy viscosity at the rear of attached cavity decreases and more levels of physical turbulent fluctuations are resolved.

2) The value of ω in the freestream significantly affects the numerical results such as time-averaged cavity and fluctuations of the lift coefficient. With decreasing f_k, the sensitivity of ω-equation on freestream becomes much weaker.

Acknowledgements
This work was supported by the National Natural Science Foundation of China (Grant Nos. 11172040 and 51239005) and Beijing Municipal Natural Science Foundation (Grant No. 3144043)

References

[1] Utturkar Y, Wu J, Wang G and Shyy W 2005 *Progress in Aerospace Sciences* **41** 558-608
[2] Wu J, Wang G and Shyy W 2005 *Int. J. Numer. Meth. Fluids* **49** 739–61
[3] Coutier-Delgosha O, Fortes-Patella R and Reboud J L 2003 *Journal of Fluids Engineering* **125** 38-45
[4] Spalart P 1999 *Engineering Turbulence Modelling and Experiments* **4** 3-17
[5] Wu J, Johansen S T and Shyy W 2004 *ASME Paper* HT-FED **2004**-56181
[6] Girimaji S and Abdol-Hami K S 2005 *AIAA paper* **2005**-502
[7] Johansen S T, Wu J and Shyy W 2004 *Int. J. of Heat and Fluid Flow* **25** 10–21
[8] Kinzel M P, Lindau J W, Peltier L J and Kunz R F 2007 *18th AIAA Computational Fluid Dynamics Conf.(Miami, FL, USA, 25-26, June 2007)* AIAA 2007-4098
[9] Ji B, Luo X, Wu Y, Peng X and Duan Y 2013 *Int J Multiphase Flow* **51** 33-43
[10] Huang B and Wang G 2011 *J. of hydrodynamics* **23** 26-33
[11] Zwart P J, Gerber A G and Belamri T 2004 *Proc. of Int. Conf. on Multiphase Flow (Yokohama Japan, May 30-June 4, 2004)*Paper No.152
[12] Lakshmipathy S and Girimaji S 2006 *44th AIAA Aerospace Sciences Meeting and Exhibit (Reno, Nevada, USA, 9-12, January, 2006)*AIAA 2006-119
[13] Huang B and Wang GY 2011 *SCIENCE CHINA Techno. Sci.* **54** 1801-12
[14] Wang G, Senocak I and Shyy W 2001 *Progress in Aerospace Sci.* **37** 551-81
[15] Coutier-Delgosha O, Reboud J L and Delannoy Y 2003 *Int. J. Numer. Meth. Fluids* **42** 527–48
[16] Girimaji S 2006 *J. of Applied Mech.* **73** 413-21

IOP Conf. Series: Materials Science and Engineering **72** (2015) 022008 doi:10.1088/1757-899X/72/2/022008

Numerical investigation of the hydroelastic response in cavitating flow around a flexible hydrofoil

Q Wu, B Huang[1], G Y Wang and J D Wang
Beijing Institute of Technology, 5 Zhongguancun South Street, Haidian District, Beijing 100081, China

E-mail: huangbiao@bit.edu.cn

Abstract. The objective of this paper is to investigate the hydroelastic response of cavitating flows around a flexible hydrofoil. The numerical simulations are performed by solving the incompressible and unsteady Reynolds Average Navier-Stokes (URANS) equations via the commercial CFD software ANSYS CFX. The k-ω SST turbulence model with the turbulence viscosity correction and the Kubota cavitation model are introduced to the present simulations. The results showed that the cavitation has significantly affected the foil deformation and the evolution of the transient cavity shape and the corresponding hydrodynamic response with time can be divided into three stages: during the development of the attached cavity, the cavity formed on the suction side of the flexible hydrofoil is much larger with a steeper slope of the cavity area, which is caused by the increase of the effective angle of attack due to the twist deformation. During the vortex-cavitation interaction process, the hydrodynamic loads for the rigid hydrofoil remain relatively flat, while that for the flexible hydrofoil fluctuates with high frequency because of the foil deformation, leading to a more complex cavitation pattern due to the interaction with the foil vibration. During the cavity shedding process, both the primary and the residual cavities shed downstream totally, together with the counter-rotational vortex structures, corresponding to a sharp drop in the hydrodynamic loads. The larger effective angle of attack leads to the advanced cavity inception of the next cavitation period.

1. Introduction

As the diameter of hydraulic machinery become larger and the material used to construct the blades become lighter, the composite material is becoming prevalent in a wide range of applications because of their reduced mass, increased strength to weight ratio and erosion resistance. Most earlier works on fluid-structure interaction problems have been on aerospace or wind engineering structures [1-2]. In addition to the dynamic response of flexible airfoils, recent works have also been conducted on the transient responses of flexible hydrofoils[3-4]. With the increasing importance of hydroelasticity in fluid structure interaction problems, the investigation of hydroelastic responses has attracted considerable interests, especially when the local pressure decreases to the saturated vapor pressure of the liquid, cavitation occurs, involving complex interaction between phase-change and vortex structures[5-7]. The unsteady breakdown and shedding of the cavities will induce strong transient loads and lead to further hydroelastic instabilities, even structure failures. Hence, it is important to study the effect of hydroelastic response on the cavitating flows to design safe and reliable hydraulic machinery.

---

[1] Corresponding author.

The objective of this paper is to investigate the hydroelastic responses of a flexible stationary hydrofoil in cavitating flows. The physical and numerical models are presented in Section 2. Detailed analysis of the hydroelastic responses of a flexible stationary hydrofoil in transient subcavitating and cavitating flow is presented in Section 3. Finally, the conclusions are summarized in Section 4.

2. Physical and numerical model

2.1. Fluid model

The Unsteady Reynolds Average Navier-Stokes (URANS) equations, in their conservative form, for a Newtonian fluid without body forces and heat transfers are presented in Ref. [8]. The current simulation solves the URANS equations using the *k-ω* SST turbulence model[9], which applies the *k-ε* model away from the wall and the *k-ω* model near the wall. In order to avoid the over-prediction of the turbulent eddy viscosity resulting to the over-prediction of the turbulent stress and improve the numerical simulations by considering the local compressibility effect of multiphase mixtures on the turbulence model, the turbulent viscosity is revised according to Ref. [10] by replacing μ_t with μ_{t_mod}:

$$\mu_{t\_\text{mod}} = \mu_t f(n), \quad f(n) = \frac{\rho_v + (1-\alpha_v)^n(\rho_l - \rho_v)}{\rho_v + (1-\alpha_v)(\rho_l - \rho_v)} \tag{1}$$

where n=3 is chosen since the validation studies made by Huang et al. [11] and Ducoin et al. [12] are performed in the same hydrofoil and show favorable agreements with the experimental results. In this work, the Kubota cavitation model [13] is used and the model constants are assumed as the default values in CFX [14].

2.2. Fluid model

The cantilevered hydrofoil with the spanwise bending and twisting flexibility is modeled as a chordwise rigid, two degree of freedom system in this work, as shown in Figure 1. Both the bending deformation h and the twisting deformation θ are defined about the elastic axis, which is located at the mid-chord of the foil. In Figure 1, b is the semi-chord and the chord length c=2b, U is the flow velocity.

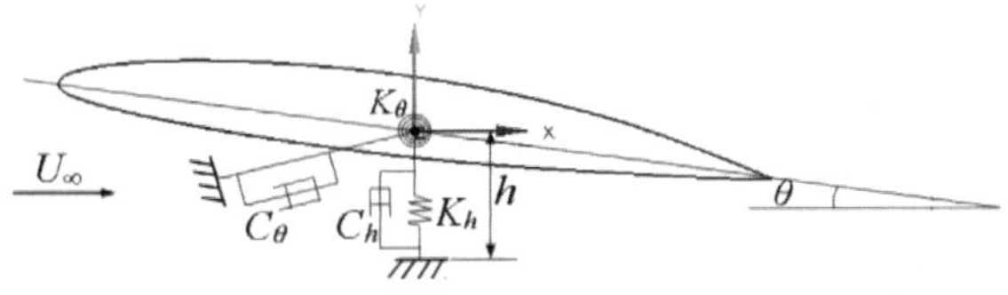

Figure 1. A two degrees-of-freedom model describing the bending h and the twisting θ of foil

The governing equation of the 2-DOF system motion can be written as:

$$\begin{bmatrix} m & S_\theta \\ S_\theta & I_\theta \end{bmatrix}\begin{Bmatrix} \ddot{h} \\ \ddot{\theta} \end{Bmatrix} + \begin{bmatrix} C_h & 0 \\ 0 & C_\theta \end{bmatrix}\begin{Bmatrix} \dot{h} \\ \dot{\theta} \end{Bmatrix} + \begin{bmatrix} K_h & 0 \\ 0 & K_\theta \end{bmatrix}\begin{Bmatrix} h \\ \theta \end{Bmatrix} = \begin{Bmatrix} F_{ex} \\ M_{ex} \end{Bmatrix} + \begin{Bmatrix} L \\ M \end{Bmatrix} \tag{2}$$

where m, S_θ, I_θ are the structural mass, the static imbalance and the moment of inertia respectively, C_h and C_θ are the structural damping values for the bending and twisting motions respectively, K_h and K_θ are the structural bending and torsional stiffness values respectively. h, $\dot{h}$ and $\ddot{h}$ are the bending displacement, velocity, and acceleration, while θ, $\dot{\theta}$, $\ddot{\theta}$ are the twisting displacement, velocity, and acceleration. F_{ex} and M_{ex} are the external excitation force and moment and set as zero in this work. L and M are the fluid lift, which is defined positive along the Y-axis, and moment, which is defined positive counter-clockwise, acting on the hydrofoil.

2.3. Hybrid coupled fluid structure interaction model

The hybrid coupled fluid structure interaction model discretizes the equation of motion, Equation (2), in the time-domain as:

$$\begin{bmatrix} m & S_\theta \\ S_\theta & I_\theta \end{bmatrix}\begin{Bmatrix}\ddot{h}\\ \ddot{\theta}\end{Bmatrix}_{n+1}^{i+1} + \begin{bmatrix} C_h & 0\\ 0 & C_\theta\end{bmatrix}\begin{Bmatrix}\dot{h}\\ \dot{\theta}\end{Bmatrix}_{n+1}^{i+1} + \begin{bmatrix}K_h & 0\\ 0 & K_\theta\end{bmatrix}\begin{Bmatrix}h\\ \theta\end{Bmatrix}_{n+1}^{i+1} - \begin{Bmatrix}L^T_{fluid}\\ M^T_{fluid}\end{Bmatrix}_{n+1}^{i+1} = \begin{Bmatrix}F_{ex}\\ M_{ex}\end{Bmatrix}_{n+1}^{i} + \begin{Bmatrix}L\\ M\end{Bmatrix}_{n+1}^{i} - \begin{Bmatrix}L^T_{fluid}\\ M^T_{fluid}\end{Bmatrix}_{n+1}^{i} \tag{3}$$

where i is the sub-iteration number within each time step n, L^T_{fluid} and M^T_{fluid} subtracted from the left and right sides of the equation are the potential flow estimate of the hydroelastic force and momentum, which is proposed by Theodorsen[15] and developed by Young[16] to account for the effects of fluid inertial, damping and restoring forces in each time step and sub-iteration, avoiding the over-estimation of the structural displacement.

2.4. Numerical setup and boundary conditions

The simulation is conducted with a 2D fluid solver and the computational domain and boundary conditions are given according to the experimental setup in Beijing Institute of Technology [17], which is shown in Fig. 2. The results shown in this paper correspond to the stationary hydrofoil with initial angle of attack of α_0=8°, which is subjected to a nominal free stream velocity of U_∞=10m/s, yielding a moderate Reynolds number of $Re=U_\infty c/v=7.5\times10^5$, where c is the chord length and v is the dynamic viscosity of the liquid (water at 25°C). The outlet pressure is set according to the cavitation number $\sigma=(p_\infty-p_v)/(0.5\rho_l U_\infty^2)=1.4$, where p_∞ is the tunnel pressure, p_v is the saturated vapor pressure.

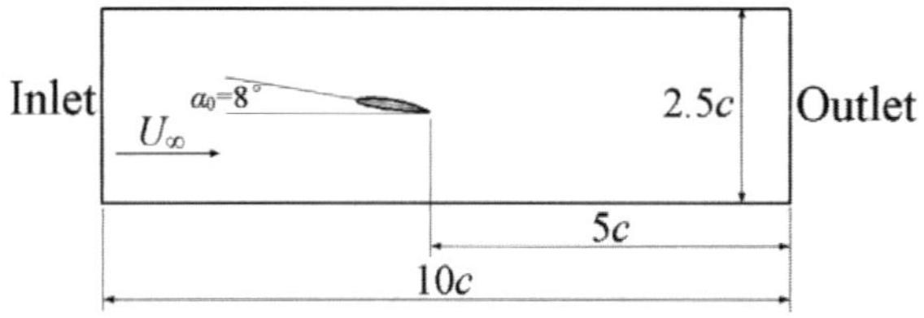

Figure 2. Computational domain and boundary conditions

The numerical mesh is shown in Fig. 3, composed of 110,000 structured elements. The mesh is refined near the leading and trailing edge of the foil and in the wake region to satisfy $y^+=yu_\tau/v\approx1$, where y is the thickness of the first cell from the foil surface, and u_τ is the wall frictional velocity. The time step chosen for the simulation is $\Delta t=1\times10^{-3}$s, which ensures that the averaged CFL= $U_\infty\times\Delta t/\Delta x\approx1$.

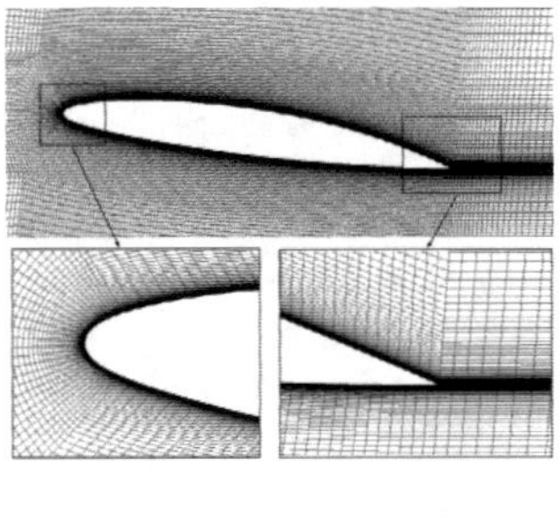

Figure 3. Fluid mesh details

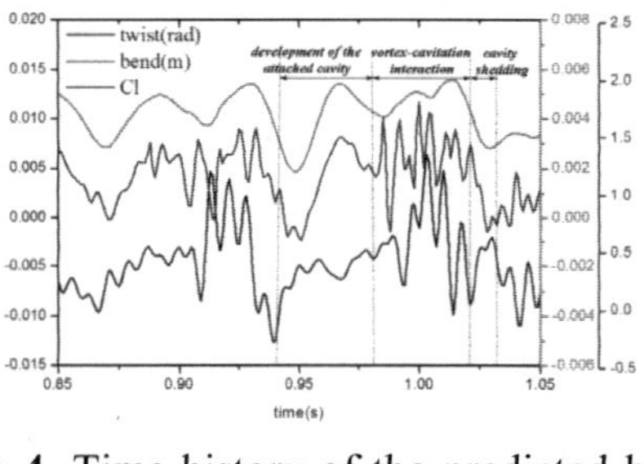

Figure 4. Time-history of the predicted bending and twisting deformations of the flexible hydrofoil in cavitating flow (σ=1.4)

3. Results and discussion

Figure 4 shows the predicted bending and twisting deformations, as well as the lift coefficient, of the flexible hydrofoil in cavitating flow at σ=1.4. From Figure 4, it is shown that the hydrofoil bends and twists with the same period time of the lift coefficient. The hydrodynamic responses within one cycle could be characterized by three stages: (I) the bending and twisting deformation increases as the evolution of the lift coefficient, while the twist angle fluctuates within a small range, which is corresponding to the development of the attached cavity, (II) the twist angle fluctuates in a large extent as the cavities break up and intensively interact with the vortex structures, (III) there is a sudden decrease of the bending and twisting deformation which is in phase with the lift coefficient, when the cavities collapse and totally shed downstream together with the vortex structures. Overall, the transient development of cavitation has affected the unsteady characteristics of the hydroelastic response, which is also closely related to the instability of the vortex structures and will be discussed following.

Figure 5 shows the evolution of the predicted lift, drag and effective angle of attack for the rigid and flexible hydrofoils with time. It is shown that the hydrodynamic load coefficients of the flexible hydrofoil fluctuate more significantly due to the disturbance caused by the deformation of the foil. The average of the lift and drag coefficients for the flexible hydrofoil are 1.067 and 0.141 respectively, which is higher than that for the rigid hydrofoil. This is because the twist deformation increases the effective angle and results to the larger hydrodynamic loads imposed on the foil.

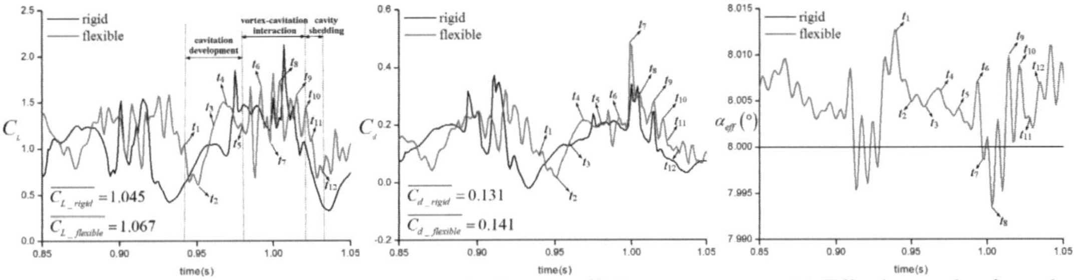

(a) Lift coefficient (b) Drag coefficient (c) Effective angle of attack

Figure 5. Comparisons of the predicted lift and drag coefficients for the rigid and flexible hydrofoils

In order to further investigate the influence of hydroelastic response on the cavitation development process and the vortex-cavitation interactions, Figures 6-7 show the evolution of vapor fraction and vorticity contours at representative times for both the rigid and flexible hydrofoils, along with the observed cavitation patterns of the flexible hydrofoil from the experimental visualizations, from which the cavity area is extracted and shown in Figure 8. The evolution of the transient cavity shapes and the corresponding hydrodynamic responses could be analyzed as follows:

3.1. development of the attached cavity (from t_1 to t_5)

During the development of the attached cavity, reasonable agreements of the cavity shapes can be obtained between the numerical results and experimental observations. As shown in Figure 5, the predicted lift and drag coefficients for both the rigid and flexible hydrofoils increase with the time, while the slope of the lift and drag coefficients for the flexible hydrofoil is larger. This is mainly because the larger effective angle of attack caused by the twist deformation between t_1 and t_5. At t_1=0.94s, the partial sheet cavity is formed at the leading edge of the foil on the suction side. Between t_2=0.95s and t_4=0.97s, the development of the small leading edge partial sheet cavity can be observed for both the rigid and flexible hydrofoil, as shown in Figures 6(b)-(c), with the cavity area increasing, as shown in Figure 8, which is corresponding to the increase of the lift and drag coefficients shown in Figure 5. Meanwhile, the re-entrant jet is formed at the rear region of the cavity, and a clockwise vortex is forming and developing on the suction side of the hydrofoil. Between t_4=0.97s and t_5=0.98s, the re-entrant jet moves toward the leading edge of the foil and the clockwise leading edge vortex develops to its maximum extent, which has an effect on the lifted cavity and leads to the decrease of the lift coefficient (shown in Figure 5(a)). Compared with the rigid hydrofoil, the cavities formed on the suction side of the flexible hydrofoil is much larger with a steeper slope of the cavity area, which is caused by the increase of the effective angle of attack due to the twist deformation.

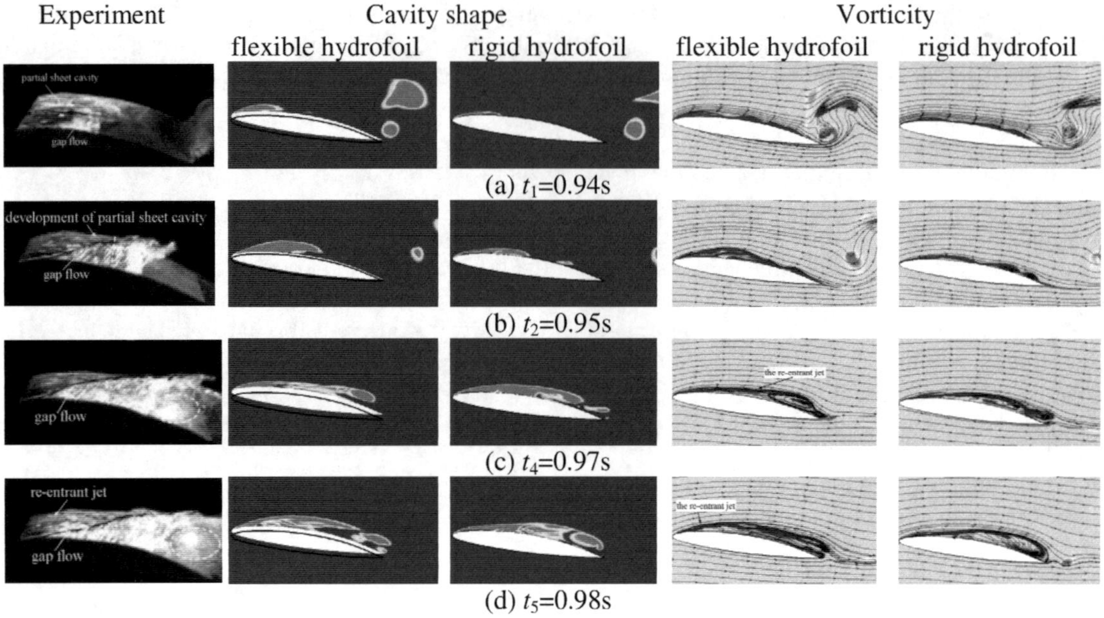

Figure 6. Predicted vapor fraction contours for the flexible and rigid hydrofoils during the development of the attached cavity, along with experimental observations of the cavitation patterns for the flexible hydrofoil.

3.2. vortex-cavitation interaction (from t_5 to t_{10})

During this stage, there is a discrepancy between the numerical results for the rigid and the flexible hydrofoil. As shown in Figure 5, the predicted lift and drag coefficients for the rigid hydrofoil remain relatively flat, while that for the flexible hydrofoil fluctuates with high frequency because the development of cavity is disturbed by the foil deformation, which leads to a more complex cavitation pattern due to the interaction with the foil vibration. At t_6=0.992s, the cloud cavity reaches to the trailing edge of the foil and is partially attached on the suction side of the foil, while the leading edge vortex is interacting with the counter-clockwise trailing edge vortex due to the reverse pressure gradient, resulting that two counter-rotating vortices with cavitating cores shed, as shown in Figure 7(b). The shedding cloud cavity moves downstream, and the cavity area reaches to its minimum value, as shown in Figure 5 and Figure 8, with the relatively lower lift coefficient and the highest drag coefficient achieved at t_7 due to the primary cavity shedding. Between t_9=1.014s and t_{10}=1.02s, governed by the main flow, the shedding cavitating vortex structures are transported towards the trailing edge of the foil, as shown in Figure 7(c), with the centers of the shedding vortex structures moving downstream along the suction side of the hydrofoil and interacting with the downstream counter-clockwise vortex again, which is corresponding to the fluctuating lift coefficients and effective angle of attack, which are shown in Figure 5.

3.3. cavity shedding (from t_{10} to t_{12})

Between t_{10}=1.02s and t_{12}=1.032s, both the primary and the residual cavities shed downstream totally and move downstream together with the primary cavity, as well as the clockwise and couter-clockwise vortex structures, which is due to the interaction between those counter-rotating vortex structures and the cavitation cores, corresponding to a sharp drop in the lift coefficient and followed by the formation of the partial sheet cavity in the next period.

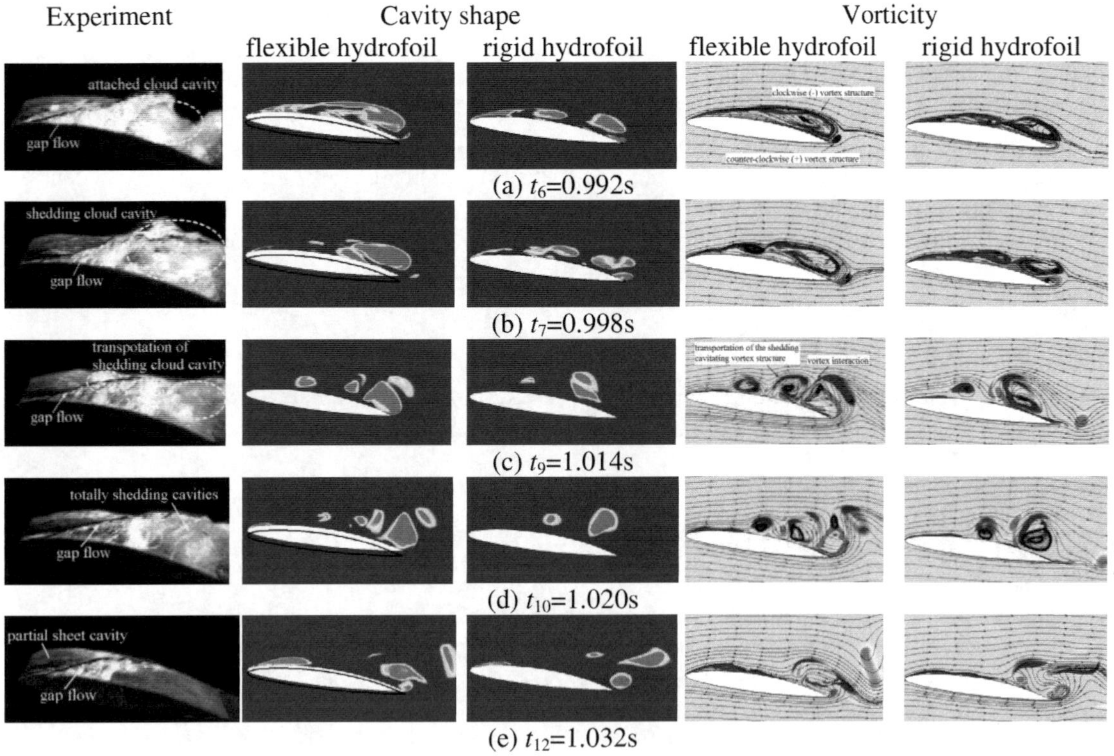

Figure 7. Predicted vapor fraction contours for the flexible and rigid hydrofoils during the vortex-cavitation interaction, along with experimental observations of the cavitation patterns for the flexible hydrofoil.

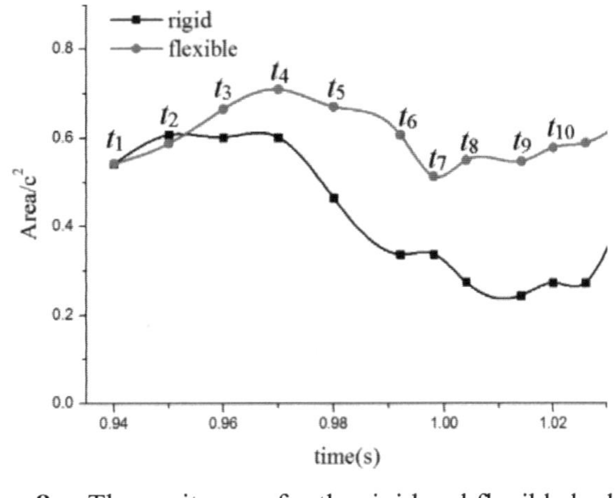

Figure 8. The cavity area for the rigid and flexible hydrofoil

4. Conclusions

In this paper, physical and numerical analysis are presented for a rigid and a flexible NACA66 hydrofoil with the angle of attack α=8° at Re=750,000 for subcavitating and cavitating conditions. The hydroelastic response has significant effect on the cavitation development process and the vortex-cavitation interactions. The evolution of the transient cavity shape and the corresponding hydrodynamic response with time can be divided into three stages: development of the attached cavity, vortex-cavitation interaction and cavity shedding. During the development of the attached cavity, the partial sheet cavity is formed with the predicted lift and drag coefficients increasing, while the cavity formed on the suction side of the flexible hydrofoil is much larger with a steeper slope of the cavity area, which is caused by the increase of the effective angle of attack due to the twist deformation. During the vortex-cavitation interaction process, the hydrodynamic loads for the rigid hydrofoil remain relatively flat, while that for the flexible hydrofoil fluctuates with high frequency because of the foil deformation, leading to a more complex cavitation pattern due to the interaction with the foil vibration. During the cavity shedding process, both the primary and the residual cavities shed downstream totally, together with the counter-rotational vortex structures, which is corresponding to a sharp drop in the hydrodynamic loads.

Acknowledgements
The authors gratefully acknowledge support by the National Natural Science Foundation of China (Grant Nos.: 51239005 and 51306020) and the Beijing Natural Science Foundation (No.: 3144034).

References
[1] Sears W R 1941 *J. Aeronaut. Sci.* **8**(3) 104–08
[2] Patil M J, Hodges D H 2006 *J. Aircr.* **43** (6) 1790–99
[3] Ducoin A, Jacques A A, Marie L G 2012 *J. Fluids Struct.* **32** 37–51
[4] Ducoin A, Young Y L 2013 *J. Fluids Struct.* **38** 40–57
[5] Wang G Y, Senocak I, Shyy W, Ikohagi T, Cao S L 2001 *Progress Aero. Sci.* **37** 551-81
[6] Huang B, Wang G Y, Zhao Y, Wu Q 2013 *Sci. China Tech. Sci.* **56** 2207-18
[7] Ji B, Luo X W, Wu Y L, Peng X X, Duan Y L 2013 *Int. J. Multiphase Flow* **51** 33–43
[8] Huang B, Ducoin A, Young Y L 2013 *Phys. Fluids* **25**(10) 1-27
[9] Menter F R 1992 *NASA Technical Memorandu* **34** 103975
[10] Coutier-Delgosha O, Fortes-Patella R, Reboud J L 2003 *J. Fluids Eng.* **125** (1) 38
[11] Huang B, Ducoin A, Young Y L 2012 *Proceedings of the 8th International Symposium on Cavitations(Singapore, 13 – 16, August, 2012)*
[12] Senocak I, Shyy W. 2002 *J. Comput. Phys.* **176** 363–83
[13] Kubota A, Kato H, Yamaguchi H 1992 *J. Fluid Mech.* **240** 59–96
[14] Zwart P, Gerber A, Belamri T 2004 *Fifth* International *Conference on Multiphase Flow (Yokohama, Japan, 30, May–4, June, 2004)*
[15] Theodorsen T 1935 *Technical Report No. 496* National Advisory Committee for Aeronautics
[16] Young Y L 2007 *J. Fluids Struct.* **23**(2) 269–95
[17] Huang B, Wang G 2011 *Sci. China Tech. Sci.* **54** 1801-1812

IOP Conf. Series: Materials Science and Engineering **72** (2015) 022009 doi:10.1088/1757-899X/72/2/022009

A simple cavitation model for unsteady simulation and its application to cavitating flow in two-dimensional convergent-divergent nozzle

Y Yamamoto[1], S Watanabe[2] and S I Tsuda[2]
[1] Graduate School of Engineering, Kyushu University, 744 Motooka, Nishi-ku, Fukuoka, Fukuoka 819-0395, Japan
[2] Department of Mechanical Engineering, Kyushu University, 744 Motooka, Nishi-ku, Fukuoka, Fukuoka 819-0395, Japan

E-mail: fmnabe@mech.kyushu-u.ac.jp

Abstract. In this paper, a simple cavitation model is developed under the framework of homogeneousone-fluid model, in which the perfect mixture of liquid and vapor phases is assumed. In most of conventional models, the vapor phase is considered as a dispersed phase against the liquid phase as a continuous phase, while in the present model, two extreme conditions are considered: for low void fraction, dispersed vapor bubbles in continuous liquid phase, while for high void fraction, dispersed droplets in continuous vapor phase. The growth of bubbles and droplets are taken into account in the mass transfer between vapor and liquid phases, and are switched according to the local void fraction. The model is applied for the simulation of cavitating flow in a two-dimensional convergent-divergent nozzle, and the result is compared with that using a conventional model. To enhance the unsteadiness of cavitation due to the instability at the cavity interphase, the turbulent shear stress is modified depending upon the continuous phases in combination with the proposed cavitation model, which drastically reduces the turbulent viscosity for high void fraction region. As a result, the unsteadiness of cavitation observed in experiments is well reproduced.

1. Introduction

The recent rapid progress of computer science has enabled us to simulate the cavitating flow and achieve qualitative, and in some extent, quantitative agreements with actual observed flows. However, such cavitation CFD (Computational Fluid Dynamics) still often fails to predict the cavitation performance even in simple cases of cavitating hydrofoil (Kato [1]). Moreover, still due to the limited computer resources, Reynolds-Averaged Navier-Stokes (RANS) simulation is often used to simulate the unsteady cavitating flow. However, the reproduction of the unsteadiness of cavitation is very limited, and key unsteady cavitation phenomena such as a re-entrant jet which develops beneath the sheet cavity and the formation of cloud cavity can never be simulated with RANS model developed based on incompressible flow. On the other hands, Reboud et al. [2]considered that the incompressible turbulence model overestimates the turbulent viscosity in the two-phase region, and has proposed the well-known Reboud correction to cut the turbulence viscosity in most of two phase flow region. With the tuned parameter, this correction method has successfully reproduced the major unsteadiness of cavitation in cases of a two-dimensional convergent-divergent nozzle [2], a two-dimensional hydrofoil [3], and a three-dimensional twisted hydrofoil [4].

The cavitating flows in two-dimensional convergent-divergent nozzles have been studied by many researchers, because of their simple structures but revealing important unsteady natures of cavitating flow such as re-entrant jet evolutions and vertical cloud cavity shedding. Stutz and Reboud [5], [6] have measured, by using a double optical probe, two-phase structures inside the cavity on the throat section of convergent divergent nozzle. Their nozzles have a common convergent angle of 43° but two different diverging angles of 4° and 8°. Keil et al. [7] have studied an aggressiveness of collapse of cloud cavitation in this kind of nozzle for the wide variety of Reynolds number. In our previous study [8], we investigated the unsteady character of cavitation in two-dimensional convergent-divergent nozzles with divergent angles of 4.0 and 8.4°. In both nozzles, we could not observe the re-entrant jet nor large scale cloud cavities, probably due to low Reynolds number flow compared to the other studies, while we observed small cloud cavity shedding from the trailing edge of sheet cavity as an essential character of cavitating flow in this kind of nozzle.

In the present study, we propose a simple cavitation model which considers virtually the interface between liquid and vapour phases in the framework of homogeneous one-fluid model, in which a subgrid-scale model of bubbles in the liquid phase and that of droplets in the vapour phase are employed. Mass source terms of liquid/vapour phases are considered through the growths and collapses of bubbles and droplets which are locally switched depending upon the local void fraction. In conjunction with this treatment of continuum phase, we also switch the eddy viscosity as well as the molecular viscosity depending upon whether the continuum phase is vapour or liquid. By doing so, it is expected to reproduce the strong unsteadiness of cavitating flow such as the generations of cloud cavities from the trailing edge of sheet cavity even ifincompressibletwo-equation RANS turbulence model is employed.In this paper, the proposed model is applied for the simulation of cavitating flow in a simple two-dimensional convergent-divergent nozzle. The effects of the choices of the mass source terms for bubbles and droplets as well as the turbulence model modification on the global motion of unsteady cavitating flow are investigated.

2. Numerical method

A numerical simulation was carried out byusing an opensource software, OpenFOAM. Several cavitation models are implemented in OpenFOAM, among which we employ interPhaseChangeFoam (IPCF) as a base solver. This solver is an incompressible Navier-Stokes one with homogeneous cavitation model considering the phase change between liquid and vapour phases.Besides the momentum equation of the mixture and the transport equations of turbulence properties, IPCF solves the following continuity equation and the mass conservation of liquid phase as follows

$$\frac{\partial u_j}{\partial x_j} = \left(\frac{1}{\rho_l} - \frac{1}{\rho_v}\right)\left(\dot{m}^+ + \dot{m}^-\right) \tag{1}$$

$$\frac{\partial \alpha_l}{\partial t} + \frac{\partial}{\partial x_j}\left(\alpha_l u_j\right) = \frac{1}{\rho_l}\left(\dot{m}^+ + \dot{m}^-\right) \tag{2}$$

where$x_j$isa Cartesian coordinate,$u_j$is velocity componentin the $x_j$direction, $\rho_l$and ρ_v are densities of liquid and vapour phases, $\alpha_l$is a volume fraction of liquid phase. The source terms in the above equations,$\dot{m}^+$and$\dot{m}^-$, are mass transfer rates between two phases due to evaporation and condensation, which are modelled in the following section.

2.1. Cavitation model

2.1.1. Schnerr-Sauer model (SS). Schnerr-Sauer (SS) model [9] implemented in OpenFOAM-2.2.1 is employed as a base model for this study, which has been used in many studies(for recent examples, [10] and [11]).In SS model, the vapour bubbles are always considered as dispersed phase in continuum liquid phase even for very large void fraction region. The SS model is based on the Rayleigh equation as follows

$$R\frac{d^2R}{dt^2}+\frac{3}{2}\left(\frac{dR}{dt}\right)^2=\frac{p-p_v}{\rho_l} \tag{3}$$

where R is a bubble radius, p is pressure, p_v is saturation pressure. If we neglect the second derivativeterm, the bubble radius growth can be represented by the following equation.

$$\frac{dR}{dt}=\text{sign}\left(p_v-p\right)\sqrt{\frac{2}{3}\frac{\left|p-p_v\right|}{\rho_l}} \tag{4}$$

In this model, the void fraction$\alpha_v = 1 - \alpha_l$represents the vapour volume per unit mixture volume and the bubble radius is derived fromα_v as follows.

$$\alpha_v=\frac{n_{nuc}\frac{4\pi}{3}R^3}{1+n_{nuc}\frac{4\pi}{3}R^3} \tag{5}$$

$$R=\sqrt[3]{\frac{\alpha_v}{\frac{4}{3}\pi n_{nuc}\left(1-\alpha_v\right)}} \tag{6}$$

where n_{nuc} is the number of bubbles per unit liquid volume.From those equations, the mass source terms of SS model can be obtained as

$$\dot{m}^+=C_c\frac{\rho_v\rho_l}{\rho}\alpha_v\left(1-\alpha_v\right)\frac{3}{R}\sqrt{\frac{2}{3}\frac{\left(p-p_v\right)}{\rho_l}}?\quad p>p_v \tag{7}$$

$$\dot{m}^-=-C_e\frac{\rho_v\rho_l}{\rho}\alpha_v\left(1-\alpha_v\right)\frac{3}{R}\sqrt{\frac{2}{3}\frac{\left(p_v-p\right)}{\rho_l}}?\quad p<p_v \tag{8}$$

where C_c and C_e are condensation and evaporation coefficients and $\rho=\rho_l\left(1-\alpha_v\right)+\rho_v\alpha_v$ is the mixture density.In the present study, the basic parameters in this model are set as $n_{nuc} = 1.6\times10^{13}\,\text{m}^{-3}$, and $C_c = C_e = 1.0$.

2.1.2. Bubble-Droplet1 model (BD1). Since applying the dispersed bubble models for large void fraction regions seems to be inappropriate, we consider another extreme case in which vapour phase contains more or less liquid droplets. In this study, this extreme condition near the void fraction is unity is taken into account in our model, at which we treat vapour phase as a continuum media. Figure 1 shows the conceptual drawing of our model.

This model virtually considers the interface between liquid and vapour phases as the iso-surface of void fraction $\alpha_{v,int}$, and in this study $\alpha_{v,int}$ is set to be 0.5 for simplicity. When the local void fraction satisfies $\alpha_v < \alpha_{v,int}$, the vapour phase is treated as dispersed phase, while $\alpha_v > \alpha_{v,int}$, the liquid phase is treated as dispersed phase. The mass transfer between vapour and liquid are dominated by that occurs at the surfaces of the bubbles/droplets, then the mass transfer rates $\dot{m}^+$ and $\dot{m}^-$ are switched depending upon the local void fraction α_v. In the model "Bubble-Droplet 1(BD1)", the SS model, i.e. equation (7) and (8), is adopted for $\alpha_v < \alpha_{v,int}$, while for $\alpha_v > \alpha_{v,int}$ the phase change at the surface of droplet is considered using Schrage's mass flux $\dot{M}$ [12], expressed as follows

$$\dot{M} = \frac{k}{\sqrt{2\pi R_g T}}(p_v - p) \tag{9}$$

where$k$ is a evaporate/condensate coefficient, $R_g$is a gas constant, and T is temperature. The value of k was set to 0.4 throughout this computation. By considering the phase change through the total surface area of droplets, the mass transfer rate can be obtained as

$$\dot{m} = \dot{m}^+ + \dot{m}^- = -\frac{k}{\sqrt{2\pi R_g T}}\frac{3(1-\alpha_v)}{R_d}(p_v - p) \tag{10}$$

where$R_d$is a radius of droplet, which can be calculated by

$$R_d = \sqrt[3]{\frac{\alpha_l}{\frac{4}{3}\pi n_{nuc}(1-\alpha_l)}} = \sqrt[3]{\frac{1-\alpha_v}{\frac{4}{3}\pi n_{nuc}\alpha_v}} \tag{11}$$

Since we do not consider any collisions and fissions of droplets/bubbles and any nucleation and destructions, n_{nuc} is constant and common for bubbles and droplets.

Since the cavity interface can be virtually treated, the mass transfer at the cavity interface is supposed to be possibly treated, which remains for our future study.

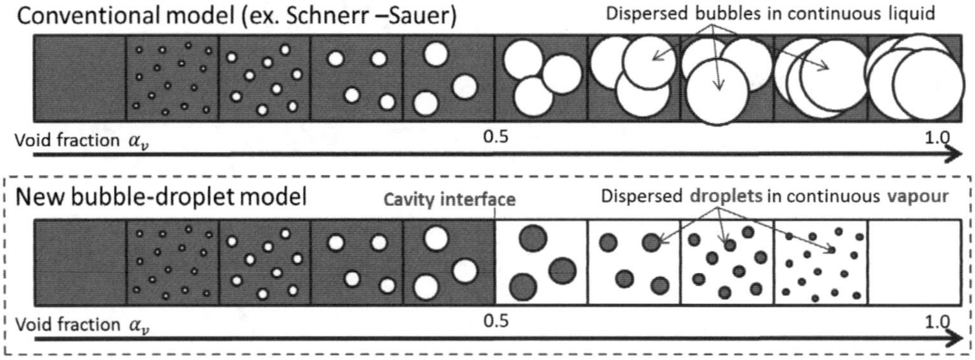

Figure 1. Conceptual drawing of bubble drop model

2.1.3. Bubble-Droplet2 model (BD2) Schrage's mass flux can also be applied for the bubbly flow region. In the model "Bubble-Droplet 2 (BD2)", the following mass transfer rate derived from the phase change through the bubble surfaces is applied for the region with $\alpha_v < \alpha_{v,int}$instead of equation (7) and (8) in SS model.

$$\dot{m} = \dot{m}^+ + \dot{m}^- = -\frac{k}{\sqrt{2\pi R_g T}}\frac{3\alpha_v}{R}(p_v - p) \tag{12}$$

2.1.4.Bubble-Droplet1 viscosity filtering model(BD1VF). Throughout the present computations, the standard k-ε model was used. However, it is known that key unsteady cavitation phenomena such as a re-entrant jet which develops beneath the sheet cavity and the formation of cloud cavity can never be simulated with incompressible RANS turbulence model.To enhance the unsteadiness due to the instability on the sheet cavity interface, we switch the eddy viscosity as well as the molecular viscosityby referring only the continuum phase. This treatment may look similar to well-known Reboud correction [2], while in this study the turbulent and molecular viscositiesare modified based on the fluid properties of continuum phase only, which is suitably applied in combination with BD1 model.

2.2. Computational method

Numerical simulations using above models were carried out for the unsteady cavitating flow in a two-dimensional convergent-divergent nozzle shownin Figure 2. The height of throat h is 5mm. The nozzle consists of a top straight wall and a bottom inclined wall with convergent and divergent angles of 43 and 8.4 degrees, respectively. This nozzle shape is similar to that used in our previous experiment [8], in which we observe a continuous cloud cavity shedding from the trailing edge of the sheet cavity.

As the boundary conditions, the velocity is fixed at the inlet ($u_{inlet} = 1.5\ \mathrm{m/s}$), and the static pressure is fixed at the outlet. The non-slip flow condition is applied on the upper and lower wall of the nozzle. The number of nodes is1,200 (along the wall) x 60 (perpendicular to the wall). For the convection scheme, second order upwind scheme is basically used except the void fraction; for the convection scheme in equation (2), TVD scheme [13] is used to well capture the cavity interface. As for the time integration, implicit Euler scheme is employed with the variable small time step satisfying local CFL number less than 0.8.

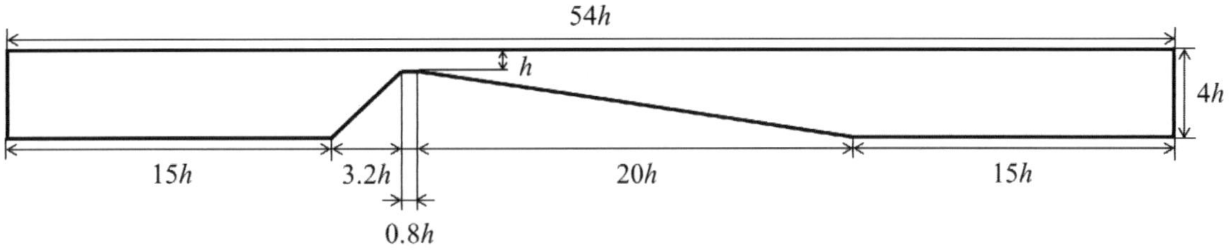

Figure 2. Computational domain (main flow direction is from left to right)

3. Result and discussion

Figure 3 shows typical cavity shapes at four instants with the time interval of 4ms, simulated by using four models described before. The cavitation number defined using the area-averaged nozzle throat velocity U_{throat} as$\sigma = 2(p_{in} - p_v)/\rho_L {U_{thorat}}^2$takes similar value of 1.63in all cases except SS model (0.49), resulting in the time averaged cavity length of $L_c/h \approx 11 - 16$. The similar cavitation observed in the previous experiment [8] is also shown in figure3 (a). By comparing the results between SS, BD1 and BD2 models, it is found that the sheet cavity is slightly longer for BD1 and BD2, but the unsteady behavior of cavitation is very similar: there are no cloud cavities observed but the sheet cavity slightly fluctuates. However, by applying the viscosity filtering model along with BD1 (BD1VF), we can clearly see the cloud cavity shedding similar to that observed in the experiment. These results suggest that the treatment of the turbulent shear stress is important rather than the cavitation model, whereas the concept of the viscosity filtering treatment in this study is conceptually suitable to the proposed cavitation model BD1 and BD2.

Figure 4(a) shows the results of the FFT analysis of pressure fluctuation measured23h downstream on the upper wall in the case with $L_c/h \approx 11 - 15$. In this figure, the strong peak around 40Hz (60Hz in SS model) and its harmonics can be clearly seen for SS, BD1 and BD2. This is associated with the cavity volume change observed in figure. 3(b)-(d), which numericallycauses surge-like oscillation of liquid column downstream of the sheet cavity. For the BD1VF model, the every frequency component is significantly larger than those for SS, BD1 and BD2, but we can clearly see the broad-banded frequency peaks around 45Hz, which is caused by the continuous cloud cavity shedding from the trailing edge of the sheet cavity. Similar tendency could be found for the other conditions with various sheet cavity lengths.

Figure 4 (b) shows the comparisons of Strouhal number $St = fh/U_{throat}$ of the cloud cavity shedding predicted by BD1VF model with that obtained by the experiment. Although the numerical data are limited to longer sheet cavities, the tendency of the predicted Strouhal number against the cavity length is very similar to the experiment. Therefore, we believe that the present model can well simulate the unsteadiness of the cloud cavity shedding in this kind of nozzle flow.

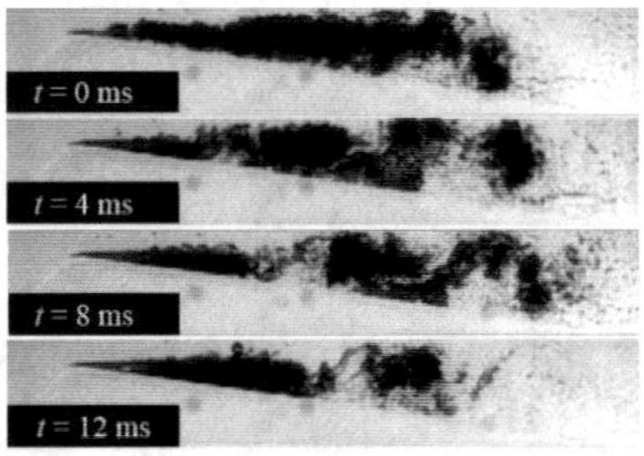

(a) Experiment [8] (σ=1.14)

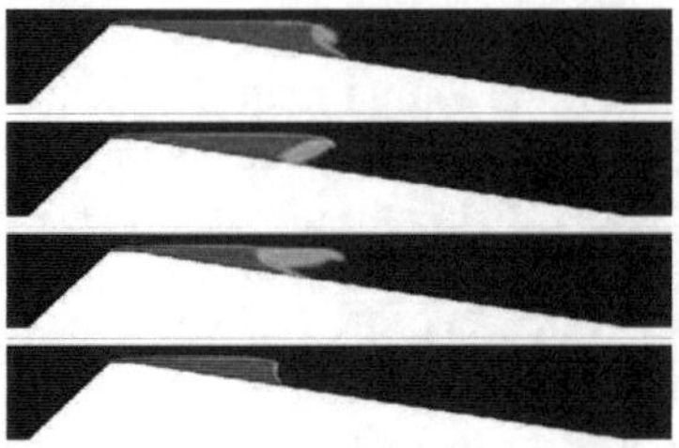

(b) SS (L_c/h=11.3, σ=0.49)

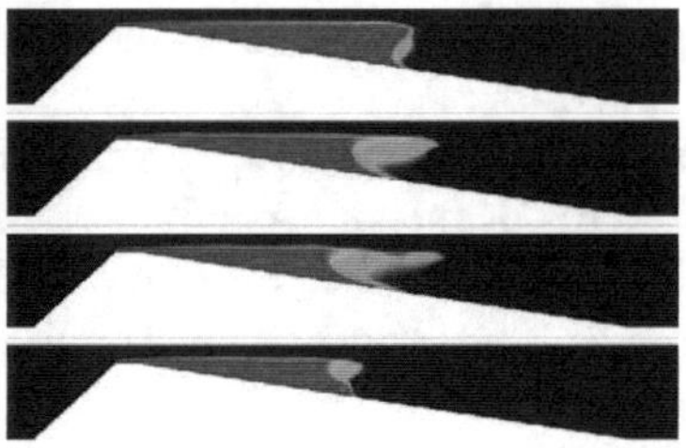

(c) BD1(L_c/h=14.4, σ=1.62)

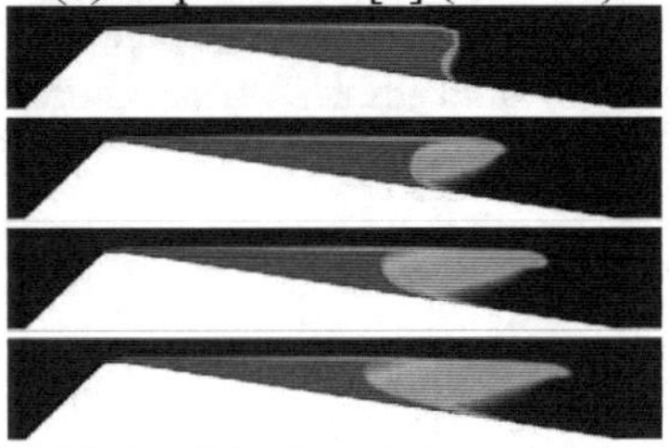

(d) BD2(L_c/h=15.7, σ=1.64)

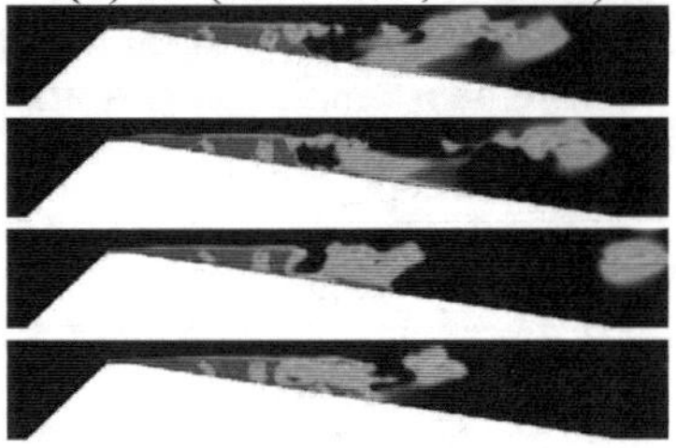

(e) BD1VF (L_c/h=11.5, σ=1.63)

Figure 3. Comparisons of cavity behaviours among four numerical models and experiment

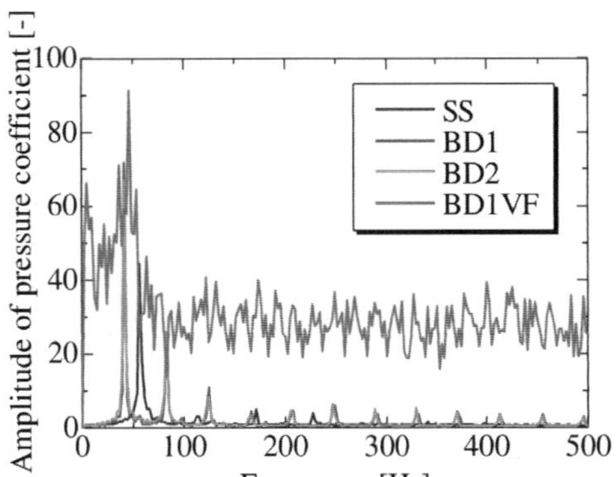

(a) Typical FFT results withL_c/h =11-16

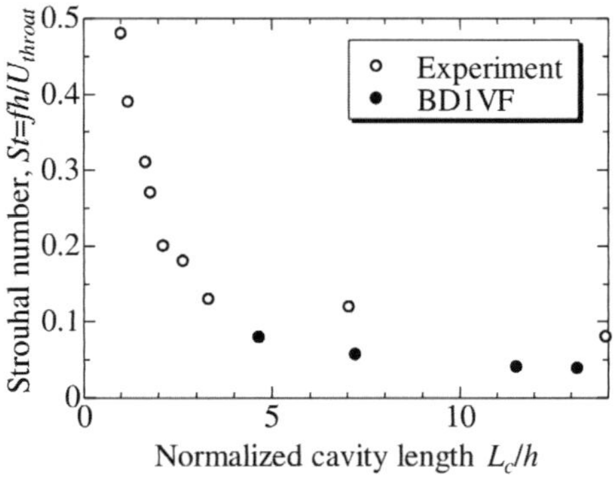

(b) Strouhal number of cloud cavity shedding

Figure 4. Results of FFT analysis of pressure fluctuation caused by unsteady cavitation

4. Concluding remarks

In this paper, a simple cavitation model was developed under the framework of homogeneous one-fluid model. The model is constructed consideringtwo extreme conditions; for low void fraction, dispersed vapour bubbles in continuous liquid phase, while for high void fraction, dispersed droplets in continuous vapour phase, and according to the local void fraction, the growth of bubbles and droplets are taken into account in the mass transfer between vapour and liquid phases. To enhance the unsteadiness of cavitation due to the instability at the cavity interphase, the turbulent shear stress is modified depending upon the continuous phases in combination with the proposed cavitation model. It is found that the proposed model can well predict the major unsteadiness of cavitation, i.e. continuous cloud cavity shedding from the sheet cavity, in a two-dimensional convergent-divergent nozzle flow.

References

[1] Kato C 2011 *Proc. ASME-JSME-KSME Joint Fluids Eng. Conf.((Hamamatsu, Shizuoka, Japan, 24-29, July,2011)* AJK2011-06084
[2] Reboud J. L., Stutz B, and Coutier O 1998 *Proc. 3rd Int. Sym. on Cavitation (Grenoble, France, 7-10, April, 1998)* 203-208
[3] Coutier-Delgosha O, Fortes-Patella R and Reboud J L 2003 *ASME J. Fluids Eng.* **125** 38-45
[4] Ji B, Luo X W, Arndt R E A and Wu Y L 2014 *Ocean Eng.* **87** 64-77
[5] Stutz B and Reboud J L 1997 *Physics of Fluids* **9** 3678-86
[6] Stutz B and Reboud J L 2000 *Experiments in Fluids* **29** 545-52
[7] Keil T, Pelz P F, Cordes U and Ludwig G 2011 *Proc. WIMRC 3rd International Cavitation Forum 2011(Warwick, UK, 4 – 6 July 2011)*
[8] Watanabe S, Enomoto K, Hara Y and Furukawa A 2012 *Proc. 5th Int. Symp. on Fluid Machinery and Fluid Eng. (Jeju, Korea, 24-27 Oct 2012,)*
[9] Schnerr G and Sauer J 2001 *Proc. 4th International Conference on Multiphase Flow (New Orleans, USA, 27 May - 1 June 2001)*
[10] Young Y L and Savander B R 2011 *Ocean Engineering* **38** 1368-81
[11] Roohi E, Zahiri A P, and Passandideh-Fard M 2013 *Applied Mathematical Modelling* **37** 6469-88
[12] Schrage R W 1953 *Theoretical Study of Interphase Mass Transfer* (Columbia: Columbia University Press)
[13] Bram V L 1974 *J. of Computational Physics* **14** 361-70

IOP Conf. Series: Materials Science and Engineering **72** (2015) 022010 doi:10.1088/1757-899X/72/2/022010

Bubble jet impact on a rigid wall of different stand-off parameters

S Li, S P Wang and A M Zhang[1]
College of Shipbuilding, Harbin Engineering University, #515 Chuanhai Building, Nantong Street No.145, Nangang District, Harbin 150001,China

E-mail: zhangaman@hrbeu.edu.cn

Abstract. One of the key features of the dynamics of a bubble near a rigid wall is the development of a high liquid jet, generating highly localized pressure on the wall. In present study, the boundary integral method is employed to simulate this phenomenon, and the vortex ring model is introduced to handle the discontinued potential of the toroidal bubble. Meanwhile, the pressure induced in the whole process is calculated by an auxiliary function. The effect of the stand-off parameter on the bubble dynamics and the pressure on the wall is investigated, and a double-peaked structure occurs in the pressure profile after the jet impact in some cases, which is associated with the jet impact and the high internal pressure inside the bubble.

1. Introduction

Bubble dynamics near a rigid wall is a particularly important subject because of the damage potential that can be caused by the impulsive pressure pulses resulting from the collapse. The growth and collapse of micron-sized bubbles near propeller blades holds the key to understanding the deleterious effects of cavitation [1]. The interaction between underwater explosion bubble and warship has important naval applications [2]. Lauterborn [3] found the velocity of the boundary-induced bubble jet is approximately 120m/s, which would cause severe damage to the structures.

Boundary integral method (BIM) is extensively used to study the dynamic of a non-equilibrium violently oscillating gas bubble for many decades [4~8]. The main features of the bubble could be simulated, such as expansion, collapse, jet and so on. Besides, the velocity and pressure in the fluid domain could also be calculated to analyze the mechanism of these phenomena. The transition of a bubble surface from singly-connected to doubly-connected induces circulation in the flow around the toroidal bubble, results in only a few works on the dynamic of toroidal bubble. Best [9] introduced a vortex sheet that moves with the fluid, rending the domain singly-connected. Wang [5] came up with a simper model consists of placing a vortex ring inside the bubble. The strength of the vortex ring is chosen to be equal to the potential difference of the jet tip and the opposite bubble surface just before the jet impacts.

In this paper, BIM is adopted to study the motion of bubble near a rigid wall, and the vortex ring model is introduced to handle the discontinued potential of the toroidal bubble. The auxiliary function method [10] is employed to calculate the pressure on the rigid wall, which avoid making finite difference of the velocity potential.

[1]Corresponding author.

2. Basic Theory

The liquid surrounding the bubble is assumed inviscid and incompressible, and the motion irrotational. The velocity potential Φ satisfies the Laplace equation:

$$\nabla^2\Phi = 0 \tag{1}$$

According to Green function, the velocity potential at any point in the domain could be expressed as an integral equation [11]

$$\iint_S \left(G(\boldsymbol{p},\boldsymbol{q}) \cdot \frac{\partial \Phi(\mathbf{q},t)}{\partial n} - \Phi(\mathbf{q},t) \frac{\partial G(\boldsymbol{p},\boldsymbol{q})}{\partial n} \right) dS(\mathbf{q}) = \varepsilon(\boldsymbol{p},t) \cdot \Phi(\boldsymbol{p},t) \tag{2}$$

where $\varepsilon(\boldsymbol{p},t)$ is the solid angle of a fixed point $\boldsymbol{p}$ on flow boundaries with the integration variable $\boldsymbol{q}$ also situated on boundaries, $\partial/\partial n$ is the normal outward derivative from the boundary S. $G(\boldsymbol{p},\boldsymbol{q}) = 1/|\boldsymbol{p}-\boldsymbol{q}| + 1/|\boldsymbol{p}-\boldsymbol{q}'|$ is the half-space Green function, with $\boldsymbol{q}'$ being the reflected image of $\boldsymbol{q}$ across the wall.

The dynamic boundary condition on the bubble surface can be written as:

$$\frac{D\Phi}{Dt} = \frac{(\nabla\Phi)^2}{2} + \frac{P_\infty}{\rho} - \frac{P}{\rho} - gz \tag{3}$$

where ρ is the density of the liquid , P_∞ is the ambient pressure of the liquid around the bubble, P is the pressure on the bubble surface, g is the gravity acceleration.

The kinematic boundary condition is:

$$\frac{d\mathbf{r}}{dt} = \nabla\Phi \tag{4}$$

The pressure inside the bubble is assumed to be uniform and consists of a constant vapor pressure and a volume-dependent noncondensable gas pressure [9]. Hence, the bubble pressure P_g as a function of the volume is described as:

$$P_g = P_c + P_{ini}(V_{ini}/V)^{\vartheta} \tag{5}$$

where the subscript *ini* denotes initial quantities, ϑ is the ratio of the specific heats for the gas, P_c is the vapor pressure.

Consider the jet impacting on the opposite bubble surface, forming a toroidal bubble. To solve this problem, a vortex ring is introduced inside the toroidal bubble [5]. The velocity induced by a circular vortex ring $\mathbf{u}_v$ can be calculated from Bio-Savart law:

$$\mathbf{u}_v(r,z) = \frac{\Gamma}{4\pi} \int_C \frac{d\mathbf{l} \times \mathbf{r}}{r^3} \tag{6}$$

More details about the vortex ring model can refer to Wang [5]. Here we have scaled length with respect to the maximum radius which the largest bubble under consideration might achieve in an infinite body of fluid R_m; pressure by $\Delta P = P_\infty - P_c$; time by $R_m\sqrt{\rho/\Delta P}$. Besides, some dimensionless parameters are given to describe a bubble:

1) Strength parameter: $\varepsilon = P_{ini}/\Delta P$
2) Stand-off parameter: $\gamma = d/R_m$
3) Buoyancy parameter: $\delta = (\rho g R_m/\Delta P)^{0.5}$

3. Results and Discussion

Three cases were calculated to study the dynamic behaviors of a bubble and the pressure induced on a rigid wall. The bubble is initialed at various stand-off parameters $\gamma = 1.3, 1.6, 2.0$, and the strength parameter is set as $\varepsilon = 20$, buoyancy parameter $\delta = 0$. The bubble motion before the jet impact has been discussed extensively before, so we just give the jet tip velocity versus time in Figure 1. Comparing the three cases, it is clear that, the jet velocity at the end of the collapse phase decreases as the bubble is initiated closer to the wall.

Figure 2 shows the toroidal bubble is rebounding for the case $\gamma = 1.3$. Because of the high-speed liquid jet, a protrusion appears at the position of jet impact, and an annular hollow move downward on the bubble surface. It can be seen that the bubble is drawn towards the rigid wall during its rebounding and the jet impacts directly on the rigid wall afterward.

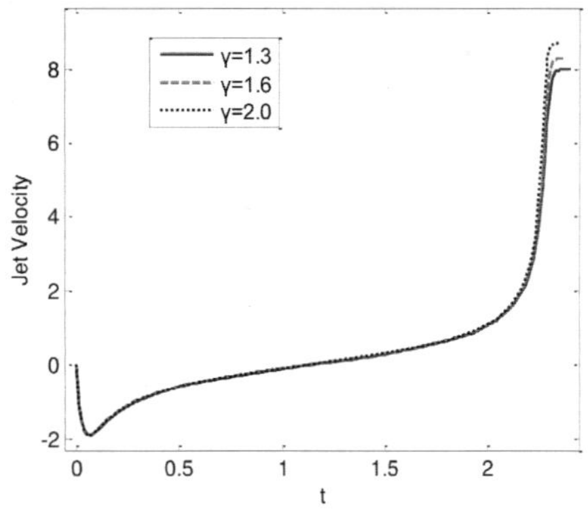

Figure 1. The jet velocity before jet impact versus time for the bubble with $\varepsilon = 20, \delta = 0$ initiated at distances $\gamma = 1.3, 1.6, 2.0$ from a rigid wall.

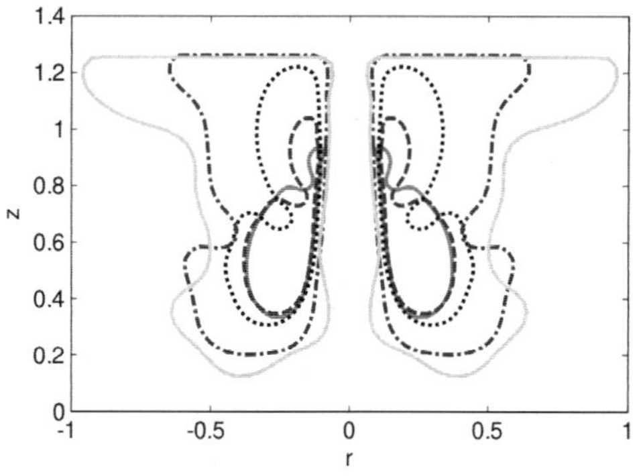

Figure 2. Evolution of the toroidal bubble with $\varepsilon = 20, \delta = 0, \gamma = 1.3$ at t=2.444 (innermost), 2.478, 2.552, 2.725, 2.873(outermost).

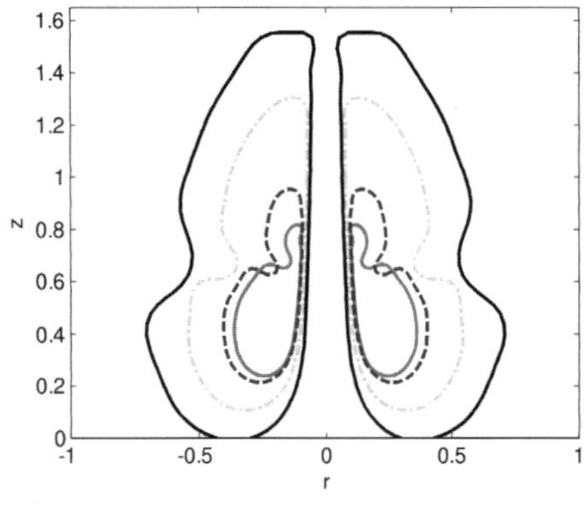

Figure 3. Evolution of the toroidal bubble with $\varepsilon = 20, \delta = 0, \gamma = 1.6$ at t=2.417 (innermost), 2.454, 2.569, 2.735 (outermost).

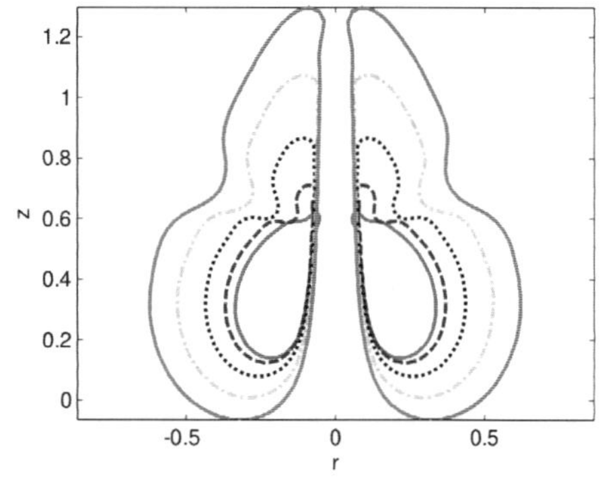

Figure 4. Evolution of the toroidal bubble with $\varepsilon = 20, \delta = 0, \gamma = 2.0$ at t=2.362 (innermost), 2.385, 2.426, 2.487, 2.565 (outermost).

Figure 3 concerns the bubble initiated at $\gamma = 1.6$. The centroid of the bubble also moves quickly toward the rigid wall. However, when the liquid jet comes into contact with the wall, the bubble reaches a large volume, so the inner-bubble pressure drops to a very low level. Meanwhile, the jet becomes very thin, so the bubble might rejoin into a singly connected form, which won't be discussed in this study. The next case relates the bubble initiated at $\gamma = 2.0$, which is shown in figure 4. The bubble rebounds in a toroidal form, moving upward, and it does not contact the rigid wall during the rebounding progress.

Figure 5 shows the kinetic energy and potential energy versus time for the bubble with different stand-off parameters. It is obvious that the bubble cycle increases as the stand-off parameter decreases, which implies the rigid wall slows down the bubble motion. As the stand-off parameter increases, the bubble will achieve greater potential energy and smaller kinetic energy when bubble reaches its minimum volume. In other words, the jet contains more energy when bubble initialled closer to the wall.

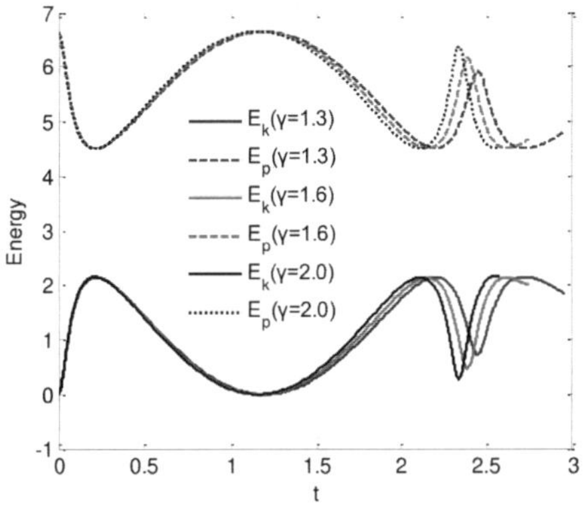

Figure 5. The kinetic energy and potential energy versus time for the bubble with $\varepsilon = 20, \delta = 0$ initiated at distances $\gamma = 1.3, 1.6, 2.0$ from a rigid wall.

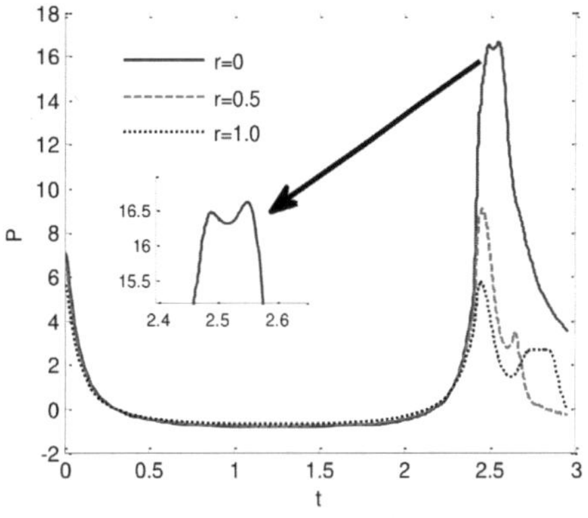

Figure 6. The pressure at different points ($r = 0, 0.5, 1.0$) of the rigid wall versus time for the bubble with $\varepsilon = 20, \delta = 0$ initiated at distances $\gamma = 1.3$ from a rigid wall.

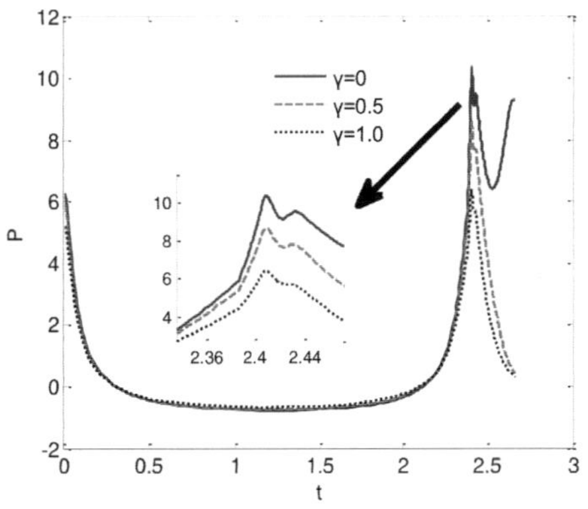

Figure 7. The pressure at different points ($r = 0, 0.5, 1.0$) of the rigid wall versus time for the bubble with $\varepsilon = 20, \delta = 0$ initiated at distances $\gamma = 1.6$ from a rigid wall.

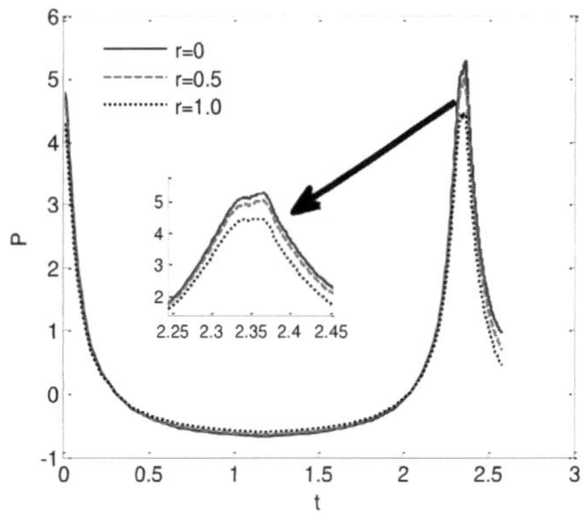

Figure 8. The pressure at different points ($r = 0, 0.5, 1.0$) of the rigid wall versus time for the bubble with $\varepsilon = 20, \delta = 0$ initiated at distances $\gamma = 2.0$ from a rigid wall.

The pressures at different position on the wall for the case $\gamma = 1.3$ are given in figure 6. At time $t = 0$, the pressure inside the bubble is very large, which induce the first maximum pressure on the wall.

However, the pressure caused by the collapsing bubble is much larger than that at $t = 0$. During the collapsing phase, two pressure peaks at the centre point of the wall, with approximately the same value, are observed. The first one occurs when the bubble reach its minimum volume, and the second one occurs when jet impact directly on the wall. As for r=0.5 and r=1.0, two small peak also turn up on the pressure profiles during the bubble rebounding phase, which are caused by the bubble protrusion pushing the liquid radially outward.

The evolution of the pressures at different position on the wall with time is shown in figure 7 for the case $\gamma = 1.6$. There are also two peaks on the pressure profile during the collapsing phase. The first one occurs when the bubble reach its minimum volume, the second one is not caused by the jet impact because the bubble surface is not contact with the wall at this time. The second peak can be explained that the bubble still maintains high pressure inside while the bubble is moving upward rapidly, so the distance between the bubble and the wall is decreasing. After $t = 2.558$, the pressure at the centre of the wall increases gradually, however, the pressures at r=0.5 and r=1.0 keeps decreasing, which implies the range of the jet impact is limited.

The evolution of the pressures at different position on the wall with time is shown in figure 8 for the case $\gamma = 2.0$. The maximum pressure during the collapsing phase is little larger than the first maximum pressure at $t = 0$. The effect of the toroidal bubble on the wall is similar to an spherical-oscillated bubble, and the liquid jet can't be reflected in the pressure profile.

4. Conclusions

Boundary integral method was employed to simulate bubble dynamics near a rigid wall, and the vortex ring model was adopted to handle the double-connected domain problem. Three different cases （$\gamma = 1.3, 1.6, 2.0$） were calculated to discuss the effect of the stand-off parameter on bubble dynamics and the pressure on the wall. As for $\gamma = 1.3$, the pressure caused by jet impact directly on the boundary is approximately the same with the time bubble reach its minimum volume, but the jet impact area is small. As for $\gamma = 1.6$, there were two peaks on the pressure profile during the bubble collapsing phase, the first one is associated with the minimum volume of the bubble, and the second one is caused by the rapidly upward movement of the bubble. As for $\gamma = 2.0$, the effect of the toroidal bubble on the wall is similar to a spherical-oscillated bubble, the jet impact on the wall could be ignored.

References

[1] Choi J, Hsiao C T, Chahine G and Ceccio S 2009 *J. Fluid Mech.* **624** 255
[2] Klaseboer E, Khoo B C and Hung K C 2005 *J. Fluid Struct.* **21** 395
[3] Lauterborn W and Bolle H 1975 *J. Fluid Mech.* **72** 391
[4] Wang Q X, Yeo K S, Khoo B C and Lam K Y 1996 *Theoret. Comput. Fluid Dynamics* **8** 73
[5] Wang Q X, Yeo K S, Khoo B C and Lam K Y 1996 *Computers & Fluids* **25** 607
[6] Blake J R, Tomita Y and Tong R P 1998 *Appl. Sci. Res.* **58** 77
[7] Tong R P, Schiffers W P, Show S J, Blake J R and Emmony D C 1999 *J. Fluid. Mech.* **380** 339
[8] Brujan E A, Keen G S, Vogel A and Blake J R 2002 *Phys. Fluids* **14** 85
[9] Best J P 1993 *J. Fluid. Mech.* **251** 79
[10] Wu G X and Hu Z Z 2004 *P Roy. Soc. A* **460** 2797
[11] Newman J N 1977 *Marine Hydrodynamics* (London: MIT Press) p 131

IOP Conf. Series: Materials Science and Engineering **72** (2015) 022011 doi:10.1088/1757-899X/72/2/022011

Numerical simulation of the interaction between two bubbles

R Han, X L Yao and A M Zhang[1]
College of Shipbuilding, Harbin Engineering University, #515 Chuanhai Building, Nantong Street No.145, Nangang District, Harbin 150001, China

E-mail: zhangaman@hrbeu.edu.cn

Abstract. Different evolution patterns of two bubbles may be observed for different values of the phase difference and the inter-bubble distance. Based on potential flow theory, a boundary element method (BEM) is adopted to simulate the interaction between two bubbles and the toroidal bubble after the jet impact is also investigated by placing a vortex ring within the bubble. Meanwhile, some numerical techniques are used to deal with problems like coalescence and collapse. Typical phenomena like coalescence, jet towards and jet away are investigated numerically in this paper. The mechanisms underlying the various phenomena are given through the analysis of the velocity and pressure fields.

1. Introduction

The dynamic character of bubble is of much importance to engineering applications, like gas bubble gun used to raise drilling speed in the field of ocean engineering, high-speed jet formed by underwater explosion bubble for the military and the jet of bubble used to remove the calculus in medical treatment. A wide range of applications make bubble dynamics deserve attention. However, complicated motion mechanisms of non-spherical bubble make the difficulties for theory research; therefore numerical simulation and experimental observations are widely used to study bubble dynamics. In aspect of numerical simulation, a boundary element method has been verified repeatedly [1-3], and a vortex ring model is introduced by Wang [4] to simulate the toroidal bubble after the jet impact; in aspect of experimental observations, researchers often use spark-generated bubble [5, 6] or laser-induced bubble [7] to study bubble dynamic behaviors, observing the phenomena like jet, collapse and toroidal bubble. With further investigations in bubble dynamics, extensive research about multiple-bubble cases has been made, such as the interaction between two bubbles and a rigid wall [7] and the interaction between a vertical column of two bubbles and free surface [8,9]. The two-bubble interactions depending on the phase differences and the inner-bubble distances have been classified in a graph by Fong et al. [5], and phenomena like coalescence, jet towards and jet away have been observed in a series of experiments involving multiple spark-generated bubbles. Rungsiyaphornrat et al. [10] have simulated the merging of two gaseous bubbles numerically and the coalescence criterion in their work is established by the measurement of explosion experiments.

Based on potential flow theory, a boundary element method (BEM) is adopted to simulate the interaction between two identical-sized bubbles in this paper and numerical techniques are also used to deal with problems like coalescence and collapse. A vortex ring model is established to further simulate the interaction between the toroidal bubble and the other singly connected one after the jet

[1] Corresponding author.

impact. Typical phenomena like coalescence, jet towards and jet away are investigated numerically in this paper, and bubble profile is presented together with the velocity and pressure fields, aiming to provide mechanisms underlying the bubble motion.

2. Theory and numerical techniques

2.1. Theory

BEM simulations are carried out in this paper to investigate the two-bubble interaction in a free field. Due to the short duration of the interaction and the large Reynolds numbers, the velocity potential in the flow field satisfies the boundary integral equation [11]

$$\varpi\Phi(\mathbf{p}) = \iint_S \left(\frac{\partial \Phi(\mathbf{q})}{\partial n} G(\mathbf{p},\mathbf{q}) - \Phi(\mathbf{q}) \frac{\partial}{\partial n} G(\mathbf{p},\mathbf{q}) \right) \mathrm{d}S \tag{1}$$

where the normal derivative is defined as $\partial/\partial n = \mathbf{n}\cdot\nabla$ and $\mathbf{n}$ is the outward normal, $\mathbf{p}$ and $\mathbf{q}$ are the fixed and the source point located at the interface, ϖ is the solid angle at the fixed point $\mathbf{p}$ with a value of $\varpi = 2\pi$, $G(\mathbf{p},\mathbf{q}) = 1/|\mathbf{r}(\mathbf{p}) - \mathbf{r}(\mathbf{q})|$ is the Green's function, S is all the boundaries of the fluid domain.

Since the influence from the motion of bubble internal gas during its overall process can be neglected, the bubble internal pressure is related to the initial state and its volume. The liquid pressure outside the bubble wall is supposed to be equal to the pressure within the bubble, it can be written as [4]

$$P = P_c + P_0 \left(\frac{V_0}{V} \right)^k \tag{2}$$

where P_c is the constant vapor pressure inside the bubble, P_0 and V_0 are the initial pressure and volume of the bubble, and k is the ratio of the specific heat for the gas related to gas component which equals 1.25 [12].

Scaling factors are used to obtain the dimensionless form of the equations. The bubble's maximum radius R_m, $P_{ref} = P_\infty - P_c$, $R_m(\rho / P_{ref})^{1/2}$, $R_m(P_{ref} / \rho)^{1/2}$ and $(\rho g R_m / P_{ref})^{1/2}$ are used for length, pressure, time, velocity potential and buoyancy, respectively. Therefore, the dimensionless form of the bubble state equation can be written as [13]

$$\frac{\mathrm{d}\Phi}{\mathrm{d}t} = 1 + \frac{|\nabla\Phi|^2}{2} - \delta^2 z - \varepsilon \left(\frac{V_0}{V} \right)^k \tag{3}$$

where $\delta = \sqrt{\rho g R_m / P_{ref}}$ is defined as buoyancy parameter with a value of $\delta = 0$ without the influence of gravity in this paper and $\varepsilon = P_0 / P_{ref}$ is defined as the strength parameter inside the bubble. In addition, we define dimensionless inter-bubble distance as $d_{in} = D_{in} / R_m$, in which D_{in} is the distance between the initial bubble centroid. The phase difference is written as $\Delta\vartheta = 2\pi \left| \frac{t_1}{t_{osc,1}} - \frac{t_2}{t_{osc,2}} \right|$ [5], in which t is the elapsed time from the first initiation of a bubble and t_{osc} is the oscillation time.

The kinematic boundary condition governing the bubble motion is [4]

$$\frac{\mathrm{d}\mathbf{r}}{\mathrm{d}t} = \nabla\Phi \tag{4}$$

The idealized impact model (one point impact) [14] is used in this paper, and a vortex ring model [4] is introduced after the jet impact. The vortex ring is placed within the toroidal bubble, the strength of which is equal to the potential difference of the jet tip and the opposite bubble surface just before the jet impacts. The total potential of the bubble surface is now decomposed into two parts

$$\Phi = \Phi_v + \Phi_r \tag{5}$$

where Φ_v is the potential due to the circulation of the vortex ring and Φ_r is a remainder. The details on obtaining Φ_v are available in Ref. [14], while Φ_r satisfies the boundary integral equation, by which the corresponding velocity $\mathbf{u}_r$ can be computed. The total velocity $\mathbf{u} = \mathbf{u}_v + \mathbf{u}_r$ is used to update the position of the node on bubble surface, and the update of Φ_r can be obtained by

$$\frac{d\Phi_r}{dt} = \mathbf{u} \cdot \nabla\Phi_r - \frac{|\mathbf{u}|^2}{2} + 1 - \delta^2 z \tag{6}$$

2.2. Numerical techniques

Three evolution patterns of two bubbles are simulated numerically in this paper. In the case of coalescence, it should be judged whether the coalescence criterion is satisfied and the jointing of nodes is performed with some numerical technique. The coalescence criterion mentioned above is based on Ref. [10] and Ref. [15], and we reckon that the minimal film thickness between bubbles satisfies the coalescence criterion when .Some nodes are now eliminated and the potential distribution over the bubble surface remain unchanged for all other nodes. In the coalescence case, the motion of the bubble can be solved by equation (1)-(4), for it only relates to the calculation of singly connected bubble.

3. Computational results

Different phenomena of two bubbles can be observed for different values of the phase difference and the inter-bubble distance [5]. Three typical phenomena- coalescence, jet towards and jet away- are investigated numerically in this paper. Meanwhile, the velocity and pressure fields at certain moment are calculated and discussions are made through the analysis of the flow field.

3.1. Coalescence

Coalescence of two bubbles is investigated with $\varepsilon = 200, R_0 = 0.1159, d_{in} = 0.9$, and the motion of two in-phase bubbles is simulated. Both the velocity and pressure fields are presented to explain the evolution patterns. Bubble profile at four different times is shown in Figure 1.

Two bubbles are incepted simultaneously and they expand rapidly due to the internal high pressure. The interplay causes the closer walls of bubbles to be flattened, as shown in Figure 1(a). At T=0.2744, the minimal film thickness between bubbles satisfies the coalescence criterion and numerical technique is used for obtaining the coalesced bubble. The pressure field at frame T=0.2744 states the bubble overexpansion, and it's noted that the subsequent expansion of the coalesced bubble will be stopped by the high pressure of flow field. The collapse of the coalesced bubble is observed at T=1.7829. At this moment, the larger pressure gradient of the flow field around the top and the bottom of bubble leads to a faster collapse of both the top and the bottom than that of the side wall. The fluid near to the axis of symmetry is rapidly drawn due to the fast contraction of both the top and the bottom, and then two high-pressure regions are formed as a result of the fluid impact on the symmetric axis. The downward jet of the top surface and the upward jet of the bottom surface impact in the middle of the coalesced bubble at T=2.2573, which are caused by the two high-pressure regions. It can be seen that a central high-pressure region is formed due to the jet impact, which can redirect the incoming fluid to form a jet pointing to the outside wall of the toroidal bubble. At the same time, the toroidal bubble proceeds to collapse. At T=2.4080, the jet impacts the outside wall and the bubble continues to collapse (the bubble profile and the flow field are shown in Figure 1(d)).

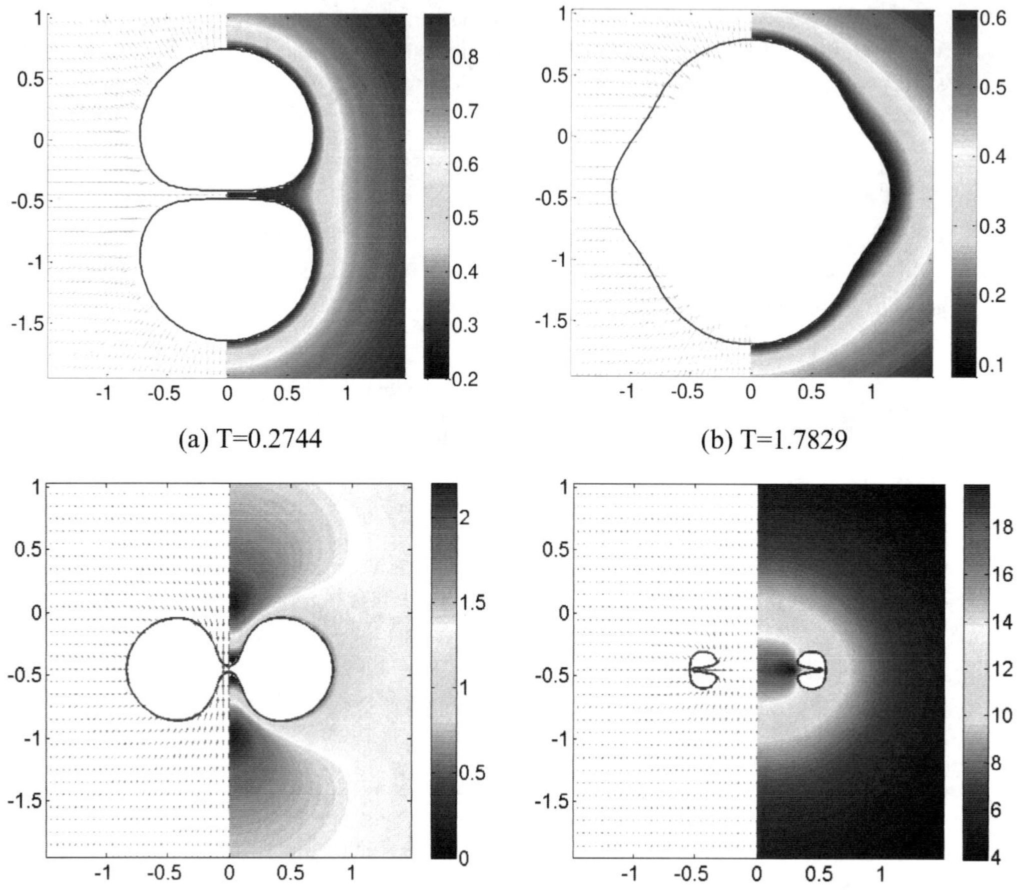

(a) T=0.2744 (b) T=1.7829

(c) T=2.2573 (d) T=2.4080

Figure 1. Coalescence case

3.2. Jet towards

The parameters for jet towards case are $\varepsilon = 20, R_0 = 0.2695, d_{in} = 3$. Two bubbles are created at the same time and calculation results of three different times are shown in Figure 2 to account for the evolution patterns of the bubbles.

Two in-phase bubbles expand due to the internal high pressure, and the pressure inside the bubble decreases gradually in the expansion phase. The expansion is about to be restrained once the internal pressure is less than the pressure of the flow field. As shown in Figure 2(a), the pressure in fluid domain is greater than that inside the bubble at T=0.4111, which restrains the expansion. It can be found that the pressure of the region between two bubbles is less than that of other regions in fluid domain, which accounts for the later collapse of the closer walls of bubbles. At T=1.2039, the bubbles attain the maximum volume with collapse of the distant walls and expansion of the closer walls (see Figure 2(b)). It can be seen that the asymmetric pressure field around one bubble causes different collapse velocities of the nodes distributed over the bubble surface, so the top of the upper bubble and the bottom of the lower bubble tend to collapse relatively faster, leading to jet towards. The bubble profile and the flow field at the moment of impact are shown in Figure 2(c), and two high-pressure regions can be obviously observed.

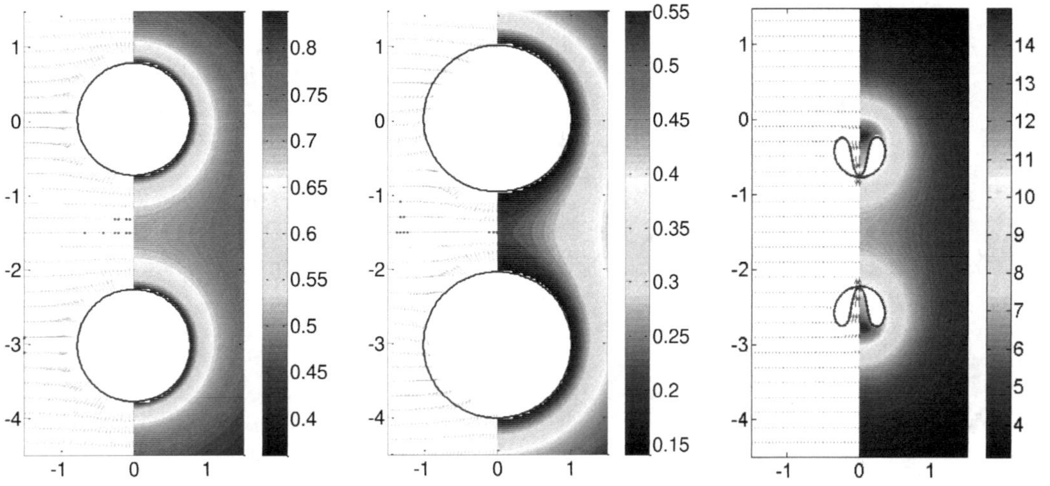

(a) T=0.4111 (b) T=1.2039 (c) T=2.3863

Figure 2. Jet towards case

3.3. Jet away

The parameters for jet away case are $\varepsilon = 20, R_0 = 0.2695, d_{in} = 2$, and the phase difference is $\Delta\vartheta = 2.9679$. The bubbles are generated sequentially with bubble 1 at T=0, and then bubble 2 at T=1.0099. Calculation results of three different times are shown in Figure 3.

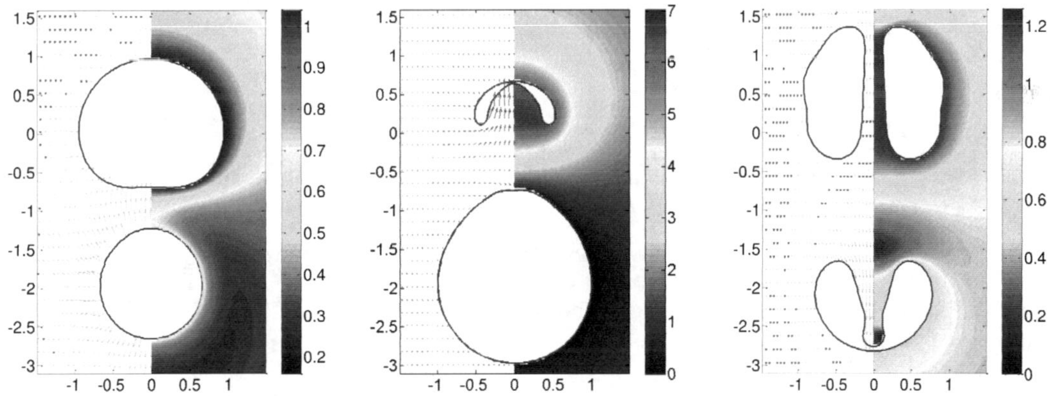

(a) T=1.2609 (b) T=1.8181 (c) T=2.5073

Figure 3. Jet away case

Bubble 1 expands rapidly after its inception at T=0 and its expansion will be restrained when the time comes to generate bubble 2 at T=1.0099. At T=1.0214, bubble 1 comes into collapse phase due to the restraint from bubble 2. As shown in Figure 3(a), the bottom of bubble 1 has the tendency to form an upward jet (a jet away from bubble 2), while the elongated shaped bubble 2 is obtained because the overexpansion of bubble 1 provides a low-pressure boundary to bubble 2. See velocity and pressure fields in Figure 3(a), the expansion of bubble 2 has been restrained by the flow fluid, however, the top of bubble 2 continues to expand due to the low-pressure boundary provided by bubble 1 and it's clear that both the collapse of bubble 1 and the elongation of bubble 2 are continued. In the collapse phase of bubble 1, fluid is rapidly drawn into the region between the two bubbles, leading to the formation of a stagnation point sometime. A high-pressure region is formed here to promote the jet of bubble 1 and

IOP Conf. Series: Materials Science and Engineering **72** (2015) 022011 doi:10.1088/1757-899X/72/2/022011

to restrain the expansion of bubble 2. At T=1.8181, the jet of bubble 1 impacts the upper wall and the obvious flattening of the top of bubble 2 can be observed, as shown in Figure 3(b). At this moment, bubble 2 is in the collapse phase and the contraction of its top surface is fastest under the effects of the high-pressure region. The interaction between the toroidal bubble 1 and bubble 2 is simulated afterwards. Two high-pressure regions are formed after the jet impact [16], so the fast contraction of the top surface of bubble 2 proceeds with fluid continually drawn into the narrow region between the two bubbles. Similar to bubble 1, the downward jet of bubble 2 is also promoted by the high-pressure region. The jet proceeds with the continued contraction of bubble 2, while bubble 1 rebounds rapidly. When it comes to T=2.5073, the jet of bubble 2 impacts the bottom surface and bubble 1 rebounds obviously (see Figure 3(c)).

4. Conclusions

Based on potential flow theory, axisymmetric numerical investigation has been carried out for the evolution patterns of two identical-sized bubbles. Three typical phenomena- coalescence, jet towards and jet away- are simulated numerically. Meanwhile, velocity and pressure fields in these cases are calculated, which accounts for the bubble evolution to some extent. The following conclusions are obtained.

1) After the coalescence of two identical-sized bubbles, the jets of the top and the bottom surface impact in the middle of the coalesced bubble. Afterwards, the bubble becomes toroidal and a jet pointing to its outside wall is formed, leading to the formation of two toroidal bubbles.

2) The existence of an in-phase bubble with the same size will restrain the collapse of the closer walls of bubbles. The asymmetric pressure field around one bubble leads to the earlier and relatively faster collapse of the distant walls, so the two bubbles jet towards each other.

3) When two bubbles are not in-phase, the later incepted bubble 2 restrains the expansion of the earlier bubble 1 and bubble 1 will jet away from bubble 2. A high-pressure region between the two bubbles is formed due to the collapse of bubble 1, which helps bubble 2 to jet away from bubble 1.

References

[1] Voinov O V and Voinov V V 1975 *Sov. Phy. Dokl.* **20** 179

[2] Blake J R and Gibson D C 1981 *J. Fluid Eng.* **111** 123

[3] Blake J R, Taib B B and Doherty G 1986 *J. Fluid Mech.* **170** 479

[4] Wang Q X, Yeo K S, Khoo B C and Lam K Y 1996 *Comput. Fluids* **25(7)** 607

[5] Fong S W, Adhikari D, Klaseboer E and Khoo B C 2008 *Exp. Fluids* **46** 705

[6] Zhang A M, Wang S P, Bai Z H and Wang C 2011 *Acta Phys. Sin.* **43(1)** 71 (in Chinese)

[7] Blake J R, Robinson P B, Shima A and Tomita Y 1993 *J. Fluid Mech.* **255** 707

[8] Robinson P B, Blake J R, Kodama T, Shima A and Tomita Y 2001 *J. Appl. Phy.* **89(12)** 8225

[9] Pearson A, Cox E, Blake J R and Otto S R 2004 *Eng. Anal. Bound. Elem.* **28** 295

[10] Rungsiyaphornrat S, Klaseboer E, Khoo B C and Yeo K S 2003 *Comput. Fluids* **32** 1049

[11] Newman J N 1977 *Marine Hydrodynamics* (1st Ed.) (London: MIT Press) 131

[12] Cole R H. 1948 *Underwater Explosions* (1st Ed.) (Prinston University Press) 164

[13] Wang C and Khoo B C 2004 *J Comput. Phys.* **194** 451

[14] Best J P 1993 *J. Fluid Mech.* **251** 79

[15] Best J P 1994 The rebound of toroidal bubbles *Bubble Dynamics and Interface Phenomena (Kluwer)* ed. Blake I R, Boulton-Stone J M, Thomas N H 405-412

[16] Li S, Zhang A M and Han R Wang C and Khoo B C 2004 *J. Comput. Phys.* **194** 451

IOP Conf. Series: Materials Science and Engineering **72** (2015) 022012 doi:10.1088/1757-899X/72/2/022012

Numerical simulation of unsteady flow characteristics for cavitation around a 3-D hydrofoil

S H Ahn, Y X Xiao and Z W Wang[1]
State Key Laboratory of Hydroscience and Engineering & Department of Thermal Engineering, Tsinghua University, Beijing 100084, China

E-mail: wzw@mail.tsinghua.edu.cn

Abstract. At present it is possible to predict more accurately by various numerical methods established for cavitation simulation around a hydrofoil. However, for the solution of the complex unsteady cavity flow, it is still marginal. In this paper, numerical method is adopted to simulate cavitation around 3-D NACA0015 hydrofoil with homogeneous two-phase flow calculation using commercial code CFX-solver with two turbulence models, the standard RNG k-ε turbulence model and the modified RNG k-ε turbulence model respectively. First, pressure coefficient for non-cavitating flow, time averaged values of unsteady cavity flow around a hydrofoil are verified to simulate more closely to an actual cavity flow. And then frequency analysis is performed with Fast Fourier Transform. The results show that the calculation results with modified RNG k-ε turbulence model agree with experimental results in terms of mean cavity length and pressure drop, but the unsteady flow characteristics of oscillating cavitation still deviate slightly in terms of unsteady cavity flow.

1. Introduction

Cavitation is a phenomenon generated from expansion of the non-condensable gas when the local pressure falls below specific value in constant temperature. This phenomenon causes complex pressure fluctuation, for hydraulic machinery, the periodic change in pressure damages to some parts in the machine. Furthermore, the cavitation bubble has an impact on surrounding liquid flow, which is relevant to the mechanical efficiency. The cavitation flow can bring out erosion, noise, vibration and so on. Thus, the cavitation is the problem which should be solved.

At present it is possible to observe cavitation phenomena more clearly and to measure more accurately by the development of experimental devices in cavitation around a hydrofoil. And also various numerical methods for cavitation calculation were already established, it is possible to more accurately predict it. Coutier-Delgosha et al. [1] presented that cavitation simulation with k-ω turbulence model including compressibility effect can be obtained better reasonable result. Kunz et al. [2] compared numerical results with turbulence modeling of between RANS and DES. Zhou et al. [3] simulated the cavitation flow on a hydrofoil considering the influence of non-condensable gas. Ji et al. [4, 5] introduced PANS method for cavitating flow and also showed the numerical results using the actual maximum density ratio. But it is still limited to solve the complex cavitation phenomena which are compressible multiphase flow. In this paper, numerical analysis is adopted to simulate unsteady cavitating flow around 3-D NACA0015 hydrofoil. The numerical result with two turbulence models

[1] Corresponding author. Tel.: +86 10 51533279.

IOP Conf. Series: Materials Science and Engineering **72** (2015) 022012 doi:10.1088/1757-899X/72/2/022012

for cavitation are compared to experimental result, first, time averaged values are compared, and then unsteady characteristics are evaluated for two turbulence models.

2. Numerical method

2.1. Governing equations

In homogeneous two phase flow, all fluids are in a flow field, share a velocity field. The homogeneous model assumes that each quantity transported for all phases are same. The homogeneous continuity and momentum equations are as:

$$\frac{\partial \rho}{\partial t}+\frac{\partial (\rho u_j)}{\partial x_j}=0 \tag{1}$$

$$\frac{\partial (u_i)}{\partial t}+\frac{\partial (u_i u_j)}{\partial x_j}=-\frac{1}{\rho}\frac{\partial p}{\partial x_i}+\nu\nabla^2 u-S \tag{2}$$

where 18, v, S are the mixture density, viscosity and the stress tensor, which are the following:

$$\rho=\sum_{a=1}^{N_p}\alpha_a\rho_a \tag{3}$$

$$\mu=\sum_{a=1}^{N_p}\alpha_a\mu_a \tag{4}$$

$$S=\frac{\partial}{\partial x_j}\left[\nu_t\left(\frac{\partial u_i}{\partial x_j}+\frac{\partial u_j}{\partial x_i}\right)-\frac{2}{3}\frac{\partial u_k}{\partial u_k}\delta_{ij}\right] \tag{5}$$

where □ is the volume fraction of each phase, N_p is the number of phases.

2.2. Cavitation model

The Rayleigh-Plesset model is typical cavitation model which control vaporization and condensation.

$$R\frac{d^2R}{dt^2}+\frac{3}{2}\left(\frac{dR}{dt}\right)^2=\frac{p_v-p}{\rho_l} \tag{6}$$

where R represents radius of bubble, 18_l is water liquid density. Equation (6) is reduced by neglecting the second order terms as following:

$$\frac{dR}{dt}=\left(\frac{2}{3}\frac{p_v-p}{\rho_l}\right)^{1/2} \tag{7}$$

The change rate of bubble volume is:

$$\dot{V}=4\pi R^2\left(\frac{2}{3}\frac{p_v-p}{\rho_l}\right)^{1/2} \tag{8}$$

The vaporization bubble initially contains non-condensable gas. The sum of volume fraction of each phase is as $\alpha_l+\alpha_v+\alpha_{ncg}=1$ and the mixture density is as $\rho=\rho_l\alpha_l+\rho_v\alpha_v+\rho_{ncg}\alpha_{ncg}$. The number of bubbles per unit volume for vaporization, N, can be written as:

$$N=\frac{3\alpha_{ncg}(1-\alpha_v)}{4\pi R^3} \tag{9}$$

and for condensation, *N* is following:

$$N = \frac{3\alpha_v}{4\pi R^3} \tag{10}$$

Cavitation process is governed by kinetics of phase change; mass transfer equation between vapor and liquid is as:

$$\frac{\partial(\rho_v \alpha_v)}{\partial t} + \nabla \cdot (\rho_v \alpha_v U) = \dot{m}_e - \dot{m}_c \tag{11}$$

where □$_v$ is vapor volume fraction. Equation (11) represents mass transfer rate of □$_v$ between evaporation and condensation.

$$\dot{m}_e = C_e \frac{3\rho_v(1-\alpha_v)\alpha_{ncg}}{R}\left(\frac{2}{3}\frac{p_v - p}{\rho_l}\right)^{1/2} \tag{12}$$

$$\dot{m}_c = C_c \frac{3\rho_v \alpha_v}{R}\left(\frac{2}{3}\frac{p_v - p}{\rho_l}\right)^{1/2} \tag{13}$$

where C_e, C_c are empirical coefficient for vaporization and condensation, which are 50 and 0.01, respectively in this paper.

2.3. Numerical model and boundary conditions

Unsteady calculation of the homogeneous two-phase flow with the cavitation model is used to obtain the results of unsteady cavity flow. The cavity flow around 3-D hydrofoil has velocity component of the spanwise direction, which is not zero by the effect of the side walls [6, 7]. Therefore, 3-D hydrofoil was adopted for this calculation domain, NACA0015 based on the experimental study by Cervone et al. [8], which is symmetry shaped model between upper and lower with chord length of 115 mm and span length of 80 mm. This simulation was used structured grids of about 1.4 million nodes, the density of grids is finer near the walls, especially around hydrofoil, where o-grid method was used as figure 1. The boundary conditions of the pressure at inlet and the mass flow rate at outlet were set according to cavitation number is defined as:

$$\sigma = \frac{p_0 - p_v}{\frac{1}{2}\rho {U_0}^2} \tag{14}$$

where p_0 and U_0 are the freestream pressure and velocity. The no slip condition was set at other walls including a hydrofoil. In this paper, the calculation condition is as following: angle of attack of 8 degree for leading edge, cavitation number of 1.8, temperature is constant of 25 °C. The initial condition in this domain was obtained from two-phase flow calculation of water liquid-vapor without the cavitation model to observe initial bubble change. Convergence criterion was specified as root mean square of 10^{-5}, time step was set and controlled by CFL number according to velocity change in calculation domain, and the commercial CFD code CFX-solver is used.

In this paper, the numerical results will be compared between RANS with RNG *k-ε* turbulence model and with the modified RNG *k-ε* turbulence model of which the eddy viscosity can be changed to the function of density proposed by Coutier-Delgosha et al. [1]. The standard RNG *k-ε* turbulence model was developed for incompressible fluid, but the cavitation is two-phase compressible flow [6]. The eddy viscosity of standard RNG *k-ε* model, .00$_t$, is as:

$$\mu_t = \rho C_\mu \frac{k^2}{\epsilon} \tag{15}$$

where C_{∞} represents turbulence model constant, *k* turbulence kinetic energy, □ turbulence dissipation rate. And modified one is as follows:

$$\mu_t = f(\rho)C_\mu \frac{k^2}{\epsilon} \tag{16}$$

$$f(\rho) = \rho_v + (\alpha_l)^n(\rho_l - \rho_v) \tag{17}$$

where α_l is the volume fraction of water liquid. And the turbulence model constant C_{∞} and $C_{\varepsilon 1}$, $C_{\varepsilon 2}$ coefficients are 0.085, 1.42 and 1.68, respectively, for RNG *k-ε* turbulence model in this paper.

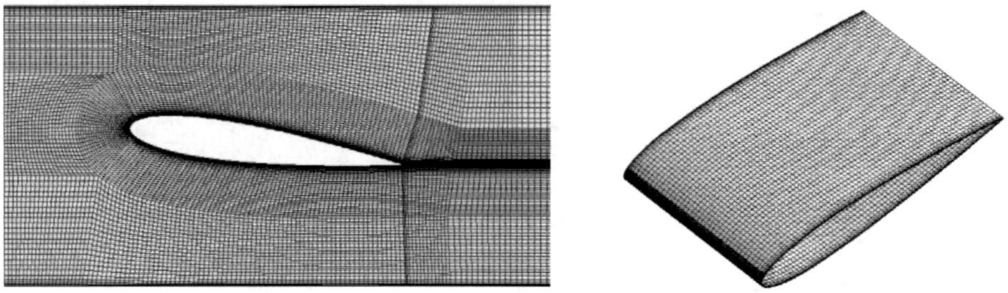

(a) Front view (b) Hydrofoil

Figure 1. Computational grids (angle of attack = 8 degree).

3. Numerical results

The results for unsteady calculation with the standard RNG *k-ε* turbulence model and the modified RNG *k-ε* turbulence model were obtained. There is no significant difference between with two different turbulence models in non-cavitating condition because this flow is almost same to single phase flow. Simulation result agrees with experimental result for pressure coefficient on a hydrofoil as figure 2. It is essential to verify calculation result of non-cavitating flow for obtaining more accurate simulation result of the cavitating flow. Table 1 shows the mean cavity length and the time averaged pressure drop around a hydrofoil of between numerical and experimental result with angle of attack of 8 degree and 1.8 cavitation number. As a result, the calculation result with modified RNG *k-ε* turbulence model is better close to experimental data, which model is changed eddy viscosity in the standard model to the function of density. The cavity flow with cavitation number of 1.8 is sufficient complex in figure 3, it shows the bubble change of vaporization and condensation with contour on the mid-plane of spanwise and iso-surface of 0.1 for vapor volume fraction, the phenomena of both bubble and cloud cavitation can be seen for this cavitation number.

Figure 2. Pressure coefficient distribution of experiment and calculation.

IOP Conf. Series: Materials Science and Engineering **72** (2015) 022012 doi:10.1088/1757-899X/72/2/022012

Table 1. Mean cavity length and pressure drop round a hydrofoil.

	l/C	$\Delta p/0.5\rho U_0^2$
EXP	0.78	0.156
RNG	0.372	0.149
RNG modified	0.782	0.161

***t* / *T* = 0.2**

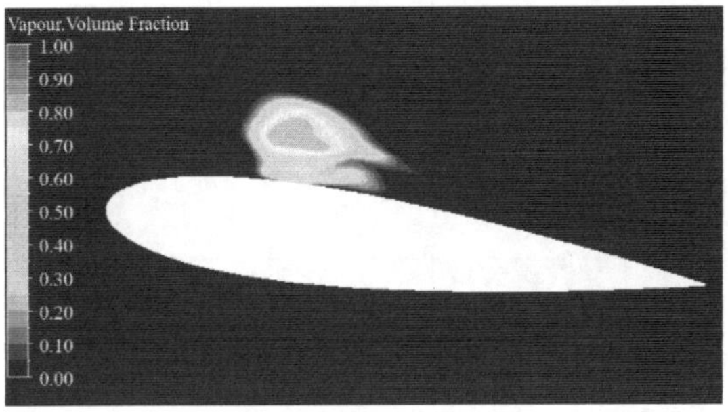

***t* / *T* = 0.4**

***t* / *T* = 0.6**

***t* / *T* = 0.8**

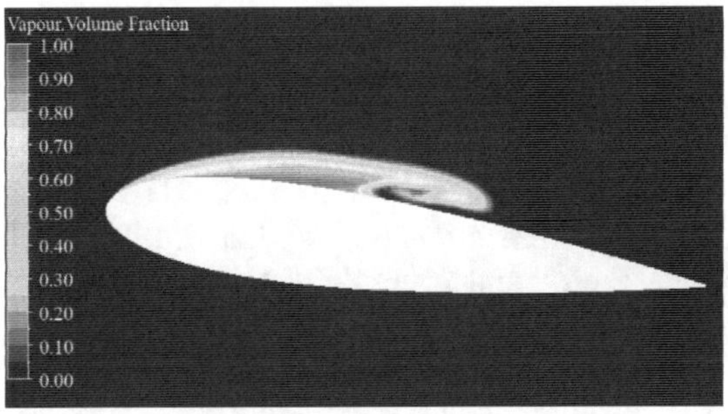

(a) Iso-surface **(b) Contour**

Figure 3. Instantaneous vapor volume fraction (σ = 1.8, 8 degree).

Figure 4 shows the result of frequency analysis with Fast Fourier Transform, the amplitude on this diagram is a non-dimensional value of pressure pulsation divided by dynamic pressure. In this result, dominant frequency of the calculation with standard RNG k-ε turbulence model, Strouhal number of 0.18, is about half of with modified one. It is still different from experimental result for oscillation cavitation and the amplitude of pressure pulsation is also larger than measurement value.

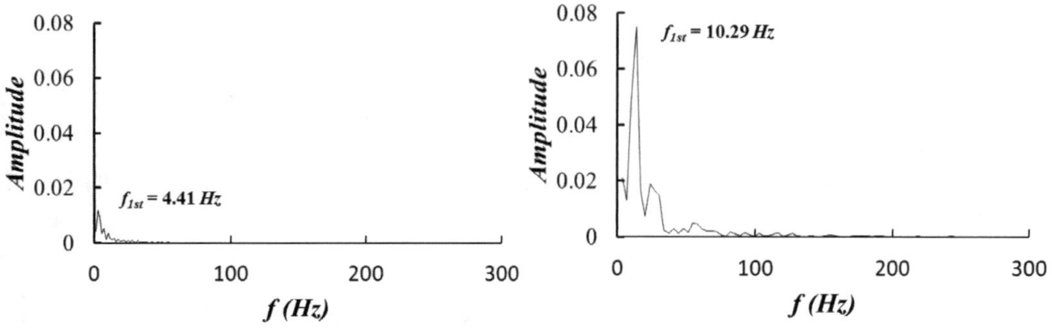

Figure 4. Frequency diagram of pressure pulsation at a hydrofoil upstream (left: standard RNG k-ε model, right: modified RNG k-ε model)

4. Conclusion

The unsteady cavitation simulation around a NACA0015 hydrofoil with two turbulence models, the standard RNG k-ε turbulence model and the modified RNG k-ε turbulence model of which turbulence viscosity is changed by local density, was calculated to evaluate an accuracy for unsteady cavitation characteristics comparing to the experimental results by Cervone et al. [8]. In this paper, a cavitation number with complex cavity flow, which has bubble and cloud cavitation simultaneously, is adopted because unsteady characteristics is obvious for this condition. To simulate more close to actual cavity flow, the calculation results of pressure coefficient for non-cavitating condition, mean cavity length and pressure drop around a hydrofoil for cavity flow were compared, and then the frequency analysis of pressure pulsation with Fast Fourier Transform was performed. The results show that modified RNG k-ε turbulence model can be simulated well through the comparison of mean cavity length and pressure drop, but the dominant frequency and amplitude of pressure pulsation still have difference in terms of unsteady cavity flow.

Acknowledgments

Special thanks are due to the State Key Program of National Science of China (Grant No. 51439002), National Natural Science Foundation of China under Contract (No. 51479093), State Key Laboratory of Hydroscience and Engineering (Grant No. 2014-KY-05) for supporting the present work.

References

[1] Coutier-Delgosha O, Fortes-Patella R and Reboud J L 2003 *J. Fluid Eng.-T. ASME* **125** 38-45
[2] Kunz R F, Lindau J W, Kaday T A and Peltier L J 2003 *Proc. the 5th international symposium on cavitation* (Osaka, Japan, 1-4 Nov 2003) pp 1-4
[3] Zhou L and Wang Z 2008 *J. Fluid Eng.* **130** 011302-1
[4] Ji B, Luo X, Wu Y, Peng X and Xu H 2012 *Int. J. Heat and Mass Transfer* **55** 6582-8
[5] Ji B, Luo X W, Wu Y L, Peng X X and Duan Y L 2013 *Int. J. Multiphase Flow* **51** 33-43
[6] Yang J, Zhou L J and Wang Z W 2011 *J. Fluid Eng.-T. ASME* **133** 081301-1
[7] He M, Yuan L L, Zhou L J, Yang J and Wang Z W 2014 *Eng. Computations* **31** 501-9
[8] Cervone A, Bramanti C, Rapposelli E and d'Agostino L 2006 *J. Fluid Eng.-T. ASME* **128** 326-31

IOP Conf. Series: Materials Science and Engineering **72** (2015) 022013 doi:10.1088/1757-899X/72/2/022013

Research on axial thrust of the waterjet pump based on CFD under cavitation conditions

Z H Shen and Z Y Pan

National Research Center of Pumps, Jiangsu University, Zhenjiang 212013, China

Email: shenzhspar@126.com

Abstract. Based on RANS equations, performance of a contra-rotating axial-flow waterjet pump without hydrodynamic cavitation state had been obtained combined with shear stress transport turbulence model. Its cavitation hydrodynamic performance was calculated and analysed with mixture homogeneous flow cavitation model based on Rayleigh-Plesset equations. The results shows that the cavitation causes axial thrust of waterjet pump to drop. Furthermore, axial thrust and head cavitation characteristic curve is similar. However, the drop point of the axial thrust is postponed by 5.1% comparing with one of head, and the critical point of the axial thrust is postponed by 2.6%.

1. Introduction

Waterjet propulsion is also known as waterjet pump. It is a device that uses reaction spraying flow to push vessel forward by a pump. Due to the advantages of waterjet pump hydraulic performance, such as superior resistance to cavitation characteristics, and structural characteristics of the device make it more and more suitable and rapid development momentum [1].

Waterjet pump power stability is of foundation to the ship stability so the research of waterjet thrust characteristics is essential. Liu Chengjiang, Wang Yongsheng et. , introducing how to obtain waterjet propulsion thrust theory, experiment and numerical calculation method, validated CFD method is feasible and credible [2]. Jules W. Lindau, Christopher Pena et. analyzed of cavitation performance AxWJ-2 type waterjet pump, discovered the relationship between waterjet thrust broken and bubbles blocked rotor runner [3]. And usually, waterjets during acceleration and cornering produces cavitation[4]. But there is rarely studies on relationship between waterjet cavitation and thrust in terms of the internal flow at home and abroad [5-7].

In this paper, the use of CFD technology, a new type of rotary axial waterjet pump cavitation was calculated at design conditions, and the predicted axial thrust and internal flow of cavitation under different conditions, and preliminary analyzed of the relationship between the two.

2. Research model and computation method

Contra-rotating axial flow pump's design parameters used in this paper are: design flow Q_d taken to 0.3 m^3 / s, first stage impeller head taken to 2 m, secondary impeller head taken to 8m. Contra-rotating axial flow pump specific speed 600. First stage impeller blade number 5, secondary impeller blades number 4, vanes leaf number is 7. Flow components joint model shown in figure 1 (figure 1 using Cartesian coordinate system),from the perspective of imports to exports, first stage impeller speed of -1450 r/min, direction is clockwise, secondary impeller speed of 1450 r/min, direction is

IOP Conf. Series: Materials Science and Engineering **72** (2015) 022013 doi:10.1088/1757-899X/72/2/022013

counterclockwise. Propeller model was three-dimensional modeling using Pro-ENGINEER, calculating area from entrance to exit and no exit tube in ordinary waterjet simulation, this study's propeller was added an export tube with special use, as shown in figure 1.

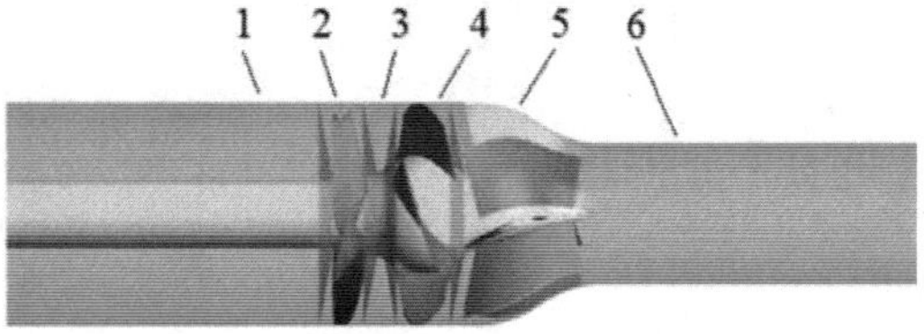

Figure 1. Model for entire pump passage: 1 Pump inlet. 2 first stage impeller. 3 axial clearance. 4 secondary impeller. 5 guide vane and nozzle. 6 pump outlet.

The grids of the three-dimensional model was generated by the ICEM CFD (integrated computer engineering and manufacturing code for computational fluid dynamics).The structured grids in computational regions are shown figure 2. The total number of grid elements is 1845627 for all the domains.

(a) first stage impeller (b) secondary impeller (c) guide vane

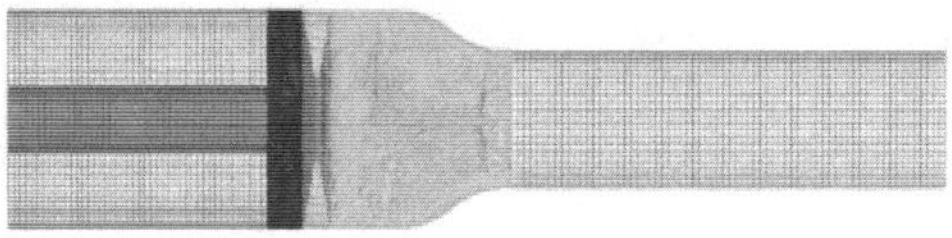

(d) grid for entire pump passage

Figure 2. Grid view in computational region.

The boundary conditions of steady flow simulation were almost the same under both cavitation and non-cavitation conditions. The total pressure at the inlet of the suction pipe was applied, and then the outlet boundary condition was mass flow rate. The impeller surfaces were set up as a rotating wall. All the other walls were stationary.

3. Governing equation and cavitation model

In the mixture model of the vapor/liquid two-phase flow, the fluid is assumed homogeneous so that the multiphase fluid components were assumed to share the same velocity and pressure [8]. A cavitation process is expressed by the mass transfer equation. Equation (1) has demonstrated the conservation equation of the vapor volume fraction. Note that the source terms and represent evaporation and condensation.

$$\frac{\partial}{\partial t}(\rho_v \alpha_v)+\frac{\partial}{\partial x_i}(\rho_v u_i \alpha_v)=R_e-R_c \quad (1)$$

$$\rho=\rho_v \alpha_v+\rho_v(1-\alpha_v) \quad (2)$$

Where α_v is volume fraction of one component.

$$R_e = F_e N_b \rho_v 4\pi R_b^2 \sqrt{\frac{2}{3}\frac{|p_v - p|}{\rho_l}}\, sign(p_v - p) \tag{3}$$

$$R_c = F_c N_b \rho_v 4\pi R_b^2 \sqrt{\frac{2}{3}\frac{|p_v - p|}{\rho_l}}\, sign(p_v - p) \tag{4}$$

For the vaporization, N_b is given by

$$N_b = \frac{3\partial_d}{4\pi R_b^3}(1 - \partial_v) \tag{5}$$

For the condensation, N_b is given by

$$N_b = \frac{3\partial_v}{4\pi R_b^3} \tag{6}$$

In equation (3) and equation (4), P_v is vapor pressure , F_e is vaporization coefficient, F_c is condensation coefficient. Empirical coefficients used to represent the vapor increasing or reducing time step. For the vaporization, $F_e = 50$. For the condensation, $F_c = 0.01$. In the present study, SST turbulence model was selected.

4. Results and discussion

Figure 3 and 4 shows pump head and pump axial thrust of the cavitation performance of the waterjet pump. With the decrease of the net positive suction head (NPSHA), the pump head decreases. NPSHA = 3.76 is approximated the net positive suction head critical (NPSHC) for the pump while head down 2.6%. Head decreases caused by increasing bubbles in the impeller. With the appearance of cavitation, the flow inside the pump state changes and the change accumulate to a certain extent on the performance of a drop head.

We find it similar to figure 4 and figure 3, i.e. the axial thrust of the pump also decreases with the decrease of the NPSHA, reaching a critical point decline intensified.

The decline-starting point of the axial thrust delays 5.1% compared with the head decline-starting point, critical point delays 2.6%. When cavitation occurs, the axial thrust decline is earlier than head fall; it is possible vacuoles might hinder the interaction between the blade and the working medium. But this does not mean cavitation is the direct cause of head drop.

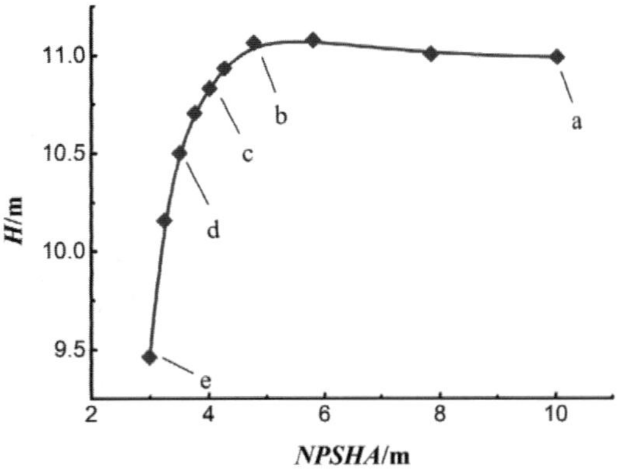

Figure 3. Cavitation characteristic curve of the pump at Q_d.

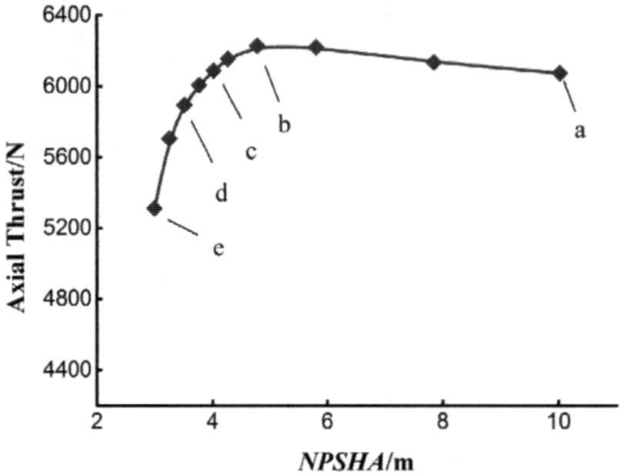

Figure 4. Cavitation characteristic curve of the axial thrust at Q_d.

4.1. Vapor bubbles distribution

This section the design point of axial thrust decline curve was selected to investigate the relationship between the development of the cavitation distribution in impeller channel and the decline of the axial thrust in conjunction with the impeller cavitation shown in Figure 5. It is analyzed the influence of cavitation based on 5 condition points (a~e) on the decreased curve of axial thrust in figure 4.

Figure 5 shows the distribution of bubbles of five operating points, while the cavitation region is composited by the volume fraction of 10% iso-surface.

When *NPSHA* was not reduced, slight cavitation occur in the suction side of the first stage impeller tip as shown in figure 5(a). When *NPSHA* was 4.78 m, the cavity area of the suction around the secondary impeller tip increased, while covered import to export side. Meanwhile, bubbles were found at pressure side near the hub of the first stage impeller, so was the suction side near the tip of the secondary impeller. When *NPSHA* dropped to 3.76 m, bubbles came out at the center of the suction side of the secondary impeller, which may be reasons and signs for the axial thrust decline beginning; When *NPSHA* was 2.99 m, bubble had covered 2/3 of the suction side of the first stage impeller, while the surface area of the bubble at the hub also further expanded, and at the same time the axial thrust had dropped 12.57%.

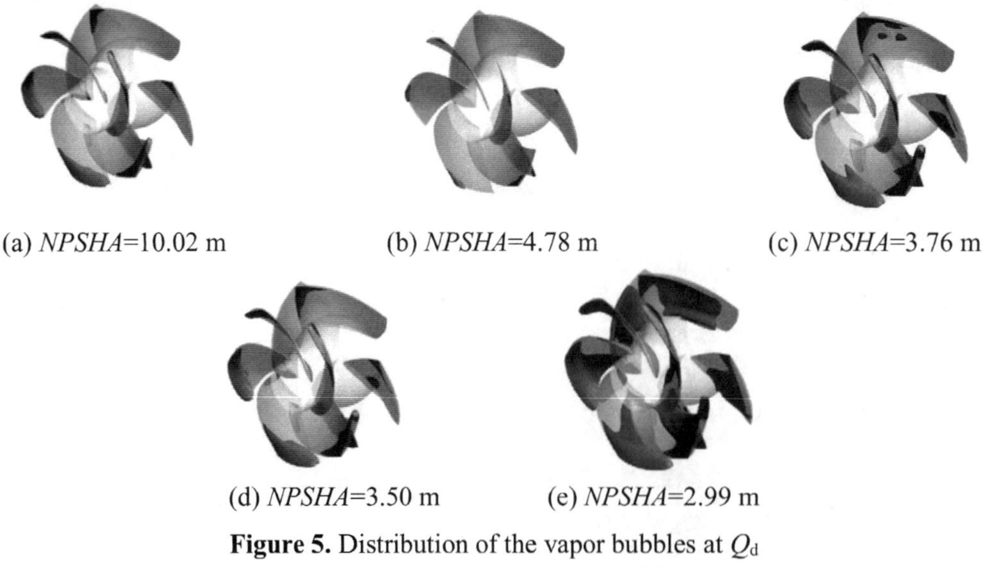

(a) *NPSHA*=10.02 m (b) *NPSHA*=4.78 m (c) *NPSHA*=3.76 m

(d) *NPSHA*=3.50 m (e) *NPSHA*=2.99 m

Figure 5. Distribution of the vapor bubbles at Q_d

4.2. Impeller surface axial thrust

Figure 6, X-axis represents *NPSHA*, where Y-axis represents the axial thrust of every part of the impeller. With the *NPSHA* decreasing, the axial thrust of the impellers each surface reduced. The difference is that the axial thrust of the first impeller rose, when *NPSHA* reached the maximum was 4.78 m, an increase of 15.26%. Then it began to decrease, when *NPSHA* was 3.76 m, the axial thrust and starting point (*NPSHA* = 10.02 m) flat, slightly decreased 0.04%. Secondary impeller axial thrust curve by a slow decline to a sharp decline in the critical point of 3.76 m, showing a similar trend in figure 4. This will also be consistent with the analysis of figure 5(c).

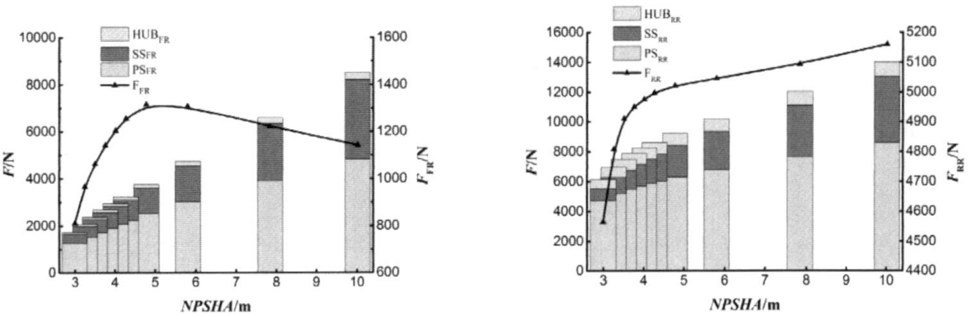

(a) first stage impeller (b) secondary impeller

Figure 6. Axial thrust of impellers.

Figure 7, the X-axis represents the position of the blade, 0 said blade inlet, 1 blade outlet; Y-axis represents the pressure coefficient of the blade. Pressure coefficient C_p is defined as

$$C_p = \frac{p - p_{in}}{\frac{1}{2}\rho v_{in}^2} \quad (7)$$

Where C_p is pressure coefficient, p is blade surface static pressure, p in is pump inlet static pressure, v_{in} is the tip peripheral speed of first stage impeller inlet side.

In figure 7(a), under conditions of no cavitation, a "low-voltage spikes "later in the work surface of impeller inlet side came out. In figure 8, the low pressure region near the inlet of the blade face appeared extending from the hub. From the above results that cavitation had occurred here, cavitation flow lines tail twisted, bubble extruded flow channel, changing the flow structure at the entrance. The suction side of the impeller blade pressure is the lowest part, high flow rate, and prone to cavitation.

With *NPSHA* = 4.78, 3.76 m, blade surface pressure decreased, but the basic trend remained unchanged. When *NPSHA* = 2.99 m, the blade face pressure was further reduced, area of "low spike" increased and figure 5(e) showed cavitation area near the hub of the first stage impeller further expansion. In this case the pressure at the suction side of the blade inlet increased, and smoothly extended to the rear of the blade.

In figure 7(b), when *NPSHA* reduced from 10.02 m to 4.78 m, pressure coefficient curve did not change significantly. *NPSHA* down to 3.50 m, the pressure coefficient decreased overall, and *NPSHA* down to 2.99 m, the pressure coefficient sharply declined.

5. Conclusion and recommendations

1) Waterjet pump axial thrust declines caused by cavitation. With NPSHA decreasing, the axial thrust and head of the pump appear a slight decline after the rise at beginning and decline intensified at the critical point. The decrease starting point of the axial thrust delays 5.1% compared with the head starting point, critical point delays 2.6%.

2) Axial thrust of the Impellers present cavitation critical point, while the initial stage of cavitation performance is of different nature. Cavitation initial stage, axial thrust of the first stage impeller rose. When NPSHA = 4.78 m it reached a maximum and then began to decline. The difference is that the initial stage of the axial thrust of the secondary impeller was gradually decreased. With NPSHA down to 3.76 m, the axial thrust appeared steep drop phenomenon.

3) We find that the curve of axial thrust presents similar trend to head cavitation characteristics, but both the starting drop point and the critical drop point is different. The next step is to explore the possible relationship between the two. In addition, this paper only preliminary reveals changes in axial thrust under cavitation conditions, but also analyzes the relationship between cavitation and axial thrust from the effects of cavitation on the flow structure.

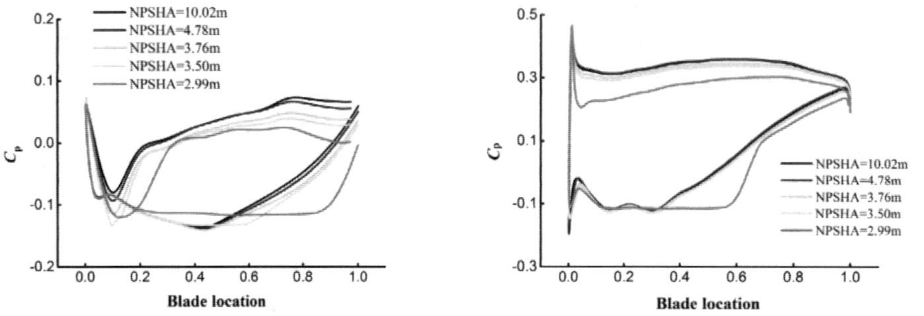

(a) first stage impeller (b) secondary impeller

Figure 7. Distribution of C_p at 0.5 span.

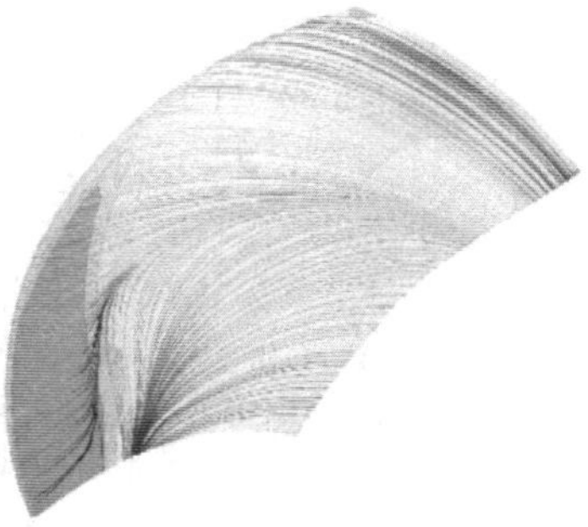

Figure 8. distribution of the pressure and velocity of first stage impeller's work side at *NPSHA*=3.25 m

Acknowledgement
This research was supported by the Stated Key Program of National Natural Science Foundation of China (Grant No. 51239005), and the Priority Academic Program Development of Jiangsu Higher Education Institutions (PAPD).

Reference

[1] Bulten N W H 2006 Numerical analysis of a waterjet propulsion system *Dissertation Abstracts Internationa* **68** (02)
[2] Liu C J, Wang Y S and Zhang Z H 2009 *Chin. J. Computational Mech.* **4** 477-82
[3] Lindau J W, Boger D A, Medvitz R B and Kunz R F 2005 *J. Fluid Eng.* **127** 995-1002
[4] Ding J M 2013 *J. Huazhong Uni. of Sci. and Tech. (Natural Science Edi.)* **41** 100-04
[5] Wang D J and Cao S L 2002 *J. Hydroelectric Eng.* **E01** 176-80
[6] Terwisgat K K and Ketal V 2003 Performance and cavitation analysis of a waterjet system in a cavitation tunnel *The 7th Int. Conf.on Oil Fast Sea Transportation* (Ischia, Italy, 07-10 Oct

2003) pp 57-62
[7] Brennen C E 2011 *Hydrodynamics of pumps* (Cambridge UK: Cambridge university press)
[8] Li X J 2013 *Mechanism and unsteady dynamic characteristics of leading edge cavitation in a centrifugal pump* (Zhenjiang: Jiangsu Uni.)

IOP Conf. Series: Materials Science and Engineering **72** (2015) 022014 doi:10.1088/1757-899X/72/2/022014

Numerical and experimental research on unsteady cavitating flow around NACA 2412 hydrofoil

M Sedlář[1,4], M Komárek[1], P Rudolf[2], J Kozák[2] and R Huzlík[3]
[1] Centre of Hydraulic Research, J. Sigmunda 190, 783 49 Lutín, Czech Republic
[2] Brno University of Technology, Faculty of Mechanical Engineering, Technická 2896/2, Brno, Czech Republic
[3] Brno University of Technology, Faculty of Electrical Engineering and Communications, Technická 3082/12, Brno, Czech Republic

E-mail: m.sedlar@sigma.cz

Abstract. This work deals with the numerical and experimental investigation of unsteady cavitating flow around a prismatic NACA 2412 hydrofoil. The main attention is focussed on the dependence of cavitation dynamics on the cavitation number at high incidence angles. The experimental research is carried out in the cavitation water tunnel the rectangular test section of which has inner dimensions 150×150×500 mm. Currently tested hydrofoils have a chord length of 120 mm and are equipped with pressure transducers at the leading edge and on the suction side. The PVDF hydrophone enables to measure high-frequency pressure pulses behind the hydrofoil trailing edge. The visualizations are based on two simultaneous high-speed cameras, recording the hydrofoil from the top and from one side. A comprehensive CFD analysis has been done with the ANSYS CFX package for a wide range of flow regimes. Different turbulence models including SAS-SST and Reynolds-stress models have been tested to capture highly unsteady phenomena on the hydrofoil. The numerical simulations show, that the dominant frequency of the cavity oscillation depends on the cavitation number and that there is a certain range of this number in which the "resonance" effect can be reached. In such regime the amplitudes of the pressure pulses on the suction side of the hydrofoil dramatically increase. The calculated results have been verified by both the visualizations and the pressure measurements carried out at the hydrofoil incidence angle of 8 degrees.

1. Introduction

Hydrodynamic pumps are very often operated in the cavitation regime. Under such conditions the pump can suffer from cavitation instabilities resulting in unwanted noise and dangerous vibrations. The cavitation instabilities have been studied intensively in the case of inducers which represent a special type of axial-flow pumps with very small incidence angles. Most of the work is based on experimental research. One of the diagnostics which is used to determinate the type of cavitation instability is the frequency analysis. We have to examine the dominant frequencies and compare them with the rotational frequency f_Ω as well as with the fundamental frequency of the inducer blade passage excitation. Typical frequencies of some cavitation instabilities in inducers can be found e.g. in [1]. Recently some studies related to the cavitation instabilities in pumps have appeared [2-5] which

[4] Corresponding author.

are based mainly on numerical simulations. In [5] it has been shown that cavitation instability in the backflow vortices in front of the pump impeller appears in a limited interval of the cavitation number which is caused by the interaction of the separated flow and cavitation regions. When the cavitation number is sufficiently high, cavities formed on the impeller blades are small and relatively stable. On the other hand when the cavitation number is very low, the cavitation regions remarkably block the blade-to blade passages of the impeller and the backflow practically vanishes.

To study experimentally interaction of cavitation structures with separated flow inside the complex geometry of a real pump is highly expensive and time-consuming and in some cases practically impossible. That is why the experimental research with a simplified geometry (prismatic NACA 2412 hydrofoil) in the cavitation tunnel has been initiated in the Centre of Hydraulic Research. This experimental research is accompanied with the numerical modelling.

2. Experimental set-up

The cavitation tunnel in the Centre of Hydraulic Research was built in 2007 especially for the investigation of erosion process triggered by cavitation [6]. The facility is a horizontal plane water tunnel for isolated hydrofoils. The closed loop is equipped with both compressor and vacuum pumps capable of creating different pressure levels while maintaining constant volume flow rate and includes the main tank with the capacity of 35 m^3. The rectangular test section has inner dimensions 150×150×500 mm and is made of organic glass to facilitate visualization from all sides (figure 1). The maximum tunnel inlet velocity 25 m/s corresponds to the maximum flow rate of the variable-speed driven axial-flow pump.

Figure 1. Cavitation tunnel in closed water loop.

The currently tested hydrofoils are prismatic with a chord length of 120 mm, which gives the span/chord ratio higher than 1 (namely S/C=1.25). The span/chord ratio influences the shape of the shedding cavities as it will be shown in following sections. The profile incidence angle can vary between 0° and ± 180°. The NACA 2412 hydrofoil has been chosen as the test case as it is close to the hydrofoils used in axial-flow and mixed-flow pumps.

The main attention is focussed on the dependence of cavitation dynamics on the cavitation number at high incidence angles. Several incidence angles have been considered nevertheless the main part of the work is devoted to the regimes with the incidence angle of 8°, which is the same as used in reference [7], though its authors used different hydrofoil (Clark Y) and different span/chord ratio (S/C=0.81). Two types of hydrofoils have been designed. The first one is equipped with pressure transducers at the leading edge and on the suction side. The second type is covered with a grid of PVDF films which enable to monitor forces resulting from violent collapses of the cavitation structures. However the results presented in this study have been obtained with the hydrofoil with only two pressure transducers at midspan, at the leading edge (pressure P0) and on the suction side (at 40% of the chord length, pressure P40), which minimizes the stimulation of cavitation by the tap holes (figure 2). The PVDF hydrophone enables to measure high-frequency pressure pulses behind the hydrofoil trailing edge (105% of the chord length). The pressures have been measured with the sampling frequency of 5 kHz, 10 seconds each record. The accuracy of the transducers is 0.5 kPa. The visualizations are based on two simultaneous high-speed cameras, recording the hydrofoil from the top and from one side. Each video-record represents two seconds with the frame rate of 1000 fps and a high resolution 1024×1024 pixels. The pressure measurements are synchronized with both cameras.

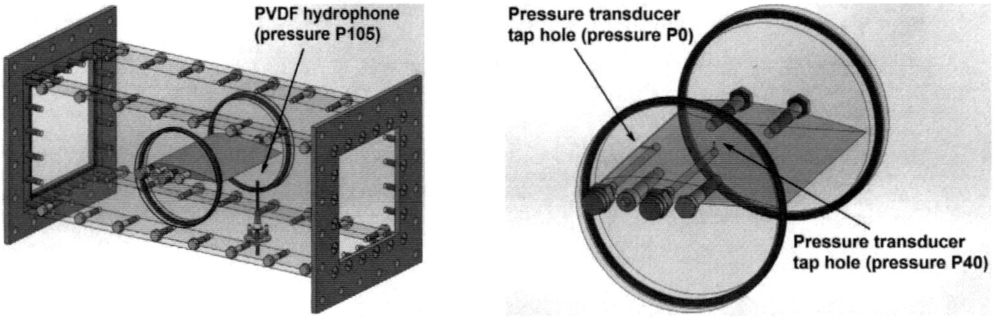

Figure 2. Scheme of test section and hydrofoil with two pressure transducers.

3. CFD tools and simulation details

In this study the ANSYS CFX package has been used to solve the URANS equations. The Rayleigh-Plesset model included in the ANSYS CFX package has been applied to describe the interphase mass transfer in the framework of the homogenous multiphase model. Though this cavitation model is based on a highly simplified Rayleigh-Plesset equation for the radius of a spherical cavitation bubble, it is quite robust and effective for problems, where the void fraction is highly important in forming flow structures but where the details of the bubble dynamics (especially the collapses and rebounds) are not of primary interest [5]. The calculations have been performed both with a full tunnel geometry (including the inlet contraction chamber, outlet diffuser and a part of the piping) and using a simplified geometry consisting of only the test section and a rectangular extension. The differences resulting from this simplification seem to be negligible, as the velocity uniformity at the front part of the test section is (based on the CFD analysis) within ± 1%, excluding the boundary layer. The structured computational grid of the test section itself represents approximately 2 million nodes in the case of the standard geometry with the span/chord ratio S/C=1.25. The computational time step has been set in the range $1\text{-}5\times10^{-4}$ s depending on the turbulence model used. The boundary conditions are based on prescribed total pressure at the inlet and the mass flow rate at the outlet. No symmetry plane at the hydrofoil midspan has been used.

To capture the highly unsteady phenomena close to the hydrofoil the widely used SST turbulence model has been tested as well as advanced turbulence models implemented in the ANSYS CFX code, namely the BSL based Reynolds-stress model, the SSG Reynolds-stress model and the SAS-SST model. No corrections of these models related to cavitation have been done.

4. Results and discussion

A comprehensive CFD study has been carried out for a wide range of flow regimes. As the main attention is focussed on the dependence of cavitation dynamics on the cavitation number, the results for a constant Reynolds number of 1.55×10^6 are discussed in this paragraph. It should be noted that the dimensionless numbers (like Reynolds and cavitation ones) in this study are based on the hydrofoil chord length and the test section inlet velocity.

The numerical simulations show, that the dominant frequency of the cavity oscillation depends strongly on the cavitation number and that there is a certain range of this number in which the "resonance" effect can be reached. In such regime the amplitudes of the pressure pulses on the suction side of the hydrofoil dramatically increase. Figure 3 shows two dominant frequencies obtained by CFD and experimentally, as a function of the cavitation number σ, and the amplitudes of the first dominant frequency. In the CFD analysis sheet cavities formed on the hydrofoil are small and relatively stable when σ is sufficiently high. There is only one dominant frequency about 21 Hz ($St \approx 0.2$) and amplitudes are very low. With decreasing cavitation number we can identify two (or more) significant frequencies. The lowest one (about 15 Hz, which corresponds to $St \approx 0.14$) is the most significant. In this regime we can see dramatically increasing amplitudes. Further drop of σ causes decrease of the amplitudes and in the end, for sufficiently low values of σ, a stable supercavitation regime starts in which it is impossible to detect any dominant frequencies. In this regime the backflow on the suction side of the hydrofoil practically vanishes. Experimentally obtained data show the same tendencies but it can be seen that the "resonance" region is quite narrow compared to the CFD results which makes it difficult to detect. The result of the FFT analysis of calculated static pressure P40 in the regime of "resonance" can be seen in figure 4.

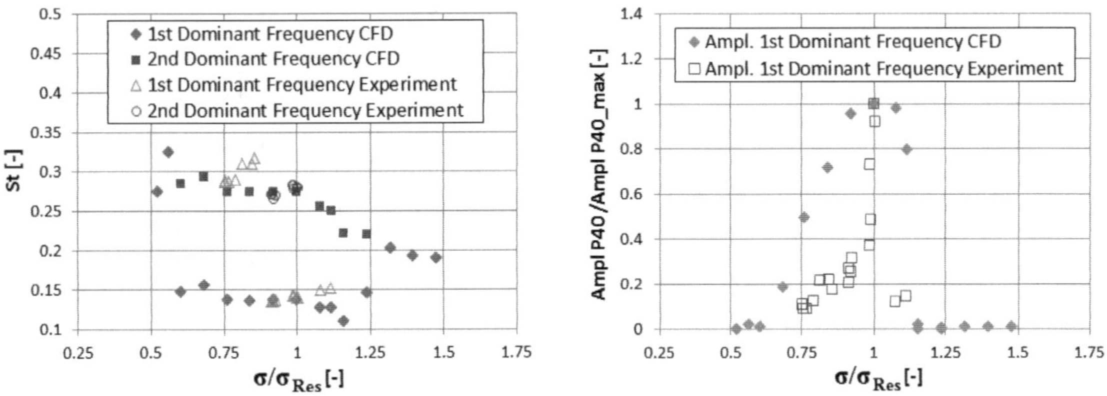

Figure 3. First two dominant frequencies (left) and pressure amplitudes of the first dominant frequency (right). Comparison of CFD and experiment.

Graphs in figure 3 are based on the normalised cavitation number (related to the cavitation number σ_{Res} in which the maximum "resonance effect" appears) and on the normalised amplitudes of the static pressure P40. The reason is that in the case of experiments the "resonance" cavitation number σ_{Res} changes according to the immediate nuclei content in water. In presented calculations σ_{Res} has a value of about 1.5; in measurements it is in the range approximately from 1.6 to 1.75. On the other hand,

IOP Conf. Series: Materials Science and Engineering **72** (2015) 022014 doi:10.1088/1757-899X/72/2/022014

values of dominant amplitudes depend on different methodology of evaluation of experimental and numerical results. Nevertheless an agreement of the calculated and measured pressure P40 in the same regime ($\sigma = \sigma_{Res}$) can be seen in figure 5.

To verify the agreement of the pressure measurements and visualizations, the image analysis of movie sequences has been carried out in several points close to the position of the pressure transducer P40 tap hole. Values from 1 to 256 were assigned to selected pixels according to the gray scale and the FFT analysis has been applied (figure 6). The transformation of the temporal image sequence to frequency domain provides the amplitude-frequency spectra, which identify dynamic behavior of the main coherent structures of the separated cavitating flow (figure 7). The image processing agrees very well with the results of CFD analysis as well as the pressure measurements for both the individual points and the point set average.

Another parameter which can affect the unsteady cavitating flow around the hydrofoil and which has been examined is the span/chord ratio. As it is very complicated to change dimensions of the test chamber in the cavitation tunnel, this investigation has been done only in a numerical way. The simplified geometry consisting of the test section itself with a rectangular extension has been applied. The span/chord ratios 0.417, 0.625, 1.25 and 1.875 have been considered. In addition a case of an infinite span has been modelled using the periodic boundary conditions on the side walls. Here only SST turbulence model and the Reynolds-stress models could be used, as the periodic (or symmetry) boundary conditions do not fit the SAS-SST turbulence model. The CFD analysis has shown that the span/chord ratio can influence the shape of the cavitation structures (figure 8) but the basic topology of the flow separation and the dominant frequencies and their amplitudes are influenced only a little. For example in the vicinity of the "resonance" cavitation number σ_{Res} the first dominant frequency changes from the value of 13.85 Hz (S/C=0.417) to the value of 14.83 Hz (infinite S/C).

All results presented in this section have been obtained with the SAS-SST turbulence model except the case with the periodic boundary conditions where the BSL Reynolds-stress model has been used.

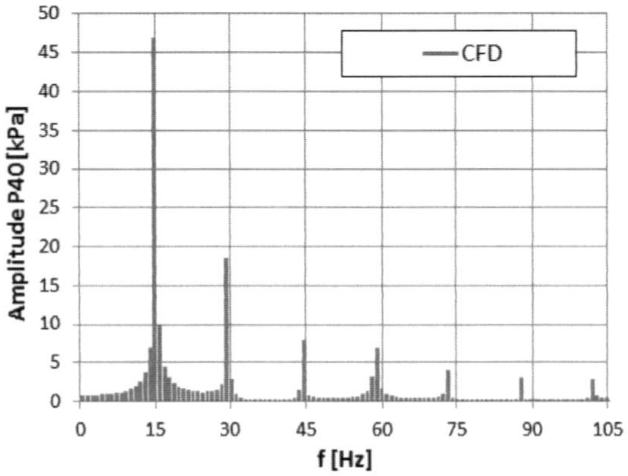

Figure 4. FFT analysis of static pressure P40. CFD, "resonance" cavitation number σ_{Res}.

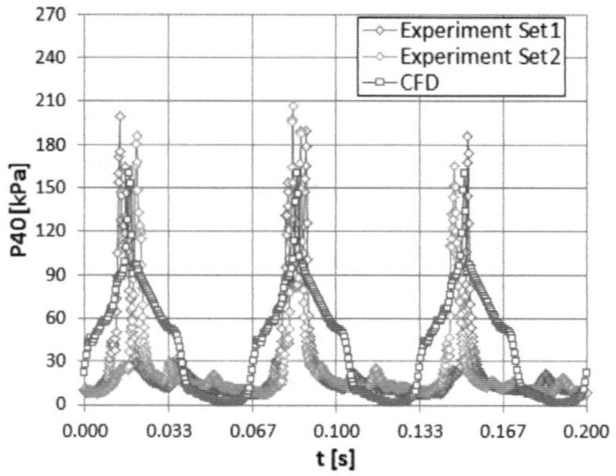

Figure 5. Dynamics of static pressure P40. 3 cycles. Experiment and CFD, $\sigma = \sigma_{Res}$.

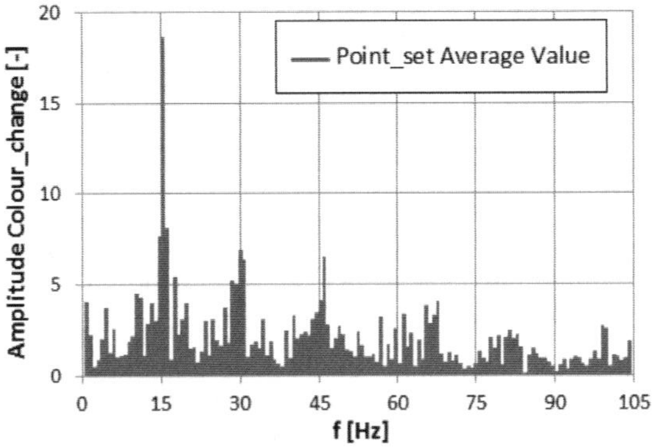

Figure 6. Image analysis of movie sequence in a set of points on profile suction side. $\sigma = \sigma_{Res}$.

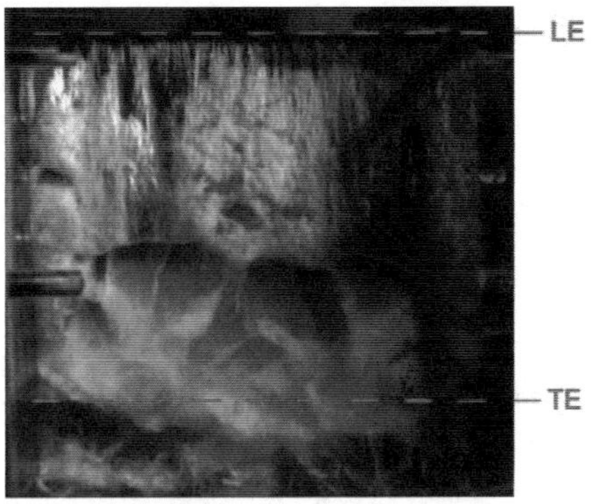

Figure 7. Instantaneous picture of separated cloud cavities in backflow region. Experiment, $\sigma = \sigma_{Res}$.

5. Conclusions

Highly unsteady cavitating flow around the NACA 2412 hydrofoil has been examined both numerically and experimentally for different flow regimes, especially for a wide range of cavitation numbers. The comparison of obtained data shows that the CFD analysis is able to predict main flow and cavitation structures (see figure 7 and figure 8 left) and numerical results agree reasonably with experiments from the quantitative point of view.

Visualizations and especially CFD analysis have provided a deep look at the development of the flow and cavitation structures around the hydrofoil in time. Unfortunately there is not enough space to present these structures in detail in this paper.

Different turbulence models have been tested during the project. Among them, the SAS-SST model has demonstrated superior results. Nevertheless future work tends to highly detailed analysis with very fine computational grids and non-URANS turbulence models like LES or DES simulations. Very promising is also the PANS model, successfully used (in the framework of the ANSYS CFX code) in [8] for very detailed analysis of flow and cavitation structures around a Delft twisted hydrofoil.

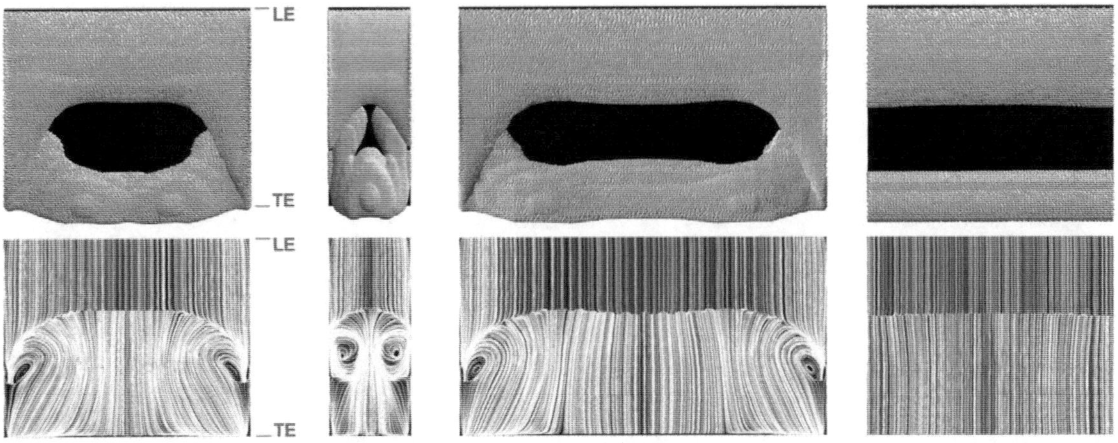

Figure 8. Separation of cloud cavities in back-flow region (top row) and surface streamlines (bottom row) on hydrofoil suction side. CFD, $\sigma = \sigma_{Res}$. Cavitation regions visualized as locations where void fraction exceeds 10%.

Acknowledgements

This work has been supported by the grant no. 101/13/23550S of the Czech Science Foundation, entitled "Experimental Research and Mathematical Modelling of Unsteady Phenomena Induced by Hydrodynamic Cavitation".

References

[1] Tsujimoto Y 2007 Cavitation instabilities in turbopump inducers *Fluid Dynamics of Cavitation and Cavitating Turbopumps* ed dAgostino L and Salvetti M V (Vienna: Springer-Verlag) pp 169-90
[2] Yamanishi N, Fukao S, Qiao X, Kato C and Tsujimoto Y 2007 *J. Fluid Eng.* **129** 587-94
[3] Yamamoto K and Tsujimoto Y 2009 *Int. J. Fluid Mach. and Systems* **2** 40-54
[4] Kobayashi K and Chiba Y 2010 *Int. J. Fluid Mach. and Systems* **3** 113-21
[5] Sedlář M, Šputa O and Komárek M 2012 *Int. J. Fluid Mach. and Systems* **5** 18-29
[6] Sedlář M, Komárek M, Vyroubal M and Müller M 2012 Experimental and numerical analysis of cavitating flow around a hydrofoil *EPJ Web of Conf. (Madrid, Spain 25-27 June 2012)* **25** 01084
[7] Watanabe S, Konishi Y, Nakamura I and Furukawa A 2011 Experimental analysis of cavitating behavior around a Clark Y hydrofoil *Proc .WIMRC 3rd Int. Cav. Forum (Warwick, UK, 4-6 July 2011)*
[8] Ji B, Luo X W, Peng Y X and Duan Y 2013 *Int. J. Multiphase Flow* **51** 33-43

IOP Conf. Series: Materials Science and Engineering **72** (2015) 022015 doi:10.1088/1757-899X/72/2/022015

Partial Averaged Navier-Stokes approach for cavitating flow

L Zhang[1] and Y N Zhang[2,3]
[1] Yalong River Hydropower Development Company, Ltd, Sichuan, China 610051
[2] Key Laboratory of Condition Monitoring and Control for Power Plant Equipment, School of Energy, Power and Mechanical Engineering, North China Electric Power University, Beijing, China 102206

Email: zhangyn02@gmail.com

Abstract. Partial Averaged Navier Stokes (PANS) is a numerical approach developed for studying practical engineering problems (e.g. cavitating flow inside hydroturbines) with a resonance cost and accuracy. One of the advantages of PANS is that it is suitable for any filter width, leading a bridging method from traditional Reynolds Averaged Navier-Stokes (RANS) to direct numerical simulations by choosing appropriate parameters. Comparing with RANS, the PANS model will inherit many physical nature from parent RANS but further resolve more scales of motion in great details, leading to PANS superior to RANS. As an important step for PANS approach, one need to identify appropriate physical filter-width control parameters e.g. ratios of unresolved-to-total kinetic energy and dissipation. In present paper, recent studies of cavitating flow based on PANS approach are introduced with a focus on the influences of filter-width control parameters on the simulation results.

1. Introduction

For studying practical engineering problems (e.g. cavitating flow, pressure fluctuation and instability inside the hydroturbine), two kinds of approaches is prevalent: (unsteady) Reynolds Averaged Navier-Stokes (RANS) and Large Eddy Simulation (LES). RANS is simple and can predict the large scale vortexes. However, the physical motions induced by small scale vortexes are usually ignored by RANS, leading to inaccuracy of prediction. LES is accurate and can successfully capture the structure of vortexes with variable scales. However, the computational cost of LES is quite intensive and sometimes it is impractical to use it for engineering simulations. Hence, a bridging method named as Partial Averaged Navier Stokes (PANS) was proposed by Girimaji and collaborators [1-3].

According to Girimaji [1], PANS is suitable for any filter width (e.g. from RANS to direct numerical simulations). The development of PANS can be divided into two steps. In the first step, one need to identify appropriate physical filter-width control parameters. Girimaji [1] suggested that the best quantitative evaluation parameters of the degree of filtering are ratios of unresolved-to-total kinetic energy and dissipation. In the second step, the PANS equations are derived from parent RANS closure by resolving more scales of motion. The scale of motion further treated in PANS is highly dependent on the values of the filter-control parameters in the first step. If we set these ratios as unity, RANS equations are obtained. If we set these ratios as zero, full Navier-Stokes equations are recovered from PANS method. Obviously, the PANS model will inherit many physical nature from parent RANS but further resolve more scales of motion in great details, leading to PANS superior to RANS. Comparing with traditional

[3] Corresponding author.

LES, PANS has distinct characteristics from three aspects [1] e.g. decomposition of velocity filed using kinetic energy content, implicit filter demarcating calculation method and grid-spacing free subfilter scale constitutive relationship.

Various kinds of PANS models have been developed based on the framework of Girimaji [1]. Ma et al. [4] proposed a low Reynolds number variant of PANS model for turbulent flow. Liu et al. [5] proposed a new nonlinear PANS model based on RNG k-ε turbulence model and validated their model through comparing experimental data with simulations (e.g a NACA hydrofoil, a curved rectangular duct, and a low specific centrifugal pump).

Currently, PANS model has been widely implemented in a wide range of engineering problems. Girimaji and Srinivasan [3] employed PANS model to simulate ramjet/scramjet mixing environment. Liu et al. [6] applied PANS model to the study of the instability of the pump turbine with misaligned guide vanes. There are also numerous studies of employing PANS model with appropriate cavitation model for the study of cavitating flow e.g. around a hydrofoil [7,8] or marine propellers [9].

In present paper, the employment of PANS model on studying of cavitating flow is introduced. The detailed discussions of principles of PANS is out the scope of present paper. For details, readers are referred to Girimaji [1].

2. Selected results of cavitating flow simulation based on PANS approach

In this section, some typical simulation results of the simulation work of cavitating flow based on PANS approach are introduced with details.

Ji et al. [9] studied the cavitating flow around a marine propeller in a uniform wake based on PANS model. Firstly, influences of two important parameters (namely, the ratio of the unresolved-to-total kinetic energy and dissipation, f_k and f_ε respectively) are discussed through simulating cavitating flow around a NACA66-mod hydrofoil. The f_k varies between 0.2 and 1 in order to fit the experimental data (as shown in Figure 1). It was found that when f_k=0.2, the predicted cavity evolution and shedding frequency agree well with the experiments. Comparing with RANS, it was found that the resolution of the simulation results predicted by PANS is better in terms of large cavity volume pulsation when the blade passing through the wake region. The choice of f_k was also found to be a crucial parameter influencing the predictions of cavitating flow. Generally, a smaller value of f_k tends to increase the acceleration of cavity pulsation while a larger value of f_k will overestimate the turbulent viscosity (referring to Figure 2). The PANS approach in Ji et al. [9] was validated by comparing with the experimental data (referring to Figure 3).

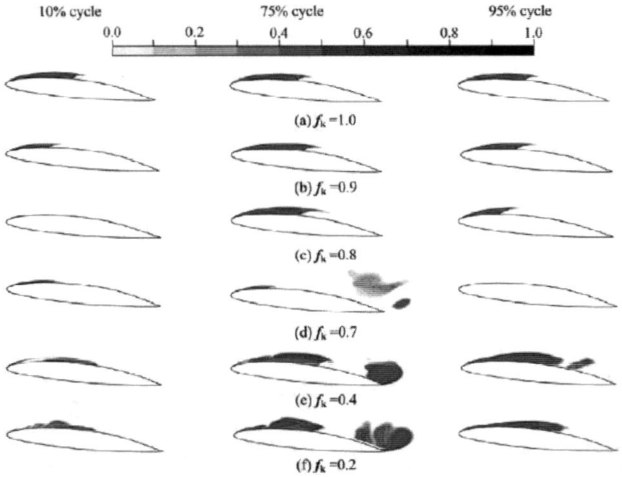

Figure 1. Influences of f_k on the instantaneous vapor volume fractions. This figure was adapted from Ji et al. [9].

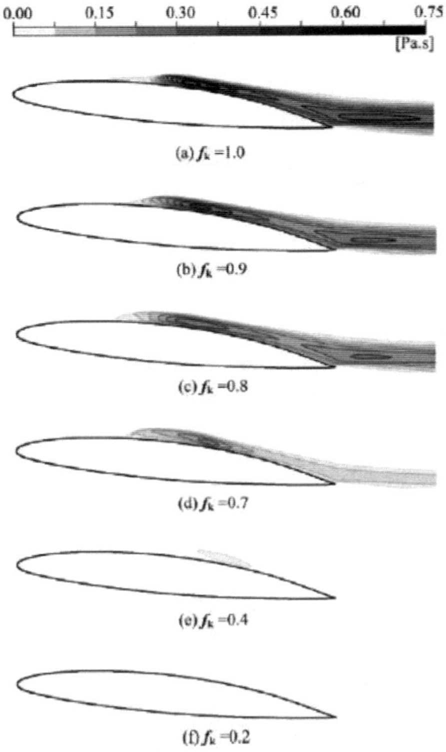

Figure 2. Influences of f_k on the time averaged turbulence viscosity. This figure was adapted from Ji et al. [9].

Huang and Wang [7] evaluated the influences of ratio of the unresolved to total kinetic energy (f_k) on the predictions of cavitating flow over a Clark-Y hydrofoil. A more rigorous comparison between predictions with the variations of f_k and the experimental data (e.g. forces, frequencies, cavity visualization and velocity distributions) are performed. With the increase of f_k, a standard k-ε model was recovered. With the decrease of the f_k, more smaller scales of the vortexes in the cavitating flow is observed (referring to more peaks in the life coefficient in Figure 4).

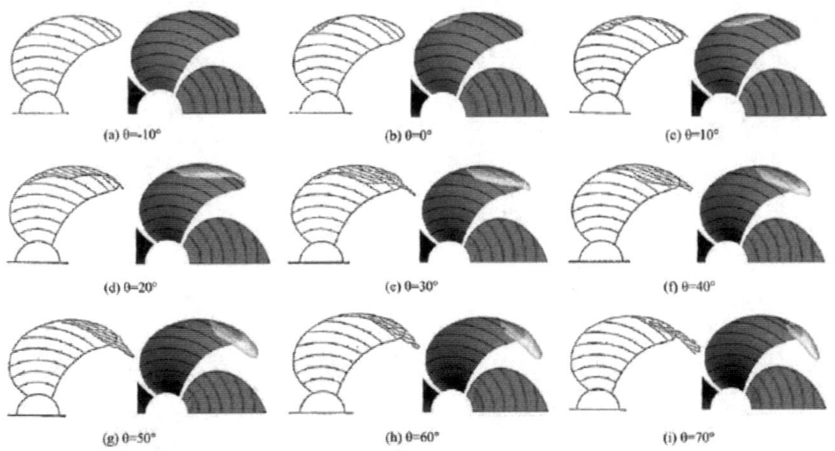

Figure 3. Comparisons between the predicted cavity patterns and the experimental evolution. This figure was adapted from Ji et al. [9].

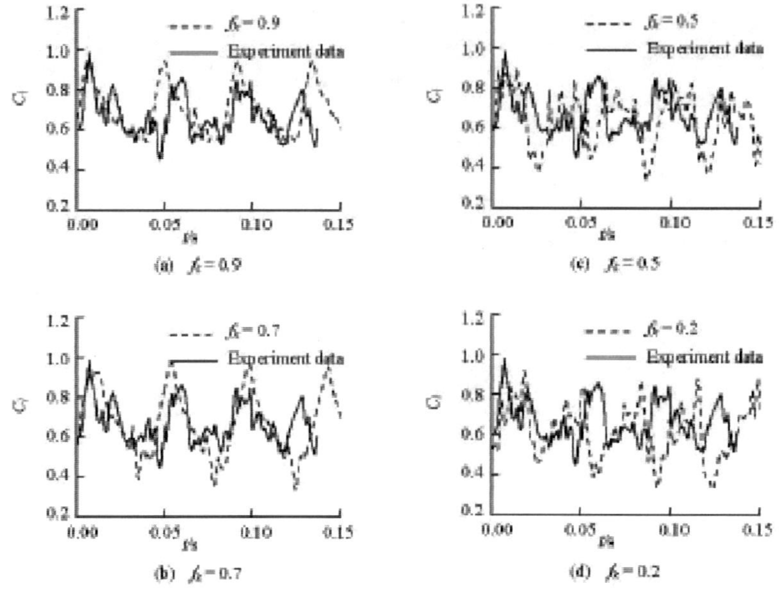

Figure 4. Influences of f_k on the variations of lift coefficient. This figure was adapted from Huang and Wang [7].

Recently, Hu et al. [10] modified the f_k as a function of the ratios between water density and mixture density in the flow as follows:

$$f_k = \tanh(\mathrm{atanh}\, C_2 \times (\rho_m / \rho_l)^{C_1} + 1 - C_2)$$

Here, ρ_m is the density of mixture; ρ_l is the density of liquid; C_1 and C_2 are two parameters newly introduced to modify the values of f_k. Figure 5 shows the influences of C_1 and C_2 on the variations of f_k.

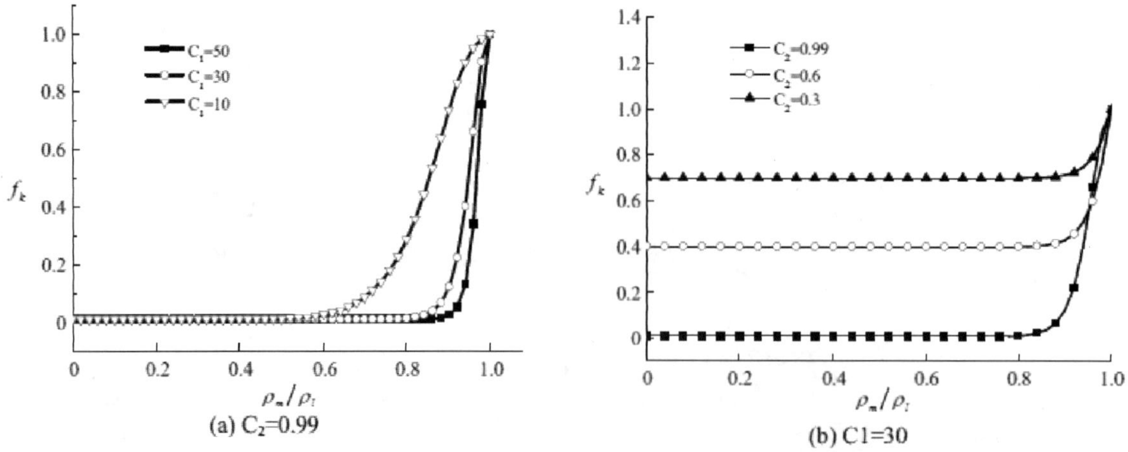

Figure 5. Influences of C_1 and C_2 on the variations of f_k. This figure was adapted from Hu et al. [10].

The modified PANS model was validated through simulating a Clark-Y hydrofoil by fitting the parameters of C_1 and C_2 with the experimental data (such as hydrofoil lift and frequency as shown in Figure 6). It was found that C_1=34.05 and C_2=0.99 show satisfactory results [10].

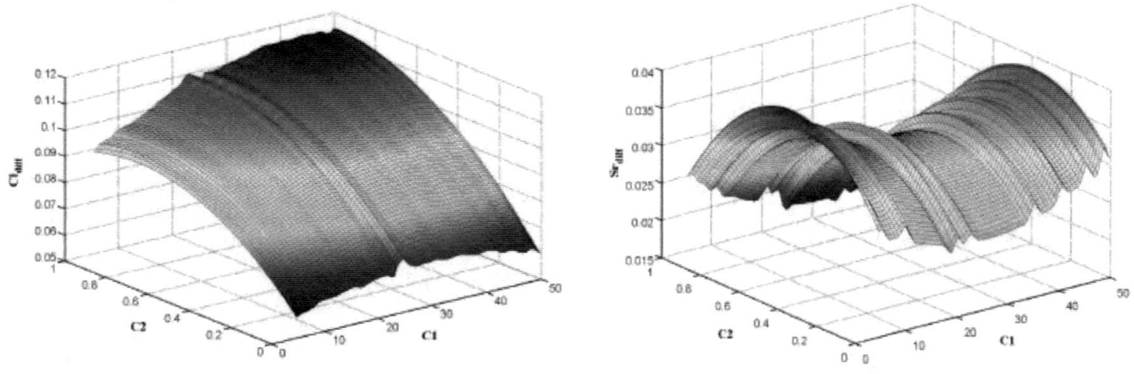

(a) (b)

Figure 6. Data fitting of hydrofoil lift and frequency using different values of C_1 and C_2. This figure was adapted from Hu et al. [10].

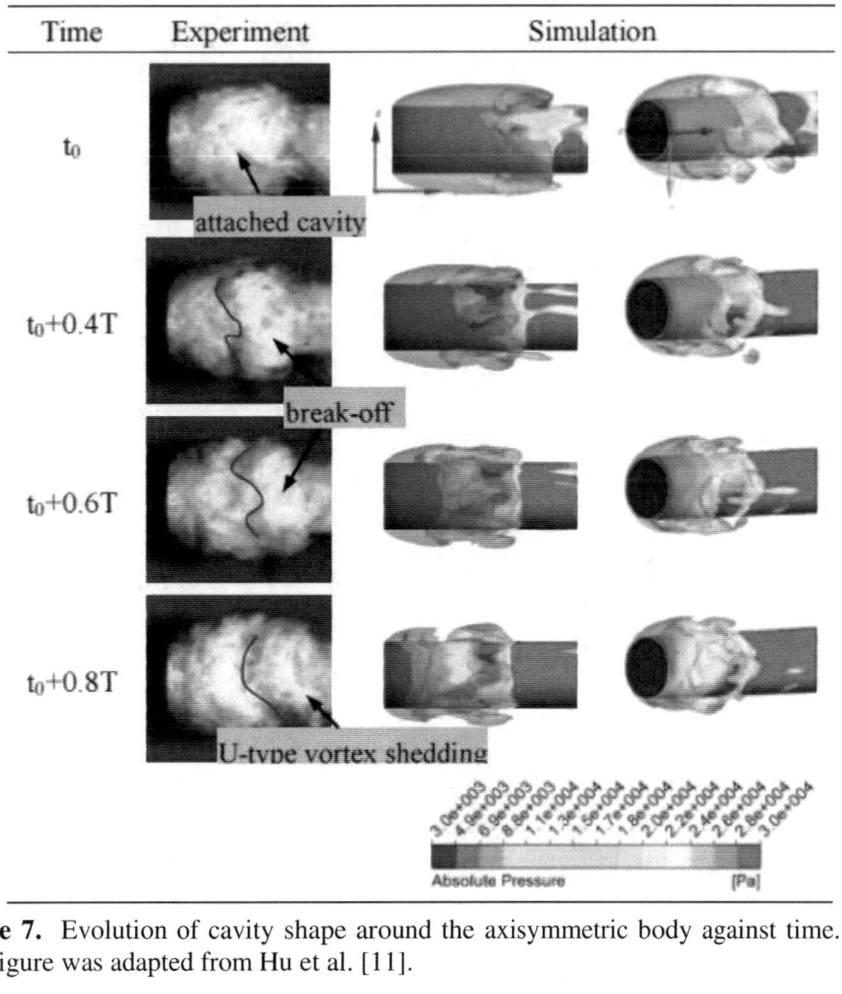

Figure 7. Evolution of cavity shape around the axisymmetric body against time. This figure was adapted from Hu et al. [11].

Hu et al. [11] further applied the modified PANS model coupled with Kubota cavitation model to simulate unsteady cavitating flows around an axisymmetric body with a blunt headform. Figure 7 shows the comparisons between experimental observations and numerical predictions. As shown in Figure 7, the simulations based on modified PANS can successfully predict the paramount phenomenon in cavitating flow (e.g. cavity break-off and U-type shedding).

3. Conclusion

Based on reported work using PANS approach, it was found that the ratio of the unresolved-to-total kinetic energy (f_k) plays an important role on the predictions (e.g. cavity pulsations, turbulent viscosity, shedding dynamics). Apart from setting it as a pure constant, variable f_k has been implemented through considering it as a function of ratio of mixture density and liquid density. Due to complex of the phenomenon, much more works on the physical origins of the paramount parameters in PANS approach on the cavitation dynamics should be investigated.

Acknowledgement

The authors acknowledge the financial support from "the Fundamental Research Funds for the Central Universities" (Project No.: 2014ZD09).

References

[1] Girimaji S 2006 *ASME J. of Applied Mechanics* **73** 413
[2] Girimaji S, Jeong E and Srinivasan R 2006 *J. of Applied Mechanics* **73** 422
[3] Girimaji S and Srinivasan R 2009 *AIAA 2009-134, 47th AIAA Fluid Dynamics Conference and Exhibit(Orlando, FL, USA, 5–8 January 2009)*
[4] Ma J M, Peng S H, Davidson L and Wang F J 2011 *Int. J. of Heat and Fluid Flow* **32** 652
[5] Liu J T, Zuo Z G, Wu Y L, Zhuang B T and Wang L Q 2014 *Computers & Fluids* **102** 32.
[6] Liu J T, Wu Y L and Wang L Q 2013 *Advances in Mech. Eng.* **2013** 710769.
[7] Huang B and Wang G.Y. 2011 *J Hydrodyn* **23** 26
[8] Ji B, Luo X W, Wu Y L and Xu H.Y. 2012 *Chinese Phys. Lett.* **29** 076401
[9] Ji B, Luo X W, Wu Y L, Peng X X and Xu H Y 2012 *Int. J. Heat Mass Transfer* **55** 6582
[10] Hu C, Wang G, Chen G and Huang B 2014 *Sci. China Physics, Mechanics & Astronomy* **57** 1967
[11] Hu C, Wang G, Chen G and Huang B 2014 *J. of Mech. Sci. Tech.* in press

IOP Conf. Series: Materials Science and Engineering **72** (2015) 032016 doi:10.1088/1757-899X/72/3/032016

Pump hump characteristic research based on mass transfer equation

D M Liu[1,2,5], Y Z Zhao[1], X B Liu[2], Y Ma[3] and W F Wang[4]

[1] Research & Development Center of Dongfang Electric Machinery CO., LTD, Deyang, 618000, China.
[2] Fluid machinery key laboratory of Xihua University, Chengdu Sichuan 610039, China
[3] Guo Dian Da Du River Repair&Installation co.,Ltd Leshan, Sichuan 614900 China
[4] Sichuan branch of China Hua-Dian Corporation, Sichuan 614900 China

E-mail: liudemin999@126.com

Abstract. The current development of modern pumped storage plants aims towards a higher flexibility in operation, an extended operation range of the hydraulic machine (especially in the pumping mode), and a higher reliability. The pumping requirements are the crucial design drivers, since, even if the turbine mode performance is very sound, the success of a project depends also on the pump turbine delivering the required maximum pump head and starting reliably in pump mode. Pump hump (pump instability working points at highest head) which is an instability source to the pump-turbine vibration is a serious damage to the pump operation on high head. So the pump hump and cavitation number based on the numerical simulation and experiment results are shown in this paper. The pump hump is sensitive affected by the cavitation number. With the cavitation number decreasing, the hump on flow characteristic curve (e.g. head-flow rate curve, *H-Q* curve) is gradually decreasing until vanished. Predicting cavitating flows with multi-phase CFD computations is still a very challenging task. Some results of ongoing work in this field are presented. The hump on *H-Q* curve with cavitation number is discussed in this paper.

1. Introduction

The hump characteristic is one of the special characteristics of a pump–turbine at pump mode with small flow. Especially, when the pump–turbine runs at high head level, the hump region is unavoidable and the performance is reduced. If a pump–turbine runs in hump region, the flow is small and it may cause large pressure fluctuations. Strong noise can be heard during the starting period and the start time is prolonged [1]. Nowadays, the stability of a pump–turbine is more and more important, and it has become essential to study the parameters cause instable characteristics. In pump operation, a positive slope of the head-flow curve occurs at operating conditions in which energy transferred to the fluid doesn't contribute to the head rise. This energy dissipates and gives rise to unsteady flow patterns such as vortices.

[5] Corresponding author.

For investigation of the stability of pump–turbines, lots of researches have been done. Genter et al. [2] makes the CFD calculation for the pump mode. According to Genter's research, the onset of fully developed pre-rotation at runner inlet shroud coincides with the instability of the head curve. The calculation shows rotating stall in the stator already in the area of stable operation. Braun et al. [3] analyzed the change in the global performance at pump mode and it was related to a change of the secondary flow pattern in the diffuser channels. Ran et al. [4] studied the unsteady flow in a pump–turbine and presented an improved runner to reduce the hump on *H-Q* curve. The researches on the hump characteristic at pump mode were limited. The mechanism of the hump characteristic was still unknown. The cavitation in a turbo-machinery may reduce the performance of the whole unit. Researches on the cavitation flow in turbo-machineries were a hot topic for the improvement of flow passage components [5-7], but the analysis of hump characteristics affected by the cavitation number have not yet been reported.

Most of recent studies on pump-turbine at pump mode were carried out by single phase, and the reason for causing the hump characteristic is unclear. In this paper, the hump characteristic affected by the cavitation number is tested by the experiment on the test stand and a CFD method with cavitation model is used to analyze the hump characteristic of a model pump-turbine by numerical simulation. The relation between hump and cavitation number is shown. The hump will become weak until vanish with cavitation number decreasing however the hump will restore if the cavitation number decreases to a certain number.

2. Pump turbine specifications

The whole configuration including spiral casing, stay vanes, guide vanes, runner and draft tube is considered in the physical model, as shown in figure 1. The computing domain enables to overcome the influence of boundary conditions, periodic interfaces, and pitch ratio of rotor-stator interface, and it especially considers the non-axisymmetric inflow from the spiral casing.

Specifications of the machine are listed in table 1.

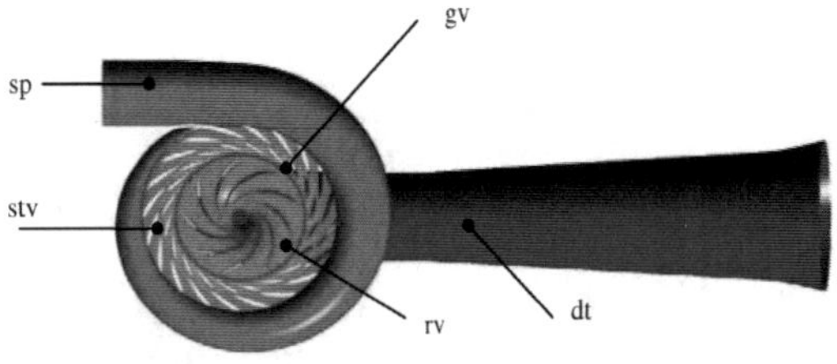

Figure 1. Computation domain of the pump-turbine model.

3. Mathematical model and computational details

Three-dimensional Reynolds-Averaged Navier-Stokes (RANS) equations are solved in conservative form on structured multi-block grids. A finite volume based discretisation scheme is used, which is up to second order accurate for the convective fluxes and truly second order accurate for the diffusive fluxes. Time dependent computations can be performed with a second order accurate time stepping scheme.

Table 1. Main characteristics of the tested pump turbine

runner diameter at inlet (mm)	560
runner diameter at outlet (mm)	250
runner blade number Z_1	7
guide vane number Z_0	20
rotational speed (rpm)	1200
specific speed (m.kW)	91

3.1. Governing equations

In homogeneous multiphase model, a common flow field is shared by all fluids, as well as other relevant fields such as temperature and turbulence. Thus, the governing equations of homogeneous model for mass, momentum, volume conservation equation can be written as

$$\frac{\partial \rho_m}{\partial t}+\frac{\partial\left(\rho_m \bar{u}_i\right)}{\partial x_i}=0 \tag{1}$$

$$\frac{\partial\left(\rho_m \bar{u}_i\right)}{\partial t}+\frac{\partial\left(\rho_m \bar{u}_i \bar{u}_j\right)}{\partial x_j}=-\frac{\partial \bar{p}}{\partial x_j}+\frac{\partial}{\partial x_j}\left(\bar{\tau}_{ij}^*-\rho_m \overline{u_i^{'} u_j^{'}}\right) \tag{2}$$

$$\frac{\partial\left(\rho_l \alpha_l\right)}{\partial t}+\frac{\partial\left(\rho_l \alpha_l u_j\right)}{\partial x_j}=\rho_l \tag{3}$$

where :

$$\bar{\tau}_{ij}^*=\begin{cases}\mu_m\left(\dfrac{\partial \bar{u}_i}{\partial x_j}+\dfrac{\partial \bar{u}_j}{\partial x_i}\right) & i \neq j \\ 2\mu_m \dfrac{\partial \bar{u}_i}{\partial x_j}-\dfrac{2}{3}\mu_m \nabla . \vec{V} & i=j\end{cases}$$

$$\bar{\Phi}=2\mu_m \overline{\left(\frac{\partial u_j}{\partial x_j}\right)^2}+\mu_m \overline{\left(\frac{\partial u_j}{\partial x_i}+\frac{\partial u_i}{\partial x_j}\right)^2}, i \neq j$$

$$\rho_m=\alpha_v \rho_v+\left(1-\alpha_v\right) \rho_l$$

$$\mu_m=\alpha_v \mu_v+\left(1-\alpha_v\right) \mu_l$$

3.2. Cavitation mass transfer model

Singhal et al. [8] has assumed as additional negative pressure drop which equals the half of the turbulence fluctuation $p_{turb}^{'}$.

$$p_{turb}^{'}=0.39 \rho k \tag{4}$$

This effect is equivalent to raise the vapor pressure $p_v^{'}$ by $p_{turb}^{'}/2$ to an equivalent vapor pressure p_v^*.

$$p_v^*=p_v+\frac{p_{turb}^{'}}{2}=p_v+0.195 \rho k \tag{5}$$

Mass transfer model for vapor–liquid mixture based on the theory of evaporation/condensation on a plane surface, and the Kinetic theory of mass transfer are used to get the source terms. To impart the above source terms a consistent appearance between the completely mushy and completely vaporous formulations, the source term of equation (6) is applied only if $p<p_v^*$. The source term of the Sharp Interfacial Dynamics Model equation (7) is employed, or vice versa. Thus, the source terms $\dot{m}$ can be expressed as:

If $p < p_v^*$

$$\dot{m} = C_1 \frac{3(1-\alpha_v-\alpha_u)}{r} \frac{2S}{2-S} \left(\frac{M}{2\pi R}\right)^{1/2} (p_v^* - p) \tag{6}$$

If $p > p_v^*$

$$\dot{m} = C_2 \frac{3\alpha_v}{r} \frac{2S}{2-S} \left(\frac{M}{2\pi R}\right)^{1/2} (p - p_v^*) \tag{7}$$

3.3. RNG k-ε model

A turbulence model is needed to close the system of equations which results from the Reynolds averaging. A variety of different turbulence models can be applied depending on the application. The turbulence model is still one of the largest error sources in modern CFD. However, methods with less or even no modeling are even nowadays not applicable on technical problems, so that the RNG $k-\varepsilon$ model is the state-of-the-art for the simulation of complex, three dimensional flows.

3.4. Grid generation

Rotating and stationary domains are matched using transient rotor stator interfaces. The global mesh encompassed 9 million nodes with smallest angles of 20 degrees. As displayed in figure 2, special refinement is applied in the runner (2.4 million nodes) and in the guide vane domain (5.2 million nodes). The mesh independent is validated before the scheme comparison. The y^+ on guide vane and runner blade is shown in figure 2(b) and (c), which are smaller than 70. The mesh number can satisfy the calculation requirement according to the comparison between simulation result and experiment result in figure 2(a).

3.5. Boundary condition

3.5.1. Inlet condition. The pressure inlet at the spiral casing (turbine regime) is used, whose profile is assumed to be uniform.

3.5.2. Outlet conditions. Pressure outlet is used at the draft tube (turbine regime), pressure is calculated according to the tail water level.

3.5.3. Other conditions. No-slip condition is assumed on all the solid walls, and standard wall function is used to calculate the turbulence kinetic energy and turbulence dispassion frequency near the wall. The rotation of runner domain is dealt with by means of transient rotator stator method in the unsteady calculation. The time step is 1degree in the unsteady calculation.

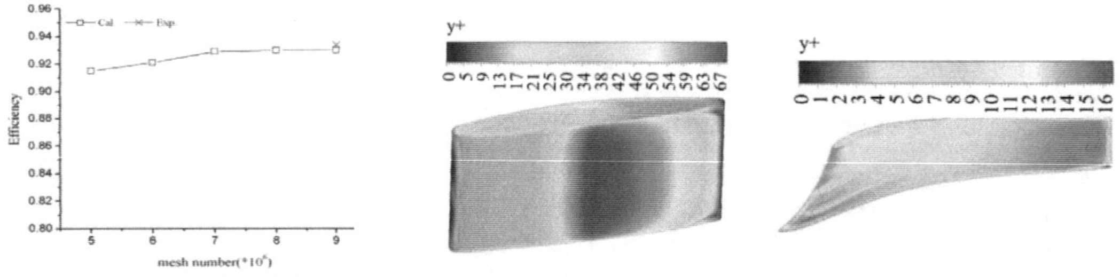

(a) Mesh independent (turbine model) (b) y^+ on guide vane (c) y^+ on runner blade

Figure 2. Mesh independent and y^+ distribution of the pump-turbine model.

4. Calculation result and analysis

The experiment results are shown in the figure 3(a), the *psi*(pressure coefficient)-*phi*(flow coefficient) curve is obtained at the same guide vane opening and different cavitation number (sigma number). The curve is different at different cavitation number. When the cavitation number equals 0.18, the hump characteristic is evidence at the *phi* = 0.212. When the cavitation number equals 0.15 and 0.12, the hump characteristic becomes weak and died away respectively. However, when the cavitation number equals 0.09, the hump characteristic becomes evidence and appears ahead of time than the cavitation number equals 0.18.

The comparison between experiment results and numerical results is shown in the figure 3(b) and (c). According to the results at cavitation number equals 0.18 and 0.09, the difference between experiment results and numerical results is in a reasonable range. The pressure coefficient at the numerical results is smaller than the experiment results. The hump position is nearly same between calculation result and experiment result.

The hump characteristic (named first hump) on *sigma* = 0.18 is caused by the guide vane inner part vortex shedding. A large number of vortexes block the guide vane flow passage from runner to spiral case. The vortex gradually appears on the guide vane flow passage when the flow coefficient decreases from working point A to C in figure 4. The centrifugal force is pushes aside by the pressure because there is not enough water to fill with the flow passage.

A little number of cavitation bubbles appears on the runner suction face which is possible reason to hump becomes weak and appears ahead of time. When the flow discharge decreases, the flow impacting becomes serious. So the cavitation bubble appears on two blades on working point C.

From the figure 6, the pressure distribution on runner suction face. On working point D, the absolute pressure is larger than the vapor pressure, so the cavitation bubble can't appear on runner suction face. On working point B, the absolute pressure is smaller than the vapor pressure, so the cavitation takes place on two runner suction faces. This is the possible reason to the hump appearing on *psi-fai* curve [9]. So in this paper, the hump mainly is caused by vortex shedding on guide vane and affected by cavitation number.

5. Conclusions

This paper mainly focuses on the cavitation number affection on the pump's hump on the smallest flow rate. The hump is compared on different cavitation numbers and same guide vane opening between experiment results and simulation results. By choosing cavitation model, mesh independent validate, hump curve are calculated by the numerical method. The following conclusions can be made:

1) Cavitation numbers make great affection on the pump's hump shape. The hump will become weak with the cavitation number decreases.

2) The hump can be calculated by the CFD method with cavitation model and the results is reasonable compared with the experiment results.

3) The hump on same guide vane opening appears ahead of time with the cavitation number decreasing.

Acknowledgments

This work is supported by the Chinese National Foundation of Natural Science (No. 51279172), the Open Research Funded Project of Key Laboratory of Fluid Machinery in Sichuan Province (Xihua University) (Grant No.Szjj2012-041, szjj2011-039).

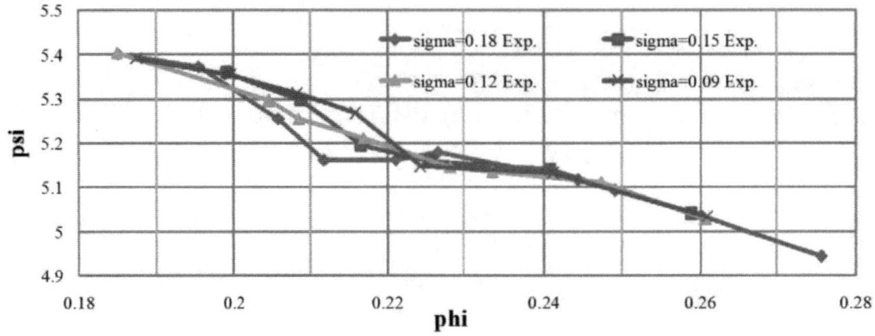

(a) Exp.

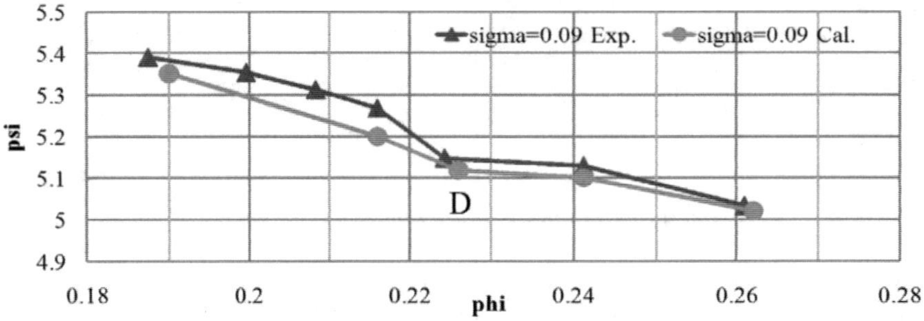

(b) Exp. and Cal. at *sigma* = 0.09

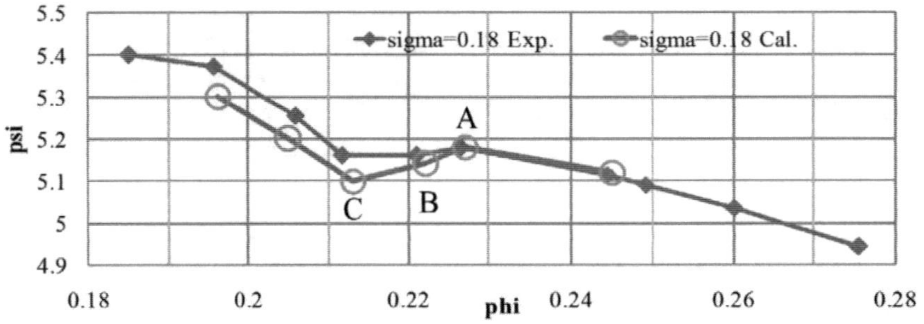

(c) Exp. and Cal. at *sigma* = 0.18

Figure 3. *Psi-phi* curve on different cavitation numbers.

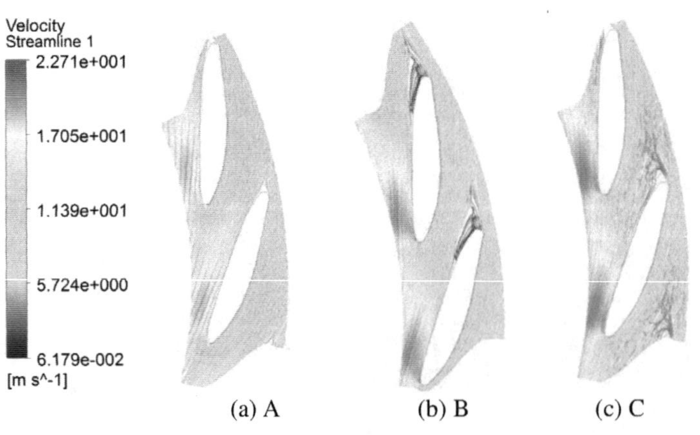

(a) A (b) B (c) C

Figure 4. Vortex sheds on guide vane flow passage at different working points.

(a) A (b) B (c) C

Figure 5. Cavitation bubbles appear on runner suction face at different working points.

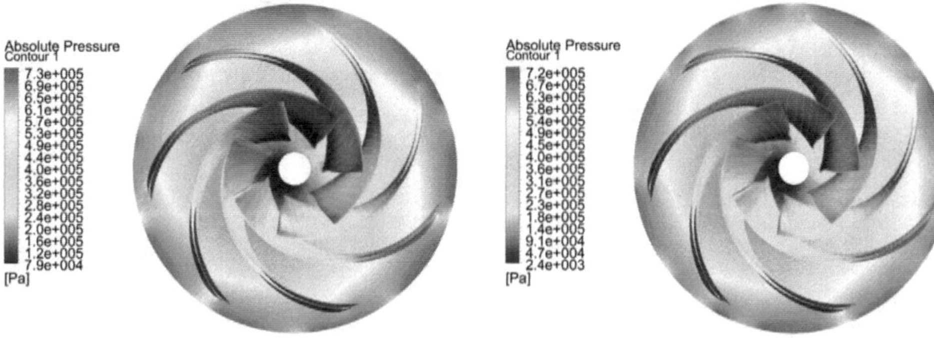

(a) D (b) B

Figure 6. Pressure distribution on runner suction face at different working points.

References

[1] Mei Z Y 2000 *Power generation technology of pumped storage power station* (Beijing: China Machine Press) p 200-201

[2] Gentner C, Sallaberger M, Widmer C, et al. *IOP Conference Series: Earth and Environmental Science* **15** 3

[3] Braun O, Kueny J L and Avellan F 2005 *Proc. ASME fluids engineering division summer meeting and exhibition* (Houston TX USA) pp 1-6

[4] Ran H J, Luo X W and Zhang Y 2008 Numerical analysis of flow phenomena related to the unsteady energy-discharge characteristic of a pump–turbine in pump mode *Proc. ASME 2008 Fluids Eng. Division Summer Meeting collocated with the Heat Transfer, Energy Sustainability, and 3rd Energy Nanotechnology Conferences (Jacksonville, Florida, USA, 10-14 Aug 2008)* pp 1115-23

[5] Fortes-Patella R, Coutier-Delgosha O, Perrin J and Reboud J L 2007 *J. Fluid Eng.* **129** 128-35

[6] Lindau J W, Kunz R F, Boger D A, Stinebring D R and Gibeling H J 2002 *J. Fluid Eng.* **124** 607–16

[7] Medvitz R B, Kunz R F, Boger D A, Lindau J W, Yocum A M and Pauley L L 2002 *J. Fluid Eng.* **124** 377-83

[8] Singhal A K, Athavale M M and Li H 2002 *J. Fluid Eng.* **124** 617-24

[9] Liu J T, Liu S H and Wu Y L 2012 *Computers & Fluids* 105-11

IOP Conf. Series: Materials Science and Engineering **72** (2015) 032016 doi:10.1088/1757-899X/72/3/032016

Nomenclature

C_1-evaporation coefficient 0.13
C_2-condensation coefficient, 0.01
C_p-Pressure coefficient
$r = 0.0001$, S is the surface tension
M-torque
N-rotational speed(r/min)
P-output power (MW)
Re-Reynolds number
p-pressure (Pa)
p_v-vapor pressure (Pa)
$r = 0.0001$
u-Velocity
α_v-Water vapor void fraction
α_u-non-dissolved gas void fraction, which takes value of 0.0005.
σ = cavitation number ($-C_p$)
ρ_m-mixture density
ρ_v-water vapor density
ρ_l-water density
μ_m-mixture viscosity
μ_v-water vapor viscosity
μ_l-water viscosity
k : Turbulent kinetic. $i, j = 1, 2, 3$.
$\eta = S\frac{k}{\varepsilon}, \eta_o = 4.38, C_\mu = 0.0845, \beta = 0.012,$
$C_{1\varepsilon} = 1.42, C_{2\varepsilon} = 1.68, \alpha_k = 1.0, \alpha_\varepsilon = 0.769.$
R-Addition term, $R = \frac{C_m r h^3 (1 - h/h_o)}{1 + bh^3} \frac{e^2}{k}$
Phi- flow coefficient, $phi = Q/\left(\pi/4D_2^2 u\right)$
Psi- pressure coefficient, $psi = 2gH/\left(u^2\right)$

Research on the characteristics of quasi-steady cavitation in a centrifugal pump

J X Lu[1,3], S Q Yuan[1], X J Li[2], Q R Si[1] and Y Luo[1]
[1] National Research Center of Pumps, Jiangsu University, Zhenjiang 212013, Jiangsu, China
[2] School of Mechanical Engineering and Automation, Zhejiang Sci-Tech University, Hangzhou 310018, Zhejiang, China

E-mail: ljx0813@gmail.com

Abstract. With the pressure decreasing, the process of cavitation in a centrifugal pump could be summarized as incipient cavitation, quasi-steady cavitation and unsteady cavitation. Quasi-steady cavitation is the condition that is between the incipient cavitation and unsteady cavitation in a centrifugal pump. Under this condition, the intensity of cavitation is relatively weak, and the head of the pump almost remains unchanged, but the cavitation exists, causing damage to the impeller by pitting and erosion. So it is important to investigate the quasi-steady cavitation. In this paper, both the numerical and experimental methods had been carried out to investigate the characteristics of quasi-steady cavitation. The internal flow in the pump, the performance of cavitation and the inlet and outlet pressure pulsation of the pump measured through experimental method have been studied under different *NPSHa* conditions. It was found that the head decreases about 0.77%-1.38% from non-cavitation condition and it could be regarded as the quasi-steady cavitation. Little change has been found from the internal flow between non-cavitation condition and quasi-steady cavitation condition. The period of inlet pressure pulsation changes from the time that the blade passes by to the period of shaft rotating with the development of cavitation. The dominant frequency of the inlet pressure pulsation is two times of shaft frequency whose amplitudes decrease firstly and then increase to a peak value, followed by a decrease to a low value in quasi-steady cavitation conditions. The dominant frequency of the outlet pressure pulsation is blade passing frequency whose amplitudes increase firstly and then decrease gradually with the decrease of *NPSHa*.

1. Introduction

Cavitation first occurs locally at the low pressure region in the impeller and develops in accordance with the decrease of suction pressure in a pump [1]. As an important energy transforming device and fluid conveying equipment, centrifugal pumps are widely used in various fields [2]. However, the life of the pump and the stability of the system are greatly affected by cavitation. Not only the noise, vibration and instability of a system can be induced, but also the erosive damage with the surface of flow channels, and the loss of performance in pumps are also produced due to cavitation [3-9].

The main processes of cavitation in a centrifugal pump could be summarized as incipient cavitation, quasi-steady cavitation and unsteady cavitation with the decrease of pressure [10]. Inception of cavitation occurs when the static pressure of the liquid at any location is lower than the saturated vapor pressure of that liquid at the prevailing temperature in the presence of nuclei [1].

[3] Corresponding author.

Unsteady cavitation include the cavitation surge (CS), rotating cavitation (RC), asymmetric cavitation (AC) and high-order instabilities [11, 12], which is responsible for the undesirable noise and vibration generation. Quasi-steady cavitation is defined as the condition that is between the occurrence of incipient cavitation and unsteady cavitation in a centrifugal pump. Under this condition, the intensity of cavitation is weak relatively, and the head of the pump almost remains unchanged, but the cavitation has been existed and damaged to the impeller by pitting and erosion. So it is difficult from the cavitation performance to find that the pump is running in the condition of quasi-steady cavitation. Therefore, it is important and necessary to investigate in detail the characteristics of the quasi-steady cavitation.

In this paper, both the numerical and experimental methods have been carried out to investigate the characteristics of quasi-steady cavitation in a centrifugal pump. The internal flow in the pump, the performance of cavitation and the inlet and outlet pressure pulsations of the pump have been studied under different *NPSHa* conditions.

2. Numerical method

2.1. Research model

Figure 1 shows the computational model for this study. The researched target is a centrifugal pump which is composed of a shrouded impeller with six blades, a single suction, and a spiral volute casing. The impeller is designed to operate at 2910 r/min. The pump's designed flow rate is Q_d=50.6 m^3/h, and the designed head *is* H_d = 20.2 m. The specific speed of the pump is n_s=132.2 at the designed point ($n_s = 3.65\ nQ^{0.5}/H^{0.75}$). The shaft frequency is f_0 = 48.5 Hz and the blade passing frequency f_d = 291 Hz of this model pump at the designed point, where $f_0 = n/60$, f_d = Z (n/60). Main geometric parameters of the pump are shown in table 1.

Figure 2 shows the structured grids in impeller regions of the pump. The grids of the three-dimensional model were generated by the ICEM CFD (integrated computer engineering and manufacturing code for computational fluid dynamics). The total number of grid nodes is 2677106 and the number of grid elements is 2518495 for all the domains. In order to make the numerical simulation more close to the real situation, import and export of the pipelines were extended appropriately.

Table 1. Main pump parameters

Parameter	Sign	Value	Unit
Inlet diameter of impeller	D_j	79	mm
Outlet diameter of impeller	D_2	140	mm
Inlet diameter of pump	D_s	65	mm
Outlet diameter of pump	D_d	50	mm
Outlet width of blade	b_2	15.5	mm
Basic circle of volute	D_3	149	mm
Total blade wrap angle	φ	140	°
Volute width	b_3	32	mm

2.2. Computational condition

In this study, the simulations were performed with the commercial software Ansys CFX 14.5, and the k-ε turbulence model was selected as turbulence model. In order to simulate the rotational effect, the multiple frames of reference were involved. The internal and external impeller surfaces were chosen as a rotating wall, the speed equals the nominal rotating speed, while all the other walls were stationary. A smooth wall condition was used for the near wall function. The grid interfaces between two stationary components were set as general grid interface (GGI). Between the rotational and stationary components GGI was also used as internal component connection. The boundary conditions of steady flow simulation were almost the same under both cavitation and non-cavitation conditions.

The total pressure at the inlet of the suction pipe was specified, while the outlet boundary condition was set up as mass flow rate. The reference pressure was set to 0 Pa. The calculations continued until the RMS (root mean square) residual dropped below 5×10^{-5}.

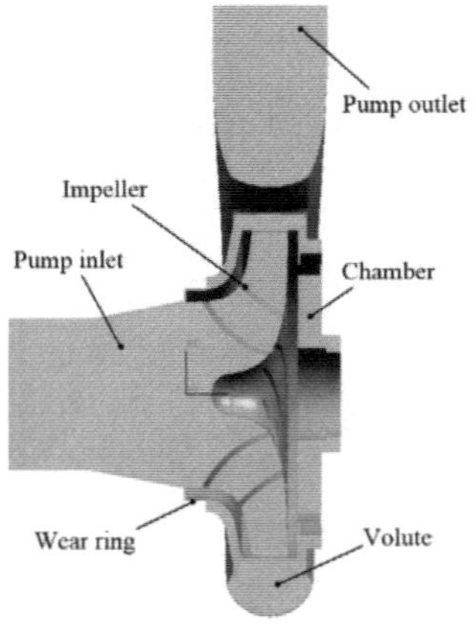

Figure 1. Model for entire pump passage.

Figure 2. Grid view in impeller region.

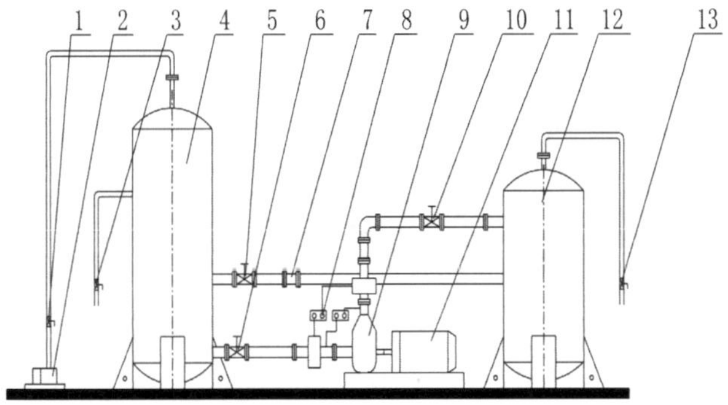

Figure 3. Schematic of the experimental test rig: 1, 3, 13) ball valve; 2) vacuum pump; 4) pressure tank; 5, 10) butterfly valve; 6) gate valve; 7) turbine flowmeter; 8) pressure transmitter; 9) test pump; 11) electromotor; 12) buffer tank.

3. Experimental methods

The predictions from the numerical model were subsequently validated with experimental data measured in a closed hydraulic test rig which was shown in figure 3. At the same time, the pump inlet and outlet pressure fluctuations were tested in the condition of quasi-steady cavitation. The pump in the rig was driven at a constant speed of 2900 r/min by an electromotor, with a blade passing frequency f_d = 290 Hz. In the experiment, a turbine flow meter installed at the downstream pipe was used to measure the flow rate. The mean static pressure was tested by two pressure transmitter with their ranges were from 0 to 4 bar installed at the outlet of the pump and -1 to 0.6 bar installed at the inlet of the pump. Two transient pressure sensors with the ranges of -100 to 100 kPa and 0 to 0.6 MPa were respectively installed at the inlet and outlet of the pump to monitor the pressure pulsations. The inlet pressure and flow rate were adjusted by the vacuum pump and the flow control valve, respectively. A PXI 4472B data acquisition module which was made by National Instruments Company was applied to capture the electric signals and convert them to digital signals.

4. Results and discussion

4.1. Quasi-steady cavitation

Figure 4 shows the power spectral density (PSD) signals of pump inlet pressure pulsations which were measured through experiment at designed flow rate (Q_d=50.6 m^3/h). Obvious changes in pump inlet pressure pulsations PSD signals have occurred starting from the *NPSHa*=6.94 m as the result of development of cavitation. The amplitudes of inlet pressure pulsation PSD signals decrease sharply at 5 times (about 250 Hz) and 6 times (blade passing frequency) of shaft frequency. However, the magnitudes increase suddenly at 3 times (about 150 Hz) and 4 times (about 200 Hz) of shaft frequency, and obvious discrete pulsations occur near the frequency. So *NPSHa*=6.94 m can be regarded as the point of cavitating inception. According to the reports of Y. Tsujimoto and T. Kimura, et al [13, 14], *NPSHa*=3.07 m can be regarded as the starting point of unsteady cavitation in the pump. Therefore, the pump is running under the condition of quasi-steady cavitation when *NPSHa* decrease from 6.94 m to 3.47 m.

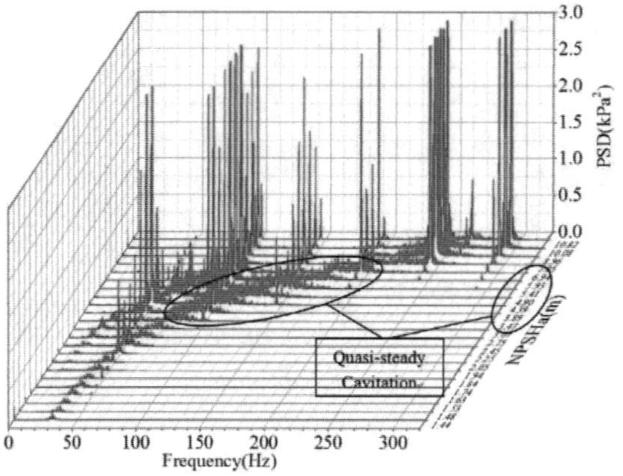

Figure 4. Waterfall plot of pump inlet pressure pulsation at Q_d.

4.2. Cavitation performance of quasi-steady cavitation

Figure 5 shows a comparison between experimental and numerical results of the suction performance curves at Q_d. The horizontal axis shows the Available Net Pressure Suction Head which is defined as $NPSHa = (p_{inlet} - p_v) / \rho g)$, where p_{inlet} is the inlet pressure and p_v is the vapor pressure. The vertical axis shows the head of the pump. A rational agreement between the experiments and the predictions can be found. The predicted accuracy was estimated to be within 5% of express for the flow regimes calculated numerically. The existed deviation between the experimental and numerical results may caused by the unavoidable testing error in the experiment. On the other hand, the roughness of the wall were not taken into account in the process of calculation may also played a critical role to the deviation. Compared with non-cavitation condition, the head of the pump decreases about 0.77% to 1.33% under the condition of quasi-steady cavitation. Hence, the head of the pump almost remains unchanged under the quasi-steady cavitation and it is difficult to find out the condition according to the suction performance.

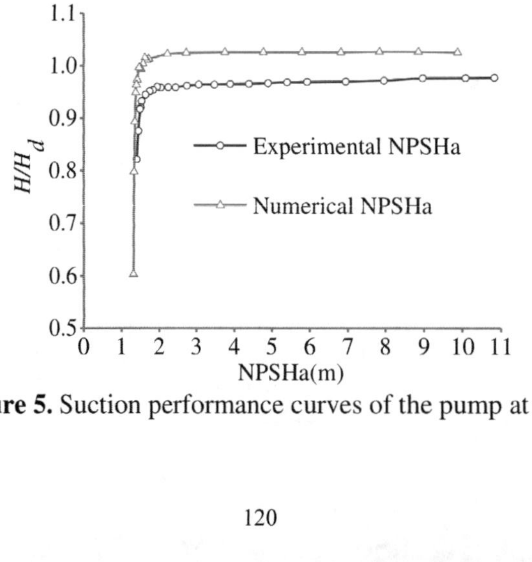

Figure 5. Suction performance curves of the pump at Q_d.

4.3. Analysis of internal flow characteristics of quasi-steady cavitation

Figure 6 has compared the distribution of water turbulence kinetic energy between non-cavitation and quasi-steady cavitation conditions in middle cross section of impeller at Q_d. It is clear that almost few changes can be found from the distribution of water turbulence kinetic energy between non-cavitation condition (*NPSHa*=9.87 m) and quasi-steady cavitation condition (*NPSHa*=3.47 m). Some turbulent kinetic energy distributes at the surface of each blade suction side which is near the outlet of the impeller, but the intensity of it in the impeller is weak. Owing to the influence of high speed rotating blades, the turbulent kinetic energy which is distributed at the interface between the blades and the volute is strong relatively. In the outlet of the volute and the tongue region, there is strong water turbulent kinetic energy existing, which indicates that the flow is relatively complex there under both non-cavitation and quasi-steady conditions.

(a) *NPSHa*=9.87m (b) *NPSHa*=3.47m

Figure 6. Distribution of water turbulence kinetic energy in middle cross section of impeller at Q_d.

The distribution of water velocity under the non-cavitation and quasi-steady cavitation conditions in middle cross section of impeller at Q_d have been shown in figure 7. Nearly few distinctions can be found from the distribution of water velocity between non-cavitation condition (*NPSHa*=9.87 m) and quasi-steady cavitation condition (*NPSHa*=3.47 m). Under the two conditions, each blade pressure surface speed is lower than that of on each blade suction side due to the pressure distributes on the blade pressure side is higher than that of on the suction side. The water velocity at the interface between each blade and the volute present a high speed owing to the high impeller tip speed.

4.4. Pump inlet pressure pulsation characteristics of quasi-steady cavitation

Figure 8 has shown the time domain of pump inlet pressure fluctuations under the quasi-steady cavitation condition at Q_d. The horizontal axis shows the time that the impeller rotates 1 revolution. The vertical axis shows the amplitude of pump inlet pressure pulsation. It is obvious that with the process of cavitation developing from non-cavitation to quasi-steady cavitation condition, not only the period of pump inlet pressure fluctuations change from the time that the blade passed by to the period of shaft rotating, but also the intensity of the fluctuations gradually become weak. It is considered that vapor bubbles in the pump play an important role in enhancing the buffering effect to the pump inlet liquid.

Figure 9 shows the PSD signals of pump inlet pressure fluctuation under the quasi-steady cavitation condition at Q_d. It is clear that the dominant frequency of pump inlet pressure fluctuation is 2 times of shaft frequency (about 97 Hz) whose magnitudes decrease first and increase to a peak value (*NPSHa*=4.39 m), then decrease to a low value. Meanwhile, the magnitudes of PSD signals of pump inlet pressure fluctuations also increase to the maximum value first at *NPSHa*=4.39 m and then decrease to a low value owing to the development of cavitation.

(a) *NPSHa*=9.87 m (b) *NPSHa*=3.47 m

Figure 7. Distribution of water velocity in middle cross section of impeller at Q_d.

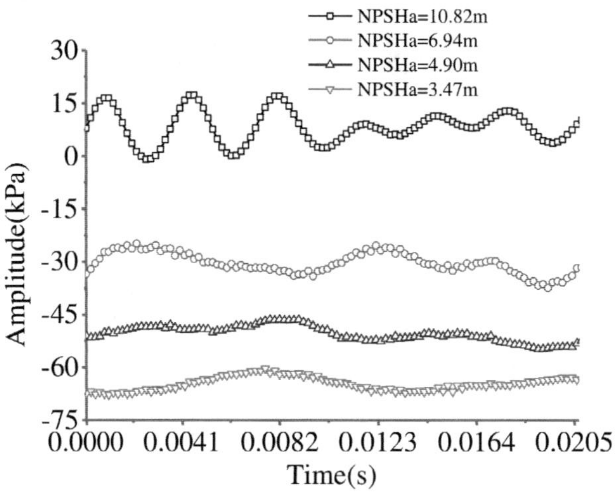

Figure 8. Time domain chart of pump inlet pressure pulsation at Q_d.

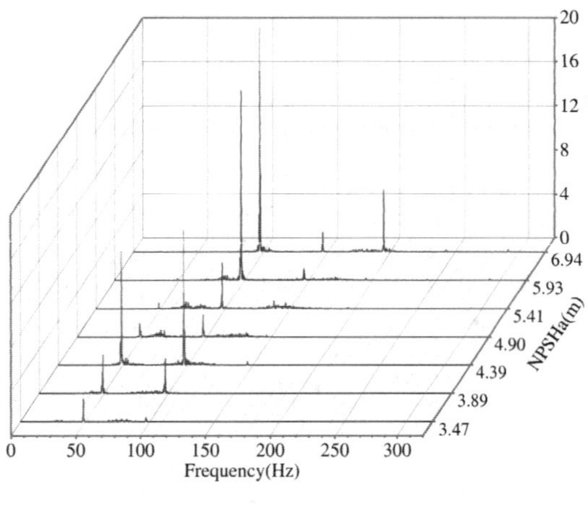

Figure 9. Power spectral density chart of pump inlet pressure fluctuation under quasi-steady cavitation condition.

4.5. Pump outlet pressure pulsation characteristics of quasi-steady cavitation

Figure 10 has shown the time domain of pump outlet pressure fluctuations under the quasi-steady cavitation condition at Q_d. The horizontal axis shows the time that the impeller rotates 1 revolution. The vertical axis shows the amplitude of pump outlet pressure pulsation. It is clear that in both non-cavitation condition and quasi-steady cavitation condition, the pump outlet pressure fluctuations present a law of sine function. It can be found that the six blades passing through the tongue cause six cycles in each impeller revolution, which is a typical pressure fluctuation distribution caused by rotor-stator interaction [10]. Compared with the cavitation-free condition, the intensity of the fluctuations gradually become strong in the quasi-steady cavitation condition due to the development of cavitation.

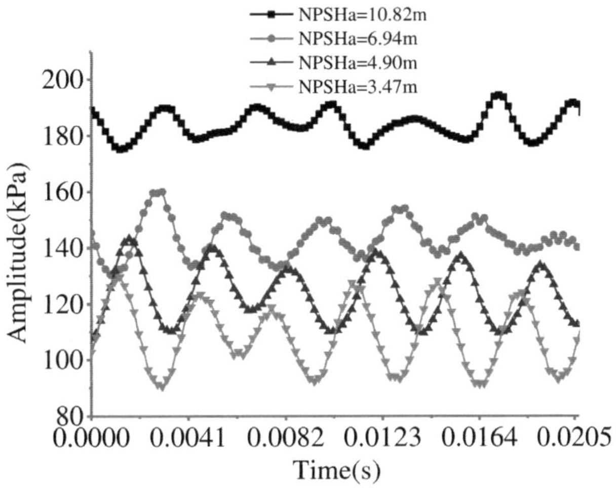

Figure 10. Time domain chart of pump outlet pressure pulsation at Q_d.

Figure 11 shows the PSD signals of pump outlet pressure fluctuation under the quasi-steady cavitation condition at Q_d. It can be found that the dominant frequency of the outlet pressure pulsation is blade passing frequency (about 290 Hz) whose amplitudes increase to the maximum value first at *NPSHa*=4.39 m, and then decrease with the decrease of *NPSHa*. At the same time, not only some discrete signals of the PSD exist, but also the subpeak occurs near the shaft frequency when *NPSHa*=4.39 m. The magnitudes of the PSD signals of outlet pressure pulsation decrease gradually near the 2 times of shaft frequency with the development of cavitation.

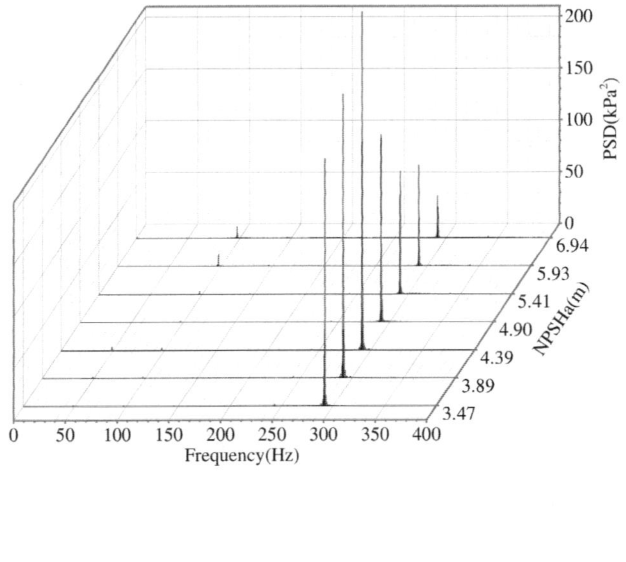

Figure 11. Power spectral density chart of pump outlet pressure fluctuation under quasi-steady cavitation condition.

5. Conclusion

The internal flow characteristics and pump inlet and outlet pressure fluctuations characteristics of the centrifugal pump under the condition of quasi-steady cavitation have been investigated through experimental method and numerical simulation. Several conclusions can be drawn as following:

1) The head of the pump decreases about 0.77% to 1.33% from non-cavitation condition can be considered as the quasi-steady cavitation condition, corresponding to the *NPSHa* decreases from 6.94 m to 3.47 m.

2) Nearly few distinctions can be found from the characteristics of internal flow between non-cavitation and quasi-steady cavitation condition.

3) Not only the period of pump inlet pressure fluctuation changes from the time that the blade passed by to the period of shaft rotating, but also the intensity of the fluctuations gradually become weak with the process of cavitation developing from non-cavitation to quasi-steady cavitation condition. In contrast, the intensity of pump outlet pressure fluctuations become strong due to the decrease of *NPSHa*.

4) The dominant frequency of pump inlet pressure fluctuation is 2 times of shaft frequency whose magnitudes decrease first and increase to a peak value at *NPSHa*=4.39 m, then decrease to a low value. The dominant frequency of pump outlet pressure pulsation is blade passing frequency whose amplitudes increase to the maximum value first at *NPSHa*=4.39 m, and then decrease gradually with the decrease of *NPSHa*.

Acknowledgements

This research was supported by the State Key Program of National Natural Science Foundation of China (Grant No. 51239005), the National Science & Technology Pillar Program of China (Grant No. 2011BAF14B04) and the Priority Academic Program Development of Jiangsu Higher Education Institutions (PAPD).

References

[1] Christopher S and Kumaraswamy S 2013 *J. Fluid Eng* .**35** 121301-1-15
[2] Yuan S Q, Si Q R, Xue F, Yuan J P and Zhang J F 2011 *J. Drainage and Irrigation Machinery Eng.* **29** (2) 93-8
[3] Ni Y Y, Yuan S Q, Pan Z Y and Yuan J P 2008 *Chin. J. Mech. Eng.* **21** (5) 46-9
[4] Chudina M 2003 *Acoustical Physics* **49** (4) 463-74
[5] Cernetic J 2009 *Mechanical Eng. Sci.* **223** 1645-55
[6] Su Y S, Wang Y S and Duan X Y 2010 *Trans. the Chin. Society for Agricultural Machinery* **41** 77-80
[7] Horiguchi H, Watanabe S and Tsujimoto Y 2000 *J. Fluid Eng.* **122** (4) 798-805
[8] Horiguchi H Watanabe S and Tsujimoto Y 2000 *J. Fluid Eng.* **122** (1) 156-63
[9] Semenov Y, Fujii A and Tsujimoto Y 2004 *J. Fluid Eng.* **126** 87-93
[10] Li X J, Yuan S Q, Pan Z Y et al. 2013 *Sci. China Tech. Sciences* **56** (9) 2156-62
[11] Subbaraman M and Burton K 2003 *Proc. Int. Cavitation Symposium* (Osaka Japan) OS-4-013
[12] Tsujimoto Y, Yoshida Y, Maekawa Y, Watanabe S and Hashimoto T 1997 *J. Fluid Eng.* **119** 775-81
[13] Fredrichs J and Kosna G 2002 *J. Fluid Eng.* **124** 356-62
[14] Kimura T, Yoshida Y, Hashimoto T and Shimagaki M 2008 *J. Fluid Eng.* **130** 051104-1-9

IOP Conf. Series: Materials Science and Engineering **72** (2015) 032018 doi:10.1088/1757-899X/72/3/032018

Cavitation performance simulation of turbine meter under different temperature water condition

Y Z Huang[1], B S Zhang[1], G Chen[2] and B L Zhu[2]
[1] China University of Petroleum-Beijing, 18 Fuxue Road, Changping, Beijing 102249, China
[2] China National Oil and Gas Exploration and Development Corporation, No.6-1 Fuchengmen Beidajie, Xicheng, Beijing 100034, China

E-mail: Huangyongzhang@cnpcint.com

Abstract. A cavitation thermodynamics model based on phase change, which is suitable for prediction of cavitation with thermal effects is developed. The cavitation characteristic at different temperature and cavitation number is investigated and analyzed. The initial cavitation of turbine flow meter generally occurs in the blade suction side. With the development of cavitation, the cavitation zone will appear on the front and the back end of the conditioner. In order to avoid the gather of cavitation, the design of the optimizing the blade structure should be adapted, and at the same time, the back pressure should be limited on the installation requirements. Expanding the measurement range and preventing cavitation occurs are the goal of the design and installation. The temperature effects on the cavitation of turbine flow meter is quite obvious and the increase of the temperature will delay the occurrence of cavitation. Pressure difference and the impeller torque will change obviously with the decrease of the cavitation number, which will cause the measurement error of the turbine meter.

1. Introduction

Turbine flow meters have been competing successfully with positive displacement flowmeters in many applications due to the economy of installation, low maintenance costs, weight, size and high flow rates per comparable connection size [1]. The turbine meters can be used to measure various flow rates, operating pressures up to 10,000 pounds per square inch, temperature range of -450° to 1000°F. Cavitation in a turbine flowmeter will take place when the local pressures fall close to or below the vapor pressure of the liquid product. The formation of bubbles and their collapse or local vaporization of product as it passes over the rotor blade surface can cause erratic behavior in the turbine flowmeter and excessive wear due to over speeding. Cavitation usually causes the rotor to speed up at the high flow rate due to the increased flow volume and causes the accuracy curve of the turbine flowmeter to be adversely affected. When turbine flow meters are used to measure high temperature water and cryogenic liquid, the cavitation characteristics in such conditions are different from that of room temperature water. This is caused by thermodynamic effects of cavitation [2].

In high temperature water, the pressures observed inside cavities are substantially lower than the vapor pressure corresponding to the free-stream temperature. The reason for this is that the temperature of the liquid in the immediate vicinity of the liquid-vapor interface is depressed below the free-stream temperature because the latent heat of vaporization must be extracted from the bulk liquid.

IOP Conf. Series: Materials Science and Engineering **72** (2015) 032018 doi:10.1088/1757-899X/72/3/032018

Thus, the growth of cavitation is suppressed and the cavity size is smaller than it would be if there were no temperature depression effects present.

Until now, there are a lot of cavitation models that have been developed. And these models can be split into two categories. One is to predict the size and shape of the cavitation bubble. For these analyses, the cavity pressure is taken as known (generally corresponding to the free-stream temperature) and thermodynamic effects are not addressed. The other is to predict the temperature depression for known cavity geometry.

For the predictions of the geometrical effects of the cavitation bubble, Furness and Hutton [3] first consider the deformation of the liquid-vapor interfaces to describe the early stage of reentrant jet formation. After that, a homogeneous two-phase mixture of liquid and vapor method has been developed in modeling the cavitation. Kubota et al. [4] proposed to relate the density evolution to the motion of bubbles in the flow. And the bubble evolution is governed by the Rayleigh-Plesset equation according to the pressure field. Recently, different authors proposed more to consider a transport equation model for the void ratio, with evaporation/condensation source terms to control the mass transfer between the two phases [5,6,7]. As the development of CFD technology, more robust numerical methodology is expected to predict and investigate the thermodynamic cavitation. In the paper, a cavitation model based on phase change is developed, suitable for prediction of cavitation with thermal effects; Building the developed cavitation model into the commercial code by using user subroutine; and the cavitation characteristic of turbine flow meter at different temperatures are analysis using the cavitation model.

2. Cavitation model considering the temperature effect

2.1. Governing equations and phase change modeling

The governing equations of mixture cavitation model are the 3D gas-liquid two-phase compressible Navier–Stokes equations, and are written as following.

Mass conservation equation of the mixture phase:

$$\frac{\partial \rho}{\partial t}+\frac{\partial}{\partial x_j}(\rho u_j)=0 \tag{1}$$

Momentum conservation equation of the mixture phase:

$$\frac{\partial(\rho u_i)}{\partial t}+\frac{\partial}{\partial x_j}\left(\rho u_j u_i\right)=\rho g_i-\frac{\partial p}{\partial x_i}+\frac{\partial \tau_{ji}}{\partial x_i} \tag{2}$$

Transport equation to account for the cavitation dynamics of vapor volume fraction:

$$\frac{\partial \alpha_v \rho_g}{\partial t}+\frac{\partial}{\partial x_j}(\alpha_v \rho_g u_j)=S \tag{3}$$

Where, ρ is the mixture density of water and vapor.

S is the mass transfer of vapor liquid phase change which is described as:

$$S=\dot{m}^{+}+\dot{m}^{-} \tag{4}$$

The former and the latter Represents the process of evaporation and condensation respectively.

As we all known, the phase change is developed in the liquid , the pressure inside the bubble is nearly same as the liquid and the pressure is the same with the vapor pressure of the liquid , so we got the mass transportation between the two phase:

$$\dot{m}=\dot{m}^{+}=\dot{m}^{-}=(\frac{2\xi}{2-\xi})\left(\frac{\mathrm{M}}{2\pi\mathrm{R}}\right)^{\frac{1}{2}}\left(\frac{p_v}{\sqrt{T}}-\frac{p}{\sqrt{T}}\right)A \tag{5}$$

For simplicity, the typical bubble size r is taken to be the same as the limiting maximum possible bubble size. Then, r is determined by the balance between aerodynamic drag and surface tension force [5]. Also, the vapor pressure need to be modification, in order to accounts for the turbulence-induced pressure fluctuations.

2.2. Physical properties of medium at different temperature
In the process of numerical simulation of the thermodynamic effect of cavitation, the physical properties of the medium must be considered.

2.2.1. Density of water. The density of water at different temperature can be determined through Tamman state equation [8]:

$$\rho_l = \frac{p + p_c}{K_1(T + T_0)} \tag{6}$$

Where, P_C = 1944.61 *MPa* and T_0 = 3837 *K* represents pressure constant and temperature constant of water respectively, K_l = 472.27 J/kg -K is a liquid constant of water.

2.2.2. The density of the vapor. According to the research of Saito [9] , the density of the vapor could be calculated according to ideal gas state equation:

$$\rho_v = \frac{p}{R_v T} = \frac{\mathrm{M}\, p}{\mathrm{R}\, T} \tag{7}$$

Where, R_v represent the vapor constant, $R_v = R / M = 461.6$.

2.2.3. Saturated vapor pressure. The relationship between the saturated vapor pressure and saturated temperature can be expressed by experience formula[9]

$$p_v = p_k \exp\{(1 - \frac{T_k}{T})(a + (b - cT)(T - d)^2\} \tag{8}$$

where: $p_k = 22.130\mathrm{MPa}$, $T_k = 647.31\mathrm{K}$, $a = 7.21379$, $b = 1.1520 \times 10^{-5}$, $c = -4.787 \times 10^{-9}$, $d = 483.16$.
2.2.4. Other physical properties. Thermal conductivity, specific heat capacity and dynamic viscosity coefficient of water and vapor, the surface tension coefficient of water, etc. must be expressed as a function of temperature through curve fitting, and dynamic update in the process of numerical simulation of when considering the effect of thermodynamic calculating process of cavitation.

3. Comparing Numerical results at different temperature

3.1. The pressure distribution of internal flow field under different temperature
Figure 1 show the pressure distribution on the impeller at different temperature (T= 25 ℃, 70 ℃ and 100 ℃) and different cavitation number (σ=1.0 and 0.5). Low pressure area is mainly located at the leading edge on the suction surface of the impeller, and as the increase of the temperature, the pressure increase which means the temperature can delay the occurrence of cavitation.

3.2. Vapor volume fraction at temperature T=25 ℃ of turbine flowmeter
According to the research results of Okita and Kajishima [10], the vapor volume fraction =10% can be regarded the gas-liquid. The vapor volume fraction$_{10}$ may reflect the bubble position and volume change. Figure 2 shows the bubble volume and position at different cavitation number when the temperature at 25℃. When σ=2.15, the cavitation begin to appear gradually, at this time, the cavitation occurs mainly in the suction side of impeller. It can be seen with the decrease of the cavitation number,

the number and volume of the bubble increase. Further reduce the cavitation number, when σ=1.2, bubble volume increase, cavitation began to appear on the front end of the front conditioner. When the cavitation number σ=0.8, the bubble volume is larger, and cavitation can be occur in the trailing edge of the impeller.

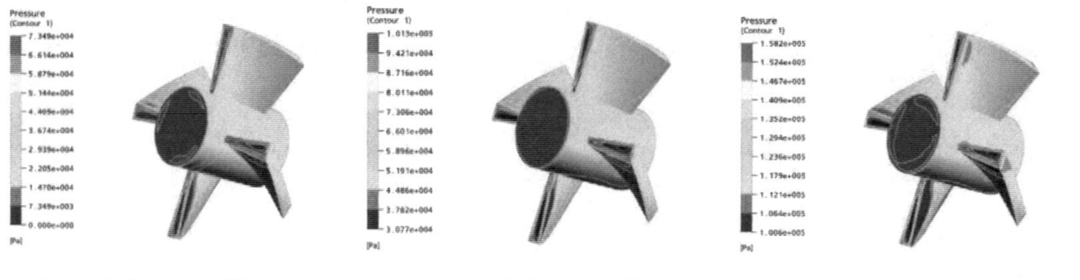

(a) σ=0.5, T=25 ℃ (b) σ=0.5, T=70 ℃ (c) σ=0.5, T=100 ℃

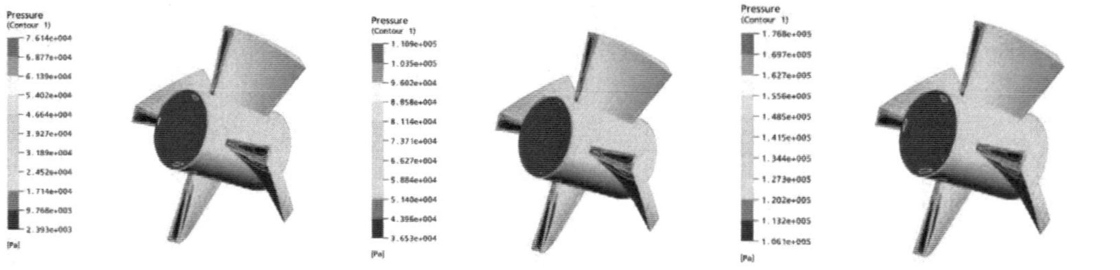

(d) σ=1.0, T= 5 ℃ (e) σ=1.0, T=70 ℃ (f) σ=1.0, T=100 ℃

Figure 1. Pressure distribution on the impeller at different temperature and cavitation number.

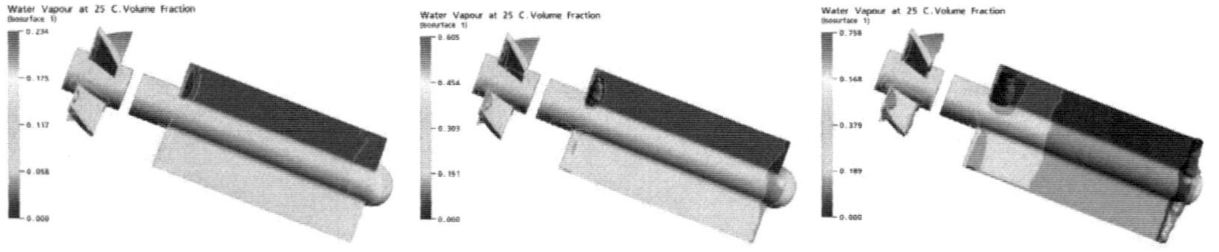

σ=2.15 σ=1.2 σ=0.5

Figure 2. Vapor volume fraction change as the decrease of cavitation number at temperature 25℃.

3.3. Temperature effects on the bubble volume fraction of turbine flowmeter

Analysis the vapor volume change at different temperature by 10% vapor volume fraction (figure 3). At the cavitation number σ=1.2, the cavitation volume decrease, as the temperature increased from 25℃ to 100℃.and cavitation appeared in the front-end and back-end of the conditioner and the suction side of the impeller. Also we can find the change rule when the cavitation number σ=0.5.so the high temperature delay the onset of the cavitation inside the turbine flowmeter.

3.4. Effect on differential pressure and torque of the cavitation number in turbine flowmeter

As we know, the measuring accuracy mainly affected through cavitation effect on the impeller torque. Figure 4 and figure 5 give the pressure and torque effect respectively with the change of cavitation number and temperature. Pressure difference and the impeller torque rise slightly with the decrease of cavitation number, which showed that under the condition of the flow rate is constant, if reduce the back pressure of the turbine flow meter, the differential pressure will rise, and the combined torque of the drive torque and the viscous drag torque will increase, which cause measurement errors. When cavitation occurs in the turbine meter, the suction side of the impeller became the cavitation area , cause the viscous friction drag torque decreased.

For the turbine flow meter is not used to doing work of rotary fluid machinery, the design requirements to minimize disturbance to the flow, so even in smaller cavitation number (σ=0.5), the presence of cavitation area on blade suction surface is lesser, also the number of blade is little, so the cavity do not blocking the flow channel generally, and falling out as the fluid flow downstream.

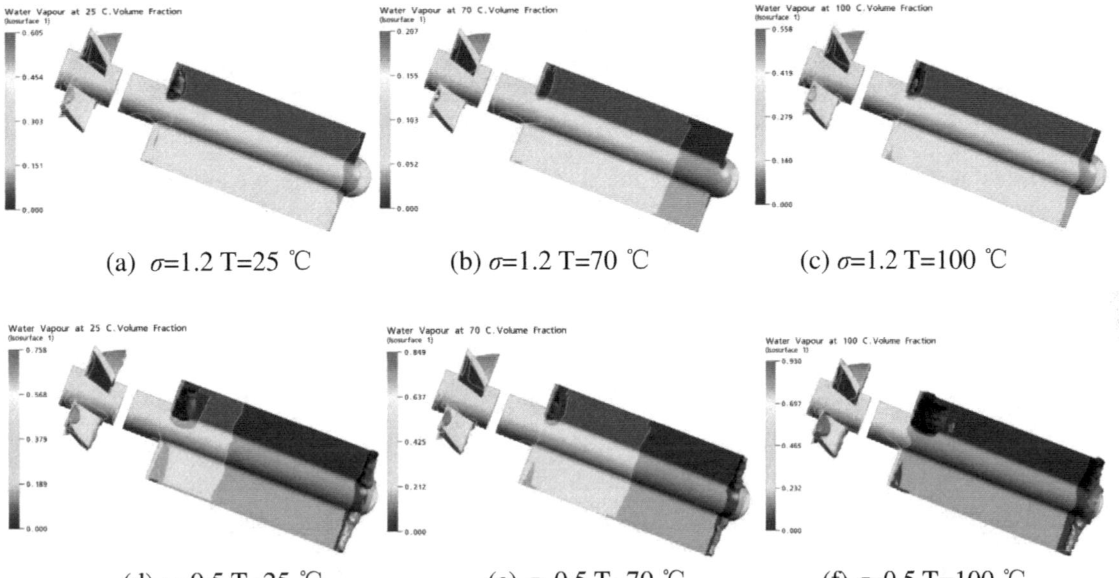

(a) σ=1.2 T=25 ℃ (b) σ=1.2 T=70 ℃ (c) σ=1.2 T=100 ℃

(d) σ=0.5 T=25 ℃ (e) σ=0.5 T=70 ℃ (f) σ=0.5 T=100 ℃

Figure 3. Temperature effects on the bubble volume fraction of turbine flowmeter.

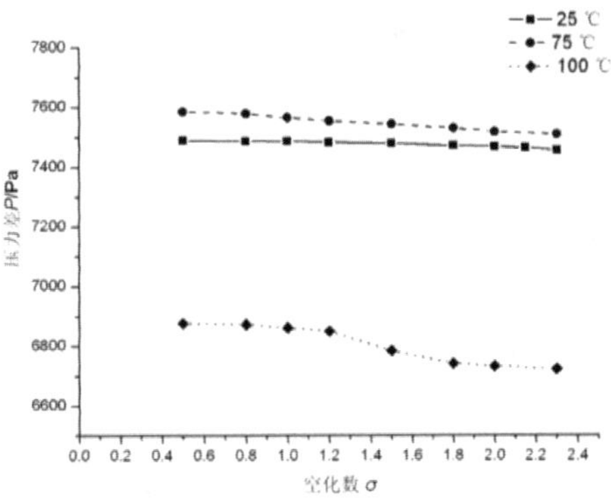

Figure 4. Pressure with cavitation number.

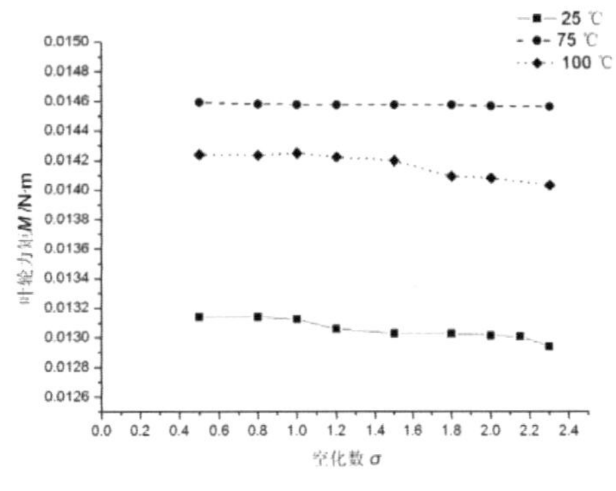

Figure 5. Impeller torque with cavitation number.

4. Conclusion

A cavitation thermodynamics turbulent model based on phase change, which is suitable for prediction of cavitation with thermal effects is developed, Building the developed cavitation model into the commercial code by using user subroutine; and the cavitation characteristic of turbine flow meter at different temperatures are analysis using the cavitation model. The cavitation characteristic at different temperature and different cavitation number is investigated and analyzed.

1) The cavitation turbulent model considering the temperature effect has been proved to be more reasonable in simulating the inner cavitation flow filed of the turbine flow meter.

2) The initial cavitation of turbine flow meter generally occur in the blade suction side, with the development of cavitation, the cavitation zone will appear on the front and the back end of the conditioner. In order to avoid the gather of cavitation, the design of optimizing the blade structure should be adapted, and at the same time, the back pressure should be limited on the installation requirements. Expanding the measurement range and preventing cavitation occurs are the goal of the design and installation.

3) The temperature effects on the cavitation of turbine flow meter is quite obvious, the increase of the temperature will delay the occurrence of cavitation. At T=25 ℃, the cavitation will occur on the suction side of the impeller at σ=2.15. But for T=70 ℃ and 100 ℃, the cavitation will occur at σ=1.5.

4) Pressure difference and the impeller torque will change obviously with the decrease of the cavitation number, which will cause the measurement error of the turbine meter.

Reference

[1] Baker R C 1991 *Flow measurement and Instrumentation* **2** 147-61

[2] Rivetti A, Martini G, Goria R and Lorefice S 1987 *Cryogenics* **27** 8-11

[3] Furness R A and Hutton S P 1975 *J. Fluid Eng.* **97** 515-22

[4] Kubota A, Kato H and Yamaguchi H 1992 *J. Fluid Mech.* **240** 59-96

[5] Singhal A K, Vaidya N and Leonard A D 1997 Multi-dimensional simulation of cavitating flows using a PDF model for phase change *ASME FEDSM* 97-3272

[6] Merkle C L, Feng J and Buelow P E O 1998 Computational modeling of the dynamics of sheet cavitation *3rd Int. Symp. Cavitation (Grenoble, France, 7-10 April 1998)* **2** 47-54

[7] Kunz R F, Boger D A, Stinebring D R et al. 2000 *Computers & Fluids* **29** 8 849-75

[8] Yogen U, Wu J Y, Wang G Y and Shyy W 2005 *Progress in Aerospace Sciences* **41** 558-608

[9] Saito Y, Takami R, Nakamori I and Ikohagi T 2007 *Comput. Mech.* **40** 85-96

[10] Okita K and Kajishima T 2002 Three-Dimensional Computation of Unsteady Cavitating Flow in a Cascade *Proc. 9th Int. Symp. on Transport Phenomena and Dynamics of Rotating Machinery* (Honolulu, USA, 10-14 Feb 2002)

Influence of blade angle distribution along leading edge on cavitation performance of a centrifugal pump

Y Xu[1], L Tan[1], S L Cao[1,3], Y C Wang[1], G Meng[1, 2] and W S Qu[1]
[1] State Key Laboratory of Hydroscience and Engineering, Tsinghua University, Beijing 100084, China;
[2] Inner Mongolia Vocational College of Chemical Engineering, Huhehaote 010070, Inner Mongolia, China

E-mail: caoshl@mail.tsinghua.edu.cn

Abstract. The influence of blade angle distribution along leading edge on cavitation performance of centrifugal pumps is analysed in the present paper. Three sets of blade angle distribution along leading edge for three blade inlet angles are chosen to design nine centrifugal pump impellers. The RNG k-ε turbulence model and the Zwart-Gerber-Belamri cavitation model are employed to simulate the cavitation flows in centrifugal pumps with different impellers and the same volute. The numerical results are compared with the experimental data, and the comparison proves that the numerical simulation can accurately predict the cavitation performance of centrifugal pumps. On the basis of the numerical simulations, the pump head variations with pump inlet pressure, and the flow details in centrifugal pump are revealed to demonstrate the influence of blade angle distribution along leading edge on cavitation performances of centrifugal pumps.

1. Introduction

Cavitation is a universal hydrodynamic phenomenon which can lead to the performance degradation and damage to the hydraulic machinery. It happens when the pressure of the liquid is less than the vaporization pressure. In centrifugal pumps, cavitation phenomenon leads to the decrease of the efficiency and impacts the stable operation. Furthermore, the cavity collapse may cause the destruction of the blades and the pump bodies. Due to the strong influence of the cavitation phenomenon on the operation of centrifugal pumps, the investigation of the cavitation flow and its relevant factors was carried out by numerical simulation.

Many studies have been developed to investigate the cavitation flow field in centrifugal pumps. Rudolf et al [1] observed the cavitation at the tongue of the volute and figured out that the 3% head drop may result from the cavitation in the tongue. Tan et al [2] introduced the compressibility into the RNG k-ε model and full cavitation model in order to simulate the cavity accurately at low flow rate. Li et al [3] calculated two pumps with and without the inducer to describe the influence on the leading edge cavitation.

The mechanisms for cavitation have been widely studied, and the relevant parameters of cavitation flow have been investigated. Wang et al [4] figured that the blade incidence angle had a significant impact on the cavitation performance, and there was an optimum value of the angle for cavitation

[3] Corresponding author.

characteristic. By using *k-ε* turbulence model and the volume of fluid cavitation model, Luo et al [5] found that the inlet geometry of impellers had great influence on the cavitation performance.

Previous studies generally indicate that the inlet geometrical parameters of impellers, such as inlet blade angle and inlet edge, have a great influence on the cavitation performance of the pumps. But the effect of the blade angle distribution along leading edge still remains unclear. The main aim of the study is to achieve the influence of blade angle distribution, and to optimize a certain centrifugal pump by changing the distribution rule.

2. Case description and numerical algorithm

2.1. Parameter of the original centrifugal pump

The centrifugal pump is made up of a suction pipe, an impeller with 7 blades and a volute. The main parameters are as follow: flow rate Q_d=25 m^3/h, pump head H=7 m and rotation speed n=1450 rpm. The wrap angle of the pump is 98° and the diameter of the outlet is 160 mm. The distribution of inlet blade angle along leading edge is shown in Fig.1.

2.2. Distribution ways of inlet blade angle

In order to investigate the influence of the blade angle distribution on pump cavitation performance, nine impellers with the same volute and suction pipe are modelled by changing the blade inlet angle and its distribution way along the blade leading edge. The blade distribution ways of the nine impellers are shown in Table 1. In order to arrange the calculation, the nine pumps are nominated according to the distribution ways of the inlet blade angle.

Table 1. Blade angle distribution

Description	β_1(hub)/°	β_1(shroud)/°	β_2/°
S1	32	46	22
S2	39	39	22
S3	46	32	22
M1(original pump)	42	56	22
M2	49	49	22
M3	56	42	22
L1	52	66	22
L2	59	59	22
L3	66	52	22

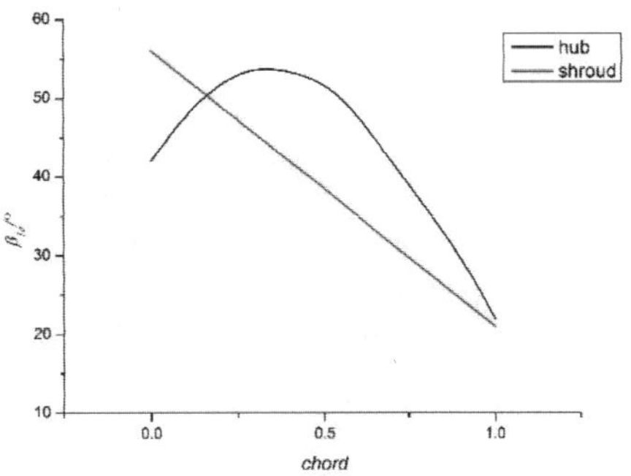

Figure 1. Distribution of blade angle along blade leading edge of the original pump.

2.3. Computational Methodology

The numerical analysis is carried out by the CFD software ANSYS-CFX. The RNG k-ε turbulence model with scalable wall functions is employed to calculate the turbulent flow. The cavitation model is Zwart-Gerber-Belamri model based on Rayleigh-Plesset Equation, which is widely used in predicting the inception, growth and collapse of the bubbles. The convergence condition is set to 1.0×10^{-5} of the RMS residual. For the non-cavitation regime, the boundary conditions are assumed: at the inlet of the domain the pressure is specified, and at the outlet, the mass flow rate is defined. Then for the cavitation condition, the inlet pressure decreases.

3. Experimental and numerical results of the centrifugal pumps

3.1. Experimental results and numerical simulation of the original pump

The experiment is conducted on an open-architectured platform which meets the national standard on centrifugal pump experiment. The electromagnetic flowmeter is used to measure the flow rate. The head of the pump is determined by the total pressure at the inlet and outlet according to the function (1) where p_1 and p_2 is the pressure of inlet and outlet of the pump; ρ is the fluid density; Z is the height of the inlet and outlet; v_1 and v_2 is the velocity of the fluid in the inlet and outlet. The efficiency is defined as function (2) where the H_t is the theoretical head value of the pump.

$$H = \frac{p_2 - p_1}{\rho g} + (Z_2 - Z_1) + \frac{v_2^2 - v_1^2}{2g} \tag{1}$$

$$\eta = \frac{H}{H_t} \times 100\% \tag{2}$$

While testing the head and efficiency, the cavitation performance of the original pump is tested by reducing the inlet pressure.

In order to insure the accuracy of numerical prediction, the performance of the original pump is calculated and compared with the experiment data. The inlet and outlet domains are extended to reduce the effect of boundary conditions. The ANSYS-TURBOGRID is used to generate the hexagonal structured grids of impeller and the hexagonal grids of volute and suction pipe are generated by ICEM. Then the grid dependence analysis is carried out, the details and the results of the four grids are in the Table 2. The three finest grids present the results that are in good agreement with the data of experiment. Grid 2 is selected for this study, and the topology is used to generate all the grids of calculations. Figure 2 shows the mesh of the impeller.

Table 2. Grid independence analysis

Grid No.	**Total grid size**	**Efficiency (%)**	**Head(m)**
Grid 1	944962	87.21	7.15
Grid 2	1110652	88.41	7.17
Grid 3	1169011	88.53	7.17
Grid 4	1246837	88.49	7.17

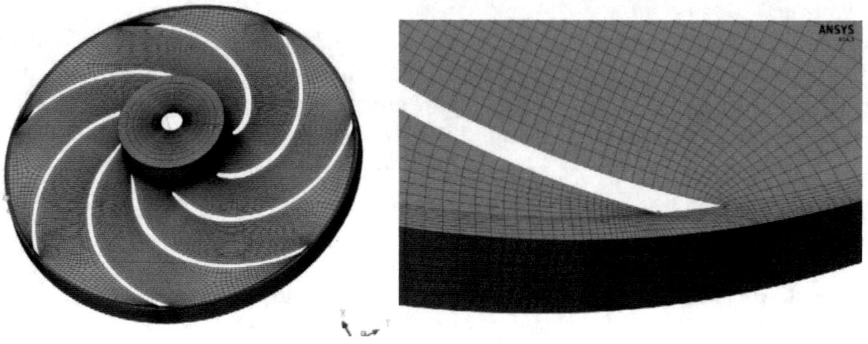

Figure 2. Mesh of impeller

The hydraulic performance and cavitation performance are calculated. The results of hydraulic performance and the comparison with experimental data is shown in Figure 3. The cavitation curves are displayed in Figure 4, and the values of $NPSH_c$ of different flow rate are compared with the data of experiment in Figure 5. At each flow rate, the numerical simulation presented a well prediction. The results indicated that the numerical simulation agreed with the experimental results, so the results are reliable.

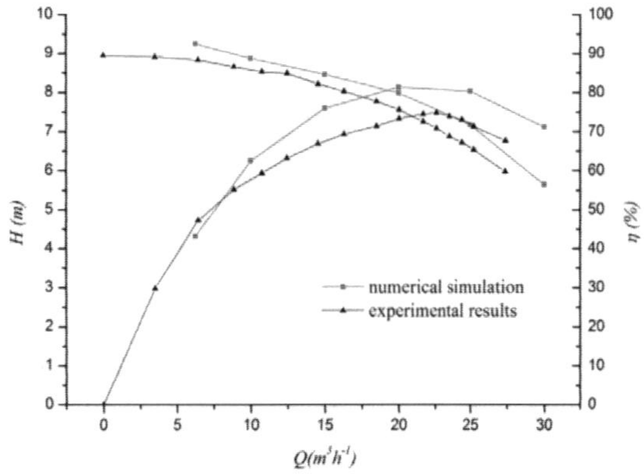

Figure 3. Comparison of hydraulic performance.

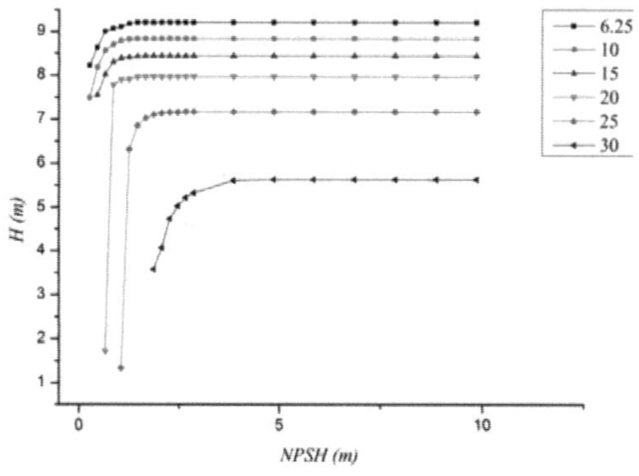

Figure 4. Cavitation curves for different flow rates.

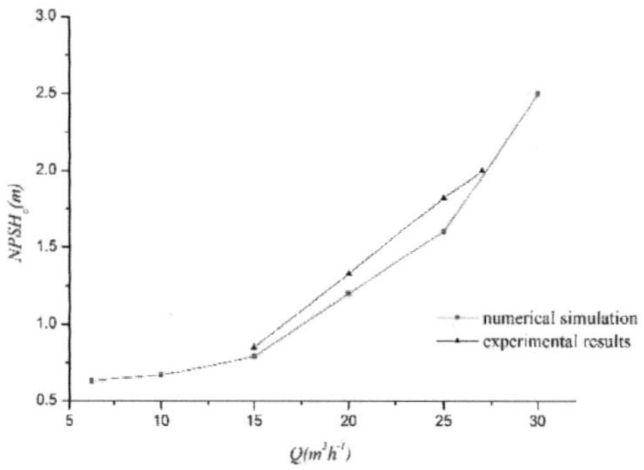

Figure 5. Comparison of cavitation performance

4. Cavitation performance of centrifugal pumps with different blade distributions

For modelling the nine impellers according to the distribution of blade inlet angle, Bladegen is employed to generate the three-dimension model of the impellers.

In order to calculate the cavitation performance of the pumps, the performance under non-cavitation regime are calculated first. Taking the results under non-cavitation condition as the initial conditions, the cavitation performances are calculated. As reducing the value of inlet total pressure, the value of $NPSH_a$ decreased and the cavitation curves are obtained.

The value of $NPSH_c$ is chosen as the evaluation criterion of the performance of cavitation. The values are obtained from the cavitation curves and are shown in Table 3 and the comparison of the nine pumps is shown in Figure 6.

It is obvious that the average value of the inlet blade angle along leading edge has significant influence on the cavitation performance of the pump.The best value is 49°, and its average value of $NPSH_c$ is 1.586m which is lower than the value of 39° for 0.05m and the value of 59° for 0.064m.

The results of three average values of inlet angle show the same conclusion that the evenly distribution way is the best distribution for cavitation performance. And the distribution ways of mounting and tapered show performance similarly. So it is helpful for optimizing cavitation performance to design pumps with evenly distribution of inlet blade angle along the leading edge.

Table 3. Values of NPSHc of the pumps

Pump No.	**$NPSH_c$ (m)**
S1	1.643
S2	1.620
S3	1.644
M1 (original pump)	1.604
M2	1.575
M3	1.580
L1	1.655
L2	1.640
L3	1.655

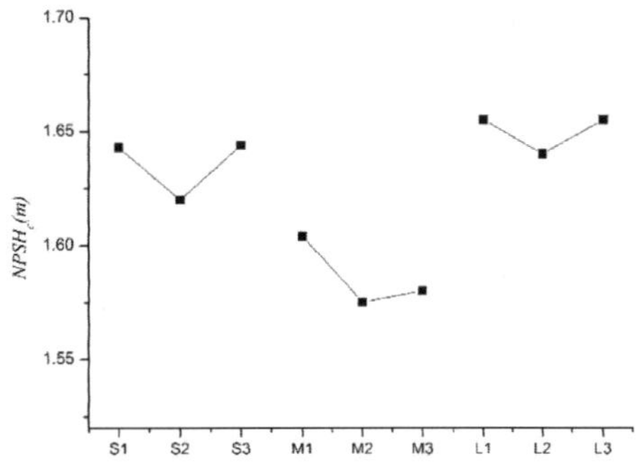

Figure 6. $NPSH_c$ of the pumps

5. Cavitation performance of the best pump

5.1. Cavitation process of the best performance pump

The analysis results indicate that the pump M2 has the best cavitation performance. In order to analyse the inception, development and collapse of cavitation, the inner flow fields of pump M2 under

IOP Conf. Series: Materials Science and Engineering **72** (2015) 032019 doi:10.1088/1757-899X/72/3/032019

different $NPSH_a$ value are investigated. It is found that the cavitation occurs in the inlet of suction side of the blade first. As decreasing the value of total pressure at inlet, the range of cavitation expands to the middle of the suction pipe, and finally occupies the passage, which reduces the efficiency and head of the pump. When the value of NPSH reduces to 1.07m, part of the bubble separate from the primary flow and run into the volute. The whole process of cavitation is shown in Figure 7.

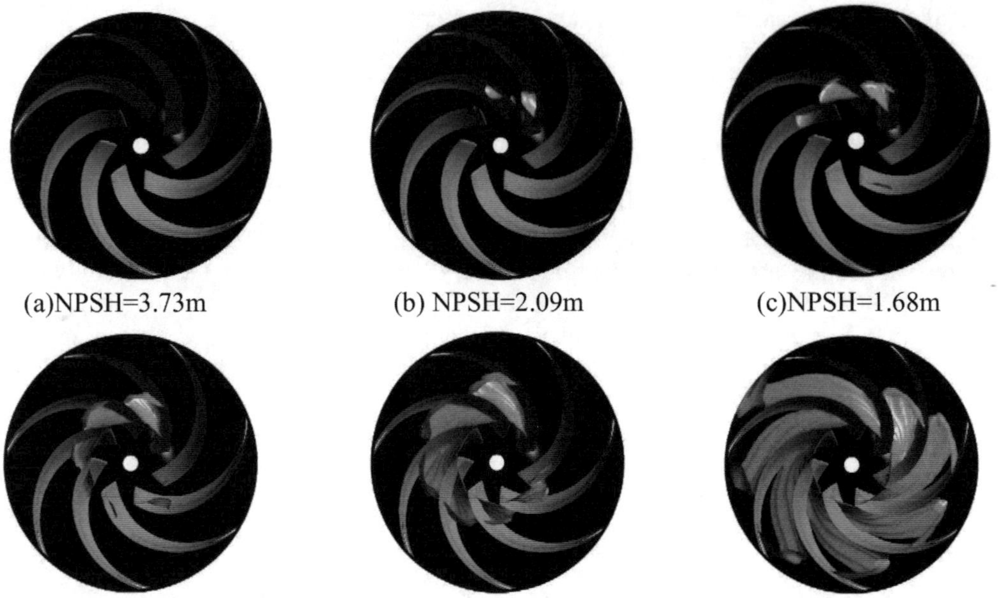

(a)NPSH=3.73m (b) NPSH=2.09m (c)NPSH=1.68m

(d)NPSH=1.48m (e)NPSH=1.27m (f)NPSH=1.07m

Figure 7. Cavitation process of the best performance pump M2

5.2. Blade load

Figure 8 shows the static pressure distributions along the blade under different value of $NPSH_a$. At decreasing the value of $NPSH_a$, the net pressure got smaller. When the value reach to 1.27m, the trend of the pressure distribution changes, and the suction effect near the leading edge reduces. The influence is significant in suction side, and the pressure mainly reduces near the inlet. When the value further reduces to 1.07m, the value of pressure decreases in both two sides near the leading edge. The distribution of the pressure coincides with the cavitation process.

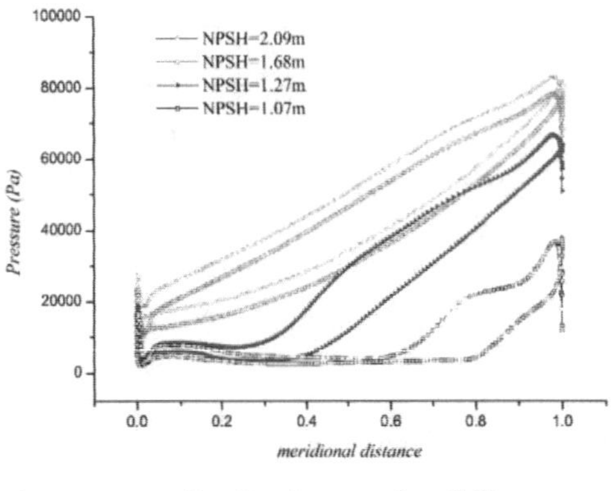

Figure 8. Static pressure distributions under different cavitation condition.

6. Conclusions

The RNG k-ε turbulence model and Zwart-Gerber-Belamri cavitation model are employed to predict the cavitation performance of nine pumps with different distributions of blade inlet angle. The values of $NPSH_c$ of the nine pumps are obtained by composing the cavitation curves. The inner flow and blade loading under different cavitation condition of the best performance pump are also analyzed. The main conclusion are as follows:

1) The best distribution way is evenly way and the best average value of the inlet blade angle is 49°. The other two ways perform similarly.

2) The cavitation first occurs near the leading edge of suction side at impeller hub, then expand to the whole passage. It is also notable that the range of cavitation moved from hub to shroud while decreasing the value of $NPSH_a$. The inlet geometry has decisive influence on the cavitation performance. To ensure the flow uniformity of the inlet by selecting proper inlet parameters is an effective way for reducing the cavitation in pumps.

References

[1] Bachert R, Stoffel B and Dular M 2010 *J. of Fluids Eng.* **132**(6) 061301.
[2] Tan L, Zhu B, Cao S and Wang Y 2013 *Chinese Sci. Bulletin* **58**(8) 949-52.
[3] Li X J, Pan Z Y, Zhang D Q and Yuan S Q 2012 *Pro. of the 26th IAHR Sym. on Hydraulic Machinery and Systems(Beijing, China, 19-23, August, 2012)*
[4] Wang Y, Liu H L, Yuan S Q, Tan M G and Wang K 2011 *Fluid Machinery* **4** 005
[5] Luo X W, Zhang Y, Peng J Q and Xu H Y 2008 *J. Tsinghua Univ. (Sci. & Techno.)* **05**

IOP Conf. Series: Materials Science and Engineering **72** (2015) 032020 doi:10.1088/1757-899X/72/3/032020

Numerical simulation and analysis of cavitation flows in a double suction centrifugal pump

G Meng[1, 2], L Tan[1], S L Cao[1, 4], W Jian[3], W W Liu[3] and D J Jiang[3]
[1] State Key Laboratory of Hydroscience and Engineering, Tsinghua University, Beijing 100084, China
[2] Inner Mongolia Vocational College of Chemical Engineering, Huhehaote 010070, Inner Mongolia, China
[3] Shanghai Eastpump (GROUP) CO., LTD, Shanghai 201906, China

E-mail: caoshl@mail.tsinghua.edu.cn

Abstract. Cavitation is an unsteady phenomenon, which is nearly inevitable in pumps. It would degrade the pump performance, generate vibrations and noises, and even erode pump flow passage components. The double suction centrifugal pump at design flow rate and large flow rate is numerically simulated using the *k-ω* turbulence model and the mass transport cavitation model. As a result, the calculated variation of pump head with pump inlet pressure agreed well with the experimental data. The results demonstrate that the numerical model and method can accurately predict the cavitation flows in a double suction centrifugal pump. The cavitation characteristics are analysed in great details. In addition, based on the calculation results, the reason that the plunge of pump head curve is revealed. It is found that the steep fall of pump head happens when the cavity reaches the blade to blade throat and the micro-vortex group appears at the back of the blade suction side. At the same time, this practice can provide guidance for the optimal design of double suction pumps.

1. Introduction

For pumps, cavitation is a very important physical phenomenon. Cavitation bubbles emerge in pump operation, where the local pressure drops below the liquid vapor pressure [1]. Cavitation has great influence on the operation performance of pumping station unit, so that many problems have occurred, such as performance degradation, vibrations, noises, pump flow passage components erosion and so on. With the rapid development of the computer technology and computational fluid dynamics (CFD), numerical simulation has become one of the most important research methods for studying cavitation flow in pumps [2-5]. Tan et al. [6] simulated the cavitation flow of centrifugal pump by using the full cavitation model and modified *RNG k–ε* turbulence model, and analysed the cavitation characteristics inside the impeller and the pressure distribution at the impeller inlet. Pouffary et al. [7] accomplished cavitation simulation on a centrifugal pump based on the cavitation model of the barotropic state law and the standard *k–ε* turbulence model. When sheet cavitation passes through the narrowest section between two adjacent blades, the pump head drops obviously and the torque decreases accordingly. At the same time, the vacuoles block the passage, which results in an uneven load inside the channel of the impeller. Chang et al. [8] adopted the *SST k–ε* turbulence model and homogeneous mixture

[4] Corresponding author.

IOP Conf. Series: Materials Science and Engineering **72** (2015) 032020 doi:10.1088/1757-899X/72/3/032020

cavitation model to simulate the cavitation flow in a mixed-flow pump. The results show that the cavitation firstly appears at suction side of blade near the leading edge and then extend to the outlet of impeller with the further reduction of net positive suction head. The serious blockage due to cavity in the flow passages results in a sharp reduction in pump head.

In the present research, the cavitation flows in a double suction centrifugal pump was predicted by using the k-ω turbulence model and the mass transport cavitation model. The calculated results are compared with the experimental data for different operating points. From these results, the cavitation characteristics and the origin of the cavitation head drop are discussed. Meanwhile, the analysis of their mechanism is proposed.

2. Computational domain and mesh

The research object is an ES8-300KPS double suction centrifugal pump as shown in Figure 1.

Figure 1. Double suction centrifugal pump.

The main parameters of the double suction centrifugal pump are given as follows: the volume flow rate Q_d=820 m^3/h, the rotational speed n=1480 r/min, the head H=64 m, the number of blades Z=6. The computational domain consists of suction casing, impeller, volute, prolongations for suction casing inlet and volute outlet as shown in Figure 2. The suction casing, impeller and volute use the tetrahedral mesh, and the extension of inlet and outlet use the hexahedral mesh. The grid independence has been verified, the coarsest mesh with 4,198,349 elements is employed for the following calculations. The mesh elements for each part are shown in table 1.

(a) Suction casing and volute

(b) Impeller

Figure 2. Computation domain of the double suction centrifugal pump.

Table 1. Computation domain and mesh of the double suction centrifugal pump.

Computation domain	Suction casing	Impeller	Volute
Mesh	2 003 664	1 084 701	1 109 984

3. Mathematical model and numerical algorithm

The fluid in the cavitation flow field for the pumps is considered as a homogeneous and compressible mixed medium of vapor and liquid. The continuity and momentum equations in the Cartesian coordinates are as follows:

$$\frac{\partial(\rho_m u_i)}{\partial x_i} = 0 \tag{1}$$

$$\frac{\partial(\rho_m u_i u_j)}{\partial x_j} = -\frac{\partial p}{\partial x_i} + \frac{\partial}{\partial x_j}\left[(\mu_m + \mu_t)\left(\frac{\partial u_i}{\partial x_j} + \frac{\partial u_j}{\partial x_i} - \frac{2}{3}\frac{\partial u_i}{\partial x_i}\delta_{ij}\right)\right] \tag{2}$$

where ρ_m and μ_m are the mixture of density and dynamic viscosity, calculated by weighted average of each phase volume fraction, u is the velocity, p is the pressure, and μ_t is the turbulent viscosity, respectively. Subscripts i, j, denote the axes directions. In the present study, the k-ω turbulence model is selected.

The mass transport cavitation model was considered in this paper, which is deduced from the Rayleigh-Plesset equation:

$$\frac{\partial(\rho_\upsilon \alpha_\upsilon u_i)}{\partial x_i} = m^+ - m^- \tag{3}$$

$$m^+ = C_{vap}\frac{3r_g(1-\alpha_\upsilon)\rho_\upsilon}{R_b}\sqrt{\frac{2}{3}\frac{\max(p_\upsilon - p, 0)}{\rho_l}} \tag{4}$$

$$m^- = C_{cond}\frac{3\alpha_\upsilon\rho_\upsilon}{R_b}\sqrt{\frac{2}{3}\frac{\max(p - p_\upsilon, 0)}{\rho_l}} \tag{5}$$

where α_v is the vapor volume fraction, m^+ and m^- represent the source terms for evaporation and condensation. C_{vap} and C_{cond} are the empirical calibration coefficients of evaporation and condensation, respectively. And r_g is the nucleation site volume fraction, R_b stands for the nucleation site radius. ρ_v is the vapor density. ρ_l is the liquid density. p_v is the water vaporization pressure. The recommended values of these coefficients are: C_{vap}=50, C_{cond}=0.01, r_g=5×10-4, R_b=10-6 m, ρ_v=0.554 kg/m^3, ρ_l=997 kg/m^3 and p_v=3574 Pa.

All the simulations were conducted by using the ANSYS-CFX commercial software. The boundary conditions are as follows: the total pressure at the inlet and the mass flow rate at the outlet based on the experimental measurements, the no-slip condition for the boundary layers imposed over the impeller blades and sidewalls, the volute casing and the inlet and outlet pipe walls. The liquid flow fields in the double suction centrifugal pump are solved first. Coupled with the cavitation model, the cavitation flow fields in the double suction centrifugal pump are solved by taking the liquid flow fields result as the initial condition. The pressure at the inlet is gradually reduced when the calculation is converged at a given pressure value.

4. Results and analysis

4.1. Pump performance curve and cavitation characteristics curve

Figure 3 shows the comparison of pump head and efficiency between the experimental and numerical results. There is a good quantitative agreement at the flow rate of 480–980 m^3/h, with the prediction errors smaller than 5.0%, which demonstrates that the numerical model and method can accurately predict the pump performance.

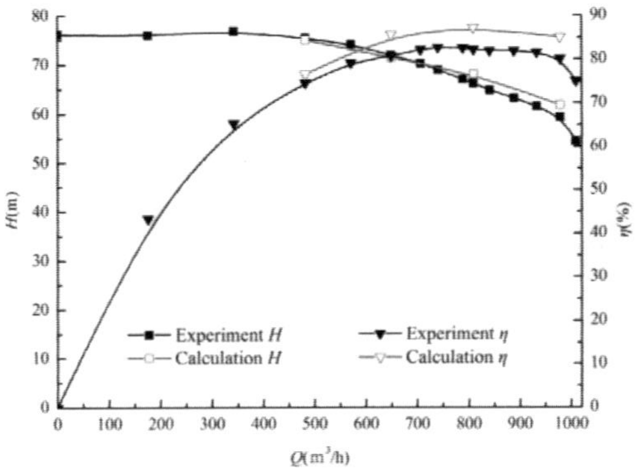

Figure 3. Comparison of pump characteristics between calculation and experiment.

The net positive suction head available (*NPSHa*) is the difference between the fluid energy of unit weight in the entrance section of the computation domain and the energy of vaporization pressure. It is defined as

$$NPSHa = \frac{p_{in}}{\rho_l g} + \frac{u_{in}^2}{2g} - \frac{p_\upsilon}{\rho_l g} \tag{6}$$

Figure 4(a) and 4(b) shows the experimental and numerical results of the head varying with *NPSHa* at $1.0Q_d$ and $1.2Q_d$ respectively. The pump head have large fluctuation at $1.2Q_d$ condition, but the trend of numerical results still coincides with the experimental data.

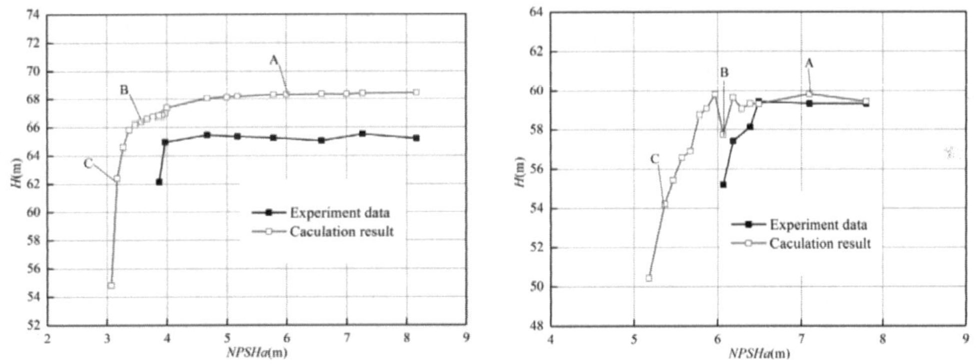

(a) $1.0Q_d$ (b) $1.2Q_d$

Figure 4. Comparison of cavitation characteristics between calculation and experiment.

4.2. The cavity development in the pump

Figure 5 and Figure 6 shows the cavity development in impeller with different *NPSHa*, corresponding to the letters A, B and C as shown in Fig.4(a) and 4(b), respectively. The cavity inception and development in the pump are clearly obtained. The cavity first appears on the suction side near the blade leading edge, as shown in Fig.5(a), at which condition there is no significant drop of the pump head. With the decrease of *NPSHa*, a large area of cavity emerges in the impeller channels and expends to the pressure side. Hence, the pump normal flow will be affected, and the pump head begins to drop, as shown in Fig.5(b). In Figure 5(c), when *NPSHa*=3.17 m, the impeller channels are seriously blocked by the cavities, which results in the pump head breaking down. On the other hand, in terms of the cavity inception and development, same tendency can be seen in large flow rate, as shown in Fig.6.

IOP Conf. Series: Materials Science and Engineering **72** (2015) 032020 doi:10.1088/1757-899X/72/3/032020

(a) *NPSHa*=6.00 m (b) *NPSHa*=3.57 m (c) *NPSHa*=3.17 m

Figure 5. Vapor volume fraction in the impeller at $1.0Q_d$.

(a) *NPSHa*=7.10 m (b) *NPSHa*=6.07 m (c) *NPSHa*=5.37 m

Figure 6. Vapor volume fraction in the impeller at $1.2Q_d$.

A section located at z=0.08 m is selected to analyse the pressure distribution of the impeller inlet. Figure 7 and Figure 8 present the distributions of pressure at the impeller inlet. With the reduction of *NPSHa*, the pressure at the impeller inlet decreases. As the pressure gradient becomes smaller, the pressure distribution becomes more even.

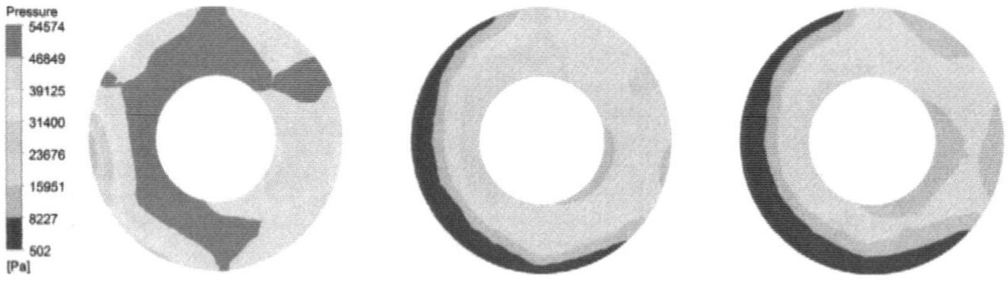

(a) *NPSHa*=6.00 m (b) *NPSHa*=3.57 m (c) *NPSHa*=3.17 m

Figure 7. Distributions of pressure at impeller inlet for $1.0Q_d$.

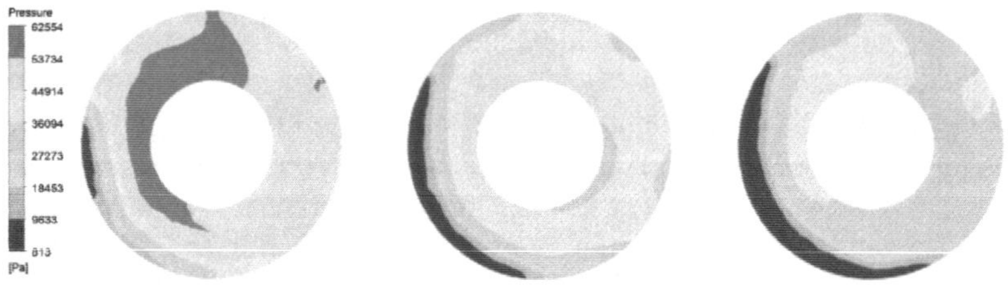

(a) *NPSHa*=7.10 m (b) *NPSHa*=6.07 m (c) *NPSHa*=5.37 m

Figure 8. Distributions of pressure at impeller inlet for $1.2Q_d$.

4.3. Internal flow characteristics analysis

Figure 9 shows the streamlines on middle span section of impeller with different *NPSHa* at $1.0Q_d$. The streamline distribution is relatively uniform, as shown in Figure 9(a). With the decrease in *NPSHa*, the streamlines disorder at the blade suction side near the trailing edge, as shown in Figure 9(b). In Figure 9(c), the streamlines at the back of the blade suction side is disorder and generate the vortex micro group. At the same time, the velocity is uneven at impeller outlet, with the decrease in *NPSHa*, the low velocity area expands gradually, as shown in Figure 10, which induces the instability of pump operation.

(a) *NPSHa*=6.00 m (b) *NPSHa*=3.57 m (c) *NPSHa*=3.17 m

Figure 9. Streamline on middle span section of impeller at $1.0Q_d$.

(a) *NPSHa*=6.00 m (b) *NPSHa*=3.57 m (c) *NPSHa*=3.17 m

Figure 10. Velocity distribution at outlet of impeller for $1.0Q_d$.

5. Conclusions

The cavitation flow in a double suction centrifugal pump is calculated by using the k-ω turbulence model and the mass transport cavitation model. Based on the results, concluding remarks as follows were obtained.

1) The calculated results of pump head varying with net positive suction head available are consistent with the measurements, which demonstrates that the numerical model and method can accurately predict the cavitation flows in a double suction centrifugal pump.

2) For the design flow rate and large flow rate, the cavity inception and development in the pump is similar.

3) The steep fall of pump head is caused when the cavity reaches the blade to blade throat and the vortex micro group appears at the back of the blade suction side.

References

[1] Ding H, Visser F C, Jiang Y and Furmanczyk M 2011 *J. Fluid Eng.* **133**(1) 011101

[2] Tan L, Cao S L, Wang Y M and Zhu B S *Transactions of the Chinese Society of Agricultural Machinery* 2012 **43**(9) 49-52 (in Chinese)

[3] Ding H, Visser F C and Jiang Y 2012 A practical approach to speed up NPSHr prediction of centrifugal pumps using CFD cavitation model *ASME 2012 Fluids Engineering Division Summer Meeting collocated with the ASME 2012 Heat Transfer Summer Conference and the ASME 2012 10th International Conference on Nanochannels, Microchannels, and*

Minichannels (Rio Grande, Puerto Rico, 8-12 July 2012)
[4] Chen T, Li S R, Li W Z, Liu Y L, Wu D Z and Wang L Q 2013 *IOP Conference Series: Materials Science and Engineering* **52**(6): 062020
[5] Tan L, Zhu B S, Cao S L and Wang B 2014 *Energies* **7**(2): 1050-65
[6] Tan L, Cao S L, Wang Y M and Zhu B S 2012 *Chinese Phys. Lett.* **29**(1) 14702-4
[7] Pouffary B, Patella R F, Reboud J L and Lambert P A 2008 *J. Fluid Eng.* **130**(6) 061301
[8] Chang S P and Wang Y S 2012 *J. of Drainage and Irrigation Machinery Eng.* **30**(2): 171-5 (in Chinese)

IOP Conf. Series: Materials Science and Engineering **72** (2015) 032021 doi:10.1088/1757-899X/72/3/032021

Numerical studies in a centrifugal pump with the improved blade considering cavitation

P F Song, Y X Zhang, C Xu, X Zhou and J Y Zhang

College of Mechanical and Transportation Engineering, China University of Petroleum, Beijing 102249, China

E-mail: spf0322@126.com

Abstract. In this paper, a centrifugal pump with the improved blade for cavitation is studied numerically. A 3D impeller with logarithmic spiral blade profile was designed by the in-house hydraulic design code using a centrifugal pump geometric parameters, and the blade profile curve of suction side on the designed impeller is replaced by a combination of tangent line and circle arc line. The cavitation flows in the centrifugal pump with designed impeller, modified impeller and centrifugal pump spectrum impeller are respectively calculated by two-phase CFD simulation at three flow rates. The tests of the centrifugal pump have been conducted to verify numerical simulation. The effects of designed impeller and modified impeller on hydraulic efficiency, critical cavitation number, cavitation length, head drop performance and vapor cavity distribution in impeller are investigated. The results show that the modification of blade profile curve of suction side can improve the cavitation performance of an impeller and hydraulic efficiency of centrifugal pump. Compared with designed impeller, the critical cavitation number of centrifugal pump with modified impeller decrease by 26.5% under the same flow rate coefficient, and the cavitation intensity in the modified impeller is weakened effectively. The hydraulic efficiency of modified impeller also increases by 4.9%.

1. Introduction

Cavitation in impeller induced instability, pressure pulsations, erosion and pump vibration have been observed in different kinds of pumps[1-3]. Cooper and Antunes[4] have explored the causes in many researches. The primary effect on the cavitation behavior is the geometry at the impeller eye for a pump. There are some geometric factors at the impeller eye have an effect on cavitation, such as, the blade inlet angles inlet and hub diameters and incidence to upstream flow, blade passage throat area and blade profile, etc.

In recent years, the research about the effect of the above factors on pump cavitation has been studied and reported[5-8]. Eugene and Ravi[6]studies the effect of blade leading edge profiles on the cavitation behavior of an impeller. Palgrave and Cooper[7], have conducted visual studies of cavitation and present a general expression for estimating $NPSH_i$ based on inlet angle and eye diameter. Schiavello and Visser[8] have studied most of the aspects of cavitation. Herbich[9] have conducted experiment of circular arc blades and it shows that the efficiency of log spiral blade is highest. Cooper[10] and Hackworth[11] proposed that an appropriate blade profile can improve effectively cavitation performance of impeller in the patent of anti-cavitation impeller design. Katz[12] and Franc[13] have studied the relationship between cavitation and boundary layer separation of stationary airfoil, the result shows that a low pressure zone is formed behind the point of boundary layer separation which induces cavitation concomitantly. Dumitrescu[16] have studied the bubble separation on the leading edge of rotating blades, and proposed that the blade profile and low pressure zone also affect the separation od leading edge.

Above researches mainly discuss the improving cavitation performance of pump are traditional methods such as changing geometry parameters of impeller inlet and inlet angle of blades, etc, however, the effect of blade profile is neglected, especially the suction surface of blade. Actually, the blade profile affects directly on the development of low pressure zone in suction surface and the motion of cavities region.

In this paper, we study the influence of the blade profile modification in suction side on the cavitation performance of a centrifugal pump. It is that the blade profile curve of suction side on designed impeller is replaced by a combination of tangent line and circle arc line. The cavitation flow in centrifugal pump with the designed impeller, the modified impeller and IS the centrifugal pump impeller were calculated by two-phase CFD simulation with the Zwart-Gerber-Belamri cavitation model. A test rig was set up to conduct the cavitation performance tests of the centrifugal pump to verify the numerical simulation. The effect of designed impeller and modified impeller on hydraulic efficiency, critical cavitation number, cavitation length, head drop performance and vapor cavity distribution in impeller are investigated.

2. Mathematical model of numerical simulation

2.1. Continuity equation and momentum equation

Since a basic assumption homogeneous multiphase transport equation model, all phases share the same velocity, the governing mixture equations for continuity and momentum and the vapor phase volume fraction of transport equation respectively are:

$$\frac{\partial \rho_m}{\partial t} + \nabla \cdot (\rho_m \vec{u}_m) = 0 \tag{1}$$

$$\frac{\partial \rho_m \vec{u}_m}{\partial t} + \nabla \cdot (\rho_m \vec{u}_m \vec{u}_m) = -\nabla p + \nabla \cdot \left[(\mu_m + \mu_t) \times \nabla \vec{u}_m \right] + \frac{1}{3} \nabla \left[(\mu_m + \mu_t) \nabla \cdot \vec{u}_m \right] \tag{2}$$

$$\frac{\partial}{\partial t}(\rho_v \alpha_v) + \nabla \cdot (\rho_v \vec{u}_v \alpha_v) = R \tag{3}$$

Where ρ_m and ρ_v are the mixture density and vapor phase density, kg/m^3; t is the time, s; $\vec{u}_m$ and $\vec{u}_v$ are the mass average velocity and the vapor phase velocity, m/s; μ_m and μ_t are the mixture dynamic viscosity and the turbulence viscosity, kg/(m·s); *P* is the pressure of flow field, Pa.

ρ_m , $\vec{u}_m$ and μ_m are the weighted average of liquid phase and vapor phase which as follows:

$$\rho_m = \rho_v \alpha_v + \rho_l (1 - \alpha_v) \tag{4}$$

$$\vec{u}_m = \vec{u}_v \alpha_v + \vec{u}_l (1 - \alpha_v) \tag{5}$$

$$\mu = \mu_v \alpha_v + \mu_l (1 - \alpha_v) \tag{6}$$

Where $\vec{u}_v$ and $\vec{u}_l$ are the velocity of liquid phase and vapor phase, m/s; μ_v and μ_l are the dynamic viscosity of vapor phase and liquid phase, Pa·s; ρ_l is the liquid phase density, kg/m^3; α_v is the vapor phase volume fraction.

The interphase mass transfer rate *R* is:

$$R = R_e - R_c \tag{7}$$

where the source terms R_e and R_c denote vapor generation (evaporation) and condensation rates.

2.2. Cavitation bubble dynamic equation

Assuming that all the bubbles in a system have the same size, Zwart-Gerber-Belamri cavitation model[14-15] is that all bubbles in system share the same radius, the interphase mass transfer rate R is:

$$R = F\frac{3\alpha_v\rho_v}{R_B}\sqrt{\frac{2}{3}\frac{|P_B - P|}{\rho_l}}sign(P_B - P) \tag{8}$$

where the source terms P_B and P denote the inside pressure and outside of cavity, Pa; the source term F is the empirical calibration coefficient; R_B is the bubble radius, m.

The final form of this cavitation model is as follows:

$$\text{if } p \le p_v \qquad R_e = F_{vap}\frac{3\alpha_{nuc}(1-\alpha_v)\rho_v}{R_B}\sqrt{\frac{2}{3}\frac{p_v - p}{\rho_l}} \tag{9}$$

$$\text{if } p > p_v \qquad R_c = F_{cond}\frac{3\alpha_v\rho_v}{R_B}\sqrt{\frac{2}{3}\frac{p - p_v}{\rho_l}} \tag{10}$$

where α_{nuc} is the nucleation site volume fraction, the source terms p and p_v denote the flow field pressure and vapor pressure, Pa; the source terms F_{vap} and F_{cond} denote evaporation coefficient and condensation coefficient.

3. Computational domain and mesh

3.1 Impeller model

This paper takes a centrifugal pump as the study object, its design flow $Q_d = 200\text{m}^3/\text{h}$, head H=20m. The parameters of impeller of centrifugal pump are listed in table 1. Based on these basic parameters, the 3D impeller with logarithmic spiral blades was designed by the in-house hydraulic design code which using two-dimension hydraulic design theory[17], as shown in figure 1 and figure 2 respectively. Keeping all other parameters as identical, the blade profile curve (AC) of suction side on designed impeller is replaced by the combination of tangent line($\overline{AB}$) and circle arc line(BC) as shown in figure 3. The tangent line($\overline{AB}$) pass through tangent point A on the blade head, and there is an appropriate transition using circular arc at point B.

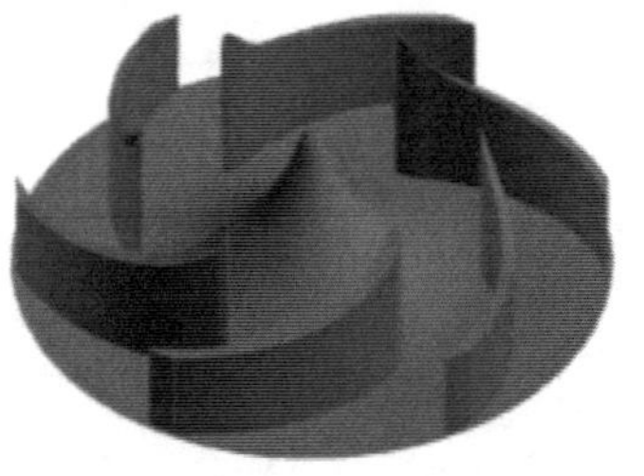

Figure 1. Design blade model.

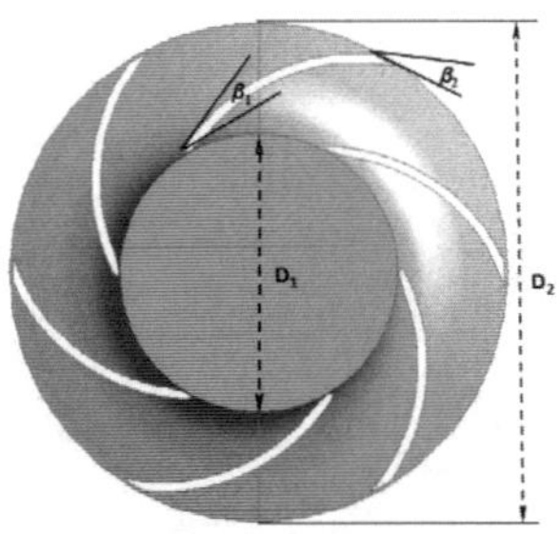

Figure 2. Design impeller model.

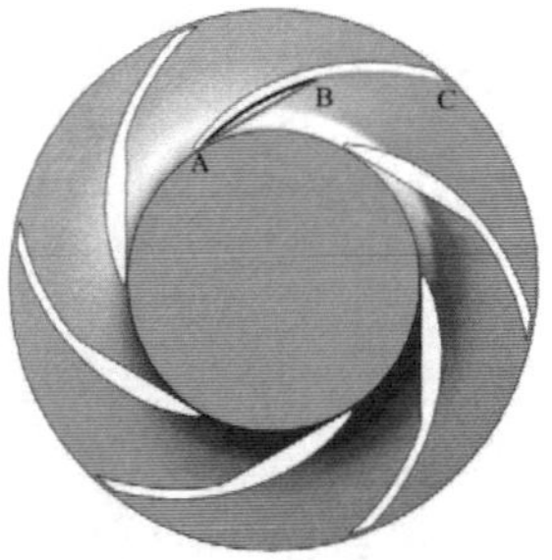

Figure 3. Modified impeller.

Table 1. Design parameters of impeller.

D_1 (mm)	D_2(mm)	B_2(mm)	β_1(°)	β_2(°)	Z	(rpm)
150	270	30	28	21	6	1450

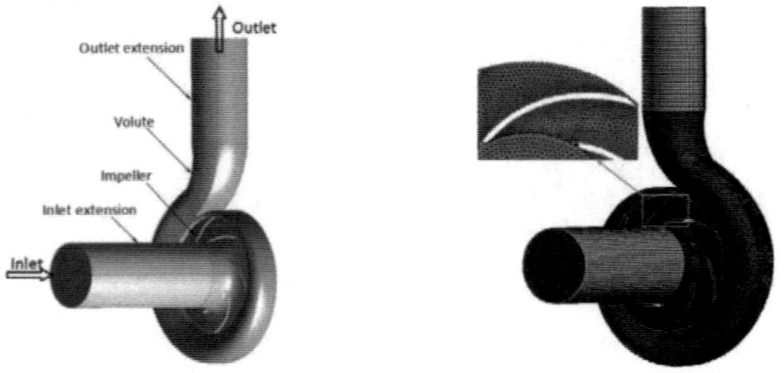

Figure 4. Computational domain and mesh of centrifugal pump

Table 2. The head coefficient of centrifugal pump at different grid numbers

	Grid number/million	Head coefficient
1	5	0.9311
2	9.1	0.9292
3	12	0.9262
4	20	0.9206

3.2 Meshing and boundary conditions

The 3D computational domain and mesh of the centrifugal pump with designed impeller and modified impeller are shown in figure 4. In order to reduce the influence of grid number on the simulation, we adopt a consistent approach of meshing for different models, and the grid independence for cavitation numerical simulation at same condition has been conducted as table 2. The calculation results indicates that the maximum head coefficient deviation is 1.12%, which shows that the influence of grid to simulation is ignored. And the number of 12 million grids are more appropriate with comprehensive consideration.

The boundary conditions of simulations are set as follows:

The inlet of the calculation domain is specified by using the averaged mass flow-rate, and the volume fraction of vapor phase is set to zero .

At the outlet plane, an average static pressure is set according to the total pressure level at the domain inlet, and the value is reduced gradually to realize the cavitation development in impeller of centrifugal pump.

All solid walls are set as the non-slip wall condition.

The coupling between impeller domain and others are set mesh motion.

The vaporization pressure $p_v = 3540\text{Pa}$, environment temperature of experiment T=20℃.

4. Results and Discussion

Three dimensionless parameters are defined to describe the cavitation flow in centrifugal pump:

Cavitation number $$\sigma = \left(p_\infty - p_v\right) / 0.5\rho_l \tag{11}$$

Head coefficient $$\psi = H / \left(U_2^2 / 2g\right) \tag{12}$$

Flow coefficient $$\phi = Q / \pi D_2 b_2 U_2 \qquad (13)$$

where p_∞ is inlet static pressure, Pa; ρ is fluid density, kg/m^3; p_v is vaporization pressure of liquid phase, Pa; H is the head of centrifugal pump, m; U_2 is the peripheral speed of impeller outlet, m/s.

4.1 The reliability verification of numerical simulation

The hydraulic performance curves and the head drop performance curves of experiment and simulated of centrifugal pump with the spectrum impeller are respectively shown in figure 5 and figure 6. It can be observed that there is a good correlation between the simulation and experiment. The simulation predictions follow the trend of the experimental data.

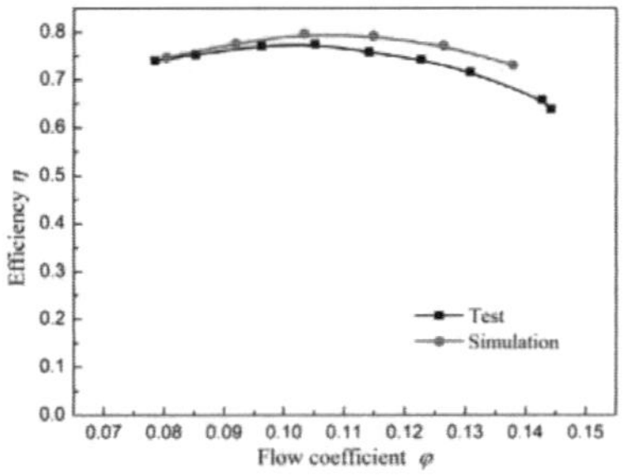

Figure 5. Hydraulic performance curves of centrifugal pump experiment and simulation

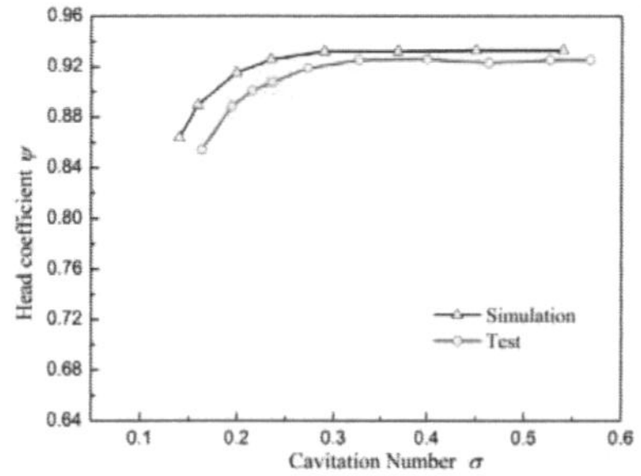

Figure 6. Head drop performance curves of centrifugal pump experiment and simulation

4.2 The analysis of hydraulic efficiency

The hydraulic efficiency curves of centrifugal pump with designed impeller and modified impeller are shown in figure 7. Note from the figure that the flow coefficient at the highest hydraulic efficiency of designed impeller and modified impeller are respectively φ=0.101 and φ=0.103, of which the deviation is 1.9%. It indicates that the modified impeller has little influence on designed condition. However, the hydraulic efficiency of modified impeller is higher than designed impeller at same flow rate, with maximum 4.9%.

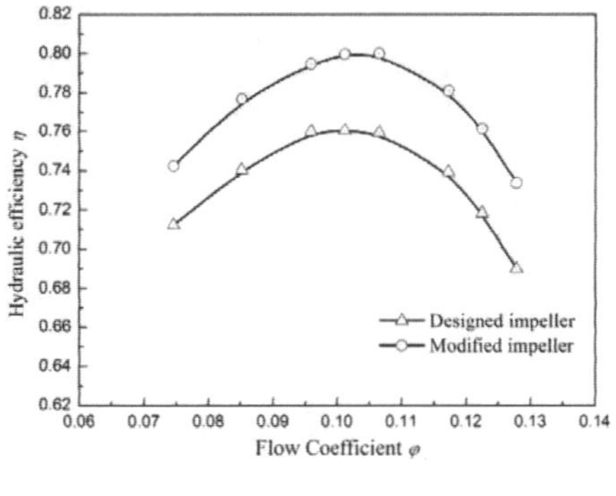

Figure 7. Hydraulic efficiency curves of centrifugal pump

4.3 The analysis of head drop curves

The head drop performances of designed impeller and modified impeller for different flow rates are shown in Figure 8. The cavitation number $\sigma_{3\%}$, $\sigma_{5\%}$ and $\sigma_{7\%}$ corresponding to the 3%, 5% and 7% drop in head are used to represent the cavitation performance of impeller. For all the flow rates, it can be observed that the $\sigma_{3\%}$, $\sigma_{5\%}$ and $\sigma_{7\%}$ of modified impeller are smaller than designed impeller obviously, and the maximal percentage of decrease are respectively 26.5%, 27.4% and 28.4%. It because that the sum of incidence loss and fraction loss in the impeller entrance decrease.

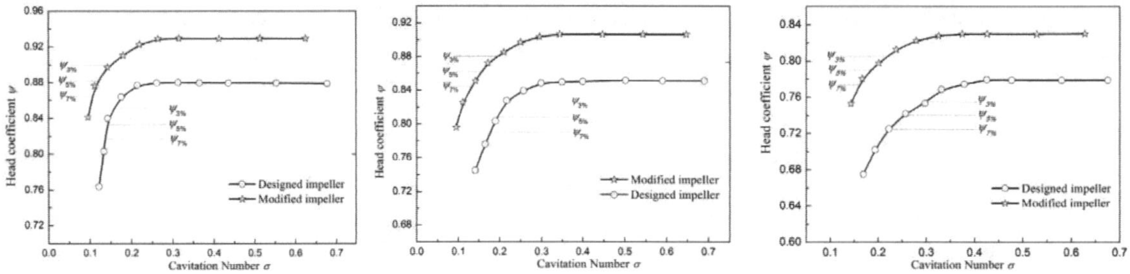

(a) $0.9Q_d$ flow rate (b) Q_d flow rate (c) $1.1Q_d$ flow rate

Figure 8. The head drop curves of centrifugal pump in cavitating condition

4.4 The analysis of cavitation flow field and cavity length

The sheet cavitation blocking the impeller channels and cavity length are appropriate representation of the amount of developed cavitation. Figure 9 shows the contours of sheet cavitation blocking the impeller channels of the impeller mid-span location at σ=0.14 for $0.9Q_d$ flow rate. The cavitation inception formed at the leading edgy of blade, and develop to the tail along blade suction surface. It can be observed that the cavitation area of modified impeller is smaller than designed impeller which will decrease the flow loss caused by vapor-liquid two phase flowing. The analysis of above indicates that the modified impeller can shrink the cavitation intensity in impeller effectively.

Figure 10 shows and the extent of vapor cavity development on the impeller mid-span location at σ=0.14. Note that the cavity lengths are non-dimensionalized by the relative distance along blade chord. From the figure, it can be noticed that the cavity length of designed impeller and modified impeller have little difference. However, the vapor volume fraction of modified impeller at same relative distance along blade chord is smaller than designed impeller.

(a) (b) (c)

Figure 9. The vapor cavity distribution in impeller mid-span: (a) Impeller mid-span location; (b) Design impeller; (c) Modified impeller.

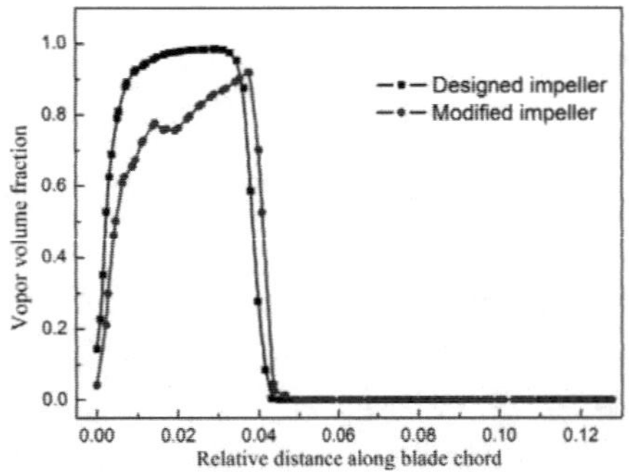

Figure 10. Vapor cavity lengths for σ=0.14 in impeller mid-span

5. Summary and Conclusion

The hydraulic efficiency, head drop performances and cavitation length at different flow rates are studied using computational fluid dynamic(CFD) analyses with the Zwart-Gerber-Belamri cavitation model. It can be concluded that the impeller of adopting the blade profile curve modification of suction side has higher hydraulic efficiency and better cavitation performance. The hydraulic efficiency of modified impeller increased by 4.9%. The head drop performance curves also have clearly demonstrated the superior $\sigma_{3\%}$ performance of modified impeller, with decreased percentage 26.5%. The growth of the vapor bubbles in modified impeller is smaller than designed impeller, and will result in lesser cavitation damage and longer impeller life.

References

[1] Brennen C E and Braisted D M. 1980 *Stability of hydraulic systems with focus on cavitating pumps*
[2] Greitzer E M 1981 J. Fluid Eng. 103(2): 193-242
[3] Cooper P, Sloteman D P , Graf E and Vlaming D J 1991 Elimination of Cavitation Related Instabilities and Damage in High-Energy Pump Impellers *Proc. of the 8th Int. Pump Users Symp. (Houston, Tex, 5-7 Mar 1991)*
[4] Cooper P and Antunes F F 1983 Cavitation in Boiler Feed Pumps Symposium *Proceedings: Power Plant Feed Pumps-State of the Art. EPRI CS-3158 (Cherry Hill, New Jersey, 2-4 June 1982)*
[5] Bakir F, Rey R, Gerber A G, Belamri T and Hutchinson B 2004 *Int. J. of Rotating Machinery* **10**(1) 15-25
[6] Balasubramanian R, Bradshaw S and Sabini E. Influence of Impeller Leading Edge Profiles on Cavitation and Suction Performance *Proc. of the 27th Int. Pump Users Symp. (Houston, Tex, 12-15 Sep 2011)*
[7] Palgrave R and Cooper P 1986 Visual studies of cavitation in pumping machinery *Proc. of 3rd Int. Pump Symp. (Houston, Tex, 20-22 May 1986)*
[8] Schiavello B and Visser F C Pump Cavitation-Various NPSHR Criteria, NPSHA Margins, and Impeller Life Expectancy *Proc. of the 25th Int. Pump Users Symp. (Houston, Tex, 23-26 Feb 2009)*
[9] Herbich J B. 1962 *Modifications in Design Improve Dredge Pump Efficiency* (Bethlehem: Fritz Engineering Lab, Lehigh University)
[10] Cooper P and Sloteman D P 1993 *Impeller for centrifugal pumps*: U.S. Patent 5,192,193[P]. 1993-3-9.
[11] Hackworth M, Eslinger D and Harrell N R 2009 *Impeller for centrifugal pump*: U.S. Patent 7,549,837[P]. 2009-6-23.
[12] Katz J 1984 *J. Fluid Mech.* **140**(4): 397-436.
[13] Franc J P and Michel J M 1985 *J. Fluid Mech.* **154** 63-90
[14] Dumitrescu H and Cardos V 2010 *J. of Aircraft* **47**(5) 1815-9
[15] Plesset M S. 1949 *J. of Applied Mechanics* **16** 228-31
[16] Zwart P J, Gerber A G and Belamri T 2004 A two-phase flow model for predicting cavitation dynamics *Fifth Int. Conf. on Multiphase Flow (Yokohama, Japan, 30 May-4 June 2004)*
[17] Zhang Y, Zhang J, Zhu H and Cai S 2014 *Advances in Mech. Eng.* **2014** 803972

The effect of gas fraction on centrifugal pump

Z T Zhu, Y Wang, L F Zhao, C Ning, S F Xie and Z C Liu
Research Center of Fluid Machinery Engine and Technology, Jiangsu University, Zhenjiang 212013, China

Email: 986853489@qq.com

Abstract: In order to study the multiphase flow field in M125 centrifugal pump, three-dimensional modeling was used for internal flow through three-dimensional software Pro/E. Then based on SST turbulence model combining with Rayleigh-Plesset cavitation model, and structured grid to simulate the hydraulic characteristics of volute and impeller within different gas conditions. The velocity, pressure and gas volume fraction distributions of the interior flow field of volute and impeller were obtained and analyzed, which revealed the effect of gas fractions on the flow characteristic of the centrifugal pump.

1. Introduction

When centrifugal pumps occur the phenomena of gas-liquid multiphase, due to the small centrifugal force caused by the small gas density, it will cause the performance degradation of centrifugal pumps, and, seriously, that the suction from impeller center cannot suck in the liquid into the pump so that zero flow occurs, and even that centrifugal pumps and its system occur serious accident[1].

With the rapid development of computational methods and numerical theory, calculation simulation has been widely used in the field of multiphase flow simulation .Many scholars have carried out a large number of technical and experimental researches on centrifugal multiphase flow simulation. Yang Minguan etc[2], Yuan Shouqi etc[3] Li bin etc[4]carried out numerical simulation analysis of three-dimensional solid-liquid two-phase in special purpose centrifugal and obtained internal flow law; Pan Zhongyong etc[5],respectively, studied the pulp suspension flow in stirred pot. Xie Peng etc[6] simulated two-phase flow field inside the pump , which initially revealed the characteristics of the pump gas-liquid two-phase flow.

This paper, basing on summarizing the results of previous studies and using centrifugal pumps with M125 models for the study and applying CFD Technology, describes the evolution of the internal flow structure of centrifugal pumps and analyses the changes of the internal gas rate of centrifugal pumps, and it is important for further studying the mixing and separation of the internal gas-liquid of centrifugal pumps and improving the performance of centrifugal pumps.

2. Numerical Simulation

2.1 Model and mesh

M125-100 centrifugal pump is selected to be the model pump. Using Pro/E software to establish a three-dimensional solid model of self-priming, the model includes four parts: impeller, volute, inlet and outlet pipelines.

The mesh was generated by software ICEM for numerical simulations. Inlet of pump and outlet of

IOP Conf. Series: Materials Science and Engineering **72** (2015) 032022 doi:10.1088/1757-899X/72/3/032022

volute were appropriately extended to increase the accuracy of calculations. Structural mesh was used in the hydraulic model. The boundary layers near stay vanes, guide vanes, and runner blades were refined. After examination, the quality of the grid meets the requirements, and the calculation model about has eight hundred thousand grid cells.

2.2 Numerical simulation settings

Numerical calculation based CFX software is conducted. Both equations of N-S and RNG k-ε are used to determine turbulent viscosity. Mixture model is used as the multiphase flow model. Using SIMPLC algorithm to solve the pressure-velocity coupling equations; Turbulence intensity is set to 5%; The total pressure and flow direction are imposed at the inlet, and the mass flow rate is given at the outlet. A no-slip wall condition is specified for the all wall boundaries . Walls are modeled using the standard wall functions.[7~8]

The flow in centrifugal pump contains the phase of water ,the phase of air bubbles and the phase of cavitation bubbles. Several simulation were conducted, and gas fractions of inlet were set to 0%, 5%, 10%, 15%, 20%, while all the air bubbles fraction were set as 0.

3. Results and analysis

3.1 Velocity and static pressure distribution of internal flow field

As shown in the Figure 1, the velocity distribution of internal flow field in impeller and volute are normal. There are no phenomenon of flow separation and the impact overall, except in the field nearby tongue. The velocity increases from the inlet to outlet of the impeller. Then the velocity decreases from impeller field to volute field while increases along the direction of volute diversion, and declines in the outlet of the pump.This is consistent with the volute design methods.Besides, the phenomenon of reflux occurs near the outer wall of the volute , which is consistent with the actual situation.

Figure 2 reveals the static pressure contours under gas fraction of 0% and water fraction of 100%. It is apparent from the static pressure contours provided that static pressure increases from the inlet to outlet of the impeller. Most of the kinetic energy has been converted into pressure energy, so the variation of static pressure is large. A small area of the blade inlet has negative pressure. At the regions of same radius, the pressures of working sides are higher than that of suction sides. In the volute, because of the different curvature radius of the volute , static pressure along the flow direction becomes larger and larger[8~9].

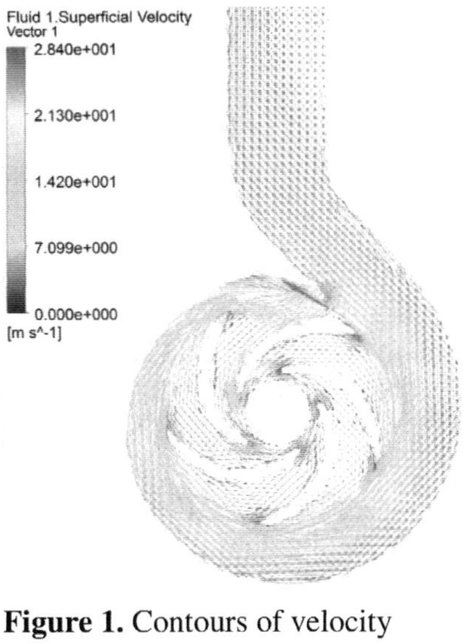

Figure 1. Contours of velocity

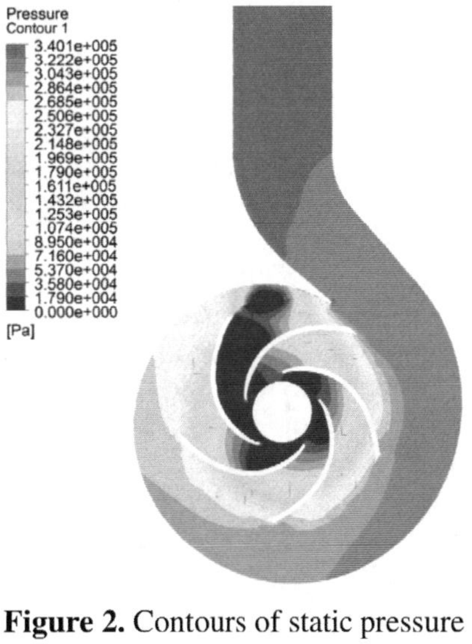

Figure 2. Contours of static pressure

3.2 Centrifugal gas-liquid two-phase simulation

3.2.1 Static pressure distribution of gas-liquid two-phase flow field
Figure 3 reveals the static pressure contours under gas conditions of 0%、5%、10%、15% and 20%. It is evident from the static pressure contours provided that static pressure increases in radial direction; Under different gas conditions, the pressure gradients along the flow path direction are different; The higher gas content, the lower the static pressure.

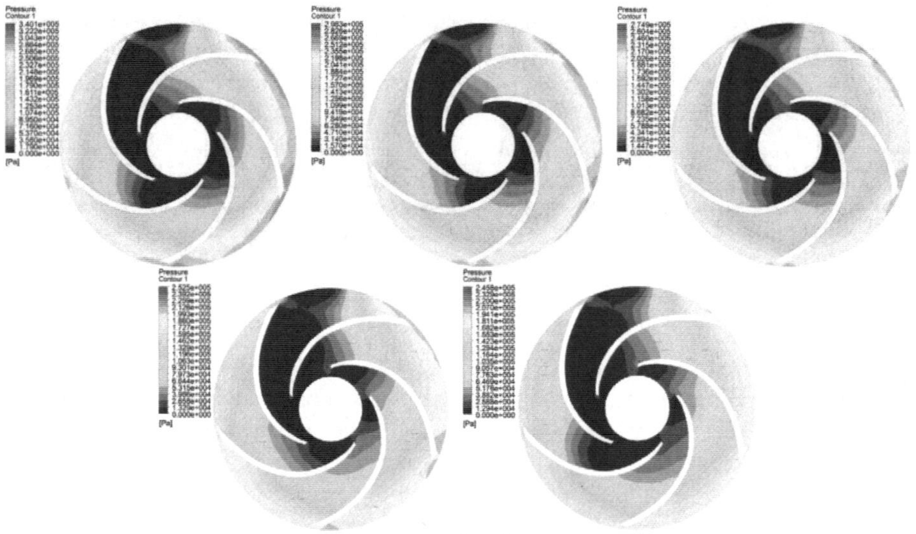

α=0% α=5% α=10% α=15% α=20%

Figure 3. Static pressure contours under the influence of different gas conditions

3.2.2 Velocity distribution of gas-liquid two-phase flow field. Figure 4 shows the velocity distribution contours under gas conditions of 0%、5%、10%、15% and 20%.It is apparent from the velocity distribution contours supplied that gas-liquid two-phase flow in centrifugal pumps under the gas conditions is very similar overall. This is because the intercoupling and mutual influence between the gas-liquid two-phase. However, with the increase of gas fraction , the velocity in outlet decreased.

3.2.3 Gas distribution of gas-liquid two-phase flow field. Figure 5 reveals the gas-phase distribution contours under gas conditions of 0%、5%、10%、15% and 20%. As shown in the figure, gas fraction on pressure side of impeller is higher than on suction side; Gas fraction on pressure side of impeller in the direction of the flow path reduced gradually; With the increase of gas fraction, gas fraction of flow channel area is increasing and high gas fraction area continues to expand. Especially when α≥15%, concentration of bubbles within the flow channel leads to the two-phase flow transition from bubbly flow to slug flow. The above phenomenon is the cause that liquid flow rate decreases as the increase in gas fraction, and then. Under the action of inertial force and centrifugal force, liquid-phase deviates from the normal flow lines track and moves to the back of the blade. At the same time, by reason of repelling effect of the liquid phase, gas-phase is forced to deviate from the normal flow lines track and moves to working surface of vane.

Gas fraction declines in radial direction in volute , and it is very low near the wall of volute. This is because the static pressure in volute radially increases. At same time, under the action of the inertia force, the liquid-phase closes to the wall of volute and repels the gas-phase. A glance at the figures of α=10%、α=15% and α=20% provided reveals large concentration of gas-phase at the outlet of the impeller.

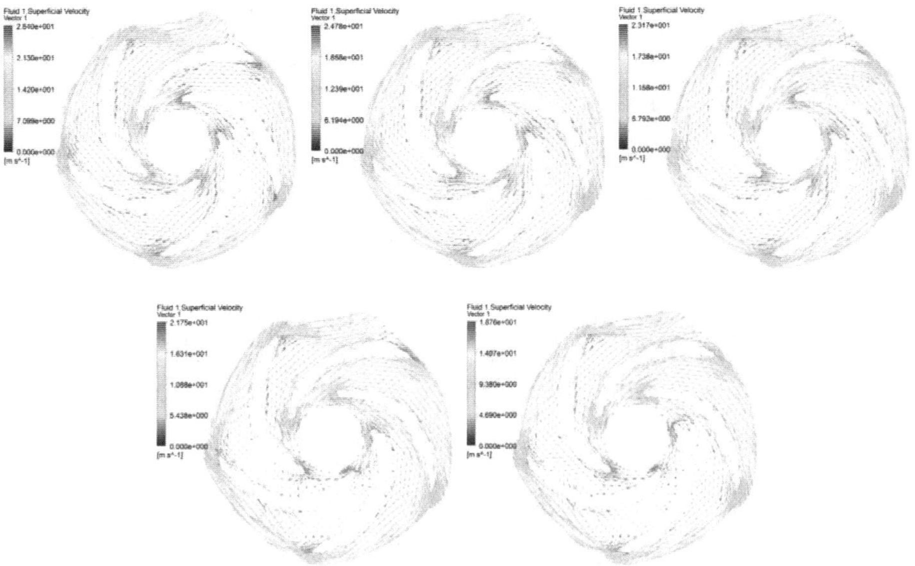

α=0% α=5% α=10% α=15% α=20%

Figure 4. Velocity distribution contours under the influence of different gas conditions

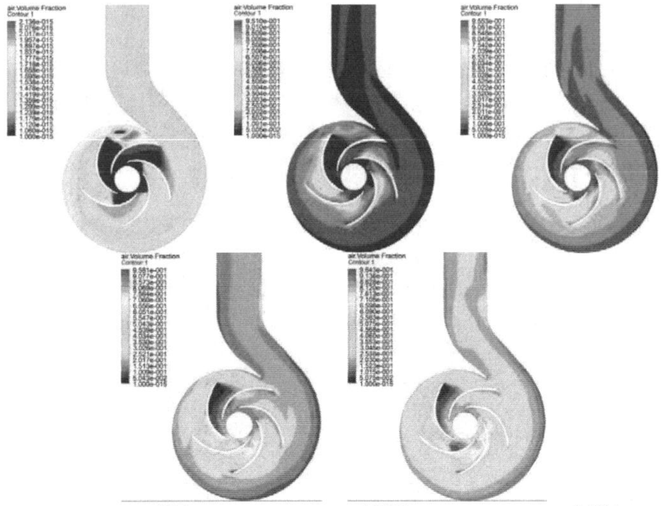

α=0% α=5% α=10% α=15% α=20%

Figure 5. Gas distribution contours under the influence of different gas conditions

3.2.4 Vapour distribution of gas-liquid two-phase flow field. Figure 6 shows the vapour distribution contours under gas conditions of 0%、5%、10%、15% and 20%. From the figure provided, it is clear that the most intensive areas of cavitation releases in the second half of the blade suction surface and decreases along the direction which opposite to the direction of rotation of the impeller. This distribution is consistent with the static pressure distribution of the flow channel. With the growth of gas fraction, the area of the cavitation region decreases and then increases. A small area of cavitation occurs on the pressure side of blades because the static pressure value is too low.

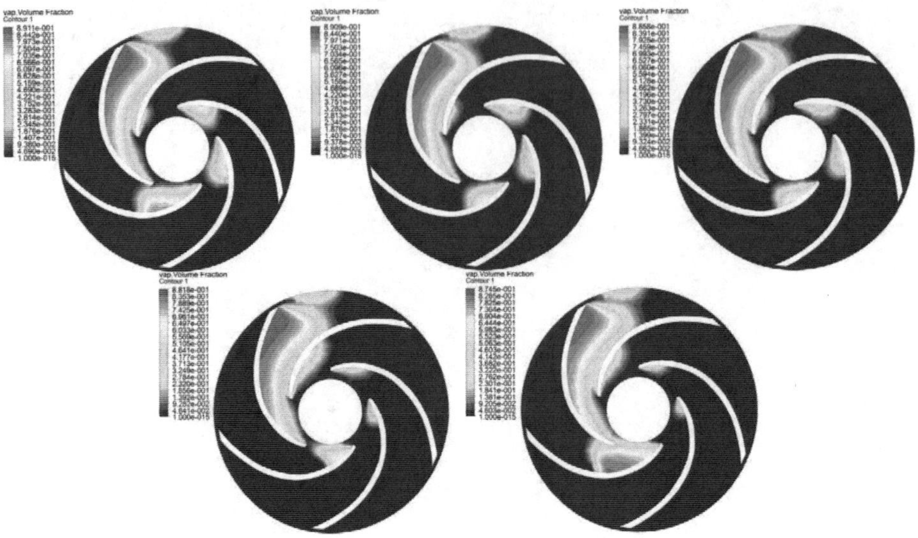

α=0% α=5% α=10% α=15% α=20%

Figure 6. Vapour distribution contours under the influence of different gas conditions

3.2.5 Changes of head in two-phase flow centrifugal pump. As shown in Figure 7, when the gas enters the pump, the head falls rapidly. Compared with single-phase conveying, two-phase flow becomes more unstable and greater flow loss, so the head declines markedly.

With the increase of gas fraction, the head still sharply drops. As shown in Figure 5, when the gas fraction is low, the small bubbles begin to gather on the region of the blade from middle of pressure surface to outlet of suction surface, and the bubbles stagnant zone is formed on the pressure side region, so the head begins to decline sharply.

With the continuing increase of gas fraction(namely as $\alpha \geq 16\%$), the head further drastically slips.Because when $\alpha \geq 16\%$, the bubble stagnant zone extends to the rear of pressure surface of the blade which further reduce the channel section area, so that both of the flow relative velocity and the flow loss increase, so the head further drastically slips. This phenomenon is probably consistent with the gas-phase distribution of centrifugal pump.

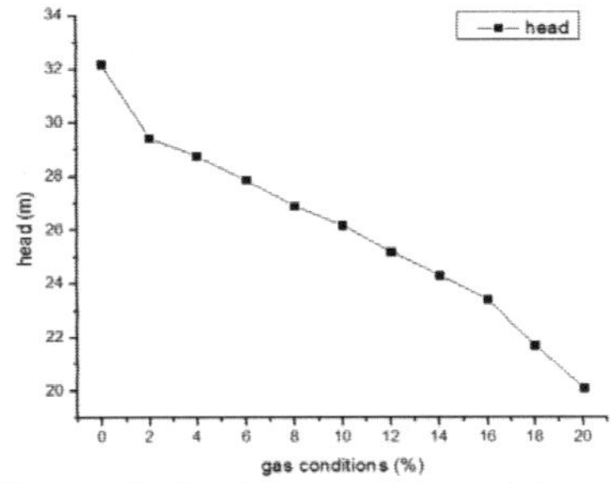

Figure 7. Head curves of model pump

4. Conclusions

1) Static pressure in impeller is increased in radial direction; With the increase of gas fraction, the pressure gradients along the flow path direction are different.

2) The distributions of gas-liquid two-phase flow under different gas conditions are very similar overall. With the increase of gas fraction , the velocity in outlet decreased.

3) With the increase of gas fraction, gas fraction of flow channel area is increasing and high gas fraction area continues to expand. Especially when $\alpha \geq 15\%$, concentration of bubbles within the flow channel.

4) The most intensive areas of cavitation releases in the second half of the blade suction surface and decreases along the direction which opposite to the direction of rotation of the impeller. A small area of cavitation occurs on the pressure side of blades.

5) The higher gas content, the lower the head. Especially, when $\alpha \geq 16\%$, the head further drastically declines.

Acknowledgement
The authors gratefully acknowledge the financial support of National Science and technology Support Plan of China (2011BAF14B01), a project funded by the Priority Academic Program Development of Jiangsu Higher Education Institutions (BK2009218).

Conflict of interests: The authors declare no conflicts of interest. The support of National Science and technology Support Plan of China (2011BAF14B01), a Project Funded by the Priority Academic Program Development of Jiangsu Higher Education Institutions (BK2009218) have no conflict of interest.

References
[1] Fu Q, Yuan S Q, Zhu R S and Wang X L 2012 *J. of Harbin Engineering University* **33**(11) 1428-34 (in Chinese)
[2] Yang M G, Gu H F, Liu D, Jia W D and Gao B 2006 *Chinese J. of Mech. Eng.* **12** 180-5 (in Chinese)
[3] Yuan S Q, Li Y and He Z H 2003 *Chinese J. of Mech. Eng.* **39**(7) 18-22 (in Chinese)
[4] Li B, Qi H and Guo Y X 2009 *Drainage and Irrigation Machinery* **27**(2) 95-9
[5] Pan Z Y, Li S, Zhang H and Shi W D 2011 *J. of Drainage and Irrigation Machinery Eng.* **29**(3) 204-8
[6] Xie P and Zhu Z C 2009 *J. of Hydraulic Eng.* 40(12) 1506-11
[7] Liu J R and Su Q Q 2009 *Transactions of the Chinese Society for Agricultural Machinery* **40**(9) 73-6
[8] Chen C C, Liu T B and Yuan S Q 1994 *Theory and Design of two-phase flow Pump* (Beijing: Ordnance Industry Press)
[9] Guan X F 1995 *Modern pumps Technical Manual* (Beijing: Astronautics Press)

Optimization of centrifugal pump cavitation performance based on CFD

S F Xie, Y Wang[1], Z C Liu, Z T Zhu, C Ning and L F Zhao
Research Center of Fluid Machinery Engine and Technology, Jiangsu University, Zhen jiang 212013, China

E-mail: 963489115@qq.com

Abstract. In order to further improve the cavitation performance of a centrifugal pump, slots on impeller blade near inlet were studied and six groups of hydraulic model were designed. Base on cavitating flow feature inside a centrifugal pump, bubble growth and implosion are calculated from the Rayleigh-Plesset equation which describes the dynamic behavior of spherical bubble and RNG κ-ε model was employed to simulate and analyze the internal two-phase flow of the model pump under the same conditions. The simulation results show that slots on blade near inlet could improve the cavitation performance and cavitation performance improvement of the second group was more obvious. Under the same conditions, the pressure on the back of blade near inlet was higher than the pressure on the back of unmodified blade near inlet, and energy distribution in the flow channel between the two blades was more uniform with a small change of head.

1. Introduction

The model pump used in this paper is a constant head pump, which has characteristics such as :steady head, high efficiency and so on. Cavitation[1] is a key issue of constant head pump which can cause noise and head fluctuation.In order to further improve the cavitation performance of a centrifugal pump, slots on impeller blade near inlet were studied .Ye Daoxing[2] investigated the effects of slots on internal flow and efficiency.Xing Gang[3] studied the impact of slotted on head.

Based on the Reynolds Average Navier-Stokes equation and RNG $\kappa-\varepsilon$ model[4] used in the CFX[5][6] software,this paper simulates internal flow of constant head pump with different impeller under different inlet pressure,and discusses the distribution of vapour volume fraction in impeller. The results could provide theory for improving the cavitation performance of centrifugal pump[7].

2. Models and Boundary Conditions

2.1. Pump geometry mode

The model pump used in present study is a low-specific-speed centrifugal pump with a specific speed of 46.3[8].Main design parameters of model pump are as follows: Q=108 m^3/h, H=140 m, n=2970 r/min,impeller suction diameter $D1$=100 mm, impeller outlet diameter $D2$=300 mm, impeller width at outlet b_2=13 mm,the number of impeller blades Z= 6,blade outlet angle β_2=90°.Three-dimensional flow models of different impellers are established by Proe software. In order to ensure uniform distribution

[1] Correspondence author. Tel.:15751010709.

IOP Conf. Series: Materials Science and Engineering **72** (2015) 032023 doi:10.1088/1757-899X/72/3/032023

of the inlet velocity, and to reduce the impact of boundary conditions on flow filed, the sections of the impeller inlet and volume outlet were carried out to extend ,as shown in figure 1.Table 1 shows the different parameters of the slotted blades ,and different impellers as shown in Figure 2.

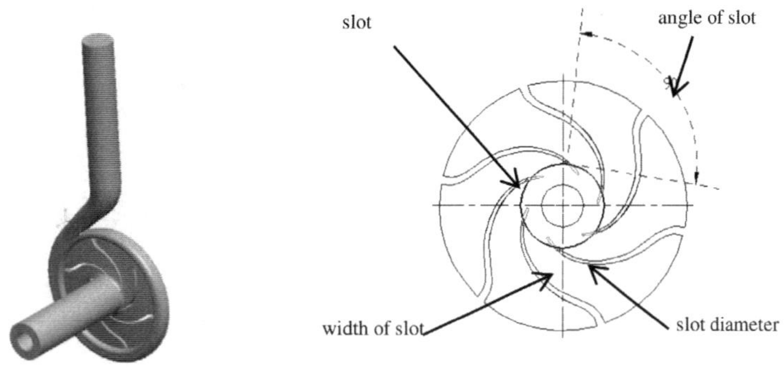

(a) 3-D models (b) 2-D models

Figure 1. Calculation models

Table 1. Parameters of the slotted blades

Parameter Number	K1	K2	K3	K4	K5	K6
Slot diameter D(mm)	109	109	109	109	109	109
Width b(mm)	3	2.65	1.31	1.18	2	1.31
Angle β(°)	90	0	0	0	90	30

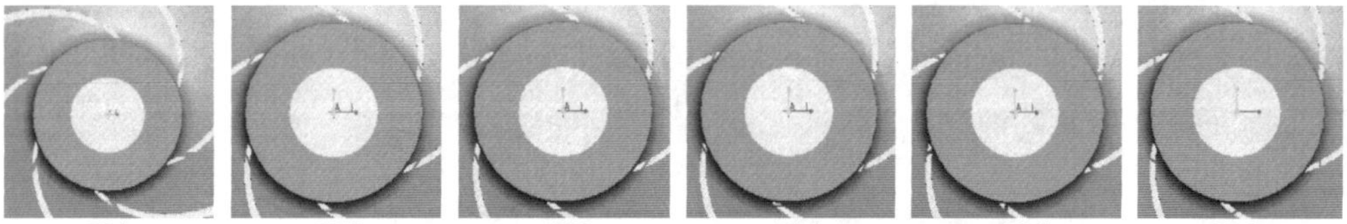

K1 K2 K3 K4 K5 K6

Figure 2. Different impellers such as table 1

2.2. Mesh generation and boundary condition

As is known， performance prediction error will be gradually decreases with the improvement of mesh quality[9] .In order to get the most economical number of grids and calculate step, this paper identifies the calculated cell number by the principle of independence of the grid, and structured hexahedral cells were used to define the inlet and outlet domain, while unstructured tetrahedral cells with strong flexibility were used for volute and impeller.Finally, the minimum mesh was determined as 1.2mm,and the total number of grid is 1 816 844.The grids number of each domain were 44352, 791495, 856379 and 124600, respectively, as shown in figure 3.

Select of the boundary conditions has a greater impact on the accuracy of the calculation and results.In this paper, the pump inlet was set as pressure inlet, outlet was set as mass flow.The no-slip condition for the boundary layers was imposed over walls and the standard wall function was used. And the interface between impeller and volute was set as frozen-rotor interface, convergence precision was set to 1 x 10^{-5}.

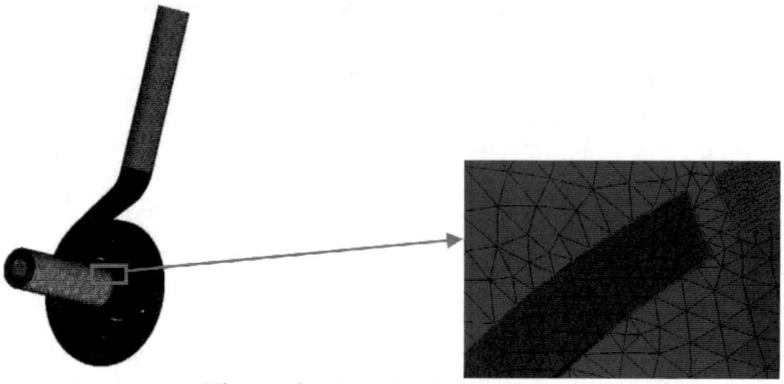

Figure 3. Calculation mesh

3. Results analysis

3.1 Vapour volume fraction distribution

Figure 4 shows that the vapour volume fraction distribution of each impeller when the inlet total pressure is set as 101325 Pa. As can be seen from the figure, cavitation is distributed in the back of the blade inlet, and vapour volume fraction of each slotted impeller are reduced compared with the original impeller. This phenomenon of figure (K2) what is that vapour volume fraction is divided into two areas and cavitation occurs in front of the slot is more obvious. Compare with figure 4 (K2), the width of figure 4 (K3) and figure 4 (K4) is reduced while the vapour volume fraction is increase. Compare figure 4 (K3) with figure 4 (K6), it can be seen that a smaller angle can be further reduced vapour volume fraction.

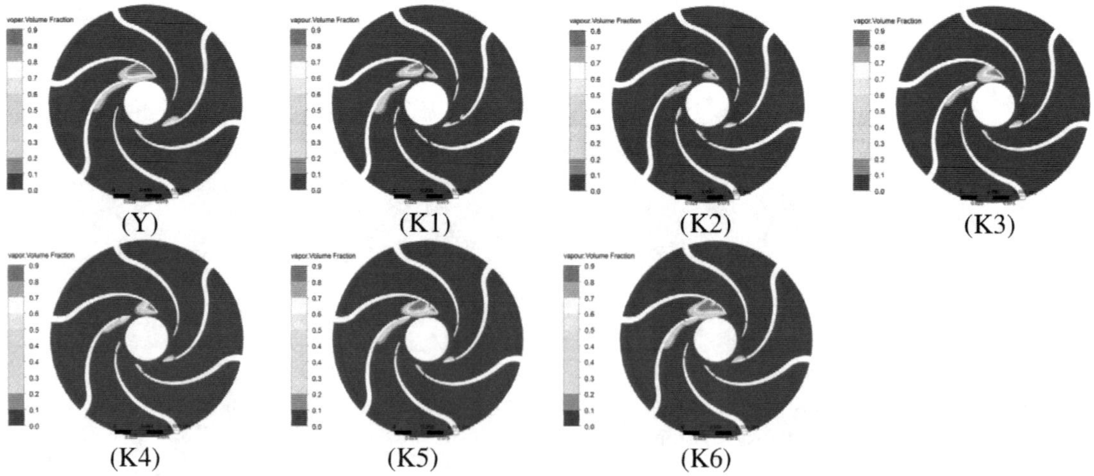

Figure 4. The vapour volume fraction distribution of different impellers under 101325 Pa

Figure 5 shows that the vapour volume fraction distribution of each impeller when the inlet total pressure is set as 50000 Pa. As can be seen from the figure, compared with the original impeller, vapour volume fraction of each slotted impeller has been significantly reduced, especially K1and K5 .Most of the cavitation area in these two impellers is behind the slot, so the impeller inlet flow channel will not blocked when cavitation occurs. From figure 5 (K1), (K2), (K3), (K4) ,you can know as the width increases, vapour volume fraction is reduced. Compare figure 5 (K1) with figure 5 (K5), it can be seen that a greater width can be further reduced vapour volume fraction.

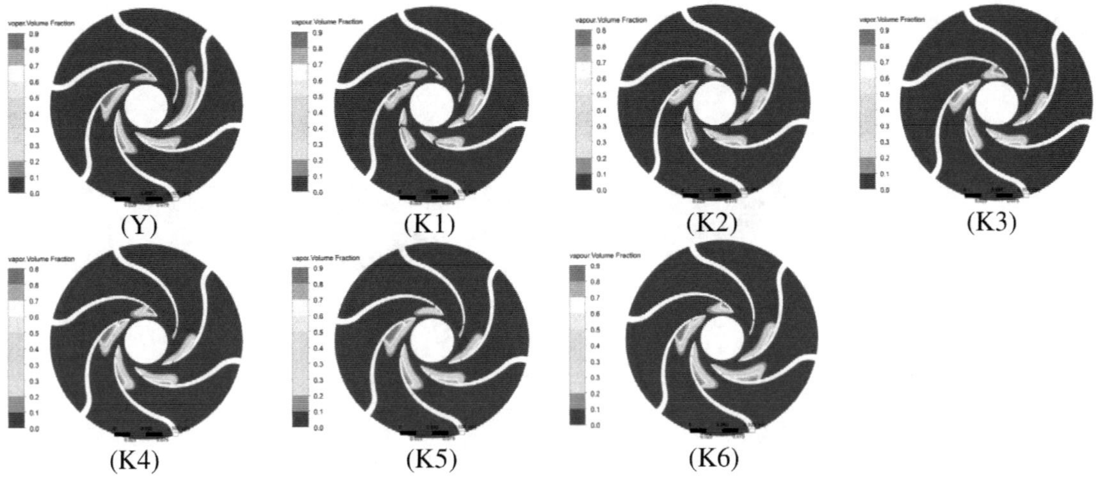

Figure 5. The vapour volume fraction distribution of different impellers under 50000 Pa

3.2 Pressure distribution

Figure 6 shows that the total pressure distribution of each impeller when the inlet pressure is set as 101325 Pa. As can be seen from the figure ,it meets the actual phenomenon that total pressure of the back of the blade is lower than the working face, and it can be seen from the figure 6 (Y), (K1), (K5), (K6), there is a low total pressure area in the impeller channel and the area near the back of the blade. As we all know that it means that there is vortex in the impeller channel, it will block the flow channel and reduce hydraulic efficiency . Compare figure 6 (K1) with figure 6 (K5), it can be seen that (K1) has a more significant area of low total pressure what meas a smaller width could cause a smaller low total pressure area . This phenomenon has also been confirmed in figure 6 (K1), (K2), (K3), (K4).

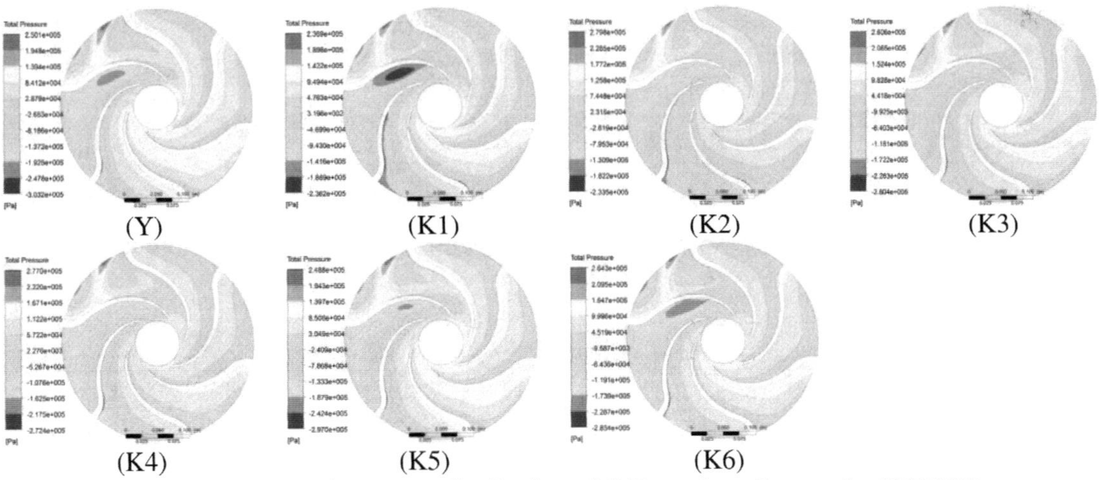

Figure 6. The total pressure distribution of different impellers under 101325 Pa

Figure 7 shows that the total pressure distribution of each impeller when the inlet pressure is set as 50000 Pa. As can be seen from the figure, The minimum value of total pressure of slotted impeller is increased compared with the original impeller, especially figure 7 (K2) and (K3) .While see intuitively from the figure, the total pressure distribution of figure 7 (K2) is better than figure 7 (K3), it has a bigger width than (K3). Compare figure 7 (K1) with figure 7 (K5) , it can also be seen that a bigger width could make uniform total pressure distribution in flow channel. Compare figure 7 (K3) with figure 7 (K6) , it can be seen that a smaller angle could also cause a uniform total pressure distribution.

In general, a larger width and a smaller angle could make a better total pressure distribution, such as width b=2.65 mm, angle β=0°.

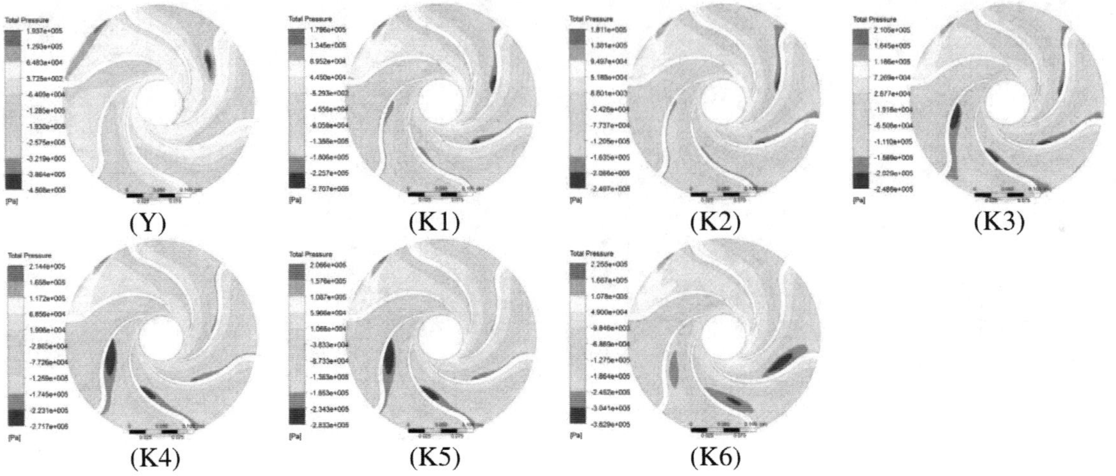

Figure 7. The total pressure distribution of different impellers under 50000 Pa

3.3 NPSH analysis

Net Positive Suction Head is a very important parameter of the pump performance.There is no accurate calculation method for NPSHR at present except to cavitation experiments.In this paper, numerical simulation of the cavitation experiment had been done by changing the inlet total pressure .Figure 8 shows that NPSH curve of numerical simulation of different impellers.

As can be seen from the figure 8, Slotted impeller has no influence on the head of pump ,while NPSH of K2 and K3 have an obvious improvement compared with the original NPSH, they are respectively 0.537 m and 0.42 m. It further evidence that the slotted impeller could improve the cavitation performance of a pump.

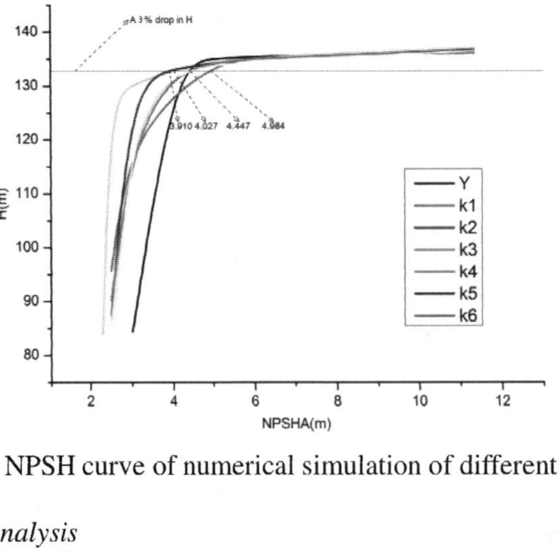

Figure 8. NPSH curve of numerical simulation of different impellers

3.4. Hydraulic efficiency Analysis

Hydraulic efficiency is another important performance parameters of the pump. As can be seen from the table 2, effect of slot on the hydraulic efficiency is little, even the slotted impeller could have a little improvement of hydraulic efficiency of a pump, such as K2, K3.

Table 2. The hydraulic efficiency of the impeller under different inlet pressure

Inlet pressure	Hydraulic efficiency(%)						
	Y	K1	K2	K3	K4	K5	K6
101325Pa	73.14	71.21	73.16	72.65	72.92	72.01	71.39
50000Pa	73.01	72.35	73.01	73.11	73.02	72.83	70.71
20000Pa	41.30	61.57	67.33	67.86	59.70	59.83	61.17

4. Conclusions

Based on the above analysis of different parameters and numerical simulation results, some conclusions are obtained:

1) In the above numerical simulation, cavitation is distributed in the back of the blade inlet and it meets the actual phenomenon.

2) Compared with the original impeller, slotted impeller can improve cavitation performance of a low-specific-speed centrifugal pump. While the parameters of slot will be different with different inlet pressure.

3) Compared with the original impeller, slotted impeller could make a more uniform pressure distribution in flow channel, such as width b=2.65 mm, angle β=0°.

4) A reasonable slot with a larger width and smaller angle has no effect on the efficiency of pump.

5) Choose the right slot position, width and angle has a significant impact on the cavitation performance, like near the blade inlet, larger width and a smaller angle.

References

[1] Brennen C E 1994 *Cavitation and bubble dynamics* (London, Oxford University Press)

[2] Ye D X 2013 *Study of impact on the performance of centrifugal pumps with slotted blade* (Zhenjiang: Jiangsu University Press)

[3] Xing G 2008 *Numerical study of the flow field of centrifugal impeller with Slotted blade* (Chongqing: Chongqing University Press)

[4] Launder B E and Spalding D B 1974 *Computer Methods in Applied Mechanics and Eng.* **3**(2) 69-89

[5] Luo X W, Zhang Y, Peng J Q and Xu H Y 2008 *J. of Tsinghua University: Sci. and Tech.* **48**(5) 836-9 (in Chinese)

[6] Wang Y, Liu H, Yuan S, Tan M and Wang K 2011 *J. of Drainage and Irrigation Machinery Eng.* **29**(2) 99-103 (in Chinese)

[7] Wang W J and Wang Y 2012 *IOP Conf. Ser.: Earth Environ. Sci.* **15**(6) 062013

[8] Guan X F *2011 Modern Pump Theory and Design* (Beijing: China Astronautic publishing House)

[9] Ferziger J H and Peric M 1996 *Computational Methods for Fluid Dynamics* (Berlin: Springer)

IOP Conf. Series: Materials Science and Engineering **72** (2015) 032024 doi:10.1088/1757-899X/72/3/032024

Numerical simulation of cavitation effects influenced by centrifugal pump inlet parameters

L F Zhao, Y Wang, C Ning, Z C Liu, Z T Zhu and S F Xie
Research Center of Fluid Machinery Engineering and Technology, Jiang Su University, Zhenjiang 212013, China

E-mail: 1020655375@qq.com

Abstract. Cavitation has great influence on performance of the centrifugal pump. However, there is still no effective design to overcome this problem. Blade leading edge of centrifugal pump impeller is the initial position of cavitation. The leading edge geometry shape not only has a great influence on the cavitation inception and its development, but also a great influence on the flow state near the impeller inlet. In this paper, the numerical simulation method is adopted. Cavitation of four different models (including rectangular-shape blade model, circular-arc-shape blade model, elliptical-shape blade model and cusp-shape blade model) are simulated under the same condition by changing the NPSHA value. The influence of different blade models on cavitation performance is analyzed. The results show that the deviation between the simulated data and experimental data is within the deviation range. The head of rectangular-shape blade model and circular-arc-shape blade model are higher than those of elliptical-shape blade model and cusp-shape blade model. However, the head of rectangular-shape blade model and circular-arc-shape blade model is smaller than the latter under the low effective cavitation margin. What's more, the head of the models with trimmed blade are higher than the head of the models with untrimmed blade under the working condition but are smaller under the low effective cavitation margin.

1. Introduction

Centrifugal pumps are widely used in chemical industry, agriculture, municipal engineering etc. Its internal cavitation has always been the major problem in fluid machinery internal flow research. Because cavitation can result in cavitation erosion, performance degradation, vibration, noise and a series of problems.so it must be considered in fluid machinery design. In recent years, the cavitation researches mainly focus on the theoretical study, experimental study and numerical analysis .Biao Huang, Guoyu Wang and Haitao Yuan proposed a density modify based cavitation model for cavitation flow computations [1]. Eric Goncalvès, Boris Charrière studied the mass transfer term between phases [2].MØRCH K. A studied some special characters of cavitation nuclei [3]. Frédéric Caupin and Eric Herbert pointed out that metastable state between liquid and vapor plays an important role in establishment of cavitation model[4]. XU Yu, WU Yulin, TANG Xuelin derived the cavitation flow control equation with the kinetics method [5]. Yuan Danqing, Chen Xiangyang, Bai Bin, Cong Xiaoqing pointed out that cavitation and cavitation erosion is the research emphasis of the future [6].What's more, the free streamline approach tells us that smooth shedding and mutation shedding also play an important role in the form of the different cavitation shape [7]. Experimental study mainly focuses on PIV, high-speed digital camera and other advanced imaging technology, which are used to study structure and characteristics of cavitation. Xu Y B has studied the resistivity imaging technology between liquid-vapor phases [8].Shi L L has studied the digital image recognition technology based on gas-liquid two phases flow parameters [9].

Blade leading edge of centrifugal pump suction side is the initial position of the cavitation, here the rapidly rotating leading edge of blade interacts strongly with the low speed liquid, which results in big pressure gradient in the blade suction surface area, and it is easier to cause cavitation under the low inlet pressure. What's more, the different leading edge of blades have great effect on the interaction method between liquid and blades, and the form of shedding vortex of the blade suction surface are different, which affects the degree of internal flow turbulence, also the performance.

This paper mainly focuses on the liquid flows around different solid obstruction and its complex cavitation flow structure under different conditions, also its effect on the performance of pump. Based on the above idea, numerical simulation is applied to study influence on the cavitation structures of four different models with different blades leading edge.

2. Model hydraulic parameters and the experimental dates

A good hydraulic performance centrifugal pump model is analyzed. The geometric parameters of the model are shown in table 1, and the experimental dates are shown in table 2.

Table 1. Geometric parameters of the pump

Impeller outer diameter(D2)	Blade width at impeller outlet(b2)	Suction diameter(Ds)
173 mm	20 mm	90 mm

Table 2. Experimental dates

Flow rate		Head	Revolving speed	Efficiency	Power	NPSHC
m^3/h	L/S	m	r/min	%	kw	m
60	16.7	37.78		70.89	8.72	2.92
100	27.8	33.98	2900	81.51	11.36	3.37
120	33.3	29.02		77.44	12.25	4.09

Figure 1 shows the four kinds of blades .ZF represents the blade with rectangular shape leading edge, YH represents the blade with circular arc shape leading edge, TY represents the blade with elliptical shape leading edge, and JJ represents the blade with cusp shape leading edge.

Figure 1. Shape of different blade leading edges.

3. Steady-state numerical calculations

3.1 Boundary conditions and mesh generation

Because the flow near inlet of impeller and volute outlet is not constant, the length of inlet pipe and the volute outlet are properly lengthened, about 3 to 4 times the diameter of the corresponding location. ICEM -CFD 14.5 is used to generate the unstructured grid, and the number of grid of the four models is about 700000, and the grid density near every tongue and back of blade suction surface of impeller are increased.

Numerical simulation shows 72 group results of four different model pumps under different inlet pressure, convergence accuracy of simulation is set to 10^{-6}.The inlet of pipe is set to pressure parameter, and volute outlet is set to velocity parameter.

3.2 Cavitation model

The tendency for a flow to cavitate is characterized by the cavitation number, defined as:

$$C_p = \frac{P - P_v}{\frac{1}{2}\rho U^2} \tag{1}$$

where P is a reference pressure for the flow, and P_v is the vapor pressure for the liquid, and the denominator represents the dynamic pressure.

The Rayleigh-Plesset equation provides the basis for the rate equation controlling vapor generation and condensation. The Rayleigh-Plesset equation describing the growth of a gas bubble in a liquid is given by:

$$R_B \frac{d^2 R_B}{dt^2} + \frac{3}{2}\left(\frac{dR_B}{dt}\right)^2 + \frac{2\sigma}{\rho_f R_B} = \frac{P_v - P}{\rho_f} \tag{2}$$

where R_B represents the bubble radius, P_v is the pressure in the bubble (assumed to be the vapor pressure at the liquid temperature), P is the pressure in the liquid surrounding the bubble, ρ_f is the liquid density, and σ is the surface tension coefficient between the liquid and vapor.

This equation gives the conditions of mechanical equilibrium of bubble in liquid .The first item of the equation represents second order vibration ,the second item is for the related item of radius change over time derivative, the third is for surface tension, and the fourth item is for the difference of inside and outside pressure of the bubble.

Neglecting the second order terms (which is appropriate for low oscillation frequencies) and the surface tension, this equation reduces to:

$$\frac{dR_B}{dt} = \sqrt{\frac{2}{3}\frac{P_v - P}{\rho_f}} \tag{3}$$

For

$$dV_B = 4\pi R_B^2 dR_B \tag{4}$$

And then

$$\frac{dV_B}{dt} = 4\pi R_B^2 \frac{dR_B}{dt} = 4\pi R_B^2 \sqrt{\frac{2}{3}\frac{P_v - P}{\rho_f}} \tag{5}$$

$$\frac{dm_B}{dt} = \rho_g \frac{dV_B}{dt} = 4\pi R_B^2 \rho_m \sqrt{\frac{2}{3}\frac{P_v - P}{\rho_f}} \tag{6}$$

If there are N_B bubbles per unit volume, the volume fraction can be expressed as:

$$\gamma_g = V_B N_B = \frac{4}{3}\pi R_B^2 N \tag{7}$$

And then the mass transfer equation can be expressed as:

$$\dot{m}_{fg} = N_B \frac{dm_B}{dt} = \frac{3\gamma_g \rho_g}{R_B}\sqrt{\frac{2}{3}\frac{P_v - P}{\rho_f}} \tag{8}$$

A Correction coefficient number F is considered for the empirical correction, the value of F may be different in process of condensation and vaporization. If the condensation process is considered, the bubble radius R_B is replaced by bubble nuclear radius R_{nuc}, the mass transfer is neglected because the mass transfer is very slow during the growing of bubble, the density of interior of bubble will decrease with growing of the bubble, γ_g is replaced by $\gamma_{nuc}(1-\gamma_g)$, and then :

$$\dot{m}_{fg} = F \frac{3r_{nuc}(1-\gamma_g)\rho_g}{R_{nuc}}\sqrt{\frac{2}{3}\frac{|P_v - P|}{\rho_f}}\,\mathrm{sgn}(P_v - P) \tag{10}$$

γ_{nuc} is the volume fraction of the nucleation sites.

In the simulation of CFX, $R_{nuc} = 1\mu m$; $\gamma_{nuc} = 5\times10^{-4}$; $F_{vap} = 50$; $F_{cond} = 0.01$.

4. Calculation results

4.1 Cavitation performance

Figure 2 shows the pump cavitation performance curve of the four different models. All the cavitation performance curves are d type curves [10], which means that the performances curve have two extreme value points (as shown in figure 2).What's more, the cavitation curve of YH have four extreme values .The head of circular arc shape blade model is higher than the head of rectangular shape blade model about 0.5 meter ,and higher than the head of elliptical shape blade model and cusp shape blade model about 1.5 meter under normal working conditions. However, the head of circular arc shape blade model decreases faster than the others under low NPSHA. In addition, the head of the models with trimmed blade are higher than the head of models with untrimmed blade under the normal working condition. However, it is smaller than the latter under the low effective cavitation margin, which means that the head of models with trimmed blade decrease faster than the head of models with untrimmed blade under the low effective cavitation margin.

The cavitation performance curve of elliptical shape blade model is very similar with the cavitation performance curve of cusp shape blade model, as shown in figure 2.

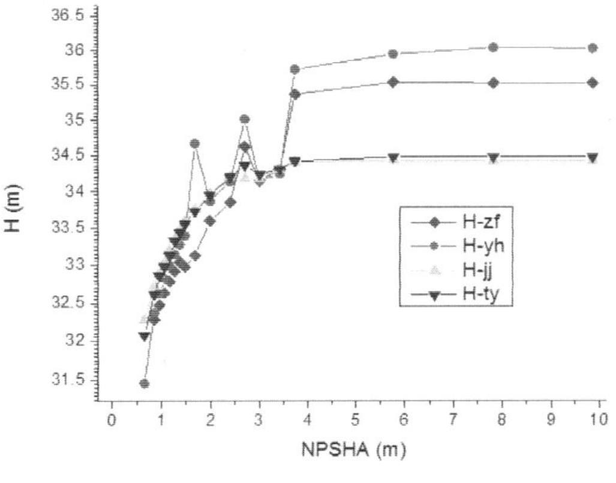

Figure 2. Cavitation performance of the four models

Figure 3 shows the changes of relevant physical parameters of the four models near the impeller inlet location about 2 mm, 25 points are selected on the straight line within the dark color area(as shown in picture a).Picture b is the pressure coefficient distribution picture, and its trend is consistent with the pressure distribution (as shown in figure d)because the value of C_p is proportional to the value of pressure. Pressure value of YH decreases firstly, and increases slowly near the suction surface of blade, while the pressure values near blade suction surface of other three models always decline. ZF model has the smallest value at the same location among the four models. The pressure distribution of TY model is almost the same with the pressure distribution of JJ model, while the pressure value of TY model is slightly bigger than the pressure value of JJ model at the same location.

Figure 3(c) gives the vapor phase volume fraction of the 25 points position which was told above. The change trend of volume fraction of ZF, TY, JJ models are almost exactly the same, but the volume fraction value of ZF is bigger than the volume fraction value of the JJ and TY at the same location. The volume fraction value of YH is bigger than the value of the other three models near blade suction surface, but smaller than the latter in the location where it is far away from the blade suction surface.

4.2 Physical parameter of the position near the impeller inlet analysis (inlet pressure P =100000Pa)

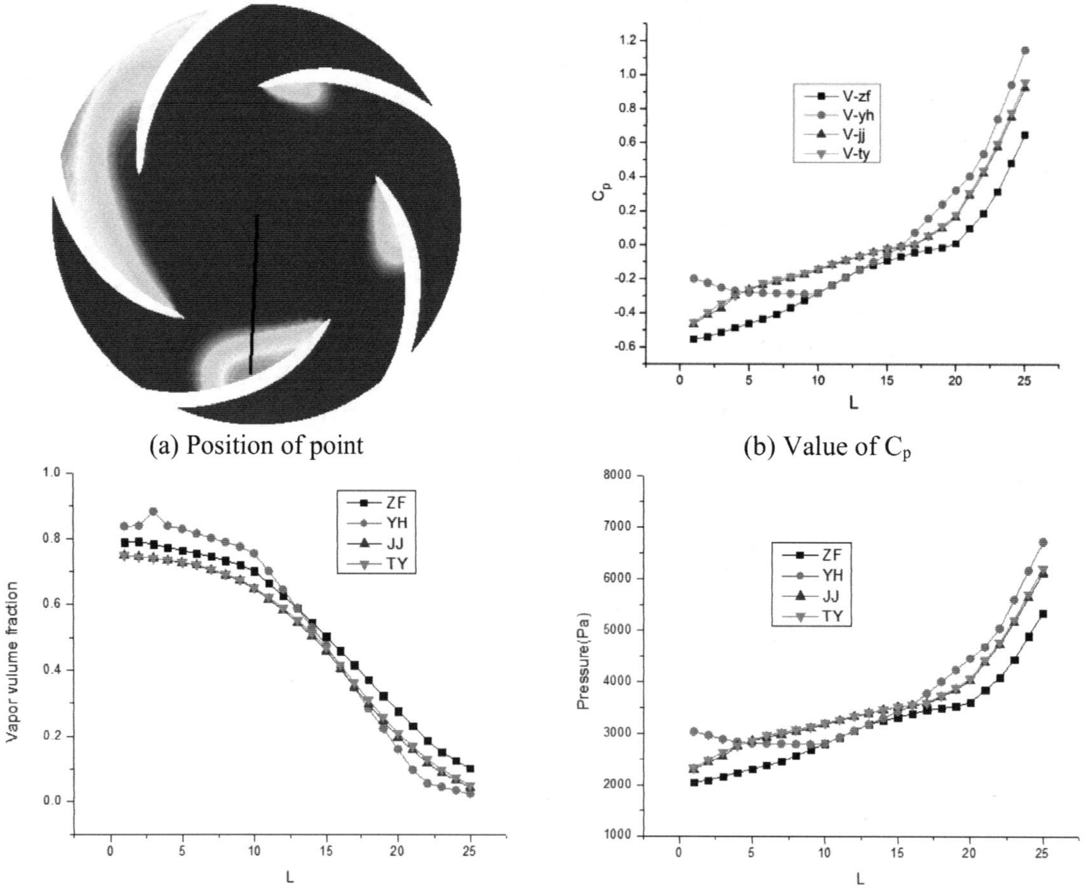

(a) Position of point (b) Value of C_p

(c) Vapor volume fraction (d) Pressure distribution

Figure 3. Values of physical parameters in the Graphical position at $P = 2\times10^4 Pa$

Figure 4 shows the pressure distribution cloud pictures of the four models under normal condition (P = 100000 pa) near the impeller inlet about 2 mm. The low pressure area of TY and JJ are relatively larger than the low pressure area of ZF and YH. The blade suction surface near the blade leading edge of ZF is easier to form low pressure area which can be seen from the picture ZF, while the blade suction surface near the blade leading edge of YH is the most difficult to form the low pressure area. However, YH has the biggest pressure value, as the picture shown in the red areas, which means higher head.

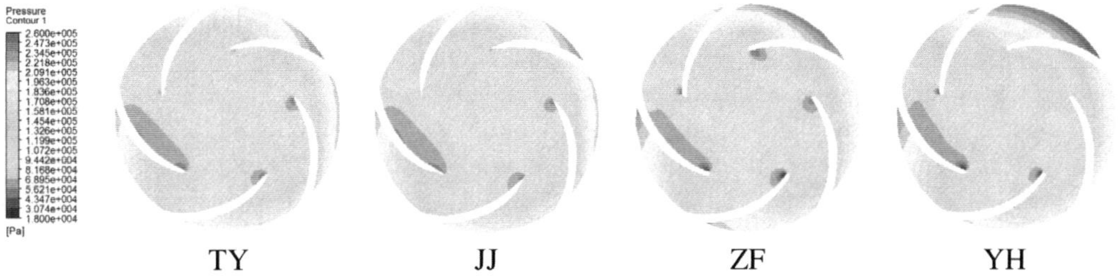

Figure 4. Pressure distribution of section near impeller inlet

4.3 Physical parameters of the section near impeller inlet analysis (inlet pressure P = 20000 Pa)

Figure 5 shows the volume fraction distribution cloud picture of the section near impeller inlet about 2 mm when the pressure of impeller inlet is 20000 Pa. As is shown in the picture, YH has the biggest volume fraction value in this section (which can be seen from the dark area).In addition, the volume fraction distribution of TY is almost the same with JJ, and the volume fraction distribution of ZF is similar to YH. In the cavitation area, the volume fraction value of YH and ZF are larger than TY and JJ. The area in the block diagram of YH and ZF has a flow structure which is formed by the interaction flow between vapor and liquid. However, this phenomenon does not occur in TY and JJ.

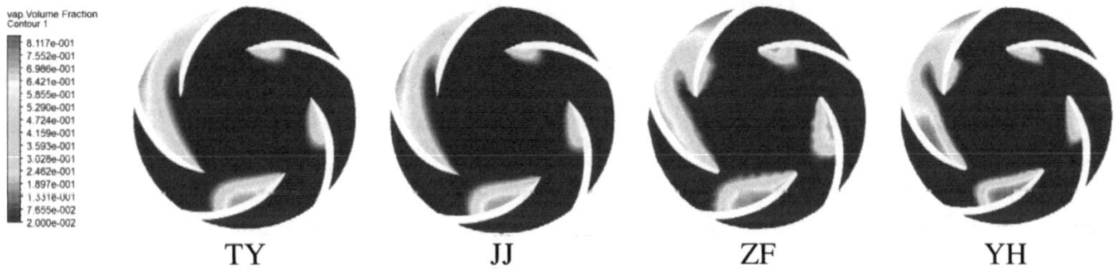

Figure 5. Vapor phase volume ratio distribution

Figure 6 shows the relative velocity distribution of the vapor phase of the corresponding location in figure 5.The relative velocity of the edge of the cavitation area is smaller, and the relative velocity of the center of the cavitation area is relatively larger. The relative velocity distributions of YH and ZF are relatively homogeneous. In block diagram of YH, the relative velocity of vapor phase is very slow, which is due to the condensation of liquid to vapor, this phenomenon also exists in ZF. However, TY and JJ do not have this phenomenon.

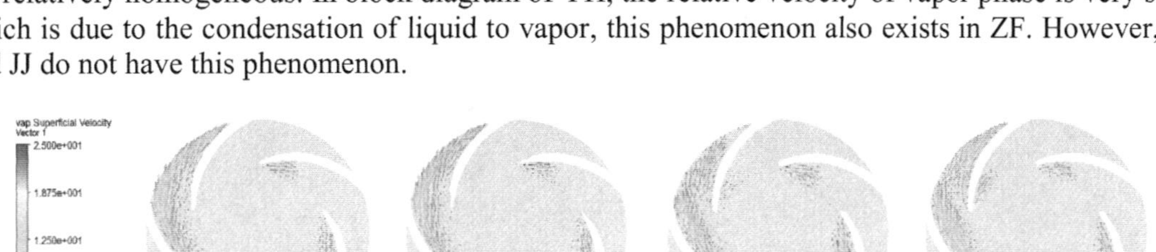

Figure 6. The relative velocity of vapor phase

5. Conclusion and discussion

This paper studies four different hydraulic models cavitation performance, we can reach the following conclusion:

1) The hydraulic performance of the models with trimmed blade is better than the model with untrimmed blade under normal working condition. Perhaps the main reason is that the shedding vortex frequency and its strength of trimmed models are different from the untrimmed models.

2) In the case of the low effective cavitation margin, the head of models with trimmed blade decrease slightly faster than the models with untrimmed blade.

3) Different hydraulic models have different cavitation performance curve shape. Although the cavitation performance curves of four kinds of models are similar to d type curve , its extreme value will change with different blades.

Acknowledgements

The authors gratefully acknowledge the financial support of National Science and technology Support Plan of China (2011BAF14B01), a project funded by the Priority Academic Program Development of Jiangsu Higher Education Institutions (BK2009218) and the .Conflict of interests: The authors declare no conflicts of interest. The support of National Science and technology Support Plan of China (2011BAF14B01), a Project Funded by the Priority Academic Program Development of Jiangsu Higher Education Institutions (BK2009218) have no conflict of interest.

The present study is also supported financially by the Research and Innovation projects of Jiangsu University (KYXX 0040) and the Research Project of Jiangsu University(13A367).

References

[1] Huang B, Wang G and Yuan H 2010 *J. of Hydrodynamics, Ser. B* **22**(5) 798-804
[2] Goncalvès E and Charrière B 2014 *Int. J. of Multiphase Flow* **59** 54 -72
[3] Mørch K A 2009 *J. of Hydrodynamics* **21**(2) 176-89
[4] Caupin F and Herbert E 2006 *Comptes Rendus Physique* **7**(9) 1000-17
[5] Xu Y, Wu Y L and Tang X L 2003 J. of tsinghua university (Sci. and Tec.) **2003**(10) 32-37
[6] Yuan D Q, Chen X Y, Bai B and Cong X Q 2009 *J. of Drainage and Irrigation Machinery Eng.* **27**(4) 269-72 (in Chinese)
[7] Brenren C E 2013 *Cavitation and bubble dynamics* (Cambridge: Cambridge University Press)
[8] Xu Y B 2008 *Resistance imaging technology study based on horizontal tube gas-liquid phases flow* (Tianjin: PhD thesis of Institute of electrical and automation engineering, Tianjin University)
[9] Shi L L 2004 *Digital image recognition technology of gas liquid two phase flow parameters* (Hanzhou: PhD thesis of Institute of information science and engineering, Zhejiang University)
[10] Pan Z Y and Yuan S Q 2013 *Fundamentals of Cavitation in Pumps* (Zhenjiang: Jiangsu University Press) chapter 5 pp 69-72

IOP Conf. Series: Materials Science and Engineering **72** (2015) 032025 doi:10.1088/1757-899X/72/3/032025

Numerical calculation for cavitation flow of inducer

C Ning, Y Wang, Z T Zhu, S F Xie, L F Zhao and Z C Liu
Research Center of Fluid Machinery Engineering and Technology, Jiangsu University, Zhenjiang 212013, China

E-mail: dream0830@163.com

Abstract: Inducer has significant effect on improving the cavitation characteristic of centrifugal pump. Several inducers were designed and modeled by Pro/E software. The mesh of flow field was done by ICEM and then was imported to ANSYS CFX to analyze the inducer's cavitation characteristic. Effects of the blade number on the performance of an inducer are investigated in the present paper. The inducers were designed on the basis of identical design flow rate and identical pressure elevation at nominal flow rate. The study focuses on the steady behavior of the inducers in cavitating conditions. Evolutions of performance, torque, mass flow rate, and amplitude of radial forces on the shaft according to the inlet pressure are considered. Furthermore, cavitation instabilities are analyzed in the study. The purpose of the present study is to investigate the pressure distribution and vapour volume fraction distribution through numerical simulations using the Navier-stokes solver with computational fluid dynamics (CFD) code.

1. Introduction

The phenomenon of cavitation in centrifugal pump is an important feature of the process of its operation and one of the concern issues of the design, manufacture, use and management of a centrifugal pump.

In the process of pump operation, for some reason, when the absolute pressure of conveying liquid medium dropped to vaporization pressure at that temperature in the flow section of the local area, liquid in this area begin to boil, produce steam, form bubbles. These bubbles flow forward along with the liquid until a certain place of high pressure, high pressure liquid around the bubbles result in the bubbles shrink sharply that broken, and at the same time, the liquid particle filling holes at high speed, hit each other and form water hammer. This phenomenon will make the flow components suffer from corrosion damage when occurred in the solid wall. The process which bubbles was formed and then ruptured and cause damage to flow components is the process of water pump cavitation.

Inducer is actually a kind of axial flow impeller with small blade load, the effect of the centrifugal force is small. It is not easy to produce the two phase separation of liquid and bubbles. It has high performance of resistance to cavitation. It also can work normally and not blocked even under the condition of local cavitation. For centrifugal pump, the model of installing inducer in front of the centrifugal impeller, the essence of which is greatly reduces the required NPSH of pump unit, was used more in order to improve its performance of resistance to cavitation. In order to improve both the cavitation resistance of the inducer itself and the head of the inducer to meet the requirement of high cavitation performance of the main impeller, the model of varying pitch inducer is often used. Under the condition of same impeller import outer diameter, the variable pitch inducer can more effectively reduce the necessary NPSH of pump unit than uniform pitch inducer.

Perfect design theory of inducer is summarized in the early 70s by NASA (National Aeronautics& Space Administration) and the former Soviet union respectively according to their work in the field of aerospace. After more than 20 years, the design theory and the development effort of inducer in our country is basically on the basis of the job.

Huang Jiande [1], further tested four different kinds of inducers on blade number and blade length, and find out the influence of the above phenomenon happened and the cavitation performance by impeller geometry, and made a preliminary study on the reason.

Zhu Zuchao, etc. [2], established a relatively perfect design method of varying pitch inducer through the analysis of the theoretical basis of inducer design, the calculation formula of its main parameters was given, the calculation method of cavitation performance and theoretical head of the inducer was deduced[3] .

Sun Jian [4], put forward that both the inducer itself no cavitation and the export head meet the requirements of import energy of main impeller of the centrifugal pump were the design principles, and further expounded the structure matching relationship between the inducer and the centrifugal pump impeller.

In this study, cavitation instabilities are analyzed. Effects of the blade number on the performance of an inducer are investigated in the present paper.

2. Design and modeling of the centrifugal pump and inducers

Selecting a centrifugal pump with the specific speed is 103 to numerical simulation and analysis of the design flow rate is 660 m^3 / h, head is 67 m, the speed is 1450 r/min, import a diameter of 250 mm, entity modeling in Pro/E, in order to avoid impact on the calculation results of velocity grads, import and export of the spiral case was extended appropriately.

To match the pump ,the flow rate of inducer is 660 m^3 / h, head is 49 m, speed is 1450r/min. NPSH is 1.2 m. As shown in figure 1.

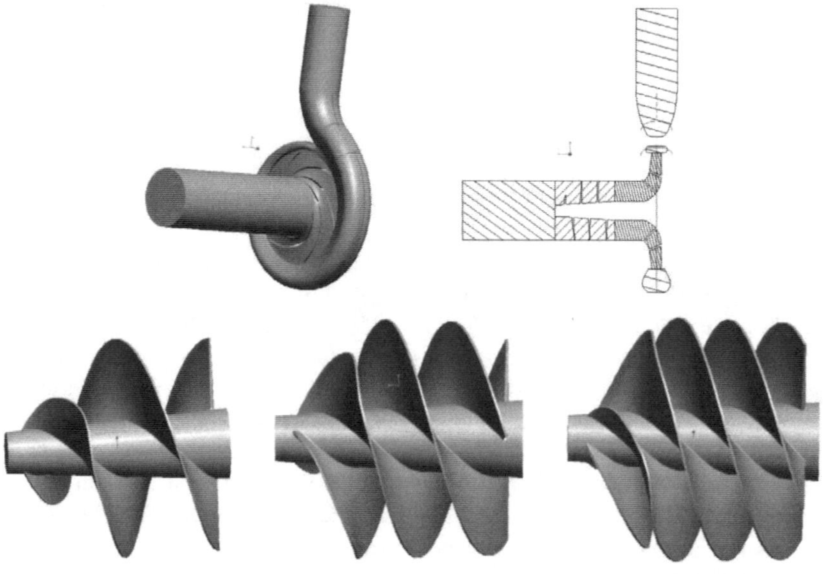

(a) 2 blades (b) 3 blades (c) 4 blades

Figure 1. Three-dimensional model of inducers

3. Calculation method of cavitation flow

3.1. Meshing

The leakage of the tip clearance and the aroused backflow should be analyzed, so when the water model was created, a tip clearance of 1 mm was considered.

Water model was meshed after imported to ANSYS-ICEM software. Due to the structure of inducer is relatively complex, adaptive unstructured grid method was used. Because the tip clearance was tiny, grids near the tip were refined. Cavitation mainly happens near the blade surface, so in order to accurately simulate the cavitation development, grids near the blade surface were also refined, and the quality of grids were above 0.3.

3.2. Calculation method

For most of cavitation flow simulation, the typical homogeneous flow model in ANSYS CFX was used. The mass transfer process between the liquid phase and vapor phase was calculated by Rayleigh-Plesset equation[5].

Continuity equation and the Reynolds averaged Navier-Stokes equations were used to simulate the flow field in the inducer, and the RNG k-ε turbulence model was used to close the equations.

The total pressure of inlet and the mass flow of outlet were selected as the boundary condition. By adjusting the inlet total pressure, the development of inducer's cavitation was controlled. The initial conditions of the cavitation calculation were depended on the results of non-cavitation results. The 25°C water was chosen as the medium, and the vapor pressure was set to be 3169 Pa. The average diameter of bubbles was set to be 2.0×10^{-6} m. The frozen-rotor interface was selected as the interface between inducer and the inlet or outlet. The convergence precision was set to be 1.0×10^{-5}.

4. Results analysis

4.1. Pressure distribution when the medium is water

Figure 2 shows that the total pressure distribution of each vane when the inlet pressure is set as 40000Pa. It can be seen from the figure 2 (1), (2), (4), there is a low pressure area in the impeller channel. Compare figure 2 (1) with figure 3 (2), (3), (4), you can see that (1) has a more significant area of low pressure. And (3) has a less significant area of low pressure.

4.2 pressure distribution when the mediums are water and vapour

Figure 3 shows that the total pressure distribution of each impeller when the mediums are water and vapour.

4.3 Vapour volume fraction distribution when the mediums are water and vapour

Figure 4 shows that the vapour volume fraction distribution of each vane when the inlet total pressure is set as 40000 Pa. As can be seen from the figure, cavitation is distributed in the inlet of blade, and vapour volume fraction of each vane in figure 4(2) are reduced compared with the original impeller. This phenomenon of figure 4(2) is that vapour volume fraction is disappear and cavitation occurs in some area of the back of the impeller .

4.4 Vapour volume fraction distribution under 30000Pa

In order to magnify the effect of cavitation, the inlet total pressure is set as 30000 Pa.

Figure 5 shows that the vapour volume fraction distribution of each vane when the inlet total pressure is set as 30000 Pa. As can be seen from the figure, the range of cavitation on the impeller is small, but the location of cavitation on the blades of inducer is different. With the increasing of the blade number of inducer, cavitation begins to occur on the back of the blades.

IOP Conf. Series: Materials Science and Engineering **72** (2015) 032025 doi:10.1088/1757-899X/72/3/032025

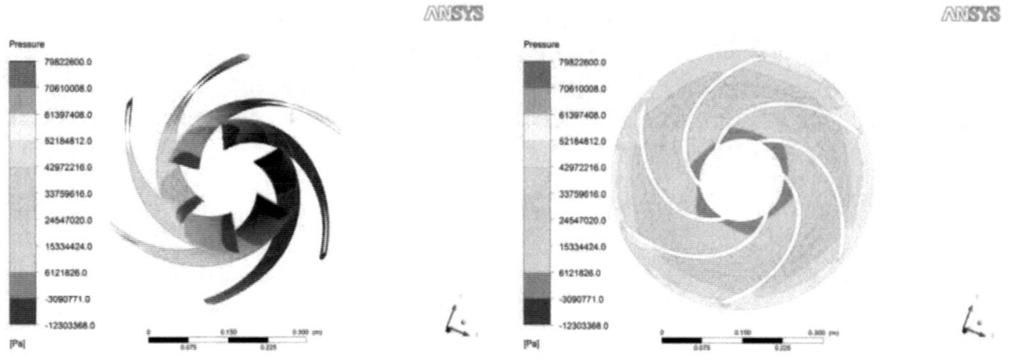

(1) no inducer

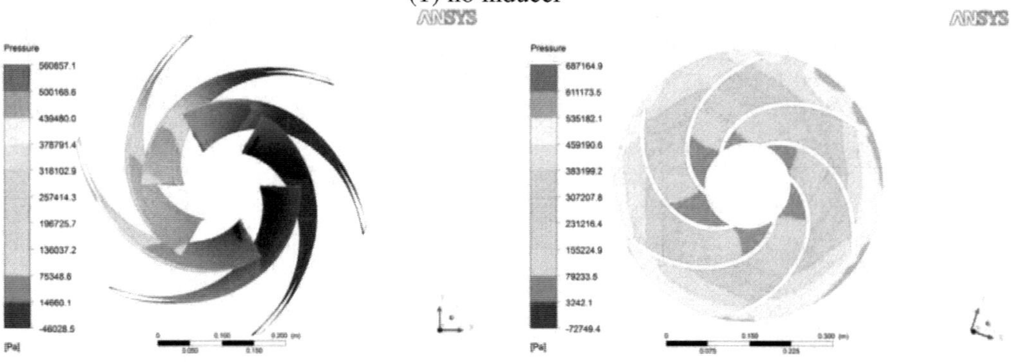

(2) with inducer of 2 blades

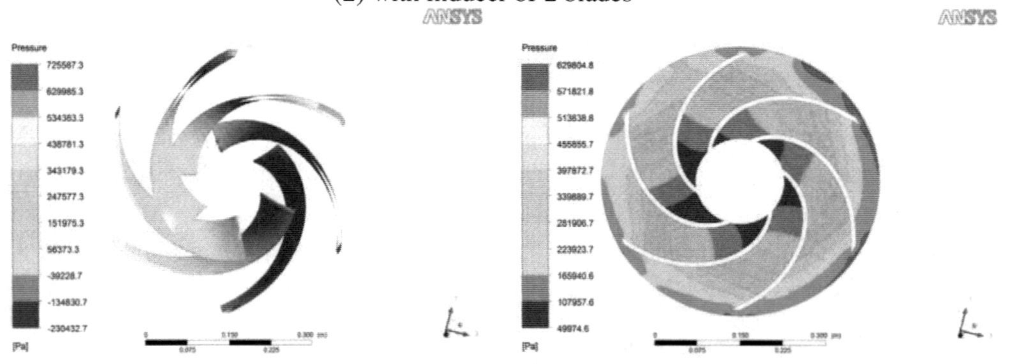

(3) with inducer of 3 blades

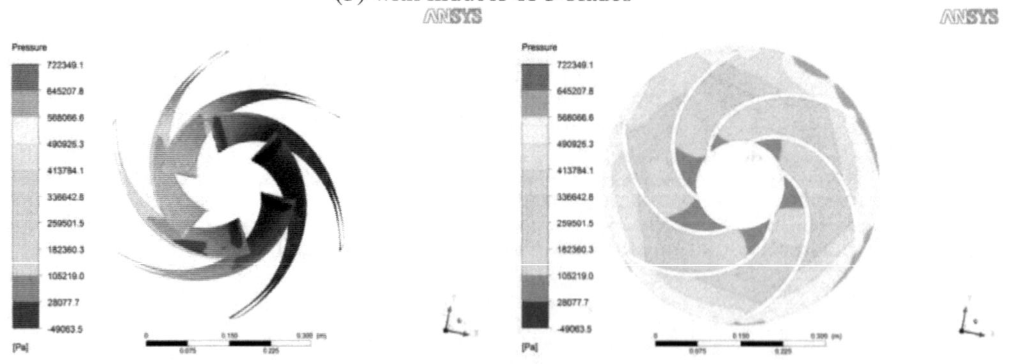

(4) with inducer of 4 blades

Figure 2. Pressure distribution under 40000Pa

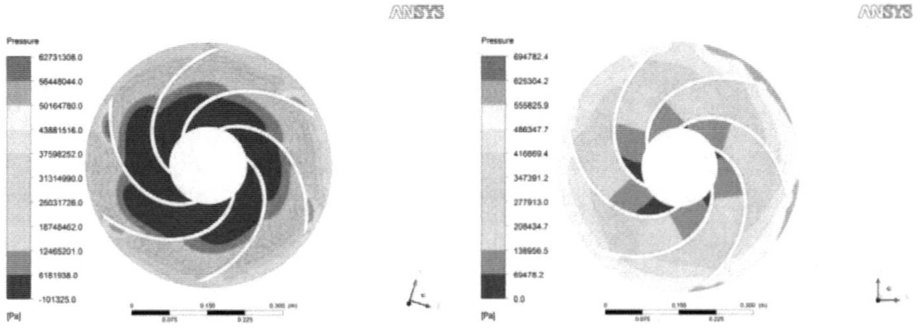

(1)no inducer (2) with inducer of 2 blades

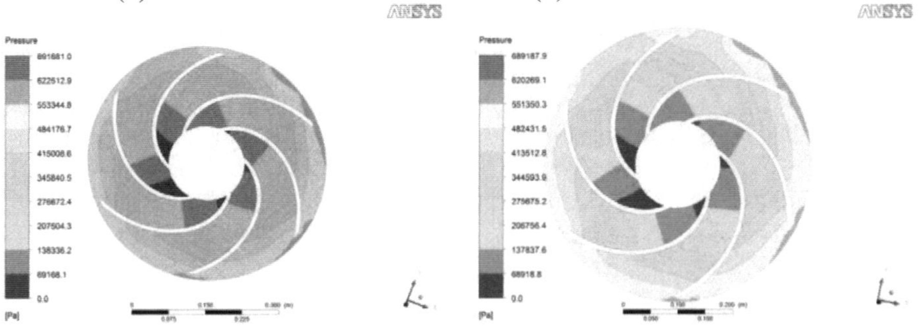

(3) with inducer of 3 blades (4) with inducer of 4 blades

Figure 3. Pressure distribution under water and vapour

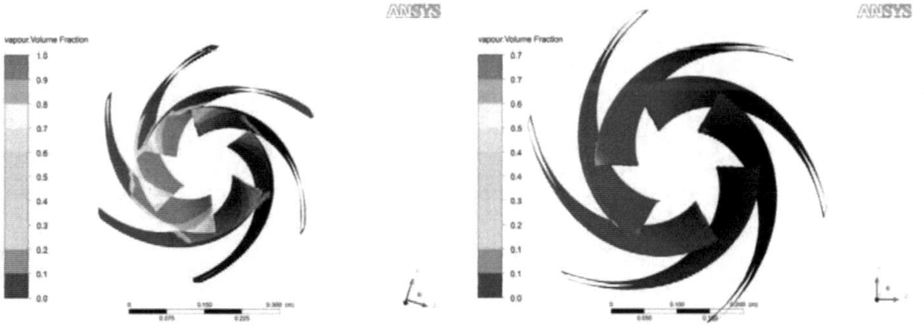

(1) no inducer (2) with inducer

Figure 4. Vapour volume fraction distribution under 40000Pa

5. Conclusions

1) The inducer has influence on cavitation performance of centrifugal pump, can greatly improve the cavitation performance of the pump.

2) With the increase of the blade number of inducer, the pressure distribution become non-uniform, so the even distribution of relative velocity is non-uniform. In order to prevent the alternate blade cavitation, hope blade for the odd number.

3) The blade number of main impeller is better equal to multiple the blade number of inducer, the flow which contains certain rotating components after come out from the inducer enters the primary impeller, thus the angle of fluid flow which enters the primary impeller is increasing.

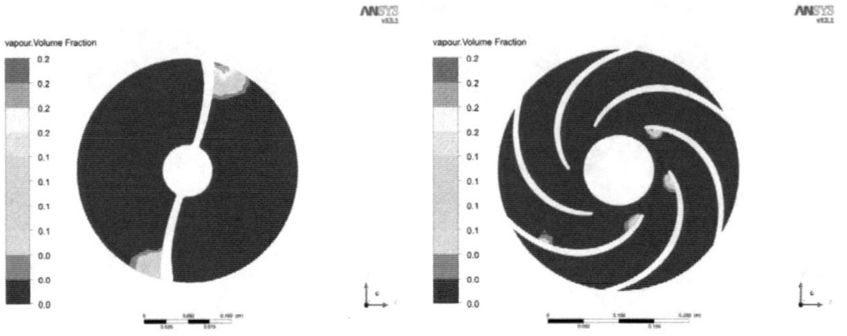

(1) with inducer of 2 blades

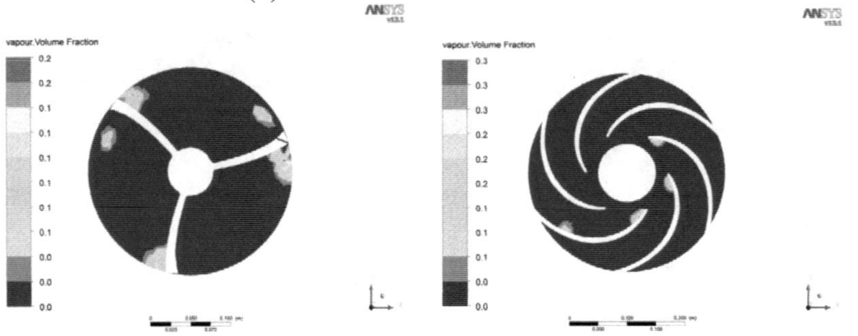

(2) with inducer of 3 blades

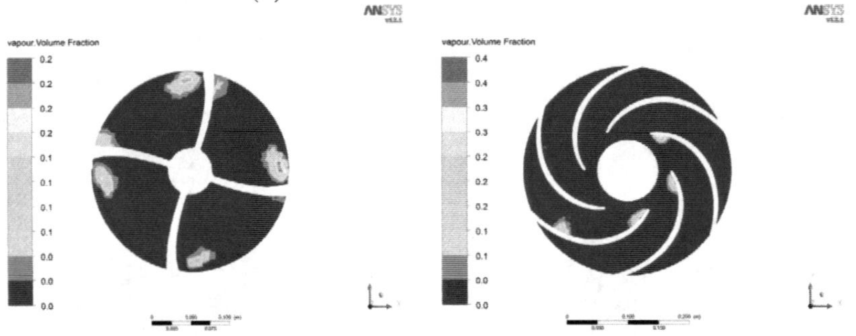

(3) with inducer of 4 blades

Figure 5. Vapour volume fraction distribution under 30000Pa

Acknowledgment

The authors gratefully acknowledge the financial support of National Science and technology Support Plan of China (2011BAF14B01), a project funded by the Priority Academic Program Development of Jiangsu Higher Education Institutions (BK2009218).Conflict of interests: The authors declare no conflicts of interest. The support of National Science and technology Support Plan of China (2011BAF14B01), a Project Funded by the Priority Academic Program Development of Jiangsu Higher Education Institutions (BK2009218) have no conflict of interest.

References

[1] Huang J D, Gu C G 2000 *J. of Aerospace Power* **2000**(6) 142-6 (in Chinese)
[2] Zhu Z C, Wang X R and Wang L Q 1997 *J. of industry machinery* **1997**(4) 102-4
[3] Zhu Z C and Wang Y Q 1998 *J. of Zhejiang University* **1998**(3) 197-200
[4] Sun J, Kong F Y and Jiao Q B 2006 *J. of fluid machinery* **2006**(3) 19-24
[5] Mejri I, Bakir F and Kouidri S 2006 *J. Power Energy* **220** 217

Numerical analysis of head degrade law under cavitation condition of contra-rotating axial flow waterjet pump

D Huang and Z Y Pan
Research Center of Fluid Machinery Engineering and Technology, Jiangsu University Zhenjiang, jiangsu 212013, China

E-mail: a08110139@126.com

Abstract. In order to study the flow-head characteristic curve, the *SST* turbulence model, homogeneous multiphase model and Rayleigh-Plesset equation were applied to simulate the cavitation characteristics in contra-rotating axial flow waterjet pump under different conditions based on ANSYS CFX software. The distribution of cavity, pressure coefficient of the blade at the design point under different cavitation conditions were obtained. The analysis results of flow field show that the vapour volume distribution on the impeller indicates that the vapour first appears at the leading edge of blade and then extends to the outlet of impeller with the reduction of Net Positive Suction Head Allowance (NPSHA). The present study illustrates that the main reason for the decline of the pump performance is the development of cavitation, and the simulation can truly reflect the cavitation performance of the contra-rotating axial flow waterjet pump.

1. Introduction

Cavitating flow are complicated because it is a rapid phase change phenomenon, which often occurs in the high-speed or rotating fluid machineries. Cavitation adversely affects the performance of hydraulic machinery, cause noise, vibrations, erosion and mechanical structure damage[1-3].So it is necessary to study caviation performance of hydraulic machinery.

Experimental methods and CFD are the main approaches to research the cavitation performance. Significant progress has been made on researching cavitation performance using experimental methods[4-6].In recent years, because of the evolution of CFD technologies and the time-consuming of experimental investigation, numerical simulation has been widely used to reach the cavitation performance and the internal flow fields in hydraulic machinery.

With the development of waterjet propulsion technology, the contra-rotating axial flow waterjet pump is considered to be an ideal form of propulsion for its compact structure, good cavitation performance and wide stable operation zone[7]. When it runs under the off-design conditions, the torque and thrust alter significantly due to cavitation[8]. Liu chengjiang found that Inlet duct cavitation lags behind the blade cavitation, and the nozzle cavitation in form of spatial cavitation occurs ahead of the blade cavitation, but there is no cavitation on nozzle wall[9]. Jules W. Lindau, Christopher Pena[10-11] explored the cavitation performance of waterjet pump, AXWJ -2,designed by Michael, then the results show the thrust breakdown coincide with cavitation choking in rotator passage. In this study, the cavitation performance of contra-rotating axial flow waterjet pump will be revealed under different conditions.

IOP Conf. Series: Materials Science and Engineering **72** (2015) 032026 doi:10.1088/1757-899X/72/3/032026

2. Physical model

The computational domain established by the Pro/E software consists of inlet flow channel, front impeller, axial clearance, rear impeller, guide vanes, outlet flow channel. Pump heads H_f=2m,H_r=8m and flow Q=0.3m^3/s were specified in pump design. The rotational speed were determined as N_f=N_r=1450 r/min. The numbers of blades were chosen as 5,4 and 7 for the front impellers, rear impellers and guide vanes respectively. The physical model of contra-rotating axial flow waterjet pump is shown in Figure 1.

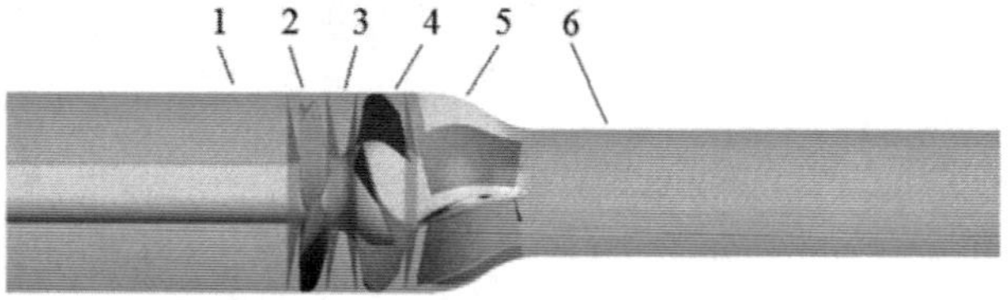

Figure 1. Physical model of contra-rotating axial flow waterjet pump: 1.inlet flow channel. 2.front impeller. 3.axial clearance. 4.rear impeller. 5.guide vanes 6.outlet flow channel.

3. Numerical simulation

3.1 Mesh generation

The grid is generated for each part by the software ANSYS ICEM CFD and the grid number and distribution are guaranteed consistently at the interface transition. In order to effectively control the mesh distribution in the contra-rotating axial flow waterjet pump, the whole hexahedral strategy is adopted. It can be observed that the results on successive grids become closer to each other and with the further refinement the results become gird independent. The final computational grid number is 1840000. Figure 2 show the grid for the computational domain: (a) Grid for front impeller, rear impeller and guid vanes (b) Grid for contra-rotating axial flow waterjet pump (c) Grid for the pump just with the front impeller.

3.2 Numerical methods

For the boundary conditions, in the cavitation simulation process, the mass flow rate was imposed at the outlet and the total pressure was specified at the inlet, where gas volume fraction of 0 and liquid volume fraction of 1.The results of noncavitating flow was used as the initial flow field of the cavitating case to improve the convergence speed and accuracy. The saturated vapour pressure is 3169pa, when the liquid temperature is 25°C.The frozen rotor interface model was employed for the interface between the rotational domains and stationary ones. The wall of impellers was set as rotational boundary conditions, while others are defined as no slip boundary conditions.

3.3 Cavitation model and turbulence model

Cavitation terms, Rayleigh-Plesset model, are used in this study. The evaporation and condensation rates are given as follows:

$$R_e = F_e N_b \rho_v 4\pi R_b^2 \sqrt{\frac{2}{3}\frac{|p_v - p|}{\rho_l}} sign(p_v - p) \tag{1}$$

$$R_c = F_c N_b \rho_v 4\pi R_b^2 \sqrt{\frac{2}{3}\frac{|p_v - p|}{\rho_l}} sign(p_v - p) \tag{2}$$

where P_v, vapour pressure; F_e, evaporation coefficient; F_c, condensation coefficient. F_e=50, F_c=0.01.The turbulence model, Shear Stress Transport model, is employed to make the control equation complete.

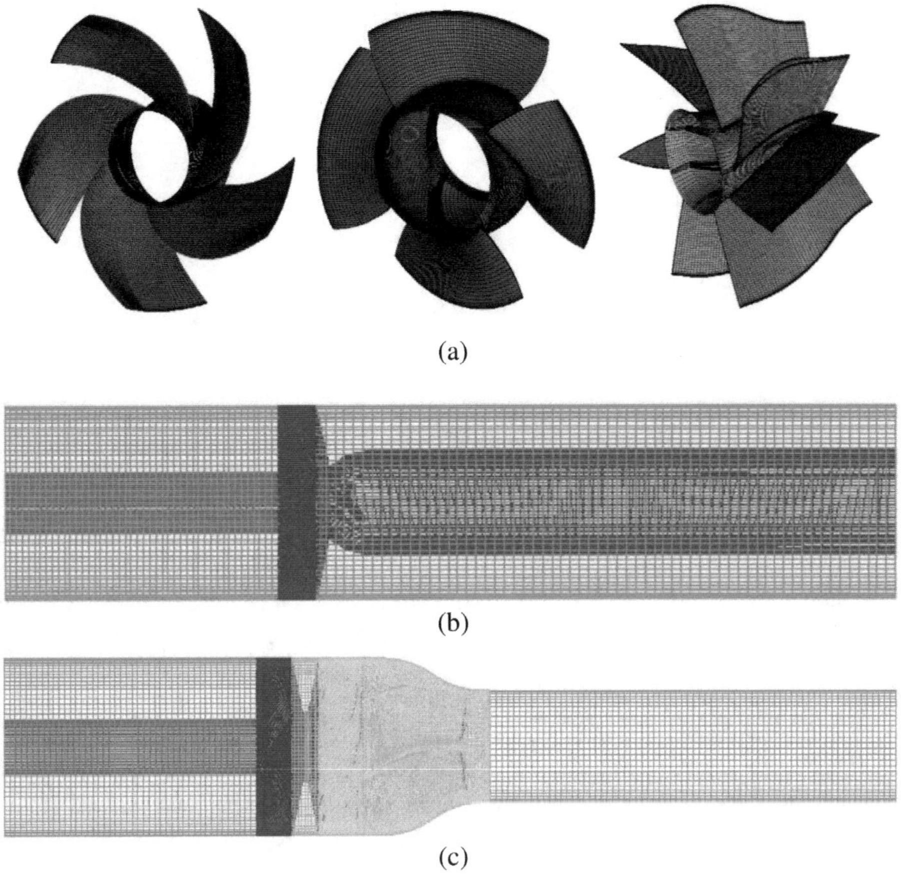

Figure 2. Grid for the computational domain: (a) front impeller, rear impeller and guid vanes. (b) axial flow waterjet pump (case 1). (c) pump just with the front impeller (case 2).

4. Results and discussion

4.1 The cavitation performance curves

Net Positive Suction Head (NPSH) is usually used to describe the problem of cavitation in pumps, which could be expressed as

$$\mathrm{NPSH} = \frac{P_{\mathrm{in}}}{\rho_l g} + \frac{v_{\mathrm{in}}^2}{2g} - \frac{P_v}{\rho_l g} \tag{3}$$

where P_{in} and v_{in} are the static pressure and velocity in the fluid close to the front impeller inlet, respectively.

The comparison of cavitation performance for the pump just with the front pump and the contra-rotating axial flow waterjet pump, called case1 and case2 respectively in the following study, is shown in Figure 3.The critical NPSH ,NPSHc, is defined as the NPSH value when the head of the pump drops by 3%. The results show that NPSHc in case2 (NPSH=3.76m) is smaller than that in case1(NPSH=4.58m),which means the cavitation phenomenon in case2 appears later than case1. The comparison of cavitation performance for case1 and case2 will be revealed in detail from three different cavitation conditions.

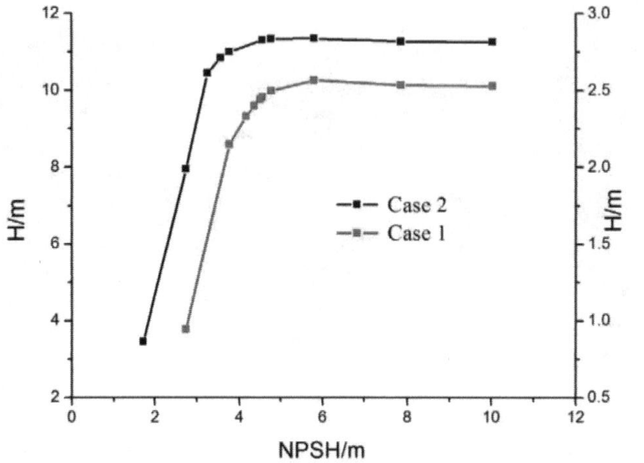

Figure 3. Cavitation performance curve

4.2 The pressure coefficient distribution under different conditions
Cavitation may alter the pressure on the pressure surface and suction surface, according to the results showed in Figure 3, some detailed investigation will be proceed. Fig.4 shows CFD-Calculated surface pressure coefficient C_p distribution around front impeller blades under different NPSHA at the design flow rate of $0.3m^3/s$. The pressure coefficient C_p is defined as

$$C_P = \frac{P - P_{in}}{0.5\rho u^2} \tag{4}$$

where P is the static pressure, P_{in} is the inlet total pressure, u is circumferential velocity of the impeller.

From Figure 4, it shows that the C_p distribution is similar at 0.3-0.7 streamwise under different NPSH before the deep cavitation occurs. The difference between case1 and case2 increases with the decrease of NPSH value at 0.8-1.0 streamwise(close to the tip of blade).Before the deep cavitation occurs, the trends of C_p distribution under different NPSH are the same and the C_p on the suction surface in case2 is always little higher than that in case1.The results show the good cavitation performance of the contra-rotating axial flow waterjet pump (case2).

4.3 The vapour volume fraction distribution
The vapour volume fraction distributions on blades of contra-rotating axial flow waterjet pump are shown in Figure 5. The cavitation phenomenon first occurs at the tip of the blades' suction surface, with the NPSH decrease, the surface covered by vapor bubbles increases. When NPSH=2.74m, the vapour region covered 75% of the suction surface of the blades, the cavitation region extended to the outlet of the impeller, at this time, the deep cavitation occurs.

Figure 6 shows the vapour volume fraction distribution in the flow passage contra-rotating axial flow waterjet pump under three conditions: the incipient cavitation condition (NPSH=4.58m), the critical cavitation condition (NPSH=3.76m), the deep cavitation condition (NPSH=1.74m).When NPSH=4.58m, the cavitation just occurs at the leading edge on the suction side of the front impeller, there is no cavitation phenomenon in the rear impeller. When NPSH=3.76m, the vapour covered on the pressure side of the front impeller and leading edge of suction side of the rear impeller. When NPSH=1.74m, the deep cavitation occurs, the flow passage is filled with bubbles, which choked the channel. Cavitation choking significantly alter the torque and the thrust. From Figure 6, it shows the cavitation first occurs in the front impeller and then extend to the rear impeller and guide vines due to the pre-pressure on the rear impeller.

IOP Conf. Series: Materials Science and Engineering **72** (2015) 032026 doi:10.1088/1757-899X/72/3/032026

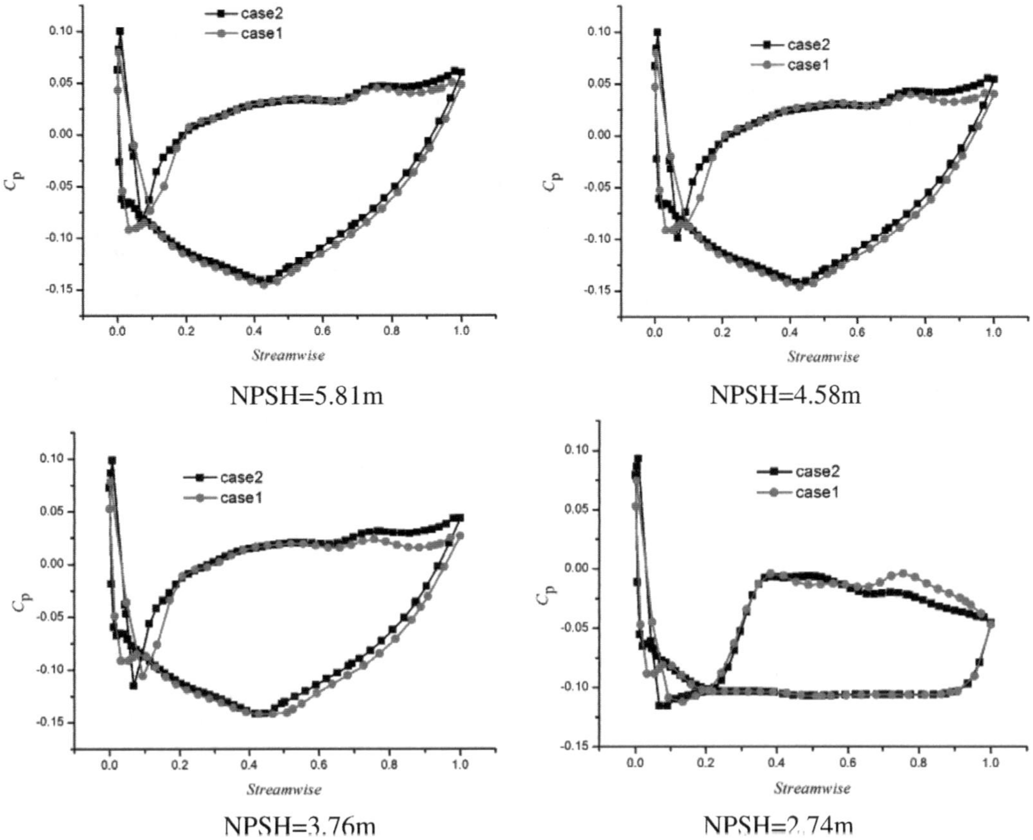

Figure 4. Pressure coefficient C_p distribution around front impeller blades

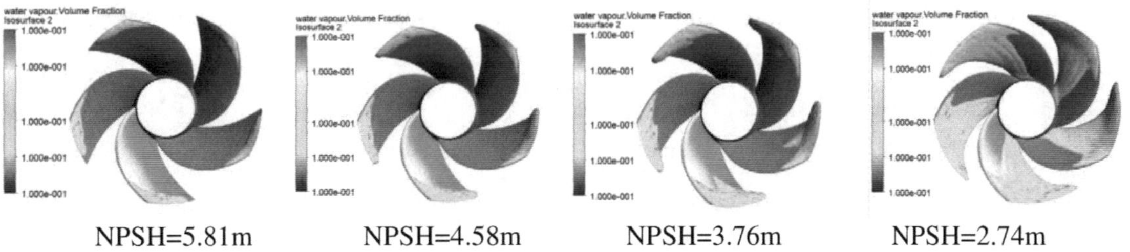

Figure 5. Vapour volume fraction distribution on blades

5. Conclusions

1) The NPSHc (NPSH=3.76m)of contra-rotating axial flow waterjet pump is smaller than that (NPSH=4.58) of the pump just with the front impeller. The contra-rotating axial flow waterjet pump has good cavitation performance.

2) With the NPSH decrease, the surface covered by vapor bubbles increases. At the deep cavitation condition, the vapour region covered 75% of the suction surface of the blades, the cavitation region extended to the outlet of the impeller

3) Cavitation first occurs in the front impeller and then extend to the rear impeller and guide vines due to the pre-pressure on the rear impeller.

NPSH=4.58m NPSH=3.76m NPSH=1.74m

Figure 6. Vapour volume fraction distribution in the flow passage

References

[1] Yongyan N, Shouqi Y, Zhongyong P and Jianping Y 2008 *Chinese J. of Mech.Eng.* **21**(5) 46-9 (in Chinese)

[2] Kumar P and Saini R P 2010 *Renewable and Sustainable Energy Review* **14**(1) 374-83

[3] Qiongfang Y, Yongsheng W and Zhihong Z 2011 *Chinese Journal of Hydrodynamics* **26**(5) 538-50 (in Chinese)

[4] Zhang D S, Shi W D, Chen B, Long F and Guan X F 2013 *Fluid Machinery* **2013**(4) 005

[5] Pereira F, Salvatore F, Felice F D and Elefante M 2002 Experimental and numerical investigation of the cavitation pattern on a marine propeller *24th Symposium on Naval Hydrodynamics (Fukuoda, Japan, 8-13 July 2002)*

[6] Bakir F, Kouidri S, Noguera R and Rey R 2003 J. Fluid Eng. **125**(2) 293-301

[7] Furukawa A, Cao Y, Okuma K and Watanable S 2000 Experimental study of pump characteristics of contra-rotating axial flow pump *Proc. 2nd Int. Symp. on Fluid Machinery and Fluid Eng. (Beijing, China, Oct 2000)* 245-252.

[8] Sun C L, Wang Y and Huang B 2011 *J. of Harbin Engineering University* **2011**(7) 008 (in Chinese)

[9] Liu Y S and Wang C 2009 *J. of Mech. Eng.* **12**(45) 76-83

[10] Lindau J W, Pena C, Baker W J, Dreyer J J, Moody W L, Kunz R F and Paterson E G 2012 Modeling of Cavitating Flow through Waterjet Propulsors *2nd International Symposium on Marine Propulsion (Hamburg, Germany, June 2011)*

[11] Lindau J W, Boger D A, Medvitz R B and Kunz R F 2005 *J. Fluid Eng.* **127**(5) 995-1002.

IOP Conf. Series: Materials Science and Engineering **72** (2015) 032027 doi:10.1088/1757-899X/72/3/032027

Numerical research on the cavitation characteristics for typical conditions of a centrifugal pump with whole flow passage

L Cao, Z W Wang [1], Y X Xiao, Y Y Yao and W Zhu

State Key Laboratory of Hydroscience and Engineering & Department of Thermal Engineering, Tsinghua University, Haidian District, Beijing 100084, China

E-mail: wzw@mail.tsinghua.edu.cn

Abstract. Cavitation is one of the key issues affecting the safe and stable operation of centrifugal pumps. This research conducted numerical simulations of the 3-D turbulent flow in the whole flow passage of a centrifugal pump using RANS method. The calculation results of cavitation characteristics agreed well with the experimental results, which were based on SST *k-ω* turbulence model and Zwart-Gerber-Belamri cavitation model. This paper analysed the cavitation development process and the corresponding pump performance for three typical conditions, namely large discharge condition, design discharge condition and small discharge condition, by changing the available Net Positive Suction Head (*NPSHa*). For large discharge condition, the incipient *NPSHa* was large, while for design discharge condition and small discharge condition, the incipient *NPSHa* values were almost the same and both small. As the flow rate decreased, the critical *NPSHa* decreased as well, and the cavitation positons gradually shift from the pressure surfaces of some blades to the suction surfaces. At the same time, the tongue has greater effect with larger flow rate and the cavitation becomes less unsteady with the decrease of flow rate. With similar vapour volume, cavitation on the blade pressure side more easily leads to the drop of pump performance. Therefore, more attention should be paid to the cavitation characteristics of centrifugal pumps in large flow conditions in hydraulic design stage.

1. Introduction

Since cavitation is one of the critical problems that restrict the promotion of pump performance, cavitation mechanism and control measures for centrifugal pumps are researching focuses all the time. Cavitation characteristics are not the same in different flow rates, and different cavitation condition will affect the pump performance to different degree. It is meaningful to probe into the cavitation characteristics in different flow rates and make out the mechanism that how cavitation affects pump performance for sake of better hydraulic design and safe and stable operation of centrifugal pumps.

Detailed information of the 3-D cavitating flows in centrifugal pumps can be conveniently obtained by numerical simulations, which is conductive to carry out researches on flow mechanism. Due to the complexity of cavitation, precise prediction of the cavitation characteristics of centrifugal pumps is always difficult. In recent years, there have been many research achievements on cavitation numerical

[1] Corresponding author.

methods and cavitation models for centrifugal pumps. Medvitz et al.0conducted quasi-3-D calculation and analysis of cavitation developments in several flow rates for a centrifugal pump and explain the issue of cavitation breakdown with the corresponding flow structure. Dijkers et al.0accurately calculated the sheet cavitation inside the centrifugal impeller with a potential-flow model. Pouffary et al.0did a quick and precise prediction on the 3-D cavitation features in a centrifugal pump and thought that it was the variation of blade load that gave rise to cavitation breakdown. Wang et al.0calculated the cavitating flows in three centrifugal pumps of different specific speed by neural network method and CFD numerical simulation method respectively, and presented that neural network method could be applied in the optimal design of centrifugal pumps since it was convenient to control the relation between input parameters and output parameters. As for turbulence models in cavitation calculations, several literatures00modified the RNG k-ε turbulence model by using density function to correct the turbulent viscosity item. In addition, some literatures0 0proposed that cavitation model should take turbulent pressure pulsation into consideration to correct the phase-change pressure. These efforts improved the prediction accuracy of cavitation characteristics to some extent and captured the cavitation features in different operation conditions of centrifugal pumps.

Up to now, explanations for mechanism of cavitation affecting pump performance mainly concentrated on two aspects. One is that different cavitation coefficients will induce cavitation of different volumes and positions, which influences the impeller discharge capacity at different levels and change the flow structure inside the impeller passage as well0 00. The other is that cavitation will change the load distribution on the blades 000. However, the pump performance change law still remains unexplained in some conditions. Recently, clearance flows in the side chambers between impeller shrouds and casing covers are usually ignored in numerical simulations. Constantin et al.0, Li et al.0, Si et al.0proposed that the unsteady flows in the side chambers should be considered in order to get a more accurate cavitation calculation results.

In this article, numerical research on the cavitation characteristics in there typical conditions of a shrouded centrifugal pump is carried out by cavitation calculation for the whole flow passage considering clearance flow. Cavitation characteristic curves and vapour distribution laws in these conditions are obtained and comparatively analysed, which is meaningful to reveal the mechanism that how cavitation affects the pump performance.

2. Physical model and numerical methodology

2.1. Geometry model and computational grid

The centrifugal pump in this article has a three-blade shrouded impeller whose outlet diameter D_2 is 310 mm. The specific speed n_s is 165. There are nine backward curved back blades on the front shroud. 3-D model of the whole flow passage was set up by software UG as shown in figure 1 and the geometry was separated to six domains including suction pipe, impeller, volute casing, back-blade region, front chamber and back chamber. The minimum axial dimension of the front chamber is 0.22 mm, and the minimum axial dimension of the back chamber is 0.78 mm.

Discretize the impeller domain with unstructured grids so that the element size at the interfaces between impeller and side chambers could be better controlled. Other domains were discretized by structured grid. Four sets of grids having 0.9 million to 2.0 million elements were used in calculation in design condition. The result with 1.63 million or more elements had no significant difference of less than 0.2%. Thus, the grid with 1.63million elements was used to balance the calculation accuracy and the cost. The grid for calculation is shown in figure 2 and the detailed information is listed in table 1.

2.2. Numerical model and computational method

The governing equations were solved by commercial CFD software Ansys CFX 14.5 with RANS method. Apply homogeneous multiphase model to calculate the liquid-vapour two-phase flow and use Reyleigh-Plesset model to deal with mass transport between water and vapour at 25 ℃. The saturated vapour pressure at 25 ℃ is 3574 Pa. In order to give a better prediction of flow separation and rotational flow, use SST k-ω turbulence model to close the governing equations. Pressure field and velocity field were coupled by SIMPLEC algorithm with high resolution scheme. Calculations were thought to get to convergence until the residual mean squares decreased to 0.0001.

Figure 1. 3D model of whole flow passage of the dredging pump.

Figure 2. Grid.

Table 1. Detailed mesh information.

Domains	Suction Pipe	Impeller	Volute Casing	Back-blade region	Front Chamber	Back Chamber	Total
Number of nodes (×10³)	70	130	680	50	90	150	1170
Number of elements (×10³)	70	670	650	40	80	120	1630

Give total pressure values at the inlet of the suction pipe and give average velocity at the outlet of the volute casing. The impeller domain and back-blade region were set as rotational domains with a rotating speed n=1000 r/min, and other domains were stationary. Frozen rotor model was used at the interfaces between rotational domain and stationary domains. The walls in side chambers, adjacent to the outside surfaces of the front shroud and the back shroud, were set as rotational walls with the same rotating speed as that of the impeller, and other walls were all set as no-slip walls. The reference pressure was set as 0 Pa. For each flow rate, calculation began from single-phase flow with high total pressure at the inlet. Gradually reduce the total pressure at the inlet. When the local pressure was lower than the saturated vapour pressure, add cavitation model to conduct two-phase flow calculation. Keep reduce the total pressure at the inlet until cavitation was seriously developed. In this paper, cavitation calculations were carried out for three flow rates, namely $1.3Q_{th}$, Q_{th} and $0.625Q_{th}$, where Q_{th} is the design flow rate. Considering the asymmetry of the volute casing and the blade-tongue interaction, each condition was calculated three times with the impeller rotated 40°each time and the arithmetic mean values of these three results were then used as the final pump performance parameters for the corresponding conditions. In addition, in order to illustrate the influence of clearance flow on the cavitation characteristics, cavitating flows were predicted without side chambers in flow rate of 1.3 Q_{th} as well.

3. Results and discussion

3.1. Validation of numerical results

The cavitation tests were conducted in a closed test rig with a measurement uncertainty of less than ±0.2% and the pump rotating speed is 1000 r/min. With flow rate of $1.3Q_{th}$, the first test point was in non-cavitation condition. Then gradually increase the vacuum degree of the pump system by a vacuum pump to reduce the avaliable Net Positive Suction Head (*NPSHa*) until the efficiency or head dropped to a certain degree. Based on the head of the initial point, H_0, calculate the increment or decrement of the head of other conditions, ΔH, and then obtain the $\Delta H/H_0$-*NPSHa* curve. Cavitation characteristic curves in $1.3Q_{th}$ obtained experimentally and numerically are both given in figure 3. The black curve represents the experimental result, while the red curve represents the numerical result with side chambers and the blue curve represents the numerical result without side chambers. It can be seen from the figure that the result considering clearance flow agrees better with the experimental result.

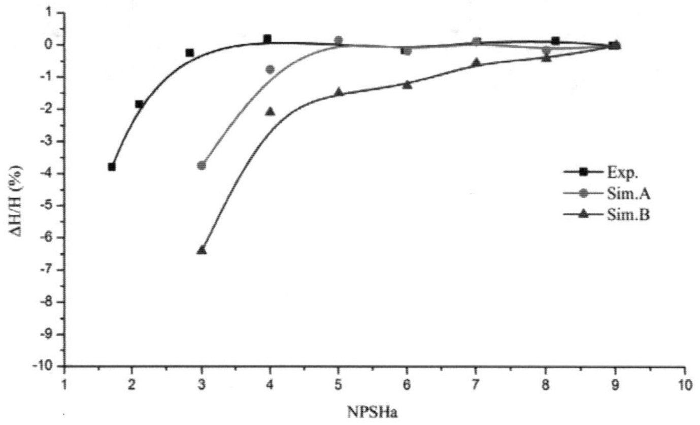

Figure 3. Comparison of experimental and numerical *H-NPSHa* curves.

3.2. Cavitation characteristics in different flow rates

It is still difficult to measure the incipient cavitation coefficient by experiment at present, while numerical simulation can conveniently predict the incipient cavitation. Figure 4 gives the cavitation characteristic curves of this pump in $1.3Q_{th}$, Q_{th} and 0.625 Q_{th}. In 1.3 Q_{th}, cavitation has occurred when *NPSHa*=9 m, while the *NPSHa* values are much smaller for Q_{th} and $0.625Q_{th}$ which is about 5 m. Regard the operation point where the head dropped by 3% as the critical cavitation point. The critical *NPSHa* decreases as the flow rate decreases. This result is consistent with the previous research findings that cavitation is apt to occur in large flow rates and has more significant effect on the pump performance.

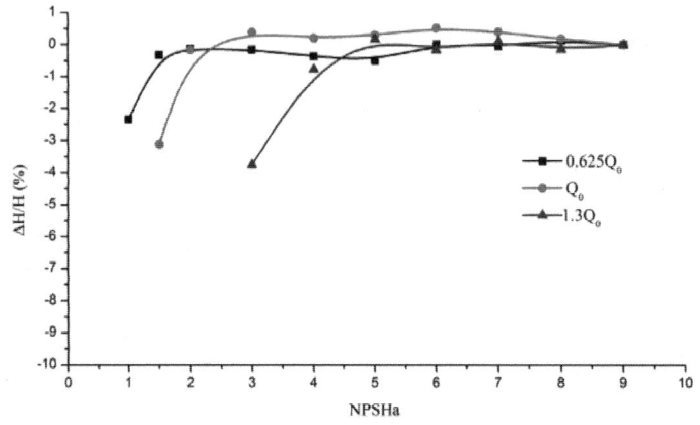

Figure 4. *H-NPSHa* curves for different flow rates.

In different flow rates, the cavitation position and its evolution process is also different. Figure 5 to figure 7 respectively show the incipient cavitation, developing cavitation and critical cavitation for the three flow rates. The blue regions represent the iso-surface of vapour volume fraction α=0.1. The walls of impeller are set as semi-transparent. So in the given orientation, vapour on the pressure surface presents faint blue due to cover of impeller walls, and the uncovered vapour presents brilliant blue.

From figure 5, in the large flow rate, cavitation first occurs on the pressure surface of one blade and near the front shroud side at the blade inlet. It is easy to cavitate at the blade leading edge which is close to the tongue. As the impeller rotates, cavitation alternately occurs at inlet of different blades. In the design flow rate, the incipient cavitation position is still related to the blade-tongue relative position. Cavitation appears on the pressure surface of the blade whose leading edge is near the tongue and at the same time appears on the suction surface of the blade whose leading edge is far away from the tongue. In the small flow rate, the incipient cavitation completely occurs on the suction surface of the blade whose trailing edge is near the tongue. As to the reason of these phenomena, it is negative attack angle at the impeller inlet for large flow rate conditions and the incoming flow impacts the suction surface of the blade which may cause flow separation at the pressure surface and induce cavitation. On the contrary, in small flow rate conditions, it is positive attack angle at the impeller inlet and the incoming flow impacts the pressure surface of the blade which may cause flow separation at the suction surface and induce cavitation. Furthermore, cavitation feature is different for the three impeller angular positions which indicates that the incipient cavitation is unsteady.

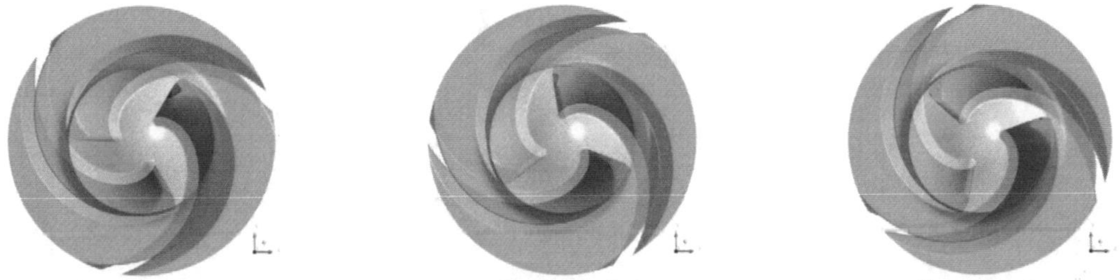

Position 1 Position 2 Position 3

(a) Q/Q_{th}=1.3, $NPSHa$=9 m

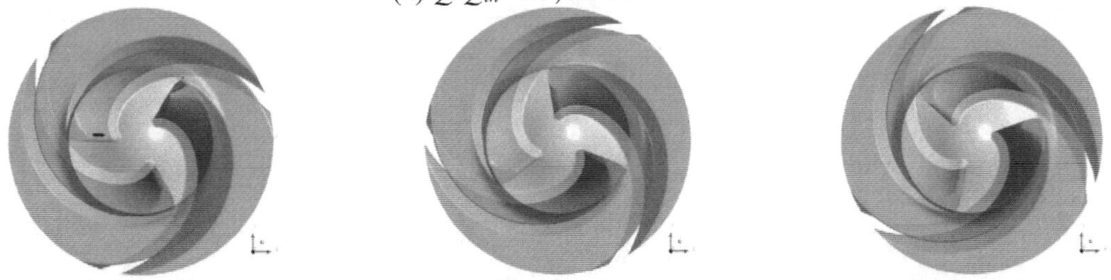

Position 1 Position 2 Position 3

(b) Q/Q_{th}=1, $NPSHa$=5 m

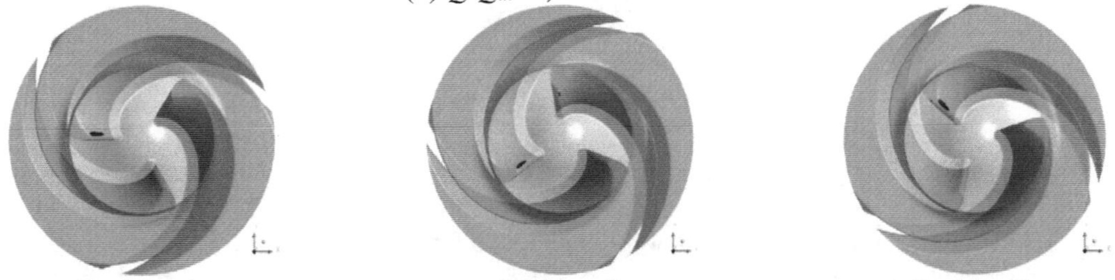

Position 1 Position 2 Position 3

(c) Q/Q_{th}=0.625, $NPSHa$=5 m

Figure 5. Iso-surfaces of vapour volume fraction α=0.1 in incipient cavitation conditions.

IOP Conf. Series: Materials Science and Engineering **72** (2015) 032027 doi:10.1088/1757-899X/72/3/032027

As the inlet total pressure falls, vapour volume increases. Figure 6 shows the developing cavitation conditions when the pump performance has not changed dramatically. For $1.3Q_{th}$, vapour still concentrates at the front shroud side of the leading edge of some blades and great influence of the tongue still exists, and cavitation has extended to the inside surface of the front shroud. Whereas for Q_{th}, the vapour mainly concentrates on the suction surface of the blades and the tongue effect has weakened. Cavitation extends from the front shroud side towards the back shroud side and occurs on the inside surface of the front shroud as well. For $0.625Q_{th}$, cavitation extends from the front shroud side towards the back shroud side similarly and vapour distribute evenly on the suction surfaces of the three blades which is hardly influenced by the tongue. Therefore, cavitation becomes less unsteady with the decrease of flow rate.

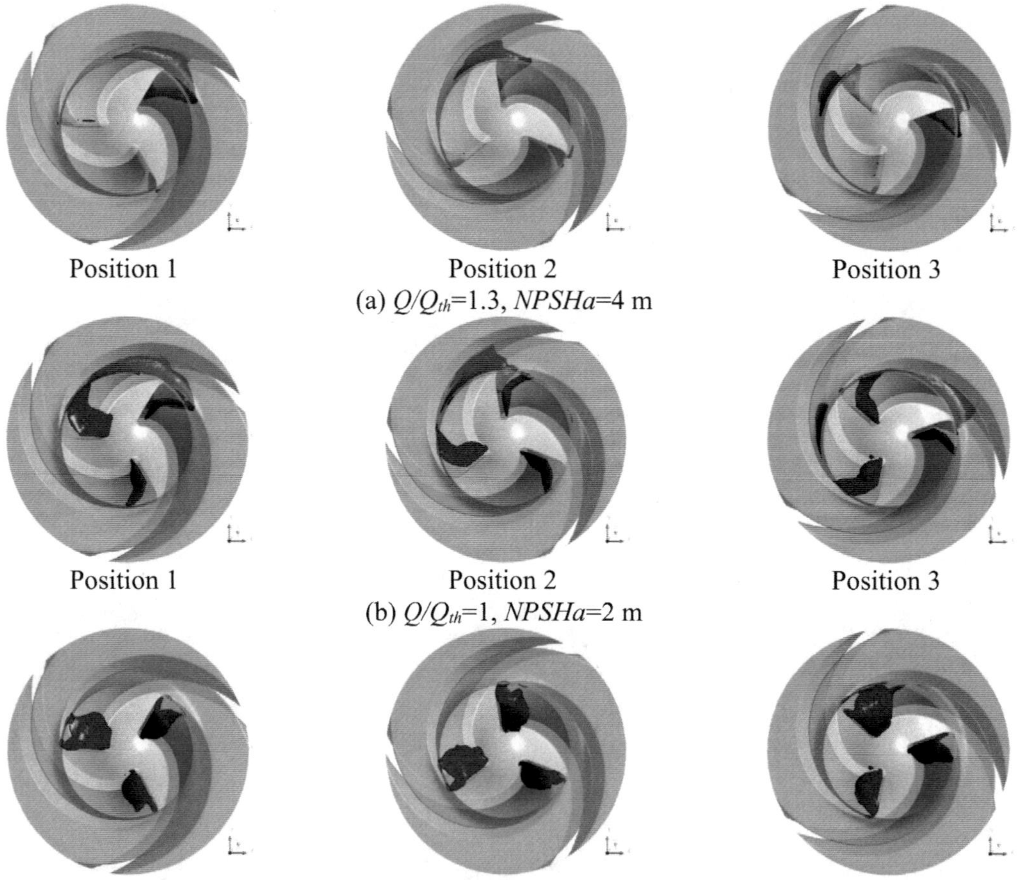

Position 1 Position 2 Position 3

(a) Q/Q_{th}=1.3, *NPSHa*=4 m

Position 1 Position 2 Position 3

(b) Q/Q_{th}=1, *NPSHa*=2 m

Position 1 Position 2 Position 3

(c) Q/Q_{th}=0.625, *NPSHa*=1.5 m

Figure 6. Isosurfaces of vapor volume fraction α=0.1 in developing cavitation conditions.

Continue to reduce the inlet total pressure, when cavitation develops to a certain extent, efficiency and head of the pump will drop remarkably. Figure 7 shows vapour distribution around the critical cavitation point. Vapour volume goes on increasing for each flow rate. The effect of tongue remains in the large flow rate condition and cavitation has develops to the suction surface. In the design condition and small flow rate condition, cavitation both develops towards the blade outlet. The increment of vapour volume is so large that has caused obvious change in the flow structure inside the blade channel, especially near the leading edge, and lead to deterioration of the pump performance.

It is worth noting that the vapour volume is similar for conditions shown in figure 7(a) and figure 6(b) but there is a big difference in their influence on pump performance. Since the blade works depending on the pressure surface, so cavitation on the pressure surface in large flow rate conditions affects the pump performance much greatly than that occurring on the suction side with less flow rate. Therefore, more attention should be paid to cavitation characteristics of centrifugal pumps in large flow conditions in hydraulic design stage.

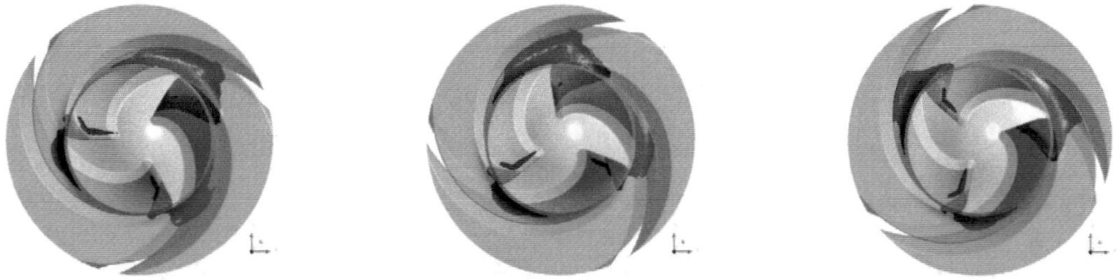

Position 1 Position 2 Position 3

(a) Q/Q_{th}=1.3, *NPSHa*=3 m

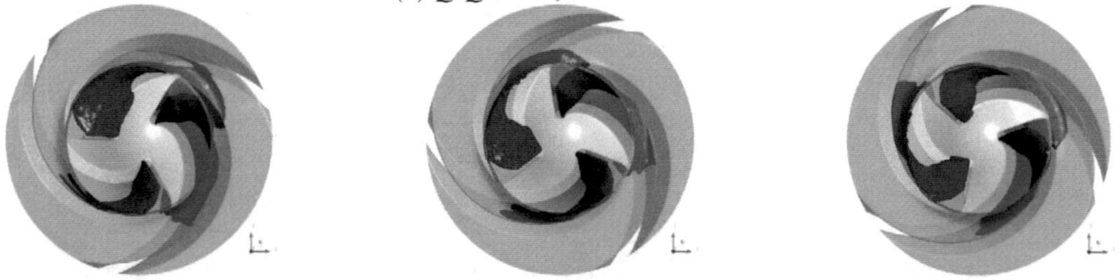

Position 1 Position 2 Position 3

(b) Q/Q_{th}=1, *NPSHa*=1.5 m

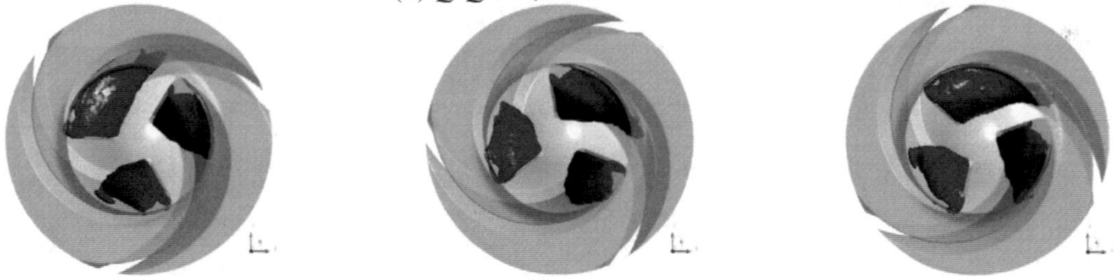

Position 1 Position 2 Position 3

(c) Q/Q_{th}=0.625, *NPSHa*=1 m

Figure 7. Isosurfaces of vapor volume fraction α=0.1 in critical cavitation conditions.

4. Conclusion

This article studies the cavitation evolution law and the effect of cavitation on pump performance via numerical simulations of cavitating flows in three typical flow rates. Compared with experimental results, it finds that considering the clearance flow can get a more accurate cavitation characteristic curve. Through comparative analysis of the cavitation characteristics for the three flow rates, it shows that cavitation is apt to occur in large flow rate conditions and meanwhile the tongue has greater effect with larger flow rate. As the flow rate decreases, the critical *NPSHa* decreases and the cavitation positons gradually shift from the pressure surfaces of some blades to the suction surfaces. With similar vapor volume, cavitation occurring on the blade pressure side is easier to lead to pump performance drop. Therefore, more attention should be paid to the cavitation characteristics of centrifugal pumps in large flow conditions in hydraulic design stage.

Acknowledgments
Special thanks are due to the State Key Program of National Science of China (Grant No. 51439002), National Natural Science Foundation of China under Contract (No. 51479093), State Key Laboratory of Hydroscience and Engineering (Grant No. 2014-KY-05) for supporting the present work.

References
[1] Medvitz R B, Kunz R F, Boger D A, Lindau J W, Yocum A M and Pauley L L 2002 *J. Fluid Eng.* **124**(2) 377-383
[2] Dijkers R J H, Fumex B, Woerd J G H O, Kruyt N P and Hoeijmakers H W M 2005 *Proc. ASME 2005 Fluids Engineering Division Summer Meeting (Houston, Texas, USA, 19-23 June 2005)* pp 1233-8
[3] Pouffary B, Patella R F, Reboud J L and Lambert P A 2008 *J. Fluid Eng.* **130**(6) 061301
[4] Wang Y, Liu H L, Yuan S Q, Tan M G and Wang K 2009 *Proc. the IEEE Int. Conf. on Computing and Intelligent Systems* **1** 137-40
[5] Liu H L, Liu D X, Wang Y and Zhuang S G 2012 *J. Huazhong Univ. of Sci. & Tech. (Natural Science Edi.)* **40**(8) 17-20
[6] Tan L, Cao S L, Wang Y M and Zhu B S 2012 *J. Chin. Phys. Lett.* **29**(1) 014702
[7] Tan L, Zhu B S, Cao S L and Wang Y M 2013 *J. Chin. Sci. Bull.* **58**(8) 949-52
[8] Liu H, Liu D X, Wang Y, Wu X F, Wang J and Du H 2013 *J. Hydrodynamics* B **25**(3) 370-8
[9] Wang W J, Wang Y, Liu R H, Li Y B, Yin G and Li G D 2014 *T. Chin. Soc. Agr. Mach.* **45**(3) 37-44
[10] Li J, Liu L J and Feng Z P 2006 *J. Xi'an Jiaotong Univ.* **40**(3) 257-60
[11] Wang X L, Yuan S Q, Zhu R S, Fu Q and Yu Z S 2012 *T. Chin. Soc. Agr. Mach.* **43**(3) 67-72
[12] Yuan J, Fu Y and Yuan S 2012 *ASME 2012 Fluids Engineering Division Summer Meeting collocated with the ASME 2012 Heat Transfer Summer Conference and the ASME 2012 10th International Conference on Nanochannels, Microchannels and Minichannels (*Rio Grande, Puerto Rico, USA, 8-12 July 2012) pp 193-202
[13] Luo X, Wei W, Ji B, Pan Z Zhou W and Xu H 2013 *J. Mech. Sci. Tech.* **27** (6) 1643-48
[14] Li X J, Yuan S Q, Pan Z Y, Yuan J P and Fu Y X 2013 *J. Sci. Chin. Tech. Sci.* **56** (9) 2156-62
[15] Wang Y, Liu H L, Yuan S Q, Tan M G and Wang K 2011 *J. Drain. Irrig. Mach. Eng.* **29** (2) 99-103
[16] He M, Fu L P, Zhou L J, Guo Q and Wang Z W 2012 *The 26th IAHR Symposium on Hydraulic Machinery and Systems* (Beijing, China, 19-23 Aug 2012) **15**(6) 062022
[17] Constantin S L 2012 *Int. Society for Optics and Photonics* 841131-10
[18] Si Q, Yuan S, Li X, Yuan J and Lu J 2014 *T. Chin. Soc. Agr. Mach.* **45**(5) 84-90

IOP Conf. Series: Materials Science and Engineering **72** (2015) 032028 doi:10.1088/1757-899X/72/3/032028

Cavitation performance and flow characteristic in a centrifugal pump with inlet guide vanes

L Tan[1], L Zha[2], S L Cao[1, 3], Y C Wang[1] and S B Gui[2]
[1] State Key Laboratory of Hydroscience and Engineering, Tsinghua University, Beijing 100084, China
[2] Changjiang Survey, Planning, Design and Research Co. Ltd, Wuhan 430010, China

E-mail: caoshl@mail.tsinghua.edu.cn

Abstract. The influence of prewhirl regulation by inlet guide vanes (IGVs) on cavitation performance and flow characteristic in a centrifugal pump is investigated. At the impeller inlet, the streamlines are regulated by the IGVs, and the axial velocity distribution is also influenced by the IGVs. Due to the total pressure loss on the IGVs, the cavitation performance of the centrifugal pump degrades. The cavitation area in impeller with IGVs is larger than one without IGVs. The specify values of total pressure loss between the suction pipe inlet and impeller inlet for three cavitation conditions show that the IGVs will generate additional pressure loss, which is related to the IGVs angles and cavitation conditions.

1. Introduction

The prewhirl regulation of IGVs is a widely used approach in industrial centrifugal compressors. It can induce a controlled prewhirl at the impeller inlet so as to regulate the pressure ratio and mass flow at the constant rotational speed. Due to the advantage of convenient installation and practical effect, the IGVs for centrifugal compressors have been sufficiently investigated for decades. The research found in the literature on IGVs for centrifugal compressors mainly fall into four categories: (1) the effect of IGVs on compressors' energy performances (pressure ratio, efficiency) [1-12]; (2) the effect of IGVs design method and structure parameter on the regulation results [2-5, 11]; (3) the unsteady interaction between the stable IGVs and the rotating impeller [6-12]; (4) the common failure modes of IGVs [13]. The focus of investigation on IGVs' effect on compressor and fan is the pressure and efficiency. For example, Fukutomi [1] experimentally investigated the effect of IGVs on a cross flow fan, and the results showed that the pressure and efficient of the fan with IGVs is higher than that without IGVs. Tan [11] investigated the effects of IGVs on the centrifugal compressor with radial and axial inlets. The results showed that the IGVs system had different performances in the radial and axial inlet models.

Though the IGVs system is a stable component of the compressor, there are still several geometric parameters of IGVs that influence the regulation results. The spherical section passage was used to eliminate the blade hub and tip clearance and the tandem vanes was presented to produce comparable swirl with significantly lower loss than flat plate vanes [2]. The significant features and effects of different settings of the original, s-cambered, and tandem inlet guide vanes on a centrifugal compressor were investigated [4]. The results showed that the tandem inlet guide vanes had superior

[3] Corresponding author.

aerodynamic performance and extended the operating range of the compressor at both positive and negative inlet swirls. Ferro [5] designed an IGVs system by using a quasi-three-dimensional method, combining the streamline curvature method for the solution of the meridional flow and a panel method for the blade-to-blade flow.

Despite the great improvement of pressure ratio and efficiency by IGVs, the operation instability of compressors resulting from IGVs should be taken into consideration. Oro [9, 10] developed an experimental openloop facility to investigate the dynamic and periodic interaction between both fixed and rotating blade rows in a single stage fan with IGVs. The wake transport through the rotor passage was influenced by the mass flow rate and the geometrical characteristics, such as the blade camber and the stagger angle. Zhou [12] used a constant-temperature hot-wire anemometer to investigate the unsteady interaction of IGVs, impeller, and diffuser in a centrifugal compressor. The unsteadiness caused by IGVs in the diffuser with negative prewhirl angle was larger than that for positive prewhirl angle.

For the centrifugal pump, the circumferential velocity at the impeller inlet directly affects the pump's efficiency and head according to the Euler equation. The main working principle of prewhirl regulation on centrifugal pump is to change the IGVs prewhirl angle to induce the circumferential velocity, thus regulating the operating condition. According to the investigation by Tan [14-16], the three dimensional IGVs can greatly improve the flow pattern at impeller inlet and widen the pump high-efficiency operation region. However, due to the friction loss on IGVs' surface the pump cavitation performance will be influenced inevitably.

Much research has been performed toward understanding the cavitation in centrifugal pumps by both means of experiment and numerical simulation. The Particle Image Velocimetry (PIV) was used to investigate the cavitation flow for the case of 3% head drop in a centrifugal pump operating at over load conditions by Bachert et al. [17]. Ding et al. [18] adopted the full cavitation model to simulate the cavitating flow in a centrifugal pump.

2. Problem statement

2.1. Parameters of centrifugal pump

The main parameters of the centrifugal pump are as follows: design flow rate Q_d=340 m^3/h, head H=30 m, rotation speed n=1450 r/min. Figure 1 shows the tested centrifugal pump and test rig.

Figure 1. Centrifugal pump and test rig.

2.2. Numerical method

The fluid in the cavitation flow field is considered a homogeneous and compressible mixed medium of liquid and vapour. The vapourization rate $\dot{m}_{vap}$ and condensation rate $\dot{m}_{con}$ in cavitation model [16] based on the mass transport are as follows:

$$\dot{m}_{vap} = C_{vap} \frac{3\alpha_{nuc}(1-\alpha_v)\rho_v}{R_b} \sqrt{\frac{2}{3}\frac{\max(p_v - p, 0)}{\rho_l}} \tag{1}$$

$$\dot{m}_{con} = C_{con} \frac{3\alpha_v \rho_v}{R_b} \sqrt{\frac{2}{3}\frac{\max(p - p_v, 0)}{\rho_l}} \tag{2}$$

where C_{vap} and C_{con} are the empirical calibration coefficients for vapourization and condensation rates, respectively; R_b is the bubble radius; p_v is the vapour pressure; α_{nuc} is the nucleation site volume fraction, α_v is the vapor volume fraction.

The computation fluid dynamics code CFX 13.0 is employed in the present calculation by combining the RNG k-ε turbulence model and mass transport cavitation model. The boundary conditions are defined as: the total pressure at pump inlet, and the mass flow at pump outlet, no slip wall at walls. Figure 2 shows the computational domain of centrifugal pump and mesh of impeller. The flow rate Q=315 m^3/h is chosen in the present study. The good agreement between the experimental and numerical results validates the reliability of the numerical model and method [16].

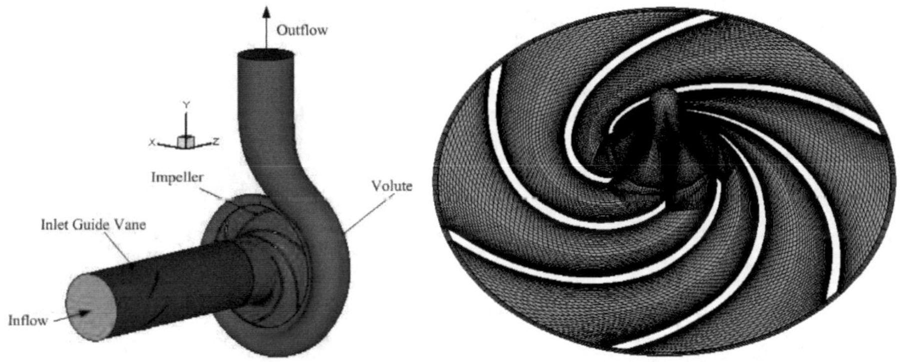

Figure 2. Computational domain of centrifugal pump and mesh of impeller.

3. Result and discussion

Figure 3 shows the streamlines in impeller for the centrifugal pump without IGVs and with IGVs at 12°, 24°, -12°, -24°. When the IGVs are set to different angles, the streamlines in suction pipe and impeller will be regulated accordingly. As shown in Fig.3, the streamlines at the impeller inlet are different for various IGVs angles, but the difference is not as obvious as that in suction pipe.

Figures 4 shows the axial velocity at impeller inlet of the centrifugal pump without IGVs and with at 12°, 24°, -12°, -24°. For both positive and negative prewhirl regulations, the axial velocity contours at impeller inlet are more even than that without IGVs. Correspondingly the high velocity region shrinks as the IGVs angle increases or decreases. Though the IGVs are circumferentially symmetrical in the suction pipe, the axial velocity distribution is still dissymmetrical at impeller inlet. The reason is that the interaction between the impeller and volute induces the uneven velocity in impeller and then influences the flow pattern at impeller inlet. In several regions close to the impeller shroud, the axial velocities are negative due to the reverse flow at partial load condition of Q=315 m^3/h.

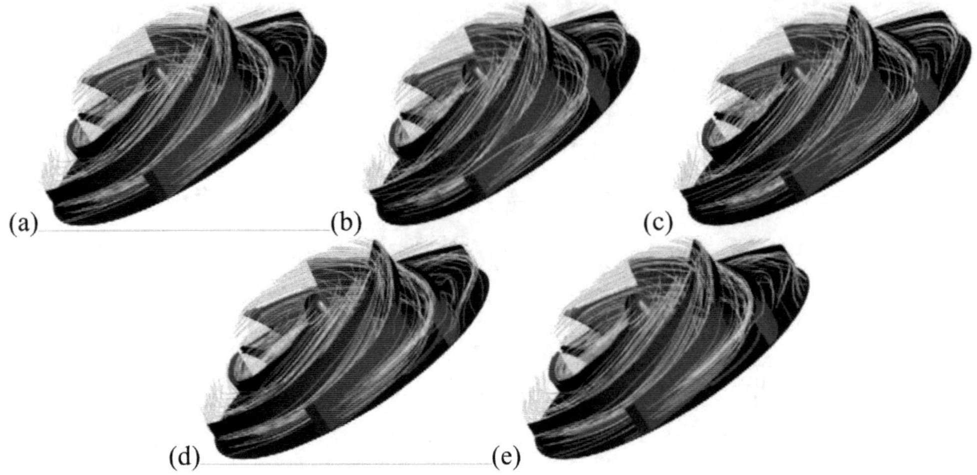

Figure 3. Streamlines in impeller: (a) no IGVs; (b) IGVs 12°; (c) IGVs 24°; (d) IGVs −12°; and (e) IGVs −24°.

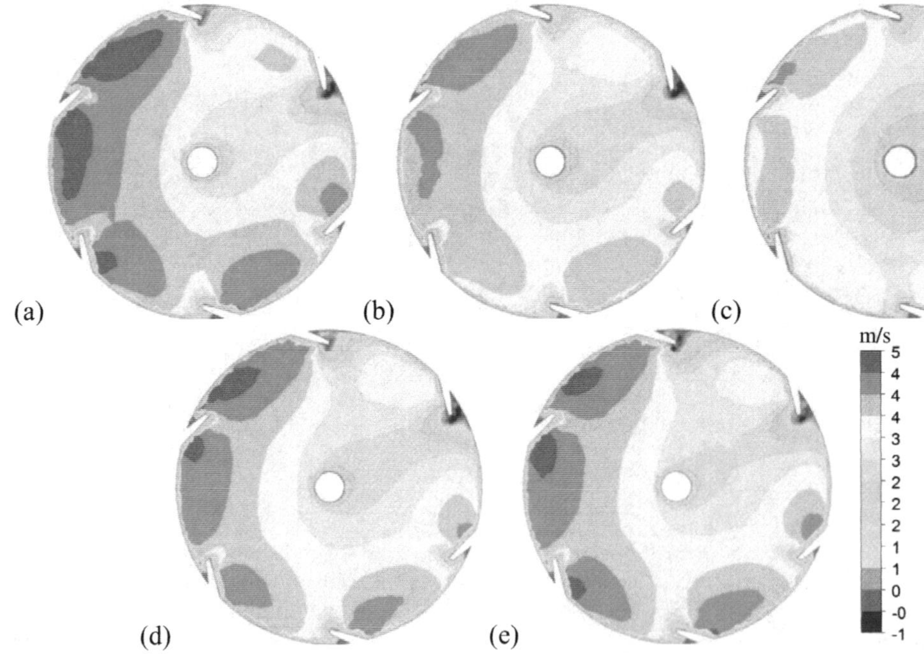

Figure 4. Axial velocity at impeller inlet: (a) no IGVs; (b) IGVs 12°; (c) IGVs 24°; (d) IGVs −12°; and (e) IGVs −24°.

Figures 5 and 6 show the cavitation structure in impeller without IGVs and with IGVs -24° for the cavitation conditions of NPSHA=4m, 3m and 2m, where the NPSHA is the difference between the total energy and vapourization energy for the unit weight of the fluid at pump inlet. With the pressure decrease at pump inlet, the cavitation develops in the impeller gradually. In comparison with the cavitation in impeller without IGVs in Fig.5, the cavitation in impeller with IGVs -24° in Fig.6 is larger, because the pressure loss increases when the IGVs are fixed in the suction pipe. Therefore, the IGVs will degrade the cavitation performance in centrifugal pump, but the influence on the cavitation structure is very tiny as shown in Figs.5 and 6.

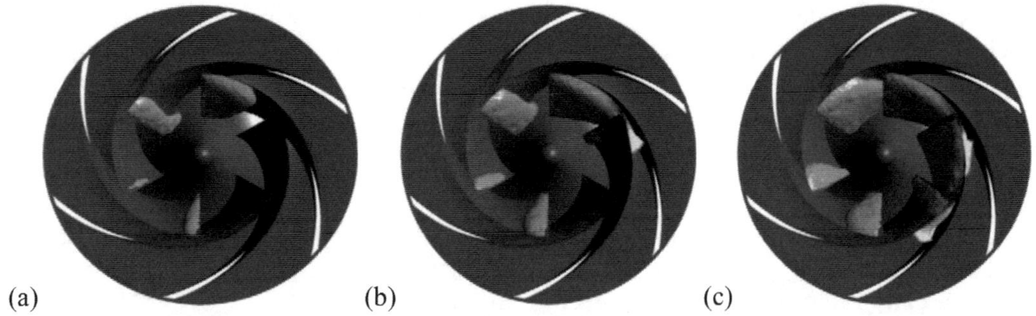

Figure 5. Cavitation structure in impeller without IGVs: (a) NPSHA=4 m; (b) NPSHA=3 m; and (c) NPSHA=2 m.

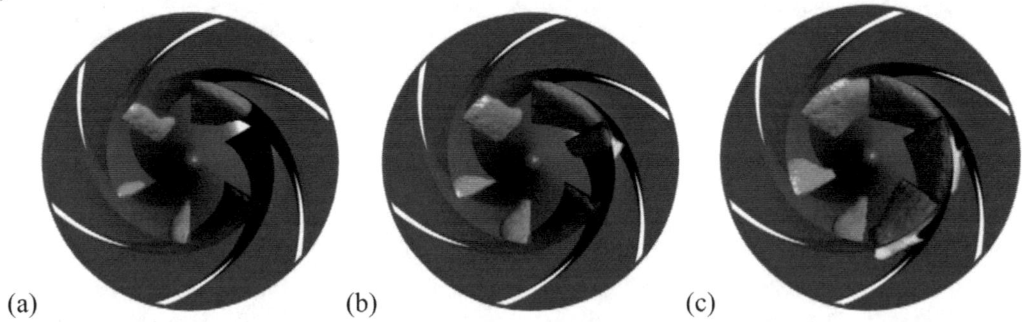

Figure 6. Cavitation structure in impeller with IGVs -24°: (a) NPSHA=4 m; (b) NPSHA=3 m; and (c) NPSHA=2 m.

To further specify the influence of IGVs on cavitation performance of the centrifugal pump, Table 1 lists the total pressure loss between the suction pipe inlet and impeller inlet. For NPSHA=4 m, the total pressure loss is 319 pa for the pump without IGVs, it increase to 2699 pa and 1199 pa when the pump installs the IGVs at 24° and -24° respectively. For NPSHA=3 m and 2 m, the difference of pressure loss for pump without and with IGVs is also very obvious. The positive prewhirl regulation has greater effect on the pump cavitation performance than the negative prewhirl regulation. The cavitation condition will also influence the pressure loss. For example, the total pressure loss for NPSHA=2 m is much larger than that for NPSHA=4 m at IGVs -24°.

Table 1. Total pressure loss between the suction pipe inlet and impeller inlet.

pump	total pressure loss (pa) for NPSHA=4 m	total pressure loss (pa) for NPSHA=3 m	total pressure loss (pa) for NPSHA=2 m
No IGVs	319	315	326
IGVs 24°	2699	2759	2729
IGVs -24°	1199	1242	1731

4. Conclusions

As an effect regulation approach of operation condition for centrifugal pumps, the IGVs can greatly widen the high-efficiency operation region. However, the pressure loss on IGVs should not be neglected for centrifugal pumps. Based on the previous experiment, the numerical research is conducted to investigate the cavitation performance and flow characteristic in a centrifugal pump without and with IGVs. The IGVs in suction pipe will generate positive or negative prewhirl on the streamlines at the impeller inlet and then influence the flow pattern in impeller. The distribution of axial velocity becomes more even after the prewhirl regulation of IGVs. Due to the pressure loss on IGVs, the cavitation performance of the centrifugal pump degrades. The cavitation area on blade surface expands when the centrifugal pump installs the IGVs. The pressure loss induced by IGVs is influenced by the IGVs angles and cavitation conditions.

Acknowledgments

This work was supported by the National Natural Science Foundation of China [Grant number 51176088 and 51179090], and the Open Research Fund Program of State key Laboratory of Hydroscience and Engineering [sklhse-2013-E-02].

References

[1] Fukutomi J and Nakamura R 2005 *JSME Int. J. Ser. B Fluids Therm. Eng.* **48** 763–69
[2] Coppinger M and Swain E 2000 *Proc. Inst. Mech. Eng. Part A J. Power Energy* **214** 153–64
[3] Liou T M, Lee H L and Liao C C 2001 *Exp. Therm. Fluid Sci.* **24** 11–23
[4] Mohseni A, Goldhahn E, van den Braembussche R A, Seume J R and Goldhahn E 2012 *J. Turbomach* **134** doi:10.1115/1.4003235
[5] Ferro L, Gato L and Falcoa 2010 *Renew. Energy* **35** 1920–8
[6] Johnston R and Fleeter S 2001 *J. Fluids Struct.* **15** 235–53
[7] Cui M M 2006 *Proc. Inst. Mech. Eng. Part G J. Aerosp. Eng.* **220** 11–28
[8] Soranna F, Chow Y C, Uzol O and Katz J 2006 *J. Turbomach.* **128** 82–95
[9] Oro J M F, Diaz K M A, Morros C S and Marigorta E B 2007 *J. Fluids Eng.* **129** 1015–29
[10] Oro J M F, Diaz K M A, Morros C S and Marigorta E B 2008 *J. Fluids Eng.* **130** doi:10.1115/1.2844578
[11] Tan J, Wang X, Qi D and Wang R 2011 *Proc. Inst. Mech. Eng. Part C J. Mech. Eng. Sci.* **225** 2089–105
[12] Zhou L, Fan H Z, Wei W and Cai Y H 2011 *Proc. Inst. Mech. Eng. Part C J. Mech. Eng. Sci.* **226** 660–80
[13] Mcalpin R L, Holm R E, Talley P L and Bernstein H L 2003 *J. Eng. Gas Turbines Power* **125** 236–40
[14] Tan L, Cao S L and Gui S B 2010 *Sci. China Technol. Sci.* **53** 2142–51
[15] Tan L, Cao S L, Wang Y M and Zhu B S 2012 *Sci. China Technol. Sci.* **55** 1037–43
[16] Tan L, Zhu B S, Cao S L, Wang Y C and Wang B B. 2014 *Energies* **7**(2) 1050-65
[17] Bachert R, Stoffel B and Dular M. 2010 *J. Fluids Eng.* **132** doi:10.1115/1.4001570
[18] Ding H, Visser F C, Jiang Y and Furmanczyk M 2011 *J. Fluids Eng.* **133**: 011101-1-14

IOP Conf. Series: Materials Science and Engineering **72** (2015) 032029 doi:10.1088/1757-899X/72/3/032029

Numerical prediction of the pressure fluctuations on small discharge condition of a pump-turbine at pump mode

Y Y Yao, Y X Xiao, W Zhu, S H An and Z W Wang[1]

State Key Laboratory of Hydroscience and Engineering & Department of Thermal Engineering, Tsinghua University, Beijing 100084, China

E-mail addresses: wzw@mail.tsinghua.edu.cn

Abstract. The operational stability of the pump turbine at the pump mode will be greatly influenced by large pressure fluctuations when operated in the small-discharge conditions. Therefore, it is significant to analyse the flow characteristic under the small discharge operating conditions deeply. Study of the internal flow in the small discharge condition has been investigate in great detail combined with model experiments in this paper. The SST k-ω turbulence model is adopted to perform three-dimensional numerical simulation of the entire pump-turbine flow passage at optimal guide vanes opening. The numerical simulation results match well with experimental data. Then internal flow under the small discharge condition is analysed. The results show that the dominant frequency inside the flow passage is a relative low frequency. In addition, there are obvious complex flow phenomena inside the draft tube, runner and diffuser domains, such as secondary flow, backflow and even vortex, leading to strong unsteady flow and significant pressure fluctuation.

1. Introduction

With the development of pumped-storage power station technology, the unstable operation issues have drawn more and more attention as pump-turbine's working condition is relatively complicated. For the pump mode of pump-turbine, the pressure fluctuations become stronger as the discharge gets further from the optimal operating condition, which can seriously affect the operating stability of the pump-turbine [1]. Therefore, it is necessary to analyse deeply on operating conditions in the small discharge conditions of the pump mode.

Study regarding to small-discharge condition's flow characteristic of pump-turbine in pump mode is not adequate so far. Qu et al [1] analyzed a mixed-flow reversible pump-turbine in pump mode through experiments, finding that the pressure fluctuation is the strongest in the area between the runner and guide vanes and the smallest in the draft tube. Wang et al [2] performed numerical simulation of a pump-turbine, exploring that the vortexes increase and their scales are enlarged as the discharge decreases. Alex et al [3] performed unsteady numerical simulation of a pump-turbine and predicted that the flow field leaving the runner is highly non-uniform under the off-design operating conditions. Ran et al [4] explored pressure fluctuations from experiments, finding that the dominant frequency under the small-discharge off-design condition is a low frequency and the low dominant frequency is believed to be induced by the rotating stall inside the flow passage. Yin et al [5] adopted the DES turbulence model to conduct unsteady numerical simulation, concluding that the low

[1] Corresponding author.

IOP Conf. Series: Materials Science and Engineering **72** (2015) 032029 doi:10.1088/1757-899X/72/3/032029

dominant frequencies under the small-discharge conditions in the pump mode are mainly caused by the flow separation inside the guide vanes and the fluctuation characteristic is induced by the generation and vanishing frequency. All the researches above lay a foundation for later related study.

In this paper, the small discharge (0.46 Q_0, where Q_0 is the optimal discharge) operating condition of pump-turbine is simulated at the optimal opening. The internal flow field was obtained through numerical calculation and the unsteady flow characteristics at the small discharge operating point were discussed.

2. Simulation model

2.1. Geometry of the model pump-turbine

The three-dimensional entire flow passage of the model pump-turbine is shown in Figure 1. The flow passage is composed of five parts, named draft tube, runner, guide vanes, stay vanes and spiral case. When operated in pump mode, the water begins from the draft tube and flows out from the spiral case. The pump-turbine consists of a runner with 9 blades and a diffuser with 20 guide vanes and 20 stay vanes (including the spiral case node). The runner diameter at outlet is 0.47 m and the optimal opening for the guide vanes is 27 mm. The rated speed of the model pump-turbine reaches 1300rpm. In addition, the discharge of the design point is 408 L/s. In this paper, the discharge is a small one with only 46% of the design discharge.

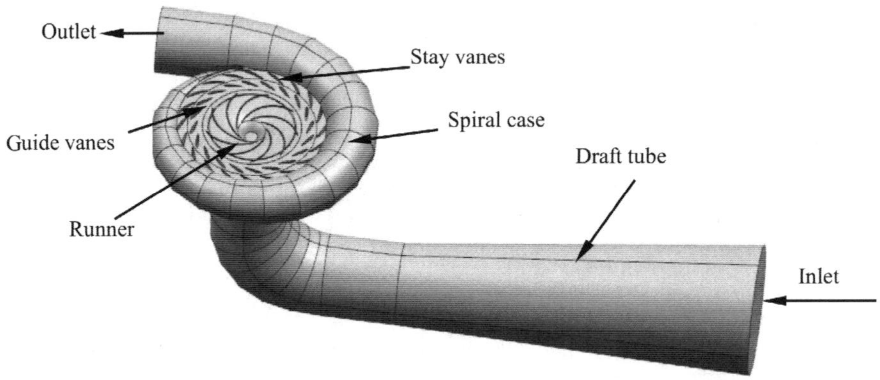

Figure 1. Computational flow domain of a pump-turbine

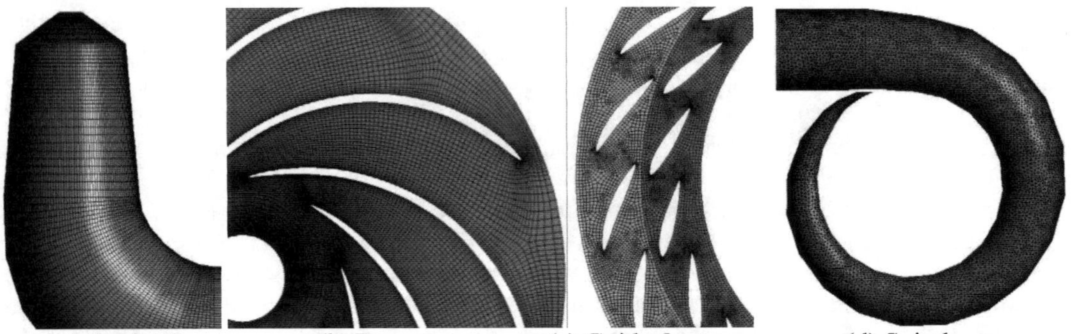

(a) Draft tube (b) Runner (c) Guide & stay vanes (d) Spiral case

Figure 2. The meshes for the model pump-turbine

2.2. Mesh generation

The hexahedral meshes were generated in stay vanes, guide vanes, runner and draft tube domains, while tetrahedral meshes were adopted in the spiral case domain. The grid generation of the entire flow passage is shown in Figure 2 and the specific cell number of each domain is listed in Table 1. It is obvious that the mesh density of runner, stay and guide vanes is much larger than the draft tube and spiral case domain. The flow speed of the former three domains is high and the internal flow conditions are complicated. Large mesh density is beneficial to capture the complex flow pattern in the three domains in order to improve the calculation precision. While the speed of draft tube and spiral case is comparatively low. It doesn't have much effect on the overall performance. Small mesh density can save computing resources and raise calculation speed.

Table 1. Mesh quantity distribution of the entire flow passage

Domain	Draft Tube	Runner	Guide vanes	Stay vanes	Spiral case	Total
Elements	418,000	1,200,000	339,000	289,000	595,000	2,841,000

2.3. Boundary conditions and corresponding simulation setup

During the simulation of pump mode, the inlet of the computational domain adopts the velocity-inlet boundary condition so as to get accurate discharge information consistent with the test data. The static pressure is specified at the outlet in order to obtain the pressure distribution consistent with the practical situation. The operating pressure is set as 1 atm. All solid walls are set as no slip. Considering that the SST turbulence model is better for turbine operating conditions [6-7], the SST turbulence model is chosen ultimately.

The unsteady numerical simulation is on the basis of the steady calculation result. During the simulation process, the time step is set according to both the blade number and the rotating speed. In this paper, 180 transient time steps make up one period calculation. A total of 16 runner revolutions are completed and the data is saved every 2 time steps.

2.4. Recording points distribution

The pressure fluctuations at certain chosen points are recorded during the unsteady numerical simulation of the pump mode. The selected recording points inside the entire flow passage are shown in Figure 3. For the draft tube, a total of 16 recording points are respectively arranged on four planes (named dt14-dt44), as shown in Figure 3(a). On each plane, there are four points uniformly circling around. For the runner domain, there are 8 points rotating with the runner between the two runner blades in the blade passage (named rv1-rv8), as shown in Figure 3(b). In addition, three sets of recording points inside the guide vanes and stay vanes, gv1 to gv4, mv1 to mv4 and sv1 to sv4, are evenly arranged along the circumferential direction on the mid-span plane, as shown in Figure 3(c). For the spiral case, there are five points inside the domain on the mid-span plane (named sp1-sp5), as shown in Figure 3(c).

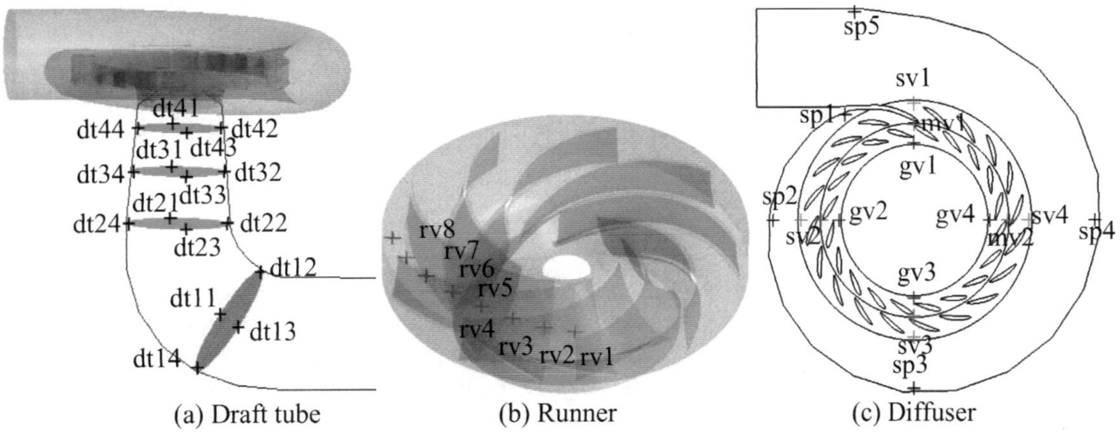

Figure 3. Recording points distribution

3. Analysis of numerical simulation results

3.1. Hydraulic performance Analysis

Operational parameters were measured at different openings to analyse the operating characteristic of pump mode through a hydraulic machinery rig. Unsteady calculation of the small discharge condition is conducted at the optimal opening in the pump mode of pump-turbine. Table 2 lists both the experimental and calculated head, power and efficiency results of the small discharge. The errors between the simulation and experiment results are also included in the table. The simulated head is close to the experimental result with the error of only -0.58%. The calculated power is relatively smaller than the experimental result with the error of -5.56%. The calculated efficiency is larger than the experimental result as the simulated power is slightly smaller than the practical condition. Generally speaking, the errors between the experimental results and the simulation results are all less than 5.6%. It is feasible to adopt the numerical simulation to analyse the internal flow characteristics under the small discharge operating condition.

Table 2. Experimental and calculated head, power and efficiency of the small discharge

Parameter	Exp.	Sim.	Err.(%)
H(m)	53.60	53.29	-0.58
P(kW)	159.00	150.16	-5.56
η(%)	62.80	66.00	5.10

3.2. Energy loss of the entire flow passage

The energy loss of each domain of the pump mode under the small discharge condition is shown in Figure 4. The energy loss is transformed into head, using m as its unit. The abscissa stands for the five domains, including draft tube, runner, guide vanes, stay vanes and spiral case, arranged according to the flow direction. The energy loss curve shapes like an inverted parabola along the flow direction. The energy loss inside the draft tube is 3.30m. The head loss inside the runner is the largest among the five domains, with the energy loss of 11.34m. The head loss inside the guide vanes is the secondly greatest among the five domains, with the loss of 10.01m, which is close to the runner's head loss. The energy loss of the stay vanes is much smaller than the guide vanes, with the loss of 2.45m. The energy loss of the spiral case is the smallest among the five domains, with the energy loss of only 0.65m.

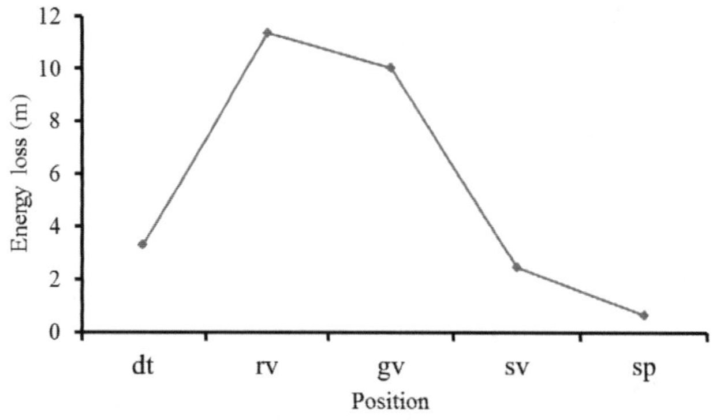

Figure 4. Energy loss of each domain

According to Figure 4, the main loss of the whole flow passage lies in the runner and guide vanes domains. The energy loss inside the draft tube is also considerable. The large energy loss is often induced by the special flow inside the passage, such as vortex and backflow, which will cause large pressure fluctuations and influence the operating stability of the pump-turbine.

3.3. *Pressure fluctuations analysis*

The relative peak-to-peak amplitudes of different points under the small discharge operating condition are shown as Figure 5. For recording points on 4 planes named dt1, dt2, dt3 and dt4 inside the draft tube, the relative peak-to-peak pressure amplitudes are almost the same, around 18%, as shown in Figure 6(a). For the 8 rotating point inside the runner, it can be seen that the amplitude gradually increases along the flow direction from 17.63% to 62.78% as the inference between the runner and guide vanes getting stronger, shown in Figure 6(b). For the 4 points on the pressure side of the guide vanes, the relative peak-to-peak amplitude at point gv1 is larger than the other 3 points with the number of 58.71%, shown in Figure 6(c). Point gv1 is more close to the casing tongue. For the 4 points on the suction side of the guide vanes, the relative peak-to-peak amplitudes are generally smaller than points on the pressure side, as shown in Figure 6(d). The results also show that the pulse amplitudes inside the stay vanes are generally smaller than the guide vanes and the amplitude at point sv2 stand out of the four points, shown in Figure 6(e). In the spiral case, the pulse amplitudes decrease gradually from sp1 to sp5, shown in Figure 6(f). Point sp5 is near the outlet of the flow passage. Its low peak-to-peak amplitude is on account of the given static pressure condition in the outlet.

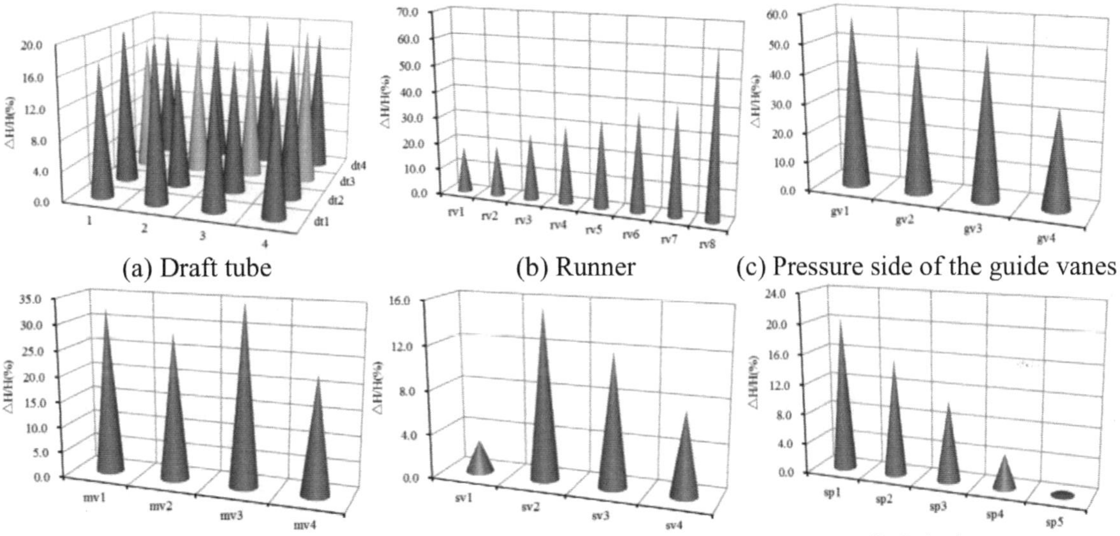

(a) Draft tube (b) Runner (c) Pressure side of the guide vanes

(d) Suction side of the guide vanes (e) Stay vanes (f) Spiral case

Figure 5. Predicted relative pulse amplitudes at different recording points

The pressure pulse spectra are analysed at various points throughout the entire flow passage. The pressure fluctuations both in the time domain and frequency domain at point dt41, rv1, rv4, rv8, gv2, mv2, sv2 and sp2 are shown in Figure 6 for the small discharge condition. The pressure pulse characteristics at point dt14 and rv1 are similar to each other, with the pressure fluctuating between -6 and 6m. The dominant frequency for dt14 and rv1 is a low frequency of $0.35f_n$. The first dominant frequency for rv4 is also $0.35f_n$ and the other dominant frequencies are on the multiple of $0.35f_n$. For point rv8, the $1.41f_n$, $0.3f_n$ and $2.81f_n$ are included and the amplitudes at these frequencies are larger than dt14 to rv4. In addition, the 20fn corresponding to the number of the guide vanes is not that obvious compared to the low frequencies, indicating that the guide vanes is not the main factor inducing the large fluctuations under the small discharge condition. At point gv2, the $0.35f_n$ and $0.88f_n$ are the dominant frequencies. For point mv2, sv2 and sp2, the first dominant frequency changes to $0.53f_n$.

As the head loss inside the runner and guide vanes is the largest among the five domains and the low frequency $0.35f_n$ is included inside the draft tube, runner and guide vanes, flow field at certain flow time is adopted and the internal flow changing characteristics is analysed. For the convenience of analysis, the 4 flow time, respectively 0.62s, 0.66s, 0.70s and 0.74s, are adopted according to the pressure fluctuation at gv2 and named t1-t4, as shown in Figure 6(e).

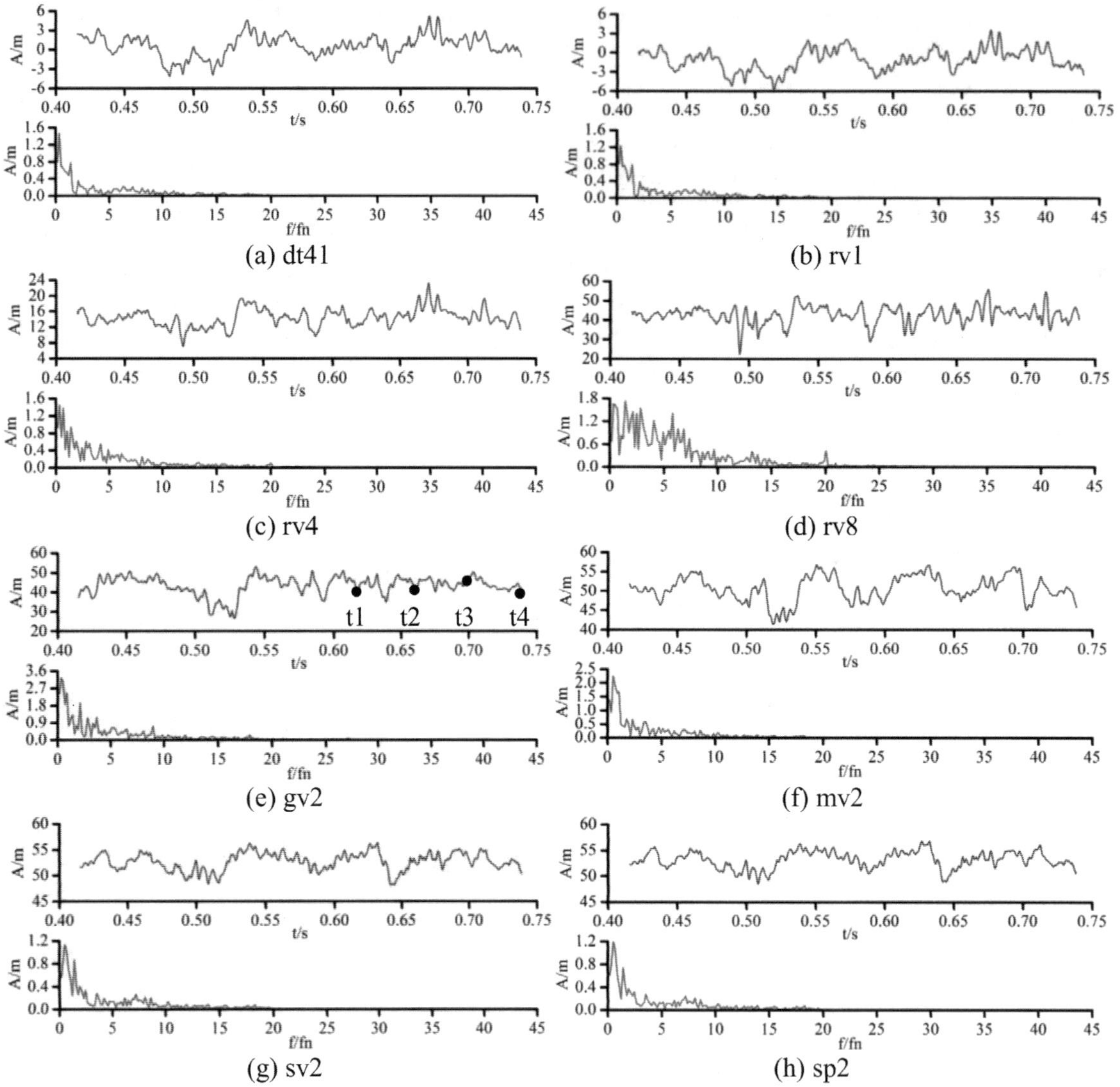

Figure 6. Pressure fluctuations both in the time and frequency domain at chosen points

3.4. Internal flow behavior analysis

The streamlines at different flow time on the mid-span plane of the runner, guide vanes and stay vanes are shown in Figure 7. The internal flow of the small discharge condition is highly disorganized. The vortex and secondary flow are obvious inside the guide vanes and stay vanes, even inside the runner. The appearance of these complex flow patterns can block the flow passage and water can't flow smoothly. In addition, part of the energy will dissipate in the vortexes. These are the factors contributing to the larger energy loss inside the runner and guide vanes. Compared the streamline at different time, it can also be dawn that the internal flow changes with the flow time. From t1 to t2, the flow in the gv2 nearby region becomes less disorganized. While, from t2 to t4, the flow in the gv2 nearby region becomes more and more disorganized. At t4, there even generates vortex flow in the gv2 nearby region. In the mv2 nearby region, the vortex and secodnary flow also experiences the strengthen and weaken process from t1 to t4. The strengthening and weakening of the vortex and secondary flow result in the low dominant frequency inside the flow passage.

Figure 8 shows the streamline on the meridional plane of the runner and draft tube. It can be found that flow separation occurs with backflow flowing from runner toward draft tube. In addition, in the elbow part of the draft tube, there also occur vortexes. The appearance of backflow and vortexes can block the flow passage and increase the flow resistance. Then the mainstream can't flow into the runner smoothly. In addition, part of the energy of the runner will dissipate in the backflow and vortexes. Therefore, the head loss inside the draft tube is relatively large although the velocity inside it is low. Compared the streamline at different time, it can also be dawn that the flow inside the draft tube changes with the flow time. From t1 to t2, the vortex on the left side migrates downwards. The two vortexes on the right side merge together. From t2 to t3, the vortex on the left side migrates upwards. The vortex on the right side splits into two parts. From t3 to t4, the vortex flow on the left side strengthens and the vortex region enlarges. The vortex on the left side weakens and the vortex region shrinks. The excursion, split and merging of the vortex and the weakening and strenghtening of the vortex are all related to the low dominant frequency inside the draft tube.

(a) t1 (b) t2 (c) t3 (d) t4

Figure 7. Streamline in the runner, guide vanes and stay vanes on the mid-span plane

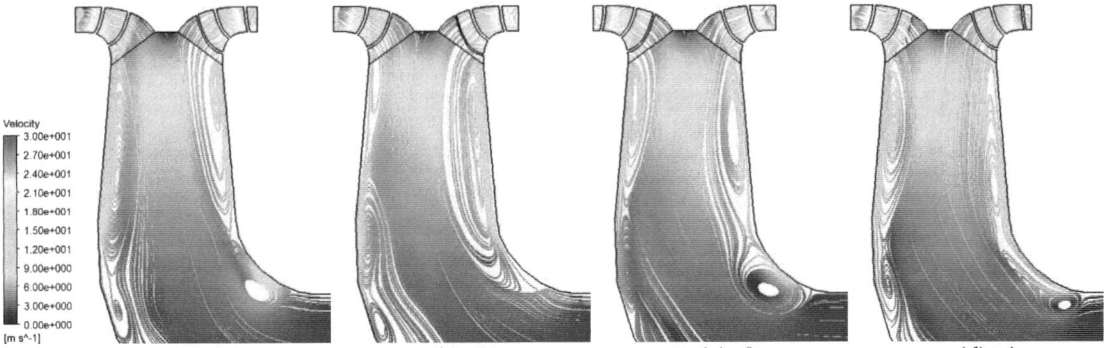

(a) t1 (b) t2 (c) t3 (d) t4

Figure 8. Streamline on the meridional plane of runner and draft tube

4. Conclusions

For the small discharge operating pump condition at the optimal opening, unsteady numerical simulation of the entire passage is conducted adopting the SST k-ω turbulence model. Conclusions drawn from the analysis are as follows.

1) The simulated results match well with experimental results. It is feasible to adopt the numerical simulation to analyze the internal flow of the small discharge condition.

2) The main energy loss is in the draft tube, runner and guide vanes domain. The large head loss inside the draft tube is induced by the backflow from the runner and the vortex inside the draft tube. The large head loss inside the runner is caused by the backflow in the runner inlet region and the

secondary flow in the runner outlet region. The large head loss inside the guide vanes is due to the large number of vortex and secondary flow inside the passage.

3) The strengthening and weakening of the vortex and secondary flow result in the low dominant frequency inside the runner downstream flow passage. The excursion, split and merging of the vortexes and the weakening and strenghtening of the vortexes are all related to the low dominant frequency inside the draft tube.

Acknowledgments

Special thanks are due to the State Key Program of National Science of China (Grant No. 51439002), National Natural Science Foundation of China under Contract (No. 51479093), State Key Laboratory of Hydroscience and Engineering (Grant No. 2014-KY-05) for supporting the present work.

References

[1] Qu L F and Wang L 1996 *Power Eng.* **06** 58-62
[2] Wang L Q, Liu J T, Zhang L F, Qin D Q and Jiao L 2011 *J. Zhejiang University Eng. Sci.* **45**(7) 1239-43
[3] Guedes A, Kueny J L, Ciocan G D and Avellan F 2002 *Proc. 21st IAHR Symp. on Hydraulic Machinery and Systems(Lausanne, Switzerland, 9-12 September)*
[4] Ran H, Luo X, Zhu L, Zhang, Y, Wang X and Xu H 2012 *Chinese J. of Mech. Eng.* **25**(6) 1205-09
[5] Yin J L, Liu J T, Wang L Q and Wei X Z 2011 *J. Eng. Thermophysics* **32**(7) 1141-4
[6] Xiao Y X, Wang Z W, Zongguo Y, Mingan L, Ming X and Dingyou L 2010 *Eng. Computations* **27(3)** 365-86
[7] Yexiang X, Zhengwei W and Zongguo Y 2011 *Eng. Computations* **28**(2) 154-71
[8] Xiao Y X, Wang Z W, Zhang J and Luo Y Y 2014 *J. Hydrodynamics Ser. B* **26**(2) 250-6

IOP Conf. Series: Materials Science and Engineering **72** (2015) 042030 doi:10.1088/1757-899X/72/4/042030

Flow analysis of NACA0015 and its application on tidal power

D M Liu[1, 2], H Chen[1], L L Yuan[1], Y Z Zhao[1] and X B Liu[2]
[1] Research & Development Center of Dongfang Electric Machinery CO., LTD, Deyang, 618000, Sichuan Province, China.
[2] Fluid machinery key laboratory of Xihua University, Chengdu 610039, China

E-mail: liudemin999@126.com

Abstract. NACA0015 is a commonly used hydrofoil for hydraulic machinery. Because the calculated flow characteristic is sensitive to the mesh density and distribution, a careful check of mesh independence should be performed before the CFD calculation. For this purpose, flow comparison was done between calculation and experimental test at the same operating condition with the Reynolds number equal to 130,000. When the span length is two times of the chord length, the ratio of lift to drag is 34.97, which is close to the test result of 36.47. The pressure coefficient and cavitation characteristic were also calculated. For example, the pressure coefficient was compared in different flow conditions with Reynolds number equal to 130,000, 200,000 and 400,000. The cavitation characteristic was calculated at different cavitation numbers, e.g. 0.5, 1.0 and 1.5. Based on these hydraulic and cavitation characteristic, a tidal turbine runner was developed by using hydrofoil NACA0015. According to our studies, the flow characteristic and cavitation index can satisfy the engineering requirements.

1. Introduction

In China and many other countries there is an emerging market for green electricity derived from renewable energy resources. Among the green energy resources is oceanic energy: for instance there are waves, currents and ocean thermal energy. To humankind, it is a serious challenging to meet the future's energy demands without harmfully impacting on the environment through getting power either by consuming more fossil fuels or relying more on nuclear energy. In this situation, developing oceanic energy resources might be a solution.

Recent studies [1, 2] indicate that marine currents have the potential to supply a significant fraction of future electricity needs. Therefore many countries have invested to develop technologies in this field. Various types of tidal turbines have been reported such as horizontal axial, vertical axial and open-center turbines. Whatever type of rotor is used, the theoretical upper limit of efficiency for extracting energy from a free stream is 59.3 percent, which can be derived from actuator disk theory-a conclusion commonly attributed to Betz [3-4]. In practice, a tidal turbine rotor, which generally will be large in relation to the flow cross-section (i.e. with its upper and lower edges close to the surface) may benefit from an effect called 'blockage'. This is where the flow is constrained by the boundaries so that it cannot diverge to the extent that it would when decelerated through a rotor in free space, and as a result a greater mass flow of fluid passes through the rotor and this manifests itself as a higher efficiency [5-6].

In this paper, a bulb-typed tidal turbine, whose blades were shaped based on NACA0015 hydrofoil, was proposed and tested by numerical simulations. This design is simple compared other horizontal turbine because of the complex seal structure for protecting the generator. This paper firstly focused on the flow analysis of NACA0015 hydrofoil in respect with hydro dynamic characteristics, and introduced some details of the mathematical model and mesh schemes used in the analysis. it then described the simulation results of hydrofoil under full wet and cavitation condition. At the end, it concluded that NACA0015 could be a better solution for tidal turbine design.

2. Mathematical model

In homogeneous multiphase model, a common flow field is shared by all fluids, as well as other relevant fields such as turbulence. Thus, the governing equations of homogeneous model for mass, momentum and volume conservation equation can be written as

$$\frac{\partial \rho_m}{\partial t}+\frac{\partial\left(\rho_m \bar{u}_i\right)}{\partial x_i}=0 \tag{1}$$

$$\frac{\partial\left(\rho_m \bar{u}_i\right)}{\partial t}+\frac{\partial\left(\rho_m \bar{u}_i \bar{u}_j\right)}{\partial x_j}=-\frac{\partial \bar{p}}{\partial x_j}+\frac{\partial}{\partial x_j}\left(\bar{\tau}_{ij}^*-\rho_m \overline{u_i^{'} u_j^{'}}\right) \tag{2}$$

$$\frac{\partial\left(\rho_l \alpha_l\right)}{\partial t}+\frac{\partial\left(\rho_l \alpha_l u_j\right)}{\partial x_j}=\rho_l \dot{m} \tag{3}$$

where $\bar{\tau}_{ij}^*=\begin{cases}\mu_m\left(\dfrac{\partial \bar{u}_i}{\partial x_j}+\dfrac{\partial \bar{u}_j}{\partial x_i}\right) & i \neq j \\ 2\mu_m \dfrac{\partial \bar{u}_i}{\partial x_j}-\dfrac{2}{3}\mu_m \nabla.\vec{V} & i=j\end{cases}$ $\quad -\rho_m \overline{u_i^{'} u_j^{'}}=\mu_t\left(\dfrac{\partial u_i}{\partial x_j}+\dfrac{\partial u_j}{\partial x_i}\right)-\dfrac{2}{3}\rho_m k \delta \quad \rho_m=\alpha_v \rho_v+\left(1-\alpha_v\right)$

$\mu_m=\alpha_v \mu_v+\left(1-\alpha_v\right)\mu_l$

All kinds of cavitation models [7-8] which is provided by many researchers are applied on cavitation phenomena. Such as Singhal（2002）[7] and Zwart-Gerber-Belamri（2004）[8] give different cavitation equations and cavitation coefficient. The source terms [9] $\dot{m}$ can be expressed as

$$\text{If } \ p<p_v^*: \ \dot{m}=C_1 \frac{3\left(1-\alpha_v-\alpha_u\right)}{r} \frac{2\sigma}{2-\sigma}\left(\frac{M}{2\pi R}\right)^{1/2}\left(p_v^*-p\right) \tag{4}$$

$$\text{If } \ p>p_v^*: \ \dot{m}=C_2 \frac{3\alpha_v}{r} \frac{2\sigma}{2-\sigma}\left(\frac{M}{2\pi R}\right)^{1/2}\left(p-p_v^*\right) \tag{5}$$

$$p_v^*=p_v+\frac{p_{turb}^{'}}{2}=p_v+0.195\rho k \tag{6}$$

3. Calculation domain and Boundary condition of hydrofoil

Figure 1 is the calculation domain of hydrofoil, the length of hydrofoil is C=0.1m and the inflow angle is 9°. To avoid the inflow condition and wall affection, the distance 5.3C and C from inflow and wall is set. The detail boundary conditions are shown in Table 1.

IOP Conf. Series: Materials Science and Engineering **72** (2015) 042030 doi:10.1088/1757-899X/72/4/042030

Table 1. Boundary condition

The chord length of the hydrofoil	0.1m	Hydrofoil wall	wall condition
The span length of the hydrofoil	0.08m	Upper plane	Wall condition
Incidence angle	9°	Lower plane	Wall condition
Fluid material	water	Cavitation number	0.5,1.0, 1.5
Inlet velocity	1.3m/s, 2m/s, 4m/s	Turbulence model	RNG $k-\varepsilon$
out	pressure	Simulation type	steady

The structure of hydrofoil mesh is finished on the *X-Y* plane as shown in Figure 1. When the node number is 400,000, the calculated ratio between lift and drag is 34.97, which is close to the test result of 36.47 as shown in Figure 2. To be insensitive to the calculation results, the mesh with the node number of 800,000 was selected in this paper.

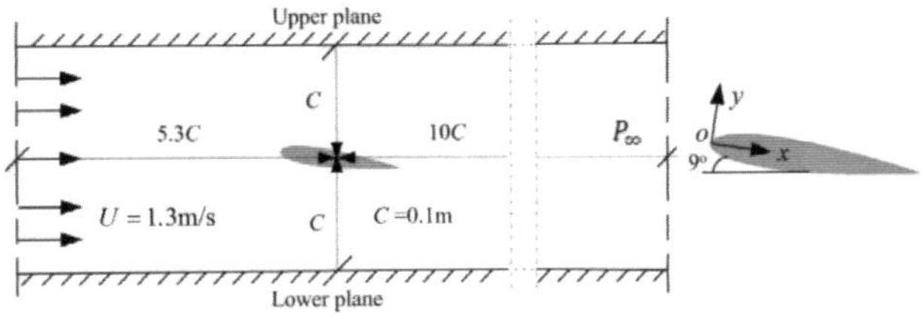

Figure 1. Model and flow conditions

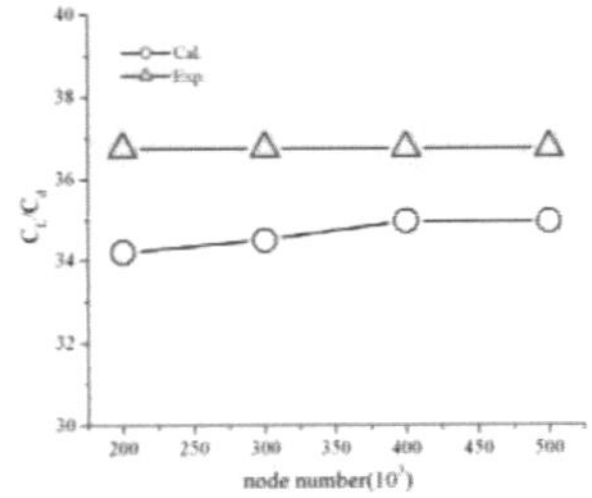

Figure 2. Lift and drag coefficient and node number

4. Calculated results and discussion of hydrofoil

The pressure coefficient C_p was compared between test and calculation as shown in Figure 3 and the consistency is excellent.

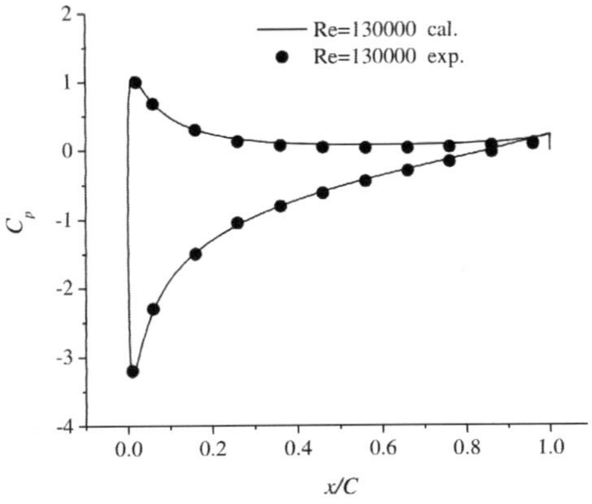

Figure 3. Pressure coefficient between test and calculation

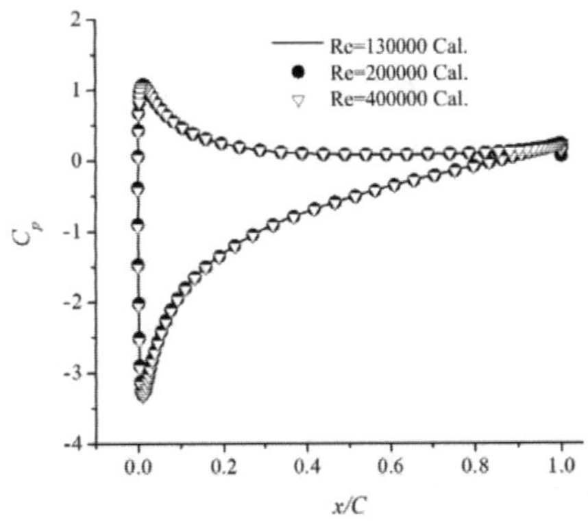

Figure 4. Pressure coefficient at different inflow conditions

Furthermore, the pressure coefficient was compared under different Reynolds number conditions of 130,000, 200,000 and 400,000. The pressure coefficient distributions are nearly identical in this inflow velocity range, as shown in Figure 4. From low velocity to high velocity, the hydrodynamics of the NACA0015 foil is excellent in wet condition.

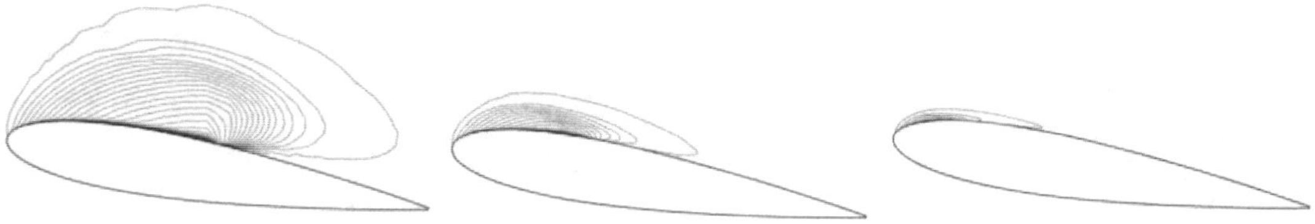

(a) (b) (c)

Figure 5. Cavitation bubble shape at different cavitation numbers (The red curve on hydrofoil means the volume of cavitation bubble): (a) $\sigma = 0.5$; (b) $\sigma = 1.0$; (c) $\sigma = 1.5$.

The cavitation bubble shape was calculated at different cavitation numbers σ ($-C_p$), e.g. 0.5, 1.0 and 1.5 (shown as in Figure 5). The cavitation bubble shape is sensitive to the cavitation number. From Figure 5 to Figure 6, we can see the volume (red circle) and length of cavitation bubbles increase with the decrease of cavitation number.

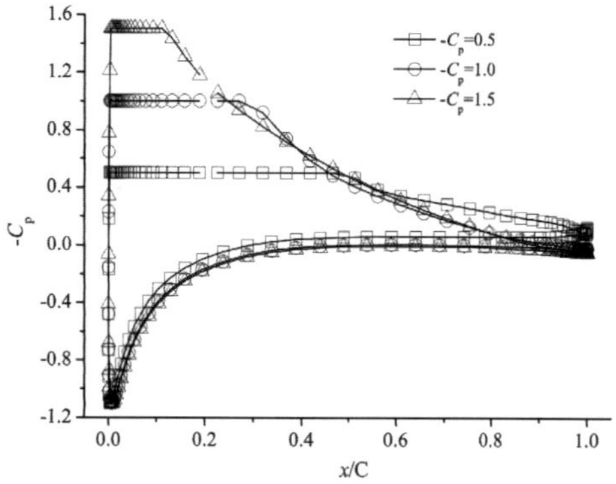

Figure 6. Pressure coefficient with cavitation numbers (The horizontal section in the curve means the length of cavitation bubble)

5. Calculated results and discussion of Tidal turbine

Based on the above research results, a new tidal turbine runner is developed by just applying the hydrofoil NACA0015. It has 10 blades, please refers to Figure 7. The calculation domain is a big box to simulate the flow domain in the ocean, please refers to Figure 7 (a). The tidal turbine domain mainly includes inlet section, rotor section and outlet section, please refers to Figure 7 (b). The rotor section is the rotation part whose rotation speed is 12r/min. The inlet section is the contraction cone, which can increase the inflow velocity to collect more energy. The outlet section is expanding cone which can recover energy. The detail boundary conditions are shown in Table 2. The blade setting angle β is the included angle between blade center line and rotor axis in Figure 7 (c).

The velocity was imposed on the inflow1 of the calculation domain as the inlet condition for the numerical simulation. And the interface boundary condition was imposed on the inflow2 of the tidal turbine as the inlet condition for the numerical simulation.

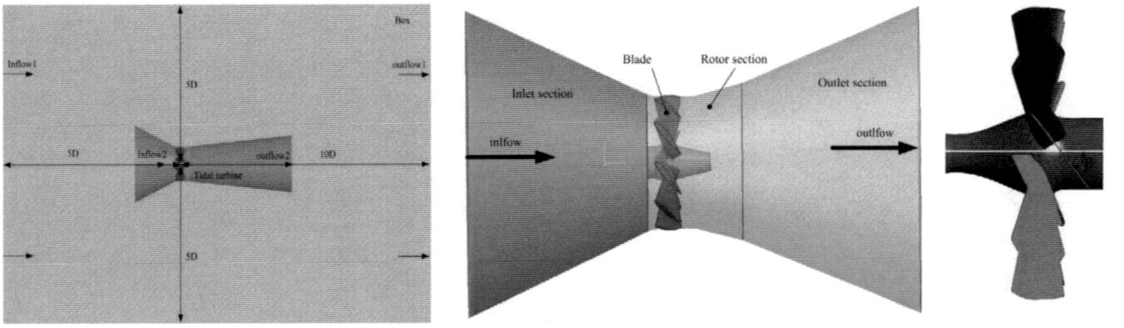

Figure 7. Calculation domain and tidal turbine structure: (a) Calculation domain; (b) Tidal turbine; (c) Blade angle.

The output power is calculated at different install angle β (Beta) in Figure 8. From the Figure 8, the output power 13KW is maximum value when the setting angle is 24°. The flow discharge is gradually decreasing with setting angle increasing in Figure 9. Because the upwind flow area increases with the blade setting angle increasing.

Table 2. Boundary condition

Fluid material	water	Cavitation number	1.5
Inletflow1 velocity	2m/s--4m/s	Moving condition	Rotor stator
D	1.5m	Calculation domain and turbine	interface
Outflow1	pressure	Turbulence model	RNG $k-\varepsilon$
blade	wall condition	Simulation type	steady

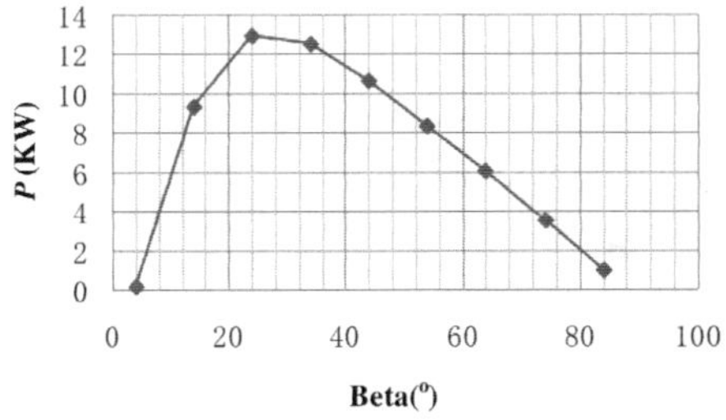

Figure 8. Power at different setting angle

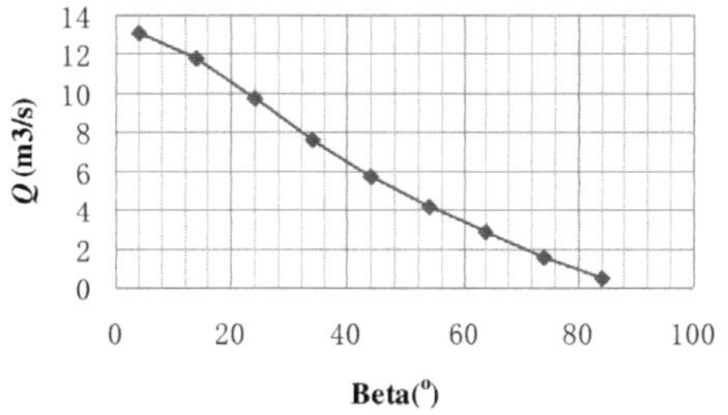

Figure 9. Discharge at different setting angle

The output power is an important parameter to determine the performance of the tidal turbine, which is calculated by Eq. (7). The power was calculated at different flow velocities of 2.0m/s, 2.5m/s, 3.0m/s, 3.5m/s and 4.0m/s. Figure 10 described the power output of the turbine in relation with the inflow velocity, taking into account the influence of the cavitation. From it we can see that the maximum output of 13KW is achieved with inflow velocity equals 4.0m/s and free of cavitation. When cavitation takes place, the outputs of the turbine runner are clearly reduced at all selected operating points; however, the maximum output power keeps still at 12KW.

$$p = \pi MN/30 \tag{7}$$

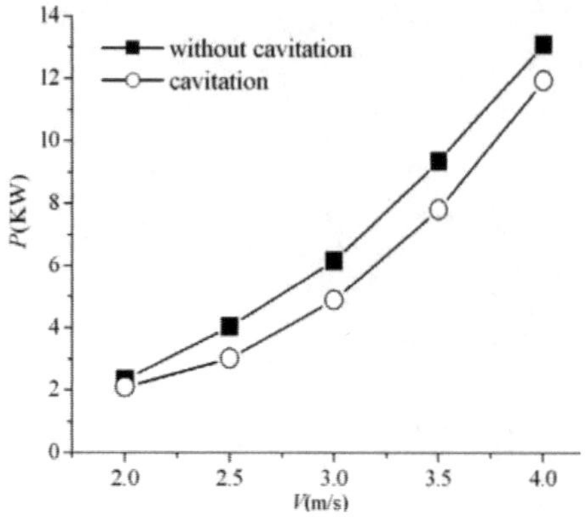

Figure 10. Power at different flow velocities

6. Conclusions

In this paper, the flow characteristics of the hydrofoil NACA0015 and a bulb-typed tidal turbine were analyzed and discussed, and got the following conclusions.

1) Simulation calculations and tests have been done to NACA0015 hydrofoil at different cavitation numbers and Reynolds numbers. The hydrodynamic characteristic and free cavitation of NACA0015 hydrofoil is adapted for the flow velocity range.

2) The tidal turbine developed by applying the hydrofoil NACA0015 is reliable and the developed tidal turbine output power is 13KW. The maximum output power appears at the blade setting angle 24o. However, the output power will further improve in the next research.

Nomenclature

C_1-evaporation coefficient 0.13
C_2-condensation coefficient, 0.01
C_p-Pressure coefficient
M-torque
N-rotational speed(r/min)
P-output power (MW)
Re-Reynolds number
p-pressure (Pa)
p_v-vapor pressure (Pa)
r =0.0001
u- Velocity
α_v- Water vapor void fraction
α_u -non-dissolved gas void fraction, which takes value of 0.0005.
σ =cavitation number (-C_p)
ρ_m -mixture density
ρ_v -water vapor density
ρ_l -water density
μ_m -mixture viscosity
μ_v -water vapor viscosity
μ_l -water viscosity
k : Turbulent kinetic. i , j =1, 2, 3.

Acknowledgments

This work is supported by the Chinese National Foundation of Natural Science (No. 51279172), the Open Research Funded Project of Key Laboratory of Fluid Machinery in Sichuan Province (Xihua University) (Grant No.Szjj2012-041, szjj2011-039).

References

[1] Binnie 1993. Tidal stream energy review: UK DTI. Report ETSU, T=05=00155=REP,

Engineering and Power Development Consultants Limited and McAlpine and Sons Limited and IT Power Limited, UK Department of Energy, UK.

[2] EU Commission 1996 The exploitation of tidal marine currents Report EUR16683EN 23-31

[3] Feasibility study of tidal current power generation for coastal waters: Orkney and Shetland. Final Report, EU contract XVII=4 1040=92-41, ICIT, March 1995.

[4] Cave P R and Evans E M 1986 Tidal stream energy systems: economics, technology and costs Third International Symposium on Wave, Tidal, OTEC and Small Scale Hydro-Energy (Brighton. England, 14-16 May 1986)

[5] HSE UK 1995 Offshore Installations: Guidance on Design, Construction and Certification (London: UK Department of Energy, HMSO)

[6] Kihoh S and Shiono M 1994 Power generation by tide and tidal current. In Proceedings of Advanced Materials '93, Vol. V=A, Eco-materials, pp. 41–44

[7] Singhal A K, Athavale M M, Li H and Jiang Y 2002 J Fluid Eng. **124**(3) 617-24

[8] Zwart P J, Gerber A G and Belamri T 2004 A Two-Phase Flow Model for Predicting Cavitation Dynamics Fifth International Conference on Multiphase Flow (Yokohama, Japan, 30 May-4 June 2004)

[9] Liu D M, Liu S H, Wu Y L and Xu H 2009 Numerical Analysis of Airfoil NACA0015 and Centrifugal Pump's Cavitation Characteristic Based On the Thermodynamic Effects Proceedings of 2009 ASME Fluids engineering conference (Vail, Colorado, USA, 2-5 August 2009)

IOP Conf. Series: Materials Science and Engineering **72** (2015) 042031 doi:10.1088/1757-899X/72/4/042031

The study of a reactor cooling pump under two-phase flow

P Wang, S Q Yuan, X L Wang and F Zhang
National research center of pumps, Jiangsu University, Zhenjiang, 212013, China

E-mail address: wangpeng2017@yahoo.com

Abstract. In this paper, the steady pressure field has been investigated numerically by computational fluid dynamics (CFD) in a nuclear reactor cooling pump. As a multiphase approach the Eulerian-Eulerian two fluid model has been applied to calculated five computational models with different kinds of blades. The analysis of inner flow field of the five model pumps shows that the pressure in the impeller increases with the increase of the gas contents and the pressure distributions are irregular at the inlet of different blades when the gas contents less than 20%. With the increase of the number of blades, the vortexes at the outlet of impeller decrease whereas the vortexes in the deep of the volute markedly increases and high velocity of the fluid huddle is generated gradually at the outlet pipes. Under the action of centrifugal force and Coriolis force, gas phase mainly concentrated at the lower velocity and lower pressure area. The radial force on the impeller gradually increases with the increase of the gas contents.

1. Introduction

Nuclear safety has been studied as one of the major issues since the inception of the nuclear industry and power generation. Establishing and maintaining core cooling to ensuring containment integrity are two main goals for nuclear safety [1-2]. As the heart of nuclear power plant, it is important for the performance of reactor cooling pump when under the two-phase flow. A.Poullikkas [3] has used the high-speed video(HSV) to observe the bubbles motion in the impeller passages and pointed out that the bubbles tend to concentrate at the blade suction side, the "high" gas contents(around 9 percent by volume) would form the passage blockage with a resulting break in the pump flow. Yuan [4] has investigated the collector discharge pressure fluctuation characteristics and the related factors in a nuclear coolant pump and pointed out the fluctuation amplitude is increased gradually from the top side wall of casing to the bottom, and the shaft rotating frequency is the major factor influencing pressure fluctuation at ever monitor when the pump is working at the design specifications.

Although a great deal of work related to traditional PWRs(Pressurized Water Reactor) has been done in the past few years[5-9], the mechanism of small break LOCA (Loss of Coolant Accident)in AP1000 has not been clearly understood, especially, the performance of coolant pump under two-phase flow(gas-liquid). To guarantee the safety operation of nuclear reactor, it is essential to study the inner gas-liquid flow in the impeller of cooling pump. In this paper, five different impellers are selected as the research subjects to investigate the performance of cooling pump under two-phase flow. The cooling pump performances and internal flow patterns are predicted by the commercial CFX 14.5. The main purpose of this paper is to detail description of the two phase flow in nuclear reactor cooling pump and discuss the relation between the gas contents and the force which load at the impeller and inner velocity flow patterns is also presented in this paper.

2. Numerical simulation methods

2.1. Model parameter and mesh generation

The design parameters are shown in Table.1.

Table 1. Main design parameters

Q (m^3/h)	H (m)	n (r/min)	Design pressure(MPa)	Specific Speed(n_s)
17886	111.3	1450	17.1	344

According to the design parameters, the computational domains are created based on the pump geometries that include 5 following fluid domains: inlet section, impeller, guide vane, volute and outlet pipes. The inlet section and outlet are extended 5x their respective diameters so that the flow can be fully developed at the two sections. After modeling in Pro/E, the 3D geometry computational domains are shown as Figure 1. The structured grids for the computational domains are generated using the grid generation tool ANSYS ICEM CFD 14.5, the impeller without shroud are partially shown in Figure 2 and the computational domains are input to the flow solver, CFX 14.5, for simulation.

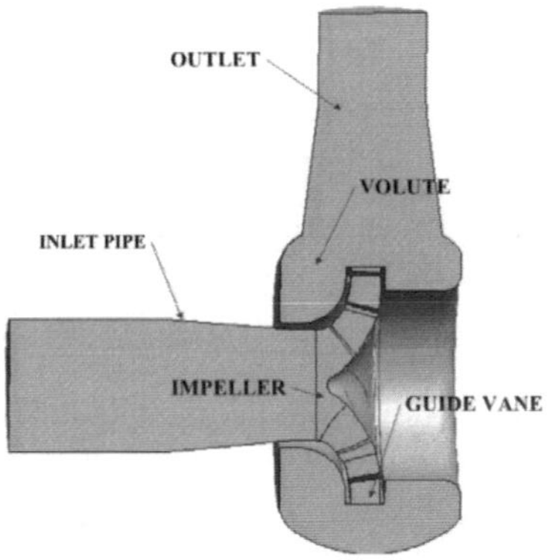

Figure 1. Geometry computational domains

Figure 2. The structure grids (5 blades)

2.2. Basic assumptions

In this paper, the Eulerian-Eulerian two-phase fluid model is selected to emphasize the two aspects of the dynamic interactions between gas and liquid phases. Gas-liquid two phase flow is affected by many factors, the situation is complex. In order to simplify this research in this paper, a simplified physical model should be used. The following basic assumptions are put forward:

1) The flow pattern is regarded as steady flow, and the liquid phase is an incompressible fluid.

2) The structure of bubbles cannot be changed when they move into the flow passages and their diameter is far less than the diameter of the flow passages.

3) The gravity of gas-liquid two phase flow is ignored.

2.3. Govern Equations

For incompressible bubbly flows without mass transfer, the instantaneous mass balance equations in the liquid and in the gas are respectively:

$$\frac{\partial(1-\chi_G)\rho}{\partial t}+\frac{\partial(1-\chi_G)\rho u_j}{\partial x_j}=0,\quad \frac{\partial\chi_G\rho_G}{\partial t}+\frac{\partial\chi_G\rho_G u_{Gj}}{\partial x_j}=0 \tag{1}$$

where u_{Gj}, ρ_G, and X_G are respectively the velocity, the density and the characteristic function of the gas phase (X_G =1 in the gas and 0 in the liquid). To simplify, the subscript G is used for the gas phase and the subscript L is omitted for the liquid phase throughout the paper. Thus u, ρ and (1-X_G) are respectively the velocity, the density and the characteristic function of the liquid phase.

At each point where the liquid phase is present, the instantaneous momentum balance is:

$$(1-\chi_G)(\rho\frac{D}{D_t}u_i = \frac{\partial}{\partial x_j}\sigma_{ij} + \rho g_i) \tag{2}$$

where $D/D_t = \partial/\partial_t + u_j(\partial/\partial x_j)$, σ_{ij} is the stress tensor and g_i is the acceleration of gravity.

For the specific case of bubbly flows, we neglect the acceleration and the weight of the gas as compared to the force exerted by the liquid on the bubbles because of the contrast of the densities $\rho_G << \rho$; so, in each point where the gas phase is present, the instantaneous momentum equation in the gas indicates that the volume density of the total force f_{Pi} exerted by the liquid on the bubbles is zero:

$$\chi_G f_{Pi} = 0 \tag{3}$$

The instantaneous momentum balance (3) represents an Eulerian description of the movement of the gas. It expresses the interfacial transfer as a local density of the force exerted by the liquid on the bubbles. In this model, the Reynolds stress tensor of the continuous phase is split into two parts, a turbulent dissipative part produced by the gradient of mean velocity and by the wakes of the bubbles and a pseudo-turbulent non-dissipative part induced by the displacements of the bubbles: each part is predetermined by a transport equation [10].

2.4. Numerical method and boundary conditions
The hydraulic passages of the modeled pump are taken in consideration as part of the computational domain. The pitch change for the interface between the rotor and stator is set as fixed rotor. The rotating speed is set as 1450rpm.

The normal velocity components are given at the inlet, and the outlet boundary is assumed to be pressure generated by the pump of the fluid. Standard wall function is adopted to predict the turbulent flow near-wall, and at the fluid physical surfaces of the pump were set to be no-slip wall,while the air physical surfaces were set to be free-slip wall. The turbulence is set to an initial intensity of 5% at the inlets. Besides, the turbulence model can predict the development of bubbly flows well. The convergence precision is set as 10^{-4}.

3. Results and discussion

3.1. Internal pressure field analysis
Due to space limitations, we only present 4 kinds of different blades schemes to analysis, which are 5, 6, 8 and 9. The pressure field distributions for 4 different blades under the calculated conditions are shown in figure 3.

From the Figure.3a) we know that the lower pressure area mainly appeared at the inlet of the impeller and also a small amount of it can be seen at the outlet edge of impeller. With the increasing of the gas contents, the lower pressure increased from the pressure side to the suction side that is the gas mainly concentrated on the suction side of the impeller. From the Figure.3b) to Figure.3d), with the increasing of the blades, the lower pressure area spread from the pressure face to the suction face. When are 9 blades, as seen at Figure.3d), appeared a large amount of high pressure area at all blades near the inlet and outlet only Figure.3a) with 5 blades have lower and more even pressure area distribution. On the other hand, the pressure in the impeller increased with the increasing of the gas contents and the pressure distributions are unregularly at the inlet of different blades when the gas contents fewer than 20%.

IOP Conf. Series: Materials Science and Engineering **72** (2015) 042031 doi:10.1088/1757-899X/72/4/042031

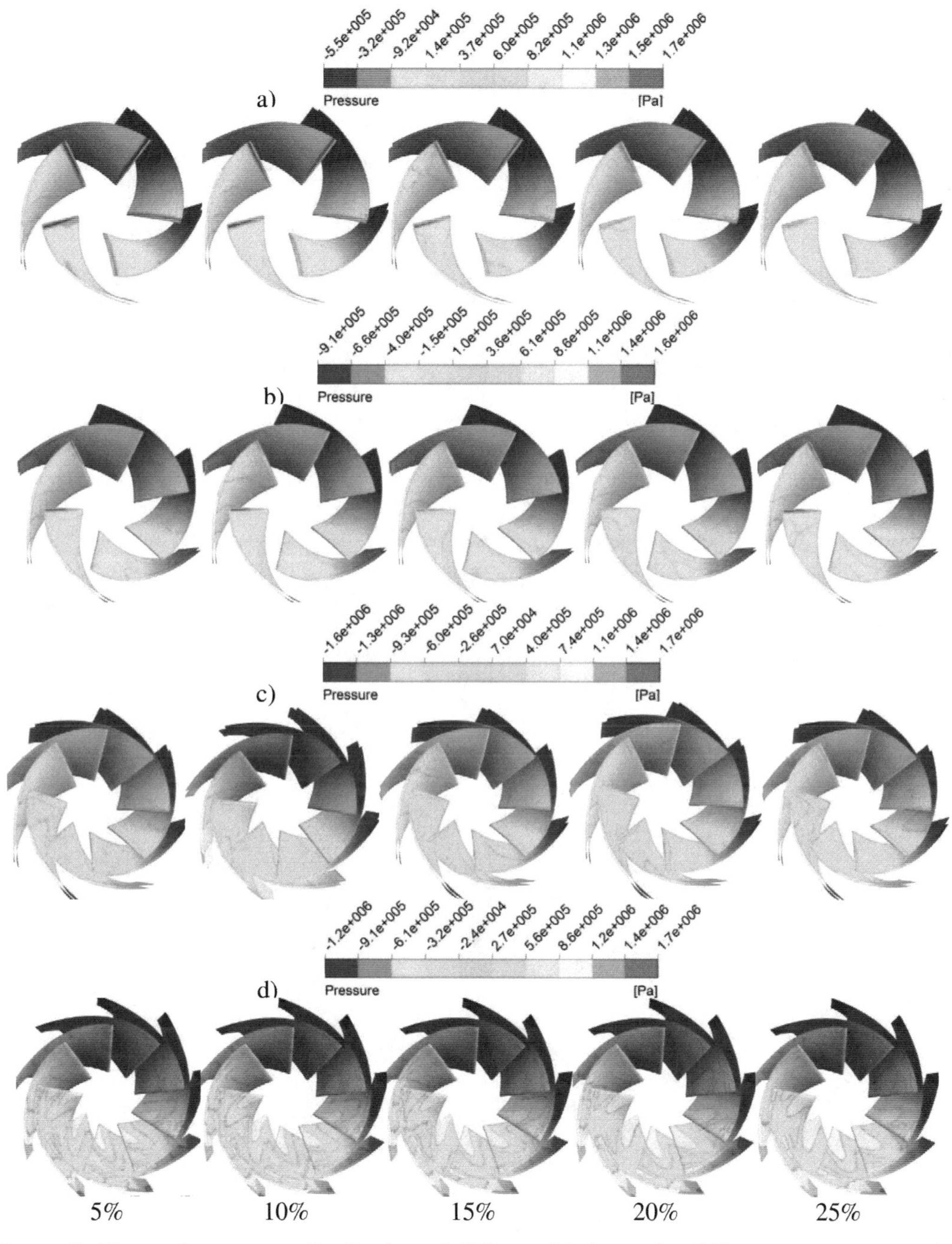

Figure 3. The static pressure distribution of different blades under different gas contents: a) 5 blades; b) 6 blades; c) 8 blades; d) 9 blades.

The distributions for gas contents of 25% are shown in Figure 3, here the high pressure area entirely fill the sides of the blades, and inevitably lower the pump performance. The main reason is probably because the larger the gas contents are, the lower inertia force of fluid has, can't keep the blades work better that poor the performance. Besides, with the increasing of blades, appeared a large amount of high pressure area that the losses within the impeller increases and the pump performance drop.

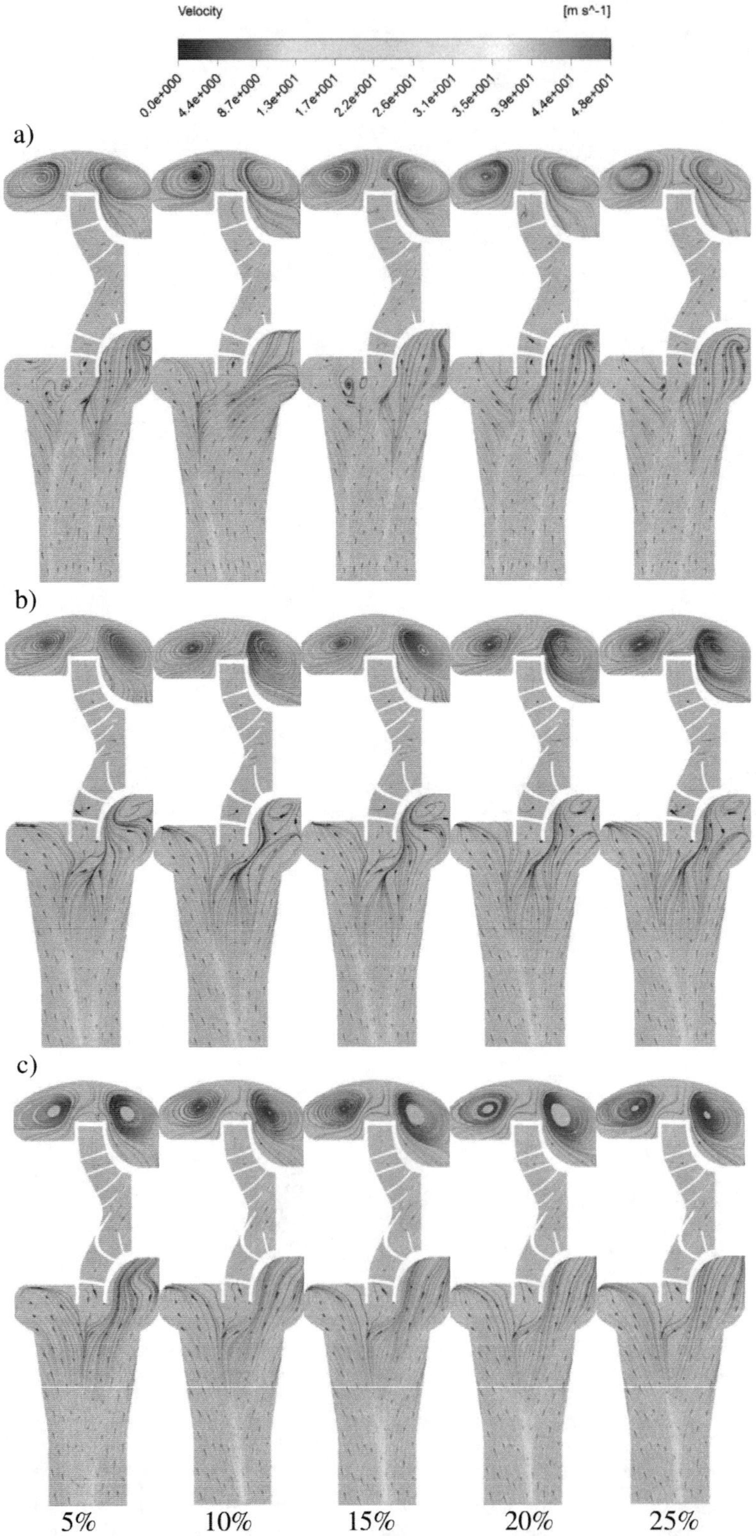

Figure 4. Streamline distribution under cross section with different gas contents: a) 5 blades; b) 7 blades; c) 9 blades.

IOP Conf. Series: Materials Science and Engineering **72** (2015) 042031 doi:10.1088/1757-899X/72/4/042031

3.2. Cross section streamlines distribution analysis

Considering space limitations, the fluid streamline distribution of cross section for the pump with 5, 7 and 9 blades are shown in figure 4, it can be seen that the distribution of streamlines within the whole section are uneven. The high velocity can be seen in the impeller; however, large amount of vortexes appeared in depths of volute and accumulated a small amount of gas wake at the end of the diffuser.

With the increase of the blades number, the vortexes of the outlet of impeller decreased, whereas the vortexes in the deep of the volute markedly increased. At the same time, the high velocity of the fluid huddle at the outlet pipes gradually. With the increase of gas contents, the vortexes in the deep of volute are direct proportional to the gas contents gradually, that is the velocity fluctuation change in the direction of large scale and irregular. The main reason is probably because the higher the gas contents, the ability of blades to do work to the fluid is decrease, gas would change the flow trend near the outlet. Besides, due to the density of the gas phase is far less than the density of the liquid phase, under the action of centrifugal force and Coriolis force gets gas phase mainly concentrated at the lower velocity and lower pressure area.

3.3 The comparative analysis of the radial force of the impeller

The pulsation curves for the radial force of the impeller as the function of the rotation angle of the impeller in one cycle under the working conditions of five kinds of gas contents from 5% to 25% are shown in figure 5. In the figure, the rotation angle increasing from the zero to 360 degree and direction of the radial force, and the projection on one cycle of the impeller each point respectively represents the axial radial force of the impeller at the moment.

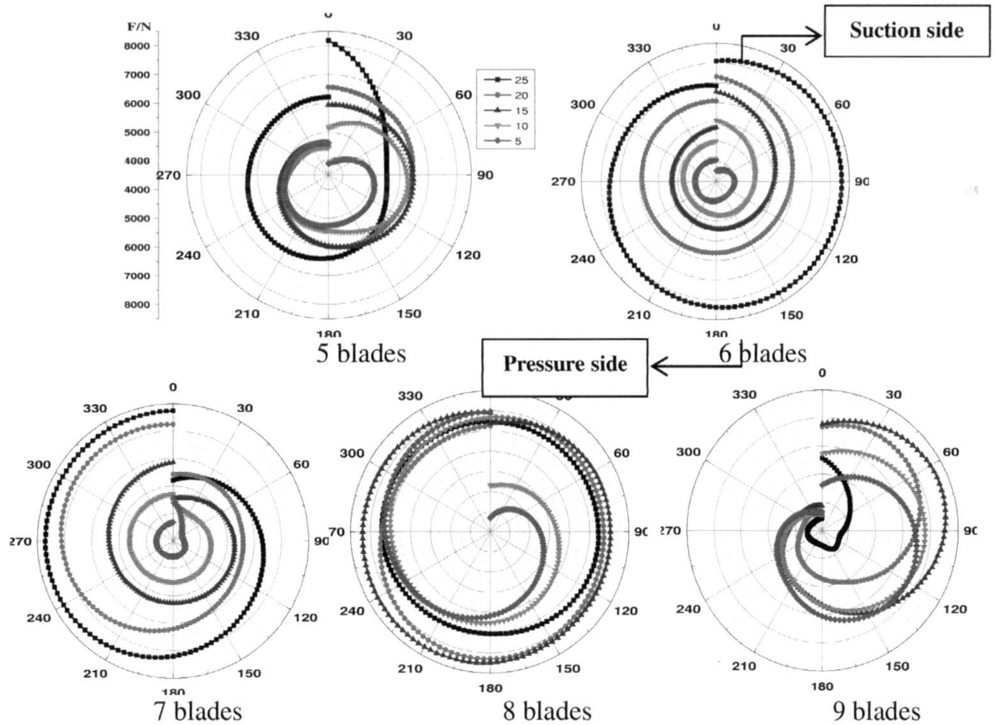

Figure 5. The curves for different blades of the impeller radial force with the impeller rotation angle at different gas contents in one cycle.

It can be seen that the radial force gradually increases with the increase of the gas contents, when the gas were 5%, the impeller with 6 and 7 blades bear the minimum force at all rotation. The force of 5, 8 and 9 blades, especially the 8 blades have the biggest force compared to the other schemes. When compared with the list five schemes, it can be seen that the 5 blades with the minimum force at the suction side, whereas the other schemes have the biggest force. Local have obvious phenomenon of circle round on the blades, however, no matter at which side of the blades, the force increasing with the gas contents. Interesting, the force on the suction side with 9 blades is minimum when compared with all schemes.

4. Conclusions

Using Euler- Euler two-phase flow model, the reactor cooling pump under of different gas contents is simulated and the analysis results show that:

1) The pressure in the impeller increased with the increasing of the gas contents and the pressure distributions are unregularly at the inlet of different blades when the gas contents fewer than 20%, with the increasing of the blades, the lower pressure area spread from the pressure face to the suction face that the losses within the impeller increases and the pump performance drop.

2) The high velocity of the fluid huddle at the outlet pipes gradually, with the increase of the blades number, the vortexes of the outlet of impeller decreased, whereas the vortexes in the deep of the volute markedly increased, with the increase of gas contents, the velocity fluctuation change in the direction of large scale and irregular.

3) Compared with the list five schemes, it can be seen that the 5 blades with the minimum force at the suction side, whereas the other schemes have the biggest force and the radial force gradually increases with the increase of the gas contents.

Acknowledgments

This research was supported by the State Key Program of National Natural Science Foundation of China (Grant No. 51239005), the National Science & Technology Pillar Program of China (Grant No. 2011BAF14B04), the National Natural Science Foundation of China (Grant No. 51379091) and Technology Support Program, Priority Academic Program Development of Jiangsu Higher Education Institutions (PAPD)

References

[1] Furukawa A 1991 *Z&E JSME Fall annual meeting* **B** 165-7
[2] Yoshida Y 1991 *JSME Fall annual meeting* **B** 168-70
[3] Poullikkas A 2000 *Progress in Nuclear Energy* **36** 123-30
[4] Yuan S Q, Zhu R S and Zheng B Y 2012 *Atomic Energy Sci. and Techno.* **46**(10) 1202-6
[5] Su X S 2000 *Foreign Nuclear Power* **2000**(6) 20-
[6] Poullikkas A 2000 *Progress in Nuclear Energy* **36** 123-30
[7] Wang W W, Su G H, Qiu S Z and Tian W X 2011 *Progress Nuclear Energy* **53**(4) 407-19
[8] Long Y, Zhu R S ,Q Fu and Yuan S Q 2014 *J. Drainage and Irrigation Machinery Eng.* **32**(4) 290-5
[9] Yang J, Wang W W, Qiu S Z, Tian W X and Su G H 2012 *Annals of Nuclear Energy* **46**(4) 81-9
[10] Chahed J, Roig A and Masbernat L 2003 *Int. J. Multiphase Flow* **29**(1) 23-49

The thermal and mechanical deformation study of up-stream pumping mechanical seal

H L Chen, C Xu[1], M Z Zuo and Q B Wu
Department of Energy and Power Engineering, Jiangsu University, Zhenjiang 212013, China

E-mail: xc15050852993@sina.com

Abstract. Taking the viscosity-temperature relationship of the fluid film into consideration, a 3-D numerical model was established by ANSYS software which can simulate the heat transfer between the upstream pumping mechanical seal stationary and rotational rings and the fluid film between them as well as simulate the thermal deformation, structure deformation and the coupling deformation of them. According to the calculation result, thermal deformation causes the seal face expansion and the maximum thermal deformation appears at the inside of the seal ring. Pressure results in a mechanical deformation, the maximum deformation occurs at the top of the spiral groove and the overall trend is inward the mating face, opposite to the thermal deformation. The coupling deformation indicate that the thermal deformation can be partly counteracted by pressure deformation. Using this model, the relationship between deformation and shaft speed and the sealing liquid pressure was studied. It's found that the shaft speed will both enhance the thermal and structure deformation and the fluid pressure will enhance the structure deformation but has little to do with the thermal deformation. By changing the sealing material, it's found that material with low thermal expansion coefficient and low elastic modulus will suffer less thermal-pressure deformation.

1. Introduction

In 1984 John Crane mechanical seals company posted a new concept about mechanical seal----up-stream pumping: open grooves in the sealing surface, and a small amount downstream leakage fluid will be pumped back. Subsequently, the company developed and produced the 8000 series of spiral groove upstream pumping mechanical seals, marking the upstream pumping mechanical seal officially put into use. Theoretically, compared with ordinary mechanical seal, upstream pumping mechanical seal with a non-contact, zero leakage, low maintenance cost and reliable operation, etc., however in fact, the research on sealing-lubricating mechanism of upstream mechanical seal and the design and application of it, etc. still need further studies.

Upstream pumping mechanical seal micro-gap lubrication problem has attracted attention of many scholars and researchers, and has made important progress. In the early time, the theory of narrow groove has been adopted [2,3]. It is assumed that the number of the spiral groove is infinite, and the geometric discontinuity and complex boundary conditions have been ignored, then the problem has been simplified to a one-dimensional model and been solved. Later, analytic methods have been used. The problem was simplified to a two-dimensional model. And the performance of mechanical seal could been predicted through solving Reynolds equation[4,5,6]. However, for the mechanical seal with

[1] Corresponding author.

complex structure such as spiral groove, analytical methods obviously was not in conformity with the pratical engineering. Recently, the numerical simulation method has been adopted. The numerical simulation has been carried on the spiral groove three-dimensional micro clearance which could effectively predict the pressure distribution in flow field inside and the flow regularity[7,8].

The mechanical seal internal flow field research has been effectively promoted. However, the thermal characteristics of the mechanical seal and the thermal - mechanical deformation has become a key point in studying the sealing mechanism and performance, especially the research about the mechanical seal with surface texturing. Li calculated the thermal deformation of mechanical seal by using the finite element method. And it is found that the thermal deformation can make the flat sealing surface deform along the radial. When the sealing surface produced dry friction, the deformation was more serious, which made the inner sealing surface more wear. So we must try to reduce the thermal deformation[9]. Jianfeng Zhou et al. established the heat transfer model of mechanical seal ring of static pressure type, and the ratio was determined that the friction heat between rotating ting and static ring accounting for the total heat. Then the coupling process between the friction heat of liquid film and the thermal deformation of the sealing surface were studied. And the rotating speed w of rotating ring, and the relationship between the film thickness of the inside diameter and the radial angle β of the sealing ring surface after deformation[10,11] were determined. Cicone er al. studied the non-adiabatic heat characteristics of the liquid film between mechanical sealing surfaces, and on the basis of this, he obtained accurate liquid film opening force, the liquid film stiffness and friction torque. But it was assumed that the viscosity of the liquid film is constant, finally it had a larger gap between the leak he obtained and the real value[12,] . Qiu yifan et al. considered the viscosity-temperature relationship of sealing medium, and solved the temperature distribution of the circular ring with spiral groove and the pressure distribution of liquid film. But they didn't further solve the thermal and mechanical deformation[13].

Based on the existing studies, this dissertation simulate the spiral groove upstream pumping mechanical face seal by combine the finite volume method(FVM) and finite element method(FEM) together, acquire the temperature and pressure field, and further imitate the thermal deformation, stress distortion and the coupling deformation. All this will provide some reference to the studying and design of upstream pumping mechanical face seal.

2. Geometry Model and Boundary Conditions

Taking the complexity of the study object, the feasibility of the modeling and the scientificity of the result into consideration, the following basic assumptions are made:

1) The performance of the sealing rings' materials does not change with temperature.

2) Ignores the impact of surface roughness.

3) Do not consider sealing ring radiation heat transfer, ignoring the heat generated by the seal ring stir and all heat is generated by the fluid film inner viscous dissipation.

4) Assumes the rings take away all the viscous dissipative energy.

5) The temperature distribution within the mechanical seal ring does not change with time, that it is a steady-state CFD model.

The 104-50 mechanical face seal was chosen and the matting rings profile parameters are shown in Figure 1. The spiral groove is set at the rotational ring end face as shown in Figure 2, and the parameters of it are shown in Table 1. Where, $\gamma=(r_g-r_o)/(r_i-r_o)$, $\beta=l_{wg}/l_{wt}$, l_{wg} is the length of groove, and l_{wg} is the length of groove plus seal dam.

The calculation model is built in the ANSYS Workbench platform, the fluid film is fluid domain and it's calculated by Fluent; the mating rings is structure domain, the heat transfer is calculated by Steady-State Thermal module, and the deformation is calculated by Steady Structure module.

As shown in Figure 3, the seal rings' temperature and stress fields are mainly produced by the fluid film. Q_f is the micro gap liquid viscous heat and it is respectively transfered to the static ring and rotational ring. Q_1 is the heat that rotational ring obtained and mainly passed to the sealing liquid by wall A-D; Q_2 is the heat that passed to the sealing liquid by stationary ring wall I-L; Q_3 and Q_4 is the heat passed to the air respectively by rotational ring and static ring. Due to heat is generated by the fluid film as well as the sealing end face's pressure is gained by the fluid film, then the 3-D micro gap need to be calculated first.

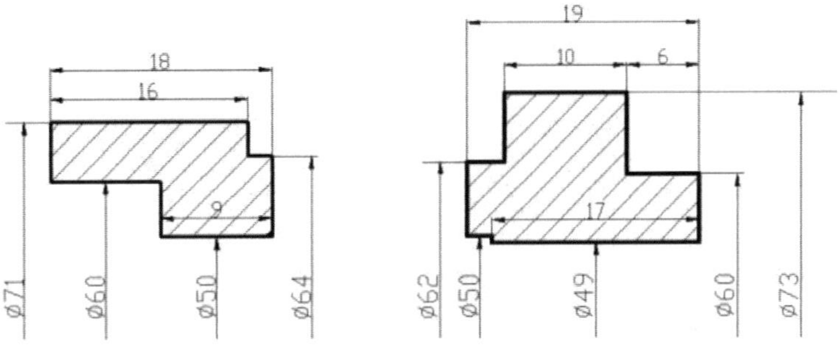

a) b)

Figure 1. geometry parameters of the a) rotational and b) stationary rings: unit:mm

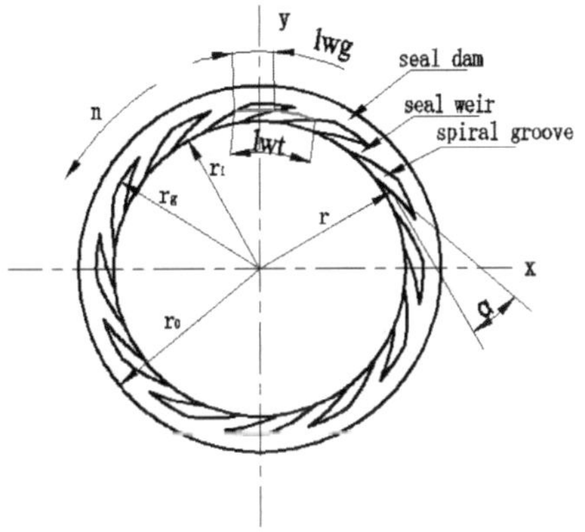

Figure 2. Rotational face geometry of upstream pumping mechanical seal

Table 1. The geometric parameters of the spiral grooves

Parameter	Value
Inner Radius r_i (mm)	25
Outer Radius r_o (mm)	32
Spiral Angle α (°)	18
Groove-to-dam Ratio γ	0.5
Groove-to-land Ratio β	0.7
Number of Grooves N_g	12
Groove depth h_c (μm)	6

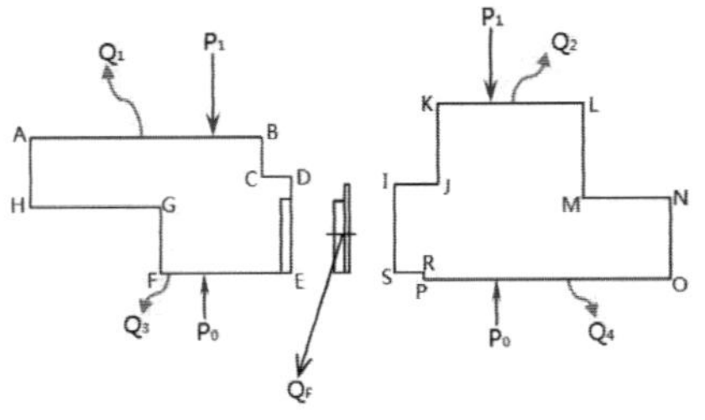

Figure 3. boundary conditions of the CFD model

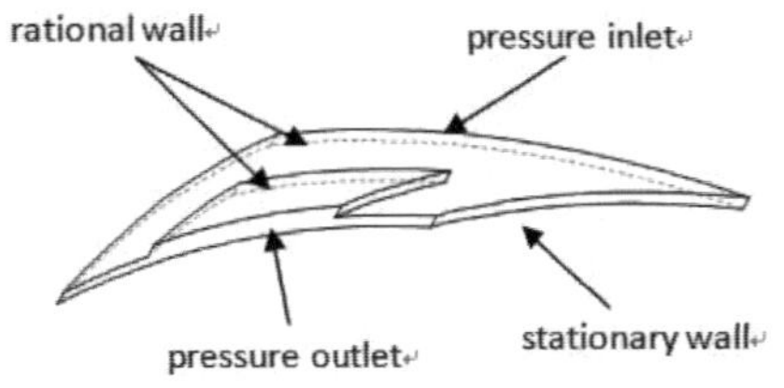

Figure 4. Computational zone and boundary conditions of one period of fluid film

As the figure 4 represents, there are two parts of the fluid film, the groove liquid which is filled in spiral groove and the gap liquid which is between seal faces. The groove liquid is set to rotation as the given rotation speed and the gap liquid is set to stationary. The end face of rotational ring, the end face and side faces of groove are set to rotational wall boundary. The end face of stationary ring is set to stationary wall boundary. The pressure of input equals the pressure of medium and the pressure of output equals atmospheric pressure. The fluid is contacted with the rotational wall and stationary wall, the heat transfer between them is set to conversation heat transfer boundary condition. α_{fs} and α_{fr} is the heat transfer coefficient between the fluid film and the rotational and stationary ring, and they can be calculated by following formula[14,15]:

$$\alpha_{fr} = \alpha_{fs} = \alpha' = 0.664 k_f pr^{0.33} \left(\frac{u_f}{v L_c} \right)^{0.5} \tag{1}$$

Where

$$L_c = \pi (r_o + r_i), \quad u_f = \frac{(r_o + r_i)}{4}, \quad pr = \frac{c_{pf} \mu_f}{k_f} \tag{2}$$

In calculating the solid domain, the use of boundary conditions set as follows: wall A ~ D and I~ L contact with the sealing liquid, high-pressure side, the pressure is medium pressure P_1; wall E ~ H and O ~ S for the low-pressure side in contact with air P_0; wall surface of the stationary ring in contact with the wall surface N ~ O, is set to a fixed constraint fix support; wall A ~ H in contact with the wall surface of the spring seat, when stabilized, the radial displacement of the spring seat as 0, so the set displacement constraints displacement, so that the axial displacement is 0. Movement loop operation, the energy obtained from the liquid film to seal liquid cooling and air cooling are set to each wall convective heat, each wall convection coefficient is calculated by the following equation [14-15]:

1) Rotational Wall A~D:

$$\alpha_{mr} = 0.135 \lambda \left[\left(0.5 \mathrm{Re}_c^{\ 2} + \mathrm{Re}_a^{\ 2} \right) P_r \right]^{0.33} / \mathrm{D}_r \tag{3}$$

where: $\mathrm{Re}_c = \omega D_r^2 / v_l$, reflect the impact of the media rotary stirring Reynolds; $\mathrm{Re}_a = v D_r / v_l$, reflect the impact of the media lateral flow around Reynolds; λ is the thermal conductivity of the sealed medium, [W/m•K]; $P_r = \mu \cdot c_{pf} / k_l$ is Plante constant, D_r is the equivalent diameter of the outer peripheral ring, [m]; v is the axial velocity of surrounding medium of the rotational ring, [m/s]; ω is the shaft speed, [rad/s].

2) Stationary Wall I~L:

$$\alpha_{ms} = 0.023 \lambda \cdot \mathrm{Re}^{0.8} \, \mathrm{Pr}^{0.4} / \mathrm{s}_s \tag{4}$$

where S_s is gap between the static ring and the seal chamber, [m].

3) Wall E~H and O~S:

Since the sealing ring inside the boundary is contacted with air, it can be seen from the above formula.4, relative to the ring, the convective heat transfer coefficient of static ring outer diameter of the inner boundary of the convective heat transfer coefficient is negligible, generally during the ANSYS simulation , its convective heat transfer coefficient only need to properly set a reasonable value.

3. Calculation and analysis

3.1. Features of Thermal Deformation

In this part, the case that the shaft speed is 1000rpm and the medium pressure(gauge pressure) is set as an example for calculation and analysis. When Calculating, the film thickness is assumed to be stable and set as 3 μ m, the choice of dynamic and static ring material parameters as shown in Table 2; the sealing liquid material parameters as shown in Table 3. μ_f variation as a function of temperature follows the exponential law as follows: $\mu_f = \mu_0 e^{-\gamma (T - T_0)}$, is implemented in Fluent by UDF programming; Table 4 shows the exact convection heat transfer coefficient in this case.

Table 2. Parameters of the sealing rings

Parameter	Rotational ring	Stationary ring
material	sic	Grp*
Density ρf (kg/m3)	3150	1810
Specific Heat cps (J/kg·K)	710	880
Thermal Conductivity ks (W/m·K)	150	45
Coefficient of Thermal Expansion α (10-6k^{-1})	4.3	6.2
Poisson's Ratio γ	0.27	0.26
Young's modulus E(GPa)	380	25

Table 3. Parameters of the sealing liquid

Parameter	Value
T∞(K)	300
Density ρf (kg/m3)	880
Specific Heat cpf (J/kg·K)	1670
Thermal Conductivity kf (w/m·K)	0.15
Viscosity μ_f (Pa·s)	0.0035
viscosity temperature coefficient γ	0.3

Table 4. The convective heat transfer coefficient of the rings

Boundary	A~B	B~C	C~D	E~H	H~A	I~J	J~K	K~L	L~S
convective coefficient $[\,w/m^2\cdot k\,]$	2091.3	1956.9	1824.7	4	1956.9	1016.1	1271.6	1629.7	4
Temperature[k]					300				

Figure 5. Temperature distribution on the seal face

Due to the liquid film is thickness, temperature distribution along the radial liquid film is almost consistent, so the temperature field distribution is mainly distribution along the direction of the seal face. Figure 5 shows the temperature distribution along static ring end face of upstream spiral groove mechanical seal The figure shows that in the inner diameter of the sealing ring, the temperature is lower. This is because here is backflow area, and the temperature close to the reflux temperature, artificially set to 300 k. And then into the spiral groove area, temperature increases, when arrives at sealing weir, the temperature reaches the maximum value. This is due to the friction of buffer fluid in the spiral groove, the temperature of the buffer fluid increases from the inner diameter to the outer diameter .While in sealing weir area, relative linear velocity is the largest, and the thickness of liquid film is relatively thinner, so the temperature reached the maximum in the sealing weir area.

Figure 6 shows the temperature profile of the stationary ring. The figure shows that the highest temperature occurs at the outer side of the seal end face, the temperature is gradually reduced from the seal end face along the axial direction, to the lowest one on the back of the sealing ring. As the interior of the seal ring touches the air, and the exterior of the ring touches the liquid, the heat transfer rate of the interior is lower than the exterior.

Figure 7 shows the thermal deformation figure of sealing ring .When calculating the thermal deformation, the sealing medium pressure and liquid film pressure were ignored, but the fixed constraint on lateral side of static ring and displacement constraint on lateral side of rotating seal ring were retained ,and then thermal deformation were figured out. The figure shows that the thermal deformation embodies in expansion with heat from the side face of sealing ring to seal face, and the maximum thermal deformation appeared in the inside of seal face. After deformation the inner side of the sealing end face is higher than the outside, the dry friction can easily occur on the end face.

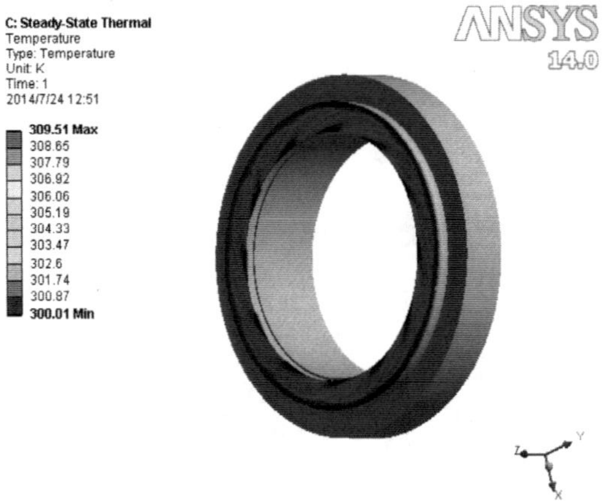

Figure 6. Temperature distribution of stationary seal ring

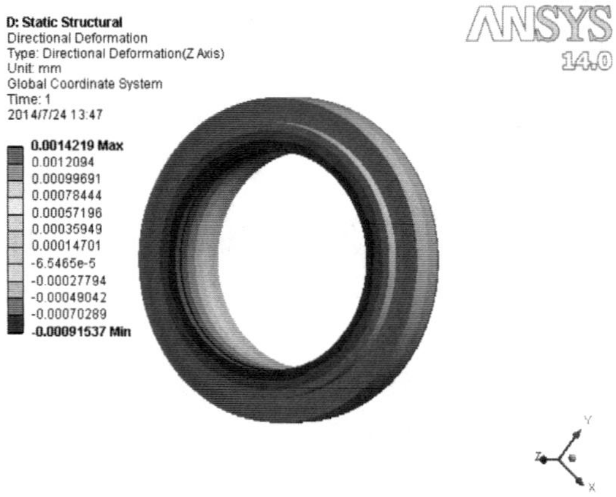

Figure 7. Temperature Deformation of stationary seal ring along axial direction

3.2. Features of Mechanical Deformation

Figure 8 shows the static pressure cloud picture of upstream pumping mechanical seal face. It depicts that the largest pressure appears on top of the spiral groove, higher than the medium pressure, and this is result of the pumping effect of spiral groove. Due to the existence of the spiral groove, leakage liquid on low pressure side will be pumped to high pressure side, causing the rise of pressure on the top of spiral groove. With minimal pressure on the inside of spiral groove, partial vacuum appears, and it is easy to produce cavitation. The figure shows that the pressure distribution on pumping mechanical seal face of spiral groove is uneven. The maximum pressure is very high(17 times of medium pressure), and the minimum pressure is very low. This big difference will obviously lead to deformation of the seal face.

Figure 9 shows deformation of seal ring end face .It depicts that the mechanical seal face is all compressive deformation, and pressure deformation and thermal deformation are in the opposite direction. The maximum deformation appears on the top of the spiral groove because of the highest pressure there. Figure shows the largest directional force deformation (Z Axis) is 0.48717 μ m, and the

IOP Conf. Series: Materials Science and Engineering **72** (2015) 042032 doi:10.1088/1757-899X/72/4/042032

maximum directional thermal deformation(Z Axis) is 1.4219 μ m. In the same working condition, the force deformation is much smaller than thermal deformation.

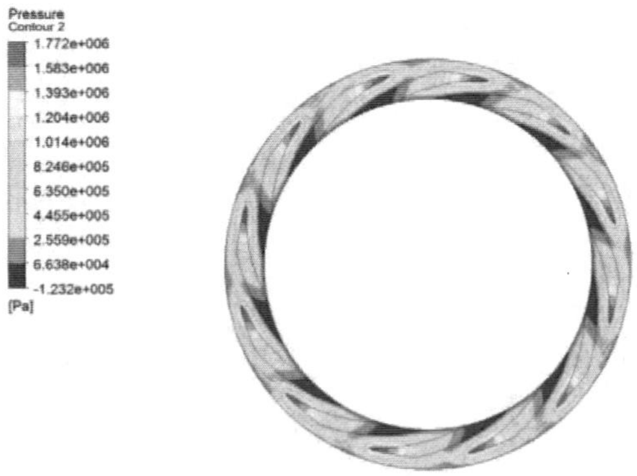

Figure 8. Pressure distribution on the seal face

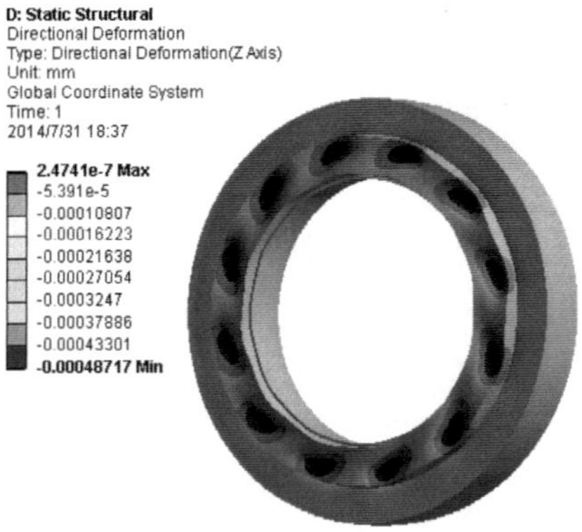

Figure 9. Pressure Deformation of stationary seal ring along axial direction

3.3. Features of Thermal-Mechanical Coupling Deformation

Figure 10. Thermal De formation along axial direction

Figure 10 is a deformation state after coupling of thermal and mechanical deformations. By contrast to Figure.8, coupling deformation and thermal deformation share the same tendency, but the amount of coupling deformation is decreased, it is because of a large thermal deformation, and the opposite direction of the mechanical and thermal deformation, which has an offset function. Therefore, as the final deformation result, the maximum deformation occurs in the inside of the end face of the seal ring, and appears to be an expansion deformation. In addition, the deformation decreases gradually outwardly.

3.4. Effect of operating conditions on the deformation

3.4.1. Effect of the sealing shaft speed on the deformation. Figure 11 shows a relationship of opening force, end face temperature and deformation amount with the shaft speed in upstream pumping mechanical sealing. According to the figure, with the constant of medium pressure, the film's opening force and end surface temperatures are increasing with the increase of shaft speed in case of film thickness remaining constant. Accordingly, the thermal deformation and mechanical deformation are also increasing with the speed increasing. Besides, it can be seen from the figure that the thermal - mechanical coupling deformation is between the thermal deformation and the mechanical deformation, which illustrates thermal deformation can be partially offset by mechanical deformation at different rotating speeds.

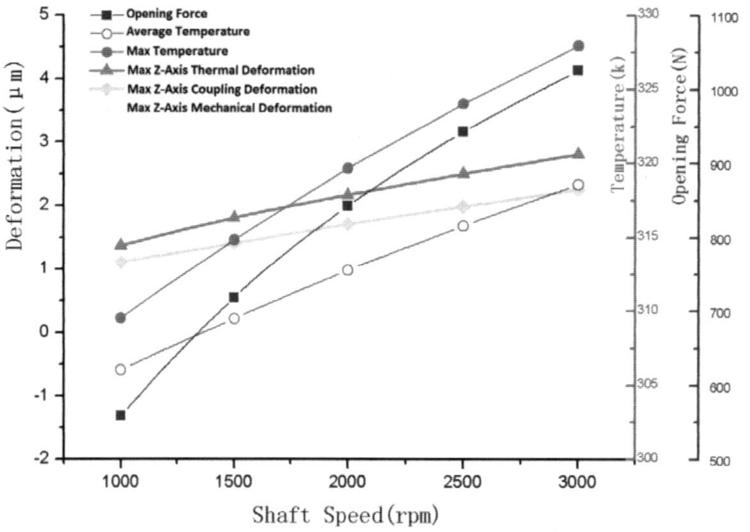

Figure 11. Shaft Speed and Deformation Relationship

3.4.2. Effect of the medium pressure on the deformation. Figure 12 shows a relationship of opening force, end face temperature and deformation amount with the medium pressure in upstream pumping mechanical sealing. According to the figure, when the rotational speed remains at 1000rpm, medium pressure increases, the opening force is increased gradually, while the temperature of the end face increases slowly. Therefore the corresponding changing amount of thermal deformation is small, and changing amount of mechanical deformation is large. However, the thermal deformation is greater than the mechanical deformation, which will increase the offset of thermal deformation to further reduce the amount of deformation.

3.5. Effect of the sealing material on the deformation

According to section 3.4, the deformation amount of the sealing ring is enlarged along with the increase of medium pressure and rotating speed, so the study will be performed in the working condition of the speed below3000rpm and the medium pressure of 0.8Mpa, to determine the effects of sealing materials on deformation amount. Table.5 shows the physical parameters of other three common sealing materials.Table.6 shows the largest deformation of sealing ring of different materials. According to Table.5 and Table.6, thermal deformation is related to the material expansion coefficient , while the mechanical deformation is related to elastic modulus E. The small thermal expansion coefficient of the seal ring produces small thermal deformation; while small elastic modulus produces a large mechanical deformation. In the working condition of large parameters, the material with small thermal expansion coefficient and low elastic modulus is selected ,to help reduce the thermal deformation, meanwhile, it will increase the offsetting effect of the mechanical deformation on the thermal deformation. Eventually coupling deformation reaches a minimum value. Table.6 Alumina ceramic material shows a verification of this conclusion.

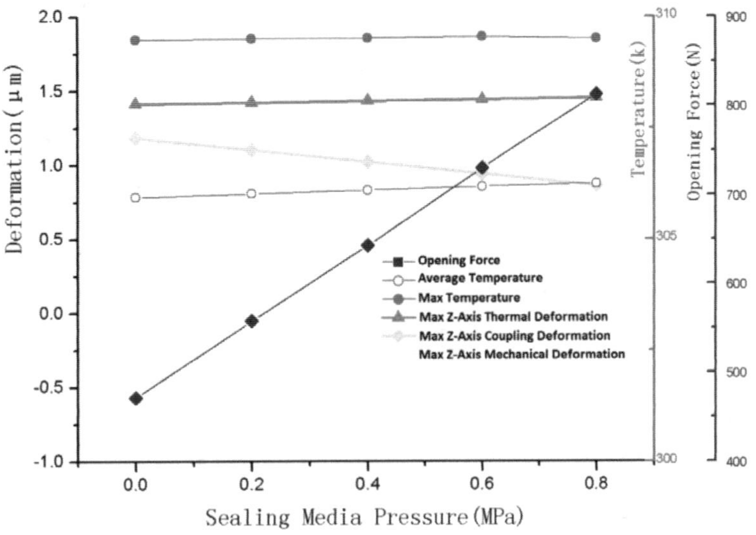

Figure 12. Sealing Medium Pressure and Deformation Relationship

Table 5. Three additional common mechanical seal materials

Parameter / Material	ρ (kg/m^3)	E (Gpa)	α (10^{-6}/K)	λ W/(m×K)	γ
Ywn8	14700	600	5.1	100	0.25
Copper alloy	8600	93	17.2	120	0.34
Alumina ceramic	3400	200	5	12	0.22

Table 6. The max deformation of each material when rpm is 3000 and the media pressure is 0.8MPa

Rotational Ring Material	Stationary Ring Material	Max Axial Thermal Deformation (μm)	Max Axial Mechanical Deformation (μm)	Max Axial Coupling Deformation (μm)
sic	SIC	2.2879	-0.069729	2.2326
	Grp*	2.8564	-1.0716	2.0212
	Ywn8	2.5939	-0.045442	2.5577
	Copper alloy	9.1747	-0.25104	8.9844
	Alumina ceramic	1.8742	-0.14142	1.7594

4. Conclusion

1) This paper established a Thermal-Mechanical Coupled model of spiral groove upstream pumping mechanical seal, and performed a numerical simulation of upstream pumping mechanical seal 3-D micro gap and sealing ring. The temperature and stress fields of the upstream pumping mechanical seal are finally obtained, along with the resulting thermal, mechanical and coupling deformation;

2) Simulation results of heat, power and coupling deformation show that the upstream pumping mechanical seal surfaces expand outward with the thermal deformation, but the mechanical

deformation compresses the surface to the inner side. The thermal deformation changes rapidly, but can be partially offset by mechanical deformation;

3) Effect of operating conditions on the deformation illustrates that the increase of shaft speed comes along with the increase of thermal deformation and mechanical deformation. Medium pressure exerts greater impact on mechanical deformation than thermal deformation. The increase of the sealing medium pressure is companied by the offsetting increase of thermal deformation by mechanical deformation at the same speed, causing the total amount of deformation to reduce;

4) Within the scope of this study, while under fast shaft speed and high sealing medium pressure condition , the deformation is mainly caused by thermal. In order to decrease the deformation, the seal material with small coefficient of thermal expansion could be selected, meanwhile, smaller elastic modulus material can be chosen to increase the mechanical deformation, which will enhance the offsets that the mechanical deformation brings to thermal deformation.

References

[1] Etsion I 1983 *J. of Tribology* **106**(3) 338-43
[2] Lai T 1994 *Lubrication engineering* **50**(8) 625-31
[3] Hao M M and Hu D M 2003 *Lubrication and seal* **2003**(1) 12-5
[4] James D D and Potter A F 1967 *J. of Tribology* **89**(4) 439-43
[5] Bonneau D, Huitric J and Tournerie B 1993 *Journal of Tribology* **115** (3) 348-54
[6] Zhou J F, Gu B Q, Chen Y 2007 *Chinese J. of Chemical Eng.* **15**(4) 499-506
[7] Zhang J F, Yuan S Q, Fu Y H and Fang Y J 2006 *Journal of Hydrodynamics, Ser. B* **18**(5) 572-7
[8] Wang Q, Chen H L, Liu T, Liu Y H, Liu Z B and Liu D H 2012 *IOP Conference Series: Earth and Environmental Science* **15**(7) 072019
[9] Chin-Hsiu L 1976 Thermal Deformation in a Mechanical Seal *ASLETrans-aetion* **19**(2) 146-52
[10] Zhou J F and Gu B Q 2007 *Nuclear Power Engineering* **2007**(2) 77-82
[11] Zhou J F and Gu B Q 2006 *Mechanical Eng.* **2006**(9) 201-7
[12] Cicone T, Pascovici M D and Tournerie B 2001 *Proceedings of the Institution of Mechanical Engineers, Part J: Journal of Engineering Tribology* 215(1) 35-44
[13] Qiu Y and Khonsari M M 2012 *J. of Tribology* **134**(2) 021703
[14] Pascovici M D and Etsion I 1992 *J. of tribology* **114**(4) 639-45
[15] Chen H L, Liu T, Ling Q L, Liu Y H, Wu R Z and Liu Z B 2012 *Fluid Machinery* **7** 008

IOP Conf. Series: Materials Science and Engineering **72** (2015) 042033 doi:10.1088/1757-899X/72/4/042033

Numerical investigation for one bad-behaved flow in a Pelton turbine

X Z Wei[1, 2, 3], K Yang[1, 2, 3, 4], H J Wang[3], R Z Gong[3] and D Y Li[3]
[1] Harbin electric power equipment co.,ltd, Harbin 150040, China
[2] Harbin Institute of Large Electrical Machinery, Harbin 150040, China
[3] Harbin Institute of Technology, School of Energy Science and Engineering, Harbin 150001, China

E-mail: woshiyangkang@126.com

Abstract. The gas-liquid two-phase flow in pelton turbines is very complicated, there are many kinds of bad-behaved flow in pelton turbines. In this paper, CFD numerical simulation for the pelton turbine was conducted using VOF two-phase model. One kind of bad-behaved flow caused by the two jets was captured, and the bad-behaved flow was analysed by torque on buckets. It can be concluded that the angle between the two jets and the value of ratio of runner diameter and jet diameter are important parameters for the bad-behaved flow. Furthermore, the reason why the efficiency of some multi-jet type turbines is very low can be well explained by the analysis of bad-behaved flow. Finally, some suggestions for improvement were also provided in present paper.

1. Introduction

The gas-liquid two-phase flow in Pelton turbines is turbulent and unsteady. Different from the reaction turbines, the hydraulic performance of a Pelton turbine is dynamic due to the unsteady flow in the rotating buckets in time and space [1]. Besides, the free surface flow should be modelled by a multiphase model [2]. So numerical simulation of the free surface flow in pelton turbines is more complex and time consuming.

In recent years, the complex free surface flow is allowed to be simulated and developed by the CFD technology. A great deal of research of the free jet from the injector was done by Zhang [3-4]. One of the first attempts to simulate the free surface flow in a pelton turbine was done by Kvinsky [5]. Numerical prediction on the efficiency of a pelton turbine was done by Jost [2]. Numerical investigation of the bad-behaved flow caused by the interaction between jet and bucket in a pelton turbine was done by Santolin [6]. On the basis of results mentioned above, the accuracy of CFD simulation on free surface flow in a pelton turbine was proved, and it can be concluded that the research on bad-behaved flow in a pelton turbine is very significant.

The bad-behaved flow mentioned in this paper can be explained as follows. There are two kinds of flow in one bucket at the same moment, one belongs to the present jet, and the other belongs to the previous jet. The fluid left in the bucket from the previous jet has not flowed out completely when the fluid from the present jet has already flowed in this bucket. So the two flows interfere with each other

[4] Corresponding author. Telephone: 18088708339.

IOP Conf. Series: Materials Science and Engineering **72** (2015) 042033 doi:10.1088/1757-899X/72/4/042033

and the interference lead to this bad-behaved flow. This bad-behaved flow exists in the multi-nozzle pelton turbine. It can reduce the power output and efficiency of the turbine.

2. Numerical simulation

In this paper, the RNG k-ε model was chosen to close the momentum equation, which is used to describe the flow in a pelton turbine. The VOF multiphase model was chosen to capture the free surface of the two phase flow. There are two domains in the simulation, one is the rotating domain, the other is the static domain, and the data is transferred through the sliding interface between the domains. Besides, the independence verification of the grid number and the boundary conditions are described as follows.

2.1. The grid independence verification

The unstructured mesh was used for the simulation. In order to get a more accurate shape of the flow, the grid at the inlet and the buckets were refined separately. The grid of the rotating domain and the static domain are shown in Figure 1. In order to verify the independence of the grid number, calculations with different grid number were done, and the calculated efficiency of different grid is shown in Figure 2. Results show that the calculation efficiency tends to be stable as the grid number increases. Finally, the grid with 15 million mesh number is selected in consideration of the calculation efficiency and calculation time.

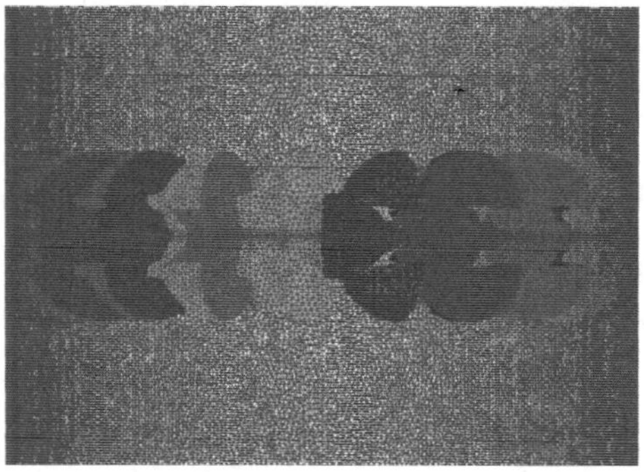

a)

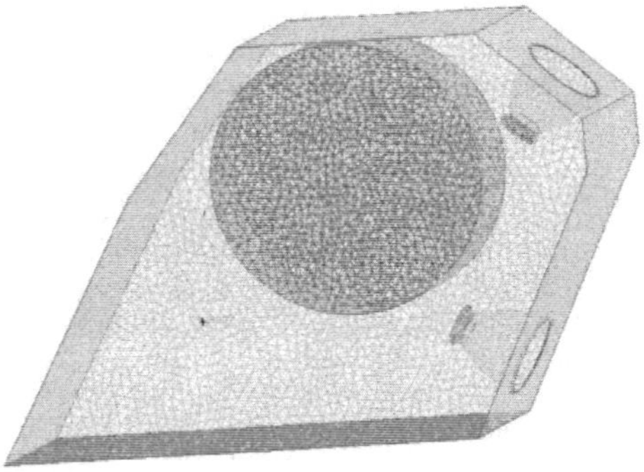

b)

Figure 1. Computational domain and grids for numerical analysis: a) rotating domain; b) static domain

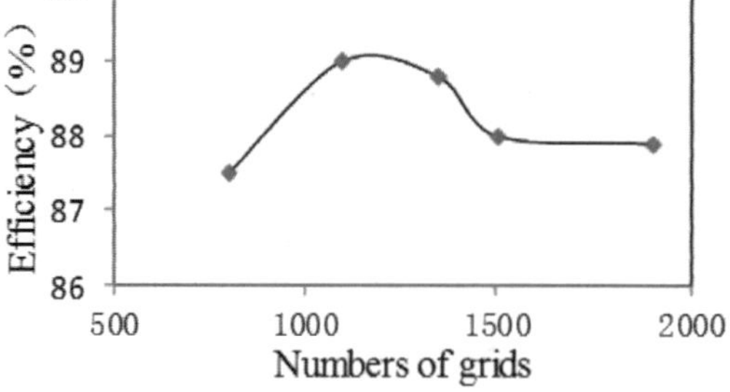

Figure 2. Independence verification on numbers of grid

2.2. Boundary conditions

As for inlet boundary conditions, the values of jet velocity and water volume fraction were prescribed. The outlet boundary condition was prescribed at the outlet of the rack with a constant pressure.

Transient calculations were performed with a time step of 0.5 degrees of runner rotation. 30 iterations were performed per each time step.

In this article, α is defined as the angle between jet 1 and jet 2, V is the jet velocity, ψ is the ratio of runner linear velocity and jet velocity, M value is the ratio of runner diameter and jet diameter. Numerical analysis was performed at 8 operating points with maximum opening in this paper as shown in Table 1.

Table 1. Operating points for numerical flow analysis

Points	α (°)	ψ	M value
1	75	0.48	9.2
2	180	0.48	9.2
3	110	0.48	9.2
4	90	0.48	9.2
5	75	0.45	9.2
6	75	0.42	9.2
7	75	0.35	9.2
8	75	0.48	11

3. Phenomenon analysis of the design condition

Operating point 1 is the design condition of this pelton turbine. Figure 3 shows the water flow within domains of point 1, it can be seen that the fluid from jet 2 is flowing into the bucket 5 while the fluid left in bucket 5 coming from jet 1 is still flowing out at this moment. So it can be concluded that the bad-behaved flow exists in bucket 5.

Figure 4 shows the value of the torque on the bucket 5 during the calculation at operating point 1, there are two torque peaks generated by the two jets. It is obvious that the torque peak generated by jet 2 is much smaller than that generated by jet 1.Then the average value of the torque generated by the two jets were calculated, the average value of the torque generated by jet 1 and 2 is 9170N·m and 7960N·m, respectively. It's the bad-behaved flow leads to the difference of the value torque generated by the two jets, which makes the torque generated by jet 2 much smaller than that generated by jet 1. In other words, the bad-behaved flow reduces the working capability of jet 2.

In order to verify this kind of bad-behaved flow, the parameter k is defined as the ratio of the average value of the torque generated by jet 1 and the average value of the torque generated by jet 2. Therefor when k=1, it can be claimed that the bad-behaved flow does not appear, and when k>1, it can be claimed that the bad-behaved flow appear. And the larger k is the more serious this kind of bad-behaved flow is. k=1.152 was calculated in operating point 1.

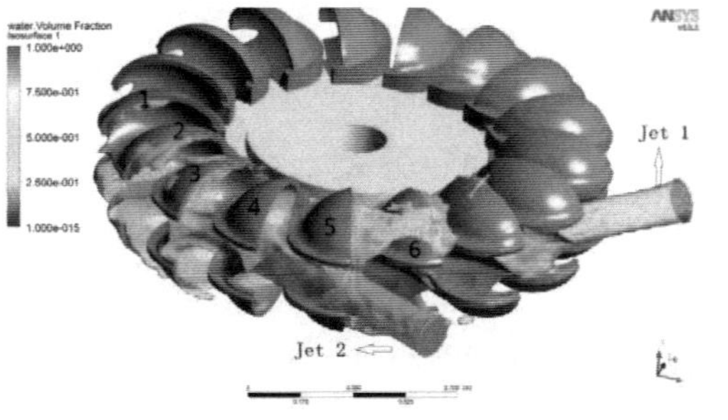

Figure 3. Water flow in domains of operating point 1

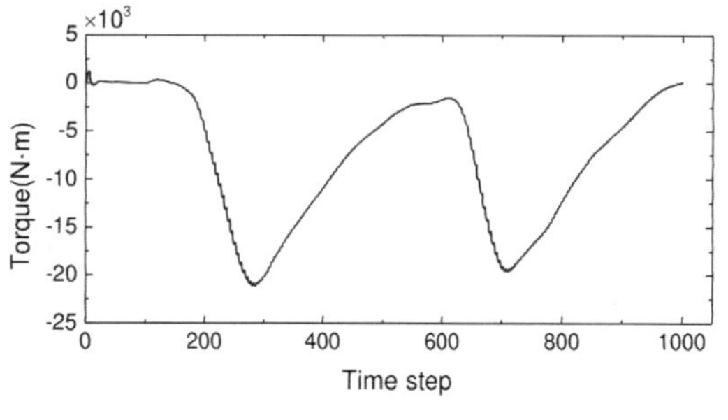

Figure 4. Torque on the bucket 5 during calculation

This bad-behaved flow is determined by two time, one time called T_1 is the time that a bucket spends rotating from jet 1 to jet 2, the other time called T_2 is the time that the fluid from jet 1 costs flowing out from the bucket completely. It can be obtained that this kind of bad-behaved flow does not appear when $T_1 > T_2$, and it appears when $T_1 < T_2$.

In this article, ω is defined as the angular velocity of the runner, therefor T_1 is the ratio of α and ω. Besides, as we know T_2 is inversely proportional to V. This kind of bad-behaved flow does not appear when $T_1 > T_2$, in order to avoid the bad-behaved flow, T_1 need to be longer while T_2 need to be shorter, in other words, α and V should be increased.

4. Schemes to avoid the bad-behaved flow

Considering the analysis above, three schemes were designed to avoid the bad-behaved flow, scheme 1 is to increase α which means to increase T_1, scheme 2 is to increase V which means to reduce T_2, and scheme 3 is to reselect the runner which also means to reduce T_2.

4.1. Scheme 1

In this scheme, the angle between jet 1 and jet 2 is increased to avoid the bad-behaved flow. Operating points 2-4 are analysed in this scheme.

The water flow within the domains of the operating point 2-4 is shown in Figure 5. It can be obtained that the α is big enough to avoid this bad-behaved flow in operating point 2 and operating point 3, fluid from jet 1 has already flowed out completely from the bucket A when bucket A rotates from jet 1 to jet 2. However it can be obtained that the bad-behaved flow appears slightly in operating point 4. In the present moment, fluid from jet 2 has already flowed into bucket A while there is still some fluid left from jet 1 flowing out from bucket A.

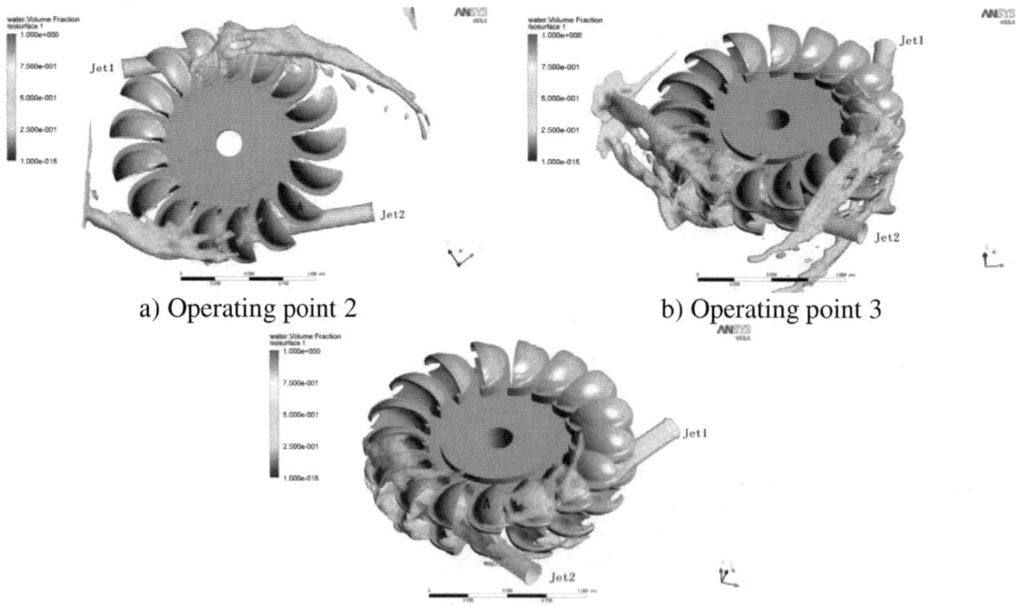

a) Operating point 2

b) Operating point 3

c) Operating point 4

Figure 5. Water flow in the domains of operating point 2-4

Then the parameter k of the operating point 2-4 is calculated, as shown in Table 2. The bad-behaved flow doesn't appear in operating point 2-3 because of the value of k is approximately equal to 1. And the value of k is 1.0132 in operating point 4, therefore the bad-behaved flow appears slightly in point 4, and 1.0132 is much smaller than the value 1.152 in design condition point, so it can be concluded that the bad-behaved flow have been greatly improved in operating point 4.

Table 2. The parameter k calculated in Scheme 1

Operating points	$\alpha(°)$	k
Operating point 2	180	1.0004
Operating point 3	110	1.0005
Operating point 4	90	1.0100

4.2. Scheme 2

In this scheme, the jet velocity is increased to avoid the bad-behaved flow. The purpose of increasing V is to reduce T_2. The linear velocity of runner is determined by the angular velocity of runner and can be seen as constant, so ψ can be used to measure the jet velocity. In order to investigate the influence the jet velocity on the bad-behaved flow, operating points 5-7 are analysed in this scheme.

The water flow in the domains of operating points 5-7 are shown in Figure 6.It can be obtained that the fluid from jet 1 has not flowed out completely when bucket A rotates from jet 1 to jet 2 at operating point 5 and operating point 6, so the bad-behaved flow still appears in operating point 5 and operating point 6. However it seems that the jet velocity is big enough to avoid the bad-behaved flow in operating point 7, the fluid from jet 1 has already flowed out completely at the moment bucket A rotates to jet 2 in this operating point, and the bad-behaved flow does not appear in operating point 7.

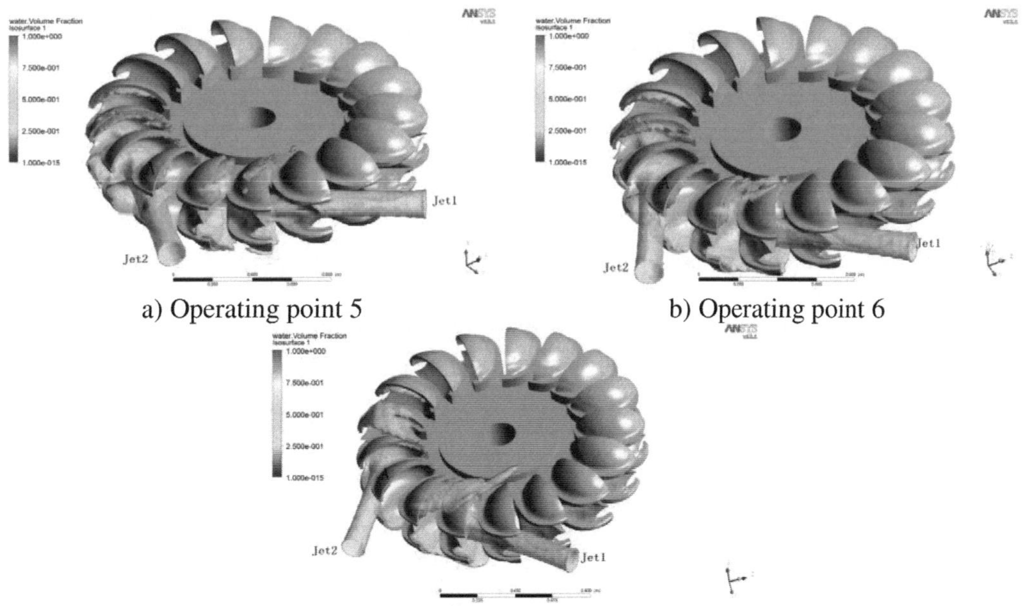

a) Operating point 5 b) Operating point 6

c) Operating point 7

Figure 6. Water flow in the domains of operating point 5-7

Then the parameter k of the operating point 5-7 is calculated. As shown in Table 3, the value of k in operating point 5 and operating point 6 is still greater than 1. It means that the bad-behaved flow still appears in these operating points, and the value of k is much smaller than that of operating point 1. It means that the bad-behaved flow has already been improved. However k=1.0008≈1 at operating 7. It means that the bad-behaved flow is avoided in operating point 7.

It is known that the efficiency of pelton turbine is determined by ψ. In this scheme, the bad-behaved flow was avoided by increasing jet velocity. But ψ was changed at the same time and the efficiency reduced, so this scheme can tell us how the jet velocity influences the bad-behaved flow, but it can't be a reasonable scheme to settle the problem of the hydropower station.

Table 3. The parameter k calculated in Scheme 2

Operating points	ψ	k
Operating point 5	0.45	1.0900
Operating point 6	0.42	1.0400
Operating point 7	0.35	1.0008

4.3. Scheme 3

In this scheme, in order to increase the jet velocity while keeping ψ constant, the runner is reselected. The M value is the ratio of the runner diameter and the jet diameter. It has a significant influence on pelton turbine. The M value of the original runner is 9.2, and the number of buckets is 17.then the M value is reselected as 11 while the number of buckets is 19. Reselection of M value can increase the jet velocity while keeping ψ constant.

The water flow within the domains of operating point 8 is shown in Figure 7. It can be obtained that the fluid coming from jet 1 has flowed out completely when bucket A rotates from jet 1 to jet 2, and the bad-behaved flow does not appear in this operating point.

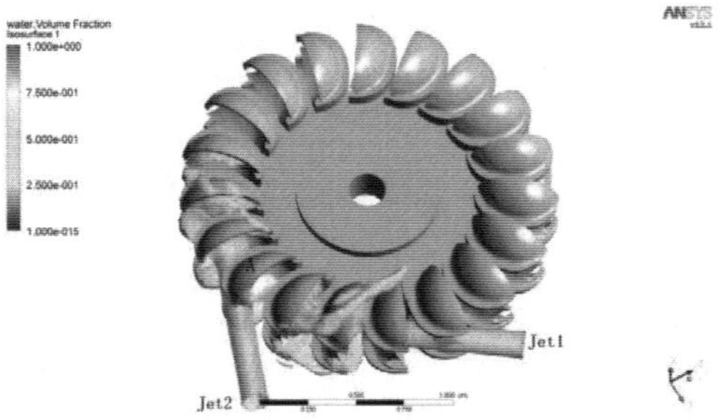

Figure 7. Water flow in the domains of operating point 8

Then the parameter k of operating point 8 is calculated. As shown in Table 4, for this new runner, k=1.0008≈1, it means that the bad-behaved flow is avoided after reselecting the runner.

Table 4 . The parameter k calculated in Scheme 3

	M value	Number of bucket	ψ	k
Point 1(Original runner)	9.2	17	0.48	1.1520
Point 8(New runner)	11	19	0.48	1.0004

5. Conclusions

The numerical analysis of flow in a Pelton turbine was performed for 8 operating points. Some conclusions can be obtained on the basis of the results:

1) There is a bad-behaved flow exists in the multi-nozzle pelton turbine, and it can decreased the working capability of some jets.

2) The bad-behaved flow is influenced by the angle of two jets, it can be avoid when the angle is big enough.

3) The bad-behaved flow is influenced by the jet velocity, it can be avoid when the jet velocity is big enough and it can also be avoid by increase the M value.

IOP Conf. Series: Materials Science and Engineering **72** (2015) 042033 doi:10.1088/1757-899X/72/4/042033

References

[1] Perrig A, Avellan F, Kueny J, Farhat M and Parkinson E 2006 *J. Fluids Eng.,Trans. ASME* **128** 350–8
[2] Jošt D and Mešnar P 2010 *Earth and Environmental Sc.* **12** 012080
[3] Zhang Z and Casey M 2007 *Power and Energy* **221** 1181-92
[4] Zhang Z 2009 *Proc. of the Institution of Mech.Eng.* **233** 609-23
[5] Kvicinsky S, Kueny J L, Avellan, F and Parkinson E 2002 Experimental and numerical analysis of free surface flows in a rotating bucket *27th IAHR Symposium on Hydraulic Machinery and Systems (Montréal, Canada, 22-26 Sep 2014)*
[6] Santolin A, Cavazzini G, Ardizzon G and Pavesi G 2009 *Proc. of the Institution of Mech. Eng., Part A: J. of Power and Energy* **233**(6) 721-8

IOP Conf. Series: Materials Science and Engineering **72** (2015) 042034 doi:10.1088/1757-899X/72/4/042034

Investigation on cavitation for hump characteristics of a pump turbine in pump mode

D Y Li[1], H J Wang[1, 3], G M Xiang[1], R Z Gong[1] and X Z Wei[1, 2]
[1] Harbin Institute of Technology, School of Energy Science and Engineering, Harbin 150001
[2] Harbin Institute of Large Electric Machinery, Harbin, 150040

E-mail: wanghongjie@hit.edu.cn

Abstract. In order to get the accurate hump characteristic curve of a pump turbine in pump mode, three dimensional steady simulations were carried out using SST k-ω turbulence model with cavitation model and without cavitation model under different operation condition points with 19 mm guide vanes opening. A refinement grids were generated to adapt the turbulence model. The results obtained with cavitation model show a better agreement with experiments. The detailed analysis was undertaken to find out the relationship between the cavitation and hump region. It is concluded that the hump characteristic is related with cavitation.

1. Introduction

Cavitation in rotating machinery is a well know phenomenon that occurs when local pressure of the working fluid approaches vapour pressure. This phenomenon is usually observed in areas of the high-speed fluid machinery such as propellers and pumps, where the flow accelerates and the pressure decreases [1]. In such a situation, there could appear strong unsteady forces, noise and erosion. These become a cause of serious accidents involving fluid machinery. Many experiments have been don't to investigate the physics of cavitation. And lots of numerical simulations have been carried out to study the cavity development inside rotating machinery.

Since the pump turbine would be switched frequently between pump mode and turbine mode, even with a few minutes, the stability shows more important than that in other rotating machinery. Hump characteristics is one of the special characteristics of a pump turbine at pump mode with small flow. Especially, when the pump turbine runs at high level head, the hump region is unavoidable and the performance is reduced [2].

Lots of research has been carried out for investigation of stability of a pump turbine at pump mode. Braun [3] analysed flow phenomena in details related to the unstable energy-discharge characteristics of a pump turbine at pump mode, he pointed out that the sudden change in the global performance parameters is related to a change of the secondary flow pattern in the diffuser channels. Yin [4] predicted performance in the vaned distributor of a pump turbine under low flow rate in pump mode and concluded that the hump characteristics can be ascribed to the special hydraulic loss characteristics of the stay vanes and guide vanes. Less information is available on the relationship between cavitation and hump characteristics. Liu [2, 5] conducted that the hump characteristics might be attributed to the appearance of cavitation at the inlet of suction side of the runner using an improved

[3] Corresponding author

cavitation model. Anciger [6] predicted rotating stall and cavitation inception in pump turbines. The analysis of hump characteristics combined with cavitation phenomenon still need be undertaken. In this paper, three dimensional steady simulations are carried out using SST k-ω turbulence model with cavitation model and without cavitation model under different guide vanes opening in order to analyze the relationship between hump characteristics and cavitation phenomenon.

2. Pump turbine modelling and numerical method

The simulation model includes draft tube (inlet), runner, guide vanes, stay vanes and spiral casing (outlet) as shown in Figure 1. Geometry configuration was established using the UG commercial software. The parameters of the reduced scale pump turbine are listed in Table 1. 19mm guide vane opening operation conditions were chosen for simulations.

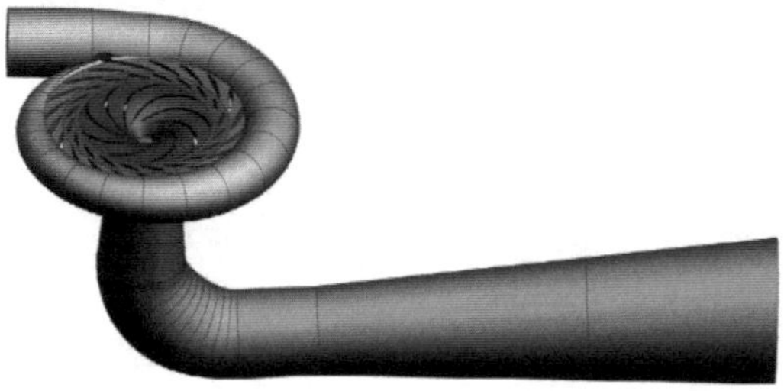

Figure 1. Computational domain

Table 1. Parameters of the reduced scale pump turbine (pump mode)

Parameters	Values
Nominal diameter of runner at inlet (mm)	274
Nominal diameter of runner at outlet (mm)	524
Number of runner blades	9
Number of stay vanes	20
Number of guide vanes	20
The height of guide vane (mm)	45.77

Structured hexahedral and unstructured (tongue as for its sharp shape) grids were generated for each part employed ANSYS ICEM. In order to capture the flow separation of guide vanes, stay vanes as well as runner using SST κ-ω turbulence model, the mesh layers are more than 10 and y+ at the wall layer is less than 2. The grids for different parts are shown in Figure 2. The total nodes are 8 million.

In this paper, the SST κ-ω turbulence model was used to close the Reynolds averaged Navier-Stokes (RANS) solver for solution of the two phase turbulence flows. The Rayleigh Plesset model was used to model cavitation in the pump turbine. The homogeneous multiphase model with water and water vapour two fluids was employed without consideration thermal phase change. In this simulation, Static pressure at draft tube inlet was specified at pump mode, and the intitial value of liquid water and water vapour was 1 and 0, respectively. The vapour pressure was set 2693 accooding to the exprimental data. Mass flow was imposed at the casing outlet. No slip wall boundary condition for the solid walls was used. High Resolution was chose for both advecition scheme and turbulence numerics. The Root Mean Square (RMS) was imposed as convergence criterion. The single phase steady flow calculation was set for initialization condition to impove the convergence rapidity.

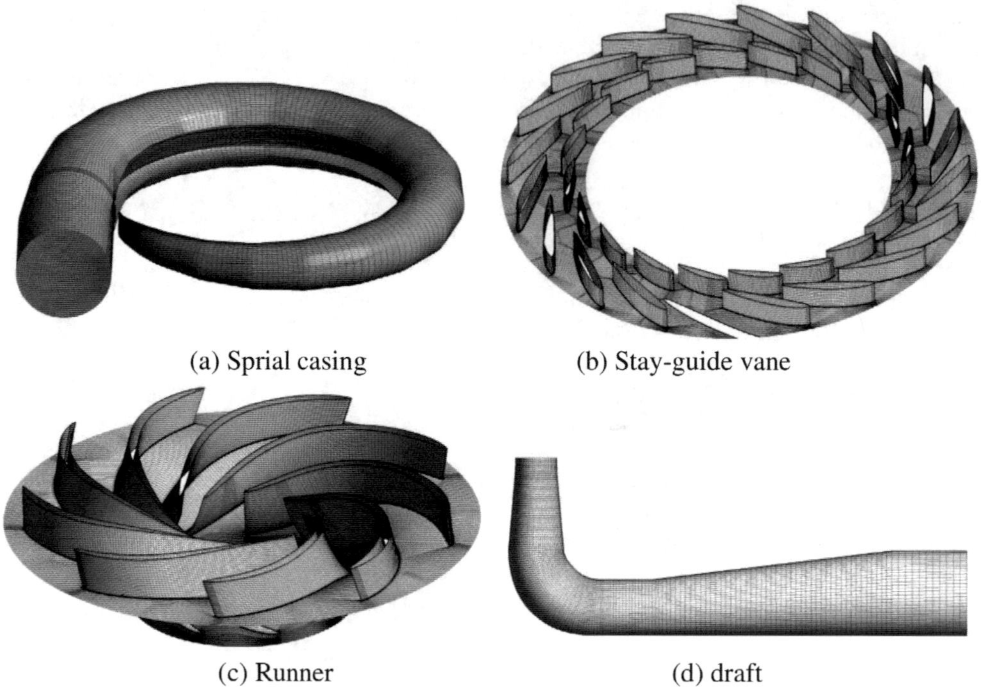

(a) Sprial casing (b) Stay-guide vane

(c) Runner (d) draft

Figure 2. Grid for different component

3. Analysis of external characteristics curve

19mm guide vanes opening was chosen for simulation. Figure 3 shows that the simulation results using SST κ-ω turbulence model with cavitation and without cavitation model. The discharge coefficient φ, head coefficient ψ, and torque coefficient λ are defined as follows.

$$\varphi=\frac{Q}{\pi R_1^{\,2}\omega} \tag{1}$$

$$\psi=\frac{2gH}{R_1^{\,2}\omega^2} \tag{2}$$

$$\lambda=\frac{M}{\pi R_1^{\,5}\omega^2\rho} \tag{3}$$

where R_1 stands for radius of runner inlet (pump mode), m. Q is volume flow rate, $m^3 \cdot s^{-1}$. ρ stands water density, $kg \cdot m^{-3}$. H is head, m.

Point D is the best efficiency point at 19mm guide vanes opening. It can be seen from Figure 3-a, as for best efficiency point, the difference between single and multiphase model is very little, but the simulation results show a better agreement with experimental data during large discharge, hump region and small discharge region and have the same peak point for head-discharge factor curve. Since the cavitation is more likely to happen during large discharge region of a pump turbine in pump mode, multiphase model results show better than that of single model.

With respect to torque-discharge curve (see Figure 3-b), the simulation results of single model show bigger than experimental data while those of multiphase shows smaller during large discharge region. As for small discharge region the value for both model shows smaller than experimental data.

It can also obtained from Figure 3-c that, as for F large discharge point, the efficiency drop very sharply. The points during the hump region are very near to best efficiency point.

Based on analyzing external characteristics curve, it can be concluded that multiphase model has some advantages to predict the performance of a pump turbine in pump mode.

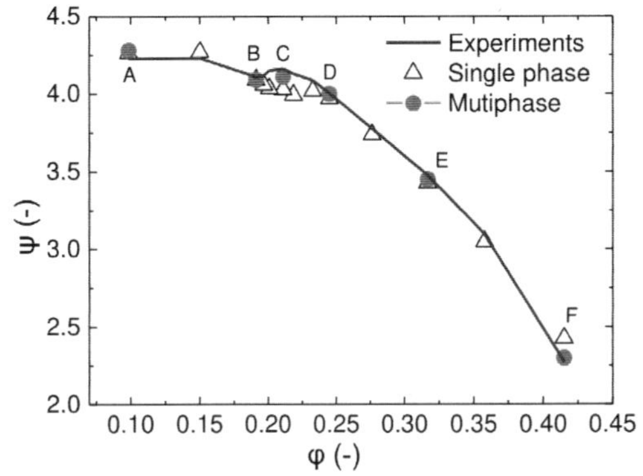

(a) Head-discharge factor curve

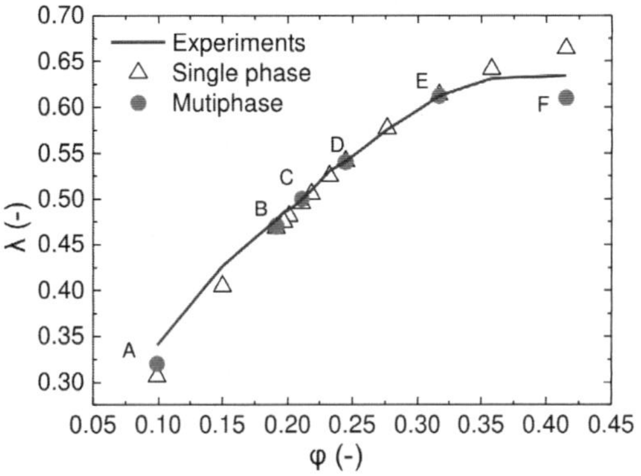

(b) Torque-discharge factor curve

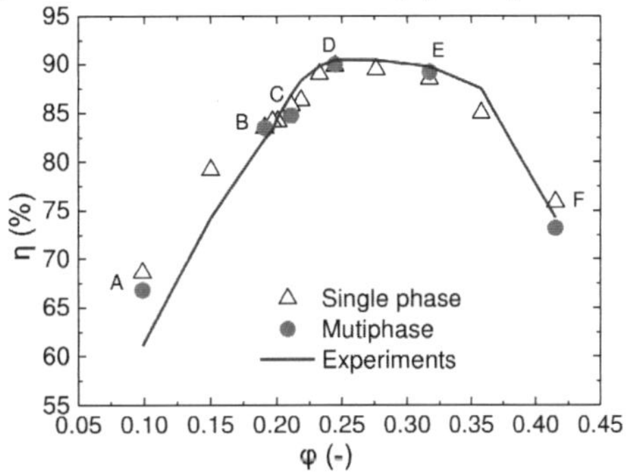

(c) Efficiency-discharge factor curve

Figure 3. Simulation results for head, torque and efficiency

4. Detailed analysis of internal flow

Since multiphase model could show obviously better agreement with experimental data during the large discharge region, F large point was chosen to analyse in detail. Figure 4 shows the comparison flow field within runner for F point. It can be seen that multiphase model could capture the flow separation which leads to cavitation while single model cannot simulate this information. As for head-discharge curve, the values obtained using single model usually are much bigger than experimental data as that single model cannot predict the loss which cavitation flow leads to, especially for pump mode.

Then small discharge point A, hump region wave tough point B and peak point C, large discharge point F were chosen to analyze the cavitation region that appears as flow separation. Figure 5 shows the cavitation region for different operation points. It can be seen that the cavitation region at the large discharge point is largest. The cavitation region of hump region wave tough point B is much larger than that at small discharge region and hump region peak point C, however which is still less than that at larger discharge point. In addition, it can also obtained that the cavitation appears at suction surface of blade inlet in pump mode.

Figure 6 shows the experimental results at the operation B and C during hump region. From the picture, it can be seen that the cavitation region at suction surface of blade inlet at hump region wave tough operation point while the cavitation region is smaller at hump region peak point C. It can be concluded that the hump characteristics has a certain relationship with cavitation. But this analysis shows cavitation influences little. Based this analysis the mechanism of hump characteristics still need to be investigated.

IOP Conf. Series: Materials Science and Engineering **72** (2015) 042034 doi:10.1088/1757-899X/72/4/042034

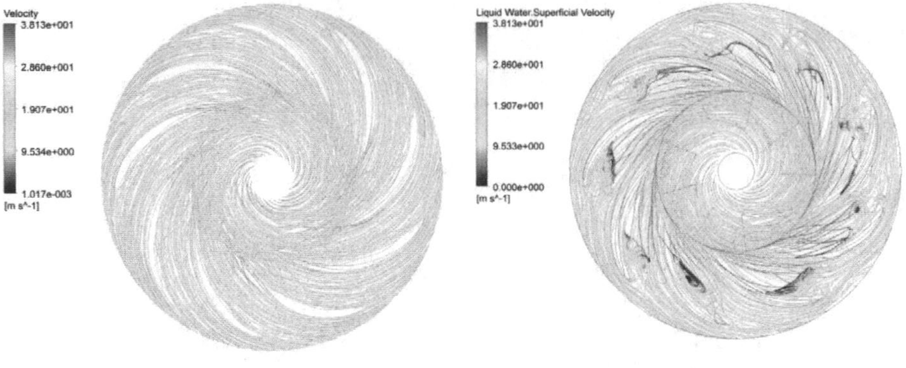

(a) Single model (b) Multiphase model

Figure 4. The comparison flow field within runner for F point

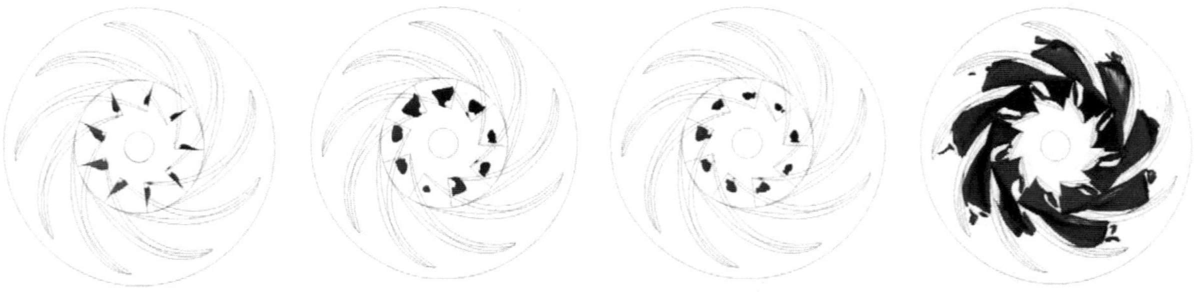

A B C F

Figure 5. The cavitation region for different operation point

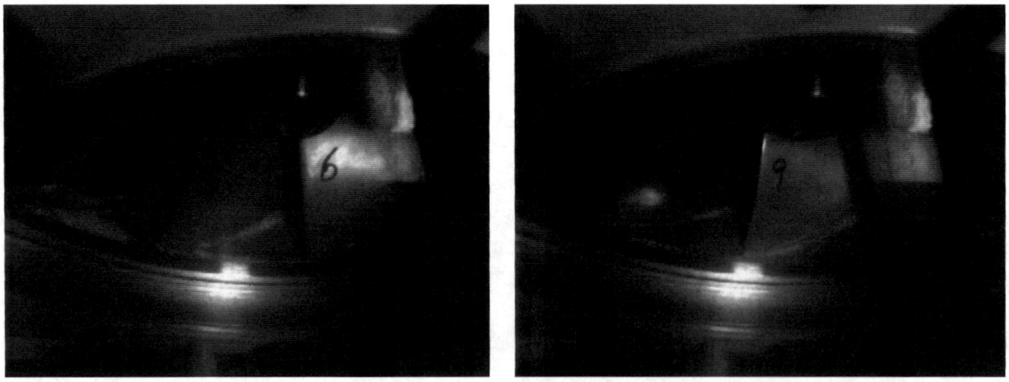

B C

Figure 6. The cavitation phenomenon of experiment

5. Conclusions

In this paper, three dimensional steady simulations were carried out using SST k-ω turbulence model with cavitation model and without cavitation model under different operation condition points at 19mm guide vanes opening. The results obtained with cavitation model show a better agreement with experiments. It can be concluded that the cavitation model could capture the cavitation phenomenon resulting from separation flow. Then, the detailed analysis is undertaken to find out the relationship between the cavitation and hump region. Analysis shows that the cavitation region appear larger at hump region wave tough point than that at peak point. In addition, the cavitation appears at suction surface of blade inlet in pump mode for both of simulation and experiments. It is concluded that the hump characteristic has a certain relation with cavitation. Based on this research, a further study could be taken in order to analyze the mechanism.

Acknowledgement
This work was supported by the National Key Technology R&G Program(Project No.2012BAF03B01-X).We thank Harbin Institute of Large Electric Machinery for its test platform.

References

[1] Iga Y, Nohml M, Goto A and Ikohagi T 2004 *J. Fluids Eng.* **126**(3) 419-29.
[2] Liu J T, Liu S H, Wu Y L, Jiao L, Wang L Q and Sun Y K 2012 *Computer & Fluids* **68** 105-11
[3] Braun O, Kueny J L and Avellan F 2005 *ASME Fluid Eng. Division Summer Meeting and Exhibition.(Houston, TX, USA, 5-11, November, 2005)*
[4] Yin J L, Liu J T， Wang L Q, Jiao L, Wu D Z and Qin D Q 2010 *SCIENCE CHINA Techno. Sci.* **53**(12) 3302-09
[5] Liu J T, Wu Y L and Liu S H 2013 *6th Int. Conf. on Pumps and Fans with Compressors and Wind Turbines (Beijing, China, 25-28, November, 2013)*
[6] Anciger D, Jung A and Aschenbrenner T 2010 *25th IAHR Sym. on Hydraulic Machinery and Systems (Timisoara, Romania, 20-24, September, 2010)*

IOP Conf. Series: Materials Science and Engineering **72** (2015) 042035 doi:10.1088/1757-899X/72/4/042035

Experiment and numerical simulation of cavitation performance on a pressure-regulating valve with different openings

W S Qu[1], L Tan[1], S L Cao[1,3], Y Xu[1], J Huang[2] and Q H Xu[2]
[1] State Key Laboratory of Hydroscience and Engineering, Tsinghua University, Beijing 100084, China
[2] Zhuzhou Southern Valve CO.,LTD., Zhuzhou 412007, China

E-mail: caoshl@mail.tsinghua.edu.cn

Abstract. As a kind of widely used device in pipe system for pressure and flow rate regulating, the valve would experience cavitation in the case when a sharp pressure drop occurs, which will induce the energy loss, noise and vibration of pipeline system, and even operational accidents. The experiment on flow resistance coefficient of a DN600 pressure-regulating valve under operation conditions from 0% to 100% openings is conducted. Based on the RNG k-ε turbulence model and the Rayleigh-Plesset cavitation equation, a set of computational model is developed to simulate the turbulent flow in the valve under operational conditions from 0% to 100% openings. The computational results of flow resistance coefficient are compared to the experimental data. And the numerical simulation is employed to predict the cavitation performance of the valve at different inlet flow conditions. The transient cavitating flow is calculated to reveal the time evolution of cavitation in the valve.

1. Introduction

Valves are commonly used in a variety of industries and play an important role in the control of flow rate, especially in hydraulic engineering. In most cases, valve acts as a throttle with the velocity increasing and the pressure dropping, therefore it is also applied to regulate the pressure in the pipeline system. Like in some other hydraulic components, cavitation appears in valves when the local pressure drops below the saturation pressure. Phase transformation occurs and bubbles formed by vapour emerge, grow and finally collapse with the pressure rising again, thus leading to vibration, noise and erosion on the valve body as well as other piping equipment. Since a pressure-regulating valve has to work under various pressure conditions such as atmosphere or even lower as much as required, it seems difficult to avoid cavitation thoroughly. However it is still very necessary to figure out the mechanism of its development and take into consideration its influence in the design of valve system.

Many works have been carried out in the investigation of flow pattern and cavitation in a valve, including both experimental measurements and numerical calculations. Vaughan et al. [1] carried out a series of simulations, and studied a range of poppet valves with different geometries and under different flow rate to predict the flow patterns and force characteristics. Recently, Valdes et al. [2] simulated the cavitating flow through a ball check valve and performed experimental tests to validate the simulation results. Tabrizi [3] et al. modelled the flow pattern in a ball valve at different opening

[3] Corresponding author.

IOP Conf. Series: Materials Science and Engineering **72** (2015) 042035 doi:10.1088/1757-899X/72/4/042035

angles by modifying the turbulent viscosity and validated the results with experiments and then discussed the pressure drop behind the valve and formation of the vortex flow downstream. Chern et al. [4] studied the cavitation flow inside a globe valve numerically and proposed a way to prevent cavitation in the valve body and at the downstream region by introducing a one-stage perforated cage. Gao et al. [5] presented a study to investigate the cavitating flow near the orifice of a poppet valve, and conducted a visualization experiment to obtain cavitation images. The results showed that to reduce the outlet area can effectively suppress cavitation. Jazi et al. [6] studied the acoustic waveform to detect the cavitation occurrence in a globe valve, and analysed the amplitude and frequency.

Many of the previous works focused on the traditional valve types such as ball valve, however, the piston valve with a perforated structure to regulate the pressure is rarely studies. Besides, the transient numerical simulation of cavitation flow in a valve has hardly been carried out.

The main aim of this study is to predict the cavitation levels under different operating conditions in a pressure-regulating valve by both steady and unsteady numerical simulation. A set of computational model is developed and validated by a series of experiments. The flow patterns, pressure distribution and vapour volume fraction inside the valve are analysed in order to explore the mechanism of the cavitation occurrence.

2. Description of the valve model and CFD method

2.1. Valve model and meshes

Figure 1 shows the schematic of valve in the present study. The fluid is supposed to flow into the valve body through a pipe with a diameter of 0.2 m. Then it is forced to jet toward the centre by a perforated cylinder structure (Figure 2) which will be named as the cage in the following passages. The internal structure as well as the operating principle of the valve is also revealed in Fig 1. The connecting rod system driven by a stepper motor pushes the cage forward and backward along a guide rail to realize different operation conditions.

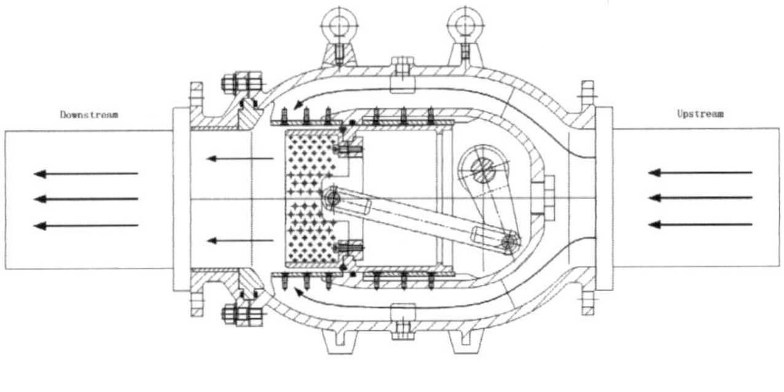

Figure 1. Schematic of valve.

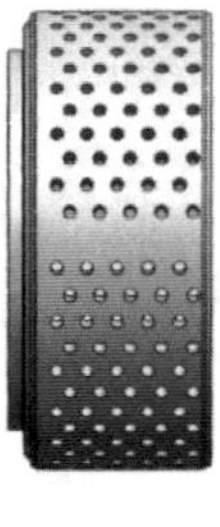

Figure 2. 3-D model of the cage structure.

The computational domain is composed of two parts. One part is the stable part under different operation conditions from 0% to 100% openings, and it is meshed in hexahedral mesh. The other part is the moving part under different operation conditions, and it contains 240 circumferential displaced holes on the cage, thus adopting the tetrahedral meshes due to the complicated structure. The mesh of both parts is shown in Fig 3. To figure out the influence of mesh number, a grid independency test is previously carried out. A variety of computational cells are utilized to simulate the flow in the valve with the opening of 40%. The result in table 1 shows that the three set of computational mesh makes little difference (error below 0.8%) in the pressure drop, especially when the cells are more than 2 million (error below 0.2%). Finally the mesh with 1,980,036 elements is chosen in the following calculation.

IOP Conf. Series: Materials Science and Engineering **72** (2015) 042035 doi:10.1088/1757-899X/72/4/042035

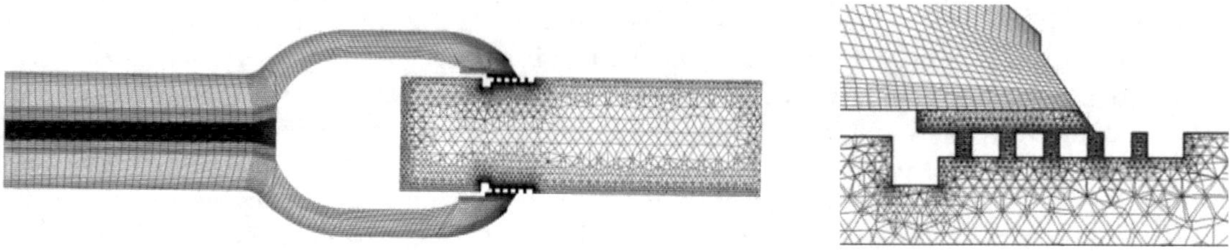

Figure 3. Computational domain and meshes (left) and meshes at the connecting holes (right)

Table 1. Result of mesh independency test.

Mesh Number	**Mass Flow Rate (kg/s)**	**P_{in} (MPa)**	**P_{out} (MPa)**	**Δ P (MPa)**
953,046	57.444	0.52	0.2534	0.2666
1,980,036	57.444	0.52	0.2520	0.2680
2,948,484	57.444	0.52	0.2514	0.2686

2.2. Numerical methods

The basic governing equation of turbulent flow in the valve is the 3-dimentional Reynolds-averaged Navier-Stokes equation, which can be denoted as

$$\frac{\partial}{\partial t}(\rho_m)+\nabla\cdot(\rho_m \boldsymbol{u})=0 \tag{1}$$

$$\frac{\partial}{\partial t}(\rho_m \boldsymbol{u})+\nabla\cdot(\rho_m \boldsymbol{uu})=-\nabla p+\nabla\cdot[(\mu_m+\mu_t)\nabla \boldsymbol{u}]+\frac{1}{3}\nabla[(\mu_m+\mu_t)\nabla\cdot \boldsymbol{u}] \tag{2}$$

where ρ_m is the mixture density; μ_t is the turbulent viscosity; μ_m is the kinetic viscosity averaged in accordance with the volume of the vapour and liquid phases; p is the pressure and $\boldsymbol{u}$ is the velocity vector. The RNG k-ε turbulence model is applied because of its advantage in disposing the flow with a high strain rate and streamline curvature.

The cavitation model applied in the present study is the transport equation based on the Rayleigh-Plesset equation. Taking into account the evaporation and condensation process in the cavitation, the transport equation is given as

$$\frac{\partial}{\partial t}(\alpha_v \rho_v)+\nabla\cdot(\alpha_v \rho_v \boldsymbol{u})=m^+-m^- \tag{3}$$

$$m^+=C_{vap}\frac{3\alpha_v \rho_v}{R_b}\sqrt{\frac{2}{3}\frac{\max(p_v-p,0)}{\rho_l}} \tag{4}$$

$$m^-=C_{cond}\frac{3\alpha_v \rho_v}{R_b}\sqrt{\frac{2}{3}\frac{\max(p-p_v,0)}{\rho_l}} \tag{5}$$

where m^+ and m^- represent the rates of evaporation and condensation, respectively. Other coefficients appeared in the equation are also specified: C_{vap}=50; C_{cond}=0.01; α_v is the vapour volume fraction; R_b is the bubble radius and set as 10^{-6} m; p_v is the saturation pressure of water at test temperature of 3574 Pa.

In consideration of the influence of turbulent pressure fluctuation to the saturation pressure, the corrected expression of saturation pressure is

$$p_v^{'}=p_v+0.195\rho_m k \tag{6}$$

The simulation of the turbulence flow and cavitation in the valve is carried out by solving the equations above with the CFD software ANSYS CFX. The pressure at inlet and mass flow at outlet are specified in accordance with the experimental measurement. The no-slip boundary conditions are imposed at the solid walls of the valve. Besides, in the unsteady calculation the time step is set as 10^{-4} s and the total time step number is 2000.

3. Results and analysis

Both steady and unsteady flows of cavitation in the valve are simulated in this study. The pressure drop results are obtained and compared with the experimental data, and then the flow patterns under five different opening conditions are analysed. On the basis of steady results, the development of cavitation is predicted in the unsteady simulation.

3.1. Experiment validation

A hydraulic test system in Fig.4 has been developed for the experiment, with a flowmeter and two pressure sensor installed to measure the flow rate and pressure at both sides of the valve. The pressure upstream is fixed by using a surge tank in the experiment, and the flow rate is controlled by an extra valve downstream. To validate the model veracity and the calculation accuracy, experiments are conducted under five different opening conditions: 20%, 40%, 60%, 80% and 100% of valve opening. Comparison has been made between the computational and experimental results. The concept of resistance coefficient is defined as

$$K = \frac{2\Delta p}{\rho v^2} \tag{7}$$

where Δp is the pressure difference between inlet and outlet ports; ρ is the fluid density; v is the cross-sectional velocity of the fluid and K is the flow resistance coefficient.

As presented in Fig 5, the experimental and numerical results agree well, except for the 20% opening, where the error may be caused by the complex flow pattern at small flow rate. The agreement in the tendencies of the two curves is good and it can be validated that the computation model has enough accuracy to be utilised in the following analysis.

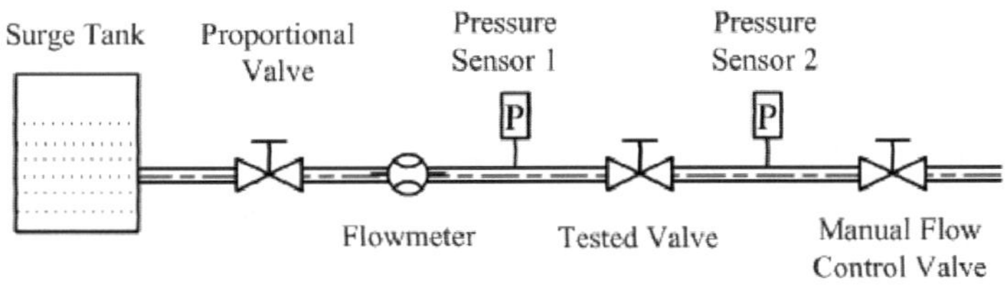

Figure 4. Diagram of the test system

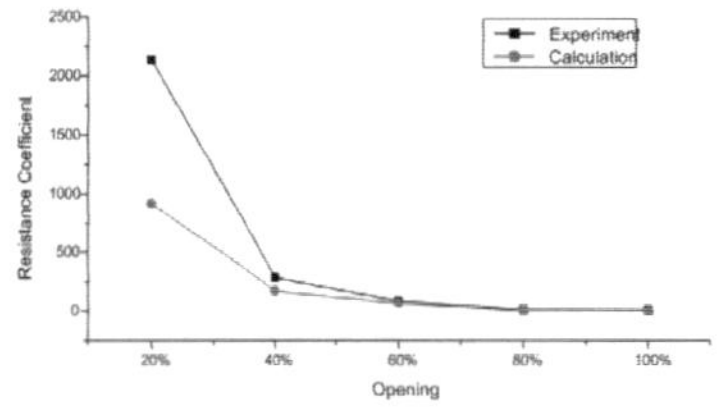

Figure 5. Resistance coefficient result

3.2. Flow analysis under different opening conditions

Figure 6 depicts the pressure contours and flow patterns in the valve with different openings. Counter-rotating vortex pairs can be seen in every figure because the flow in this region is stopped and reversed by the wall. At small openings below 60%, a sharp pressure drop is caused when the fluid flows through the holes on the cage. So the pressure in the moving part is much lower than that in the stable part. With the increase of the opening, the pressure loss in the holes decreases, and the pressure in the moving part rebounds. For the opening above 80%, the flow resistance is further reduced and the pressure loss becomes unconspicuous because the fluid passes the wide channel rather than the holes on the cage.

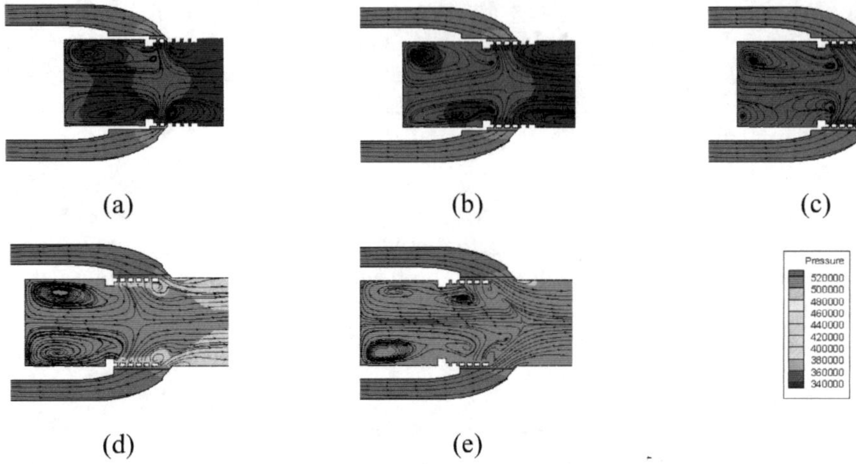

Figure 6. Flow pattern under different openings: (a) 20%; (b) 40%; (c) 60%; (d) 80% and (e) 100%.

Basically speaking, smaller valve opening leads to greater pressure decrease. Therefore, when the inlet pressure is low enough, the cavitation will occur. Figure 7 demonstrates the cavitation 3D structure by using the iso-surfaces with the vapour fraction of 0.1. Distributions of vapour differ when different inlet pressure conditions are given. Vapour cavities hardly appears with an inlet pressure higher than 0.4 MPa as indicated in Fig.7 (a). They emerge at the inlet of the holes with the inlet pressure around 0.39 MPa as shown in Fig.7 (b). When the set value approaches 0.37 MPa as shown in Fig.7 (c), vapour almost fills the hole walls, which can be defined as the cavitation inception because erosion on the valve body and influence on the downstream flow come to be significant. With a further drop in inlet pressure as shown in Fig.7 (d), cavitation becomes more serious, vapour expands and joins together in the flow passage. Cavitation shape grows more disordered and severe damage can be caused under this circumstance.

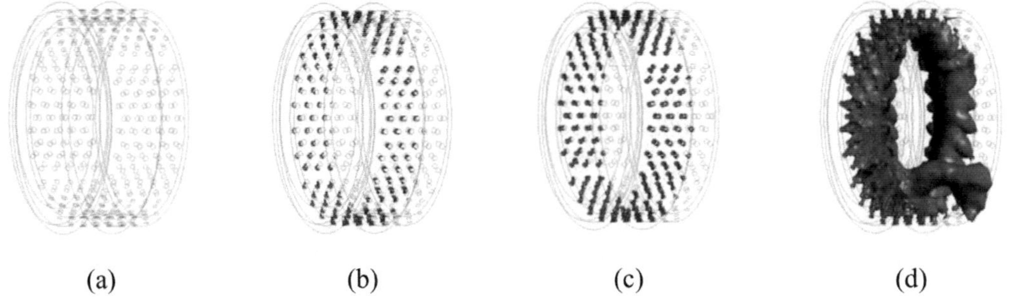

Figure 7. Cavitation under different pressure conditions (40% opening): (a) P_{in}=0.52MPa; (b) P_{in}=0.39MPa; (c) P_{in}=0.37MPa and (d) P_{in}=0.31MPa.

3.3. Unsteady calculation of cavitation

The transient cavitation flow is also simulated based on the previously developed cavitation model in this study. The operating condition with an opening of 40% and inlet pressure of 0.31 MPa has been taken as an instance. The evolution of the cavitation shape inside the valve is shown in Fig.8, with a time interval of 0.05 s (i.e. 500 time steps) between two sequent captures. At first the cavities are relatively scattered, with several parts distributing in different regions along the flow passage. The largest cavity is situated at the outlet of the holes and forms a ring in the circumferential direction. This annular cavity gradually extends and unites downstream. The shape turns more ordered and the volume increases after that.

(a) (b) (c) (d)

Figure 8. Time evolution of cavitation shape for transient cavitation flow (40% opening, P_{in}=0.31MPa): (a) t=0.02s; (b) t=0.07s; (c) t=0.12s and (d) t=0.17s.

4. Conclusion

A series of numerical simulations of the turbulent flow through a pressure-regulating valve has been performed. The cavitation is studied by developing and applying a set of cavitation model based on the transport equation. A batch of experiments on the flow resistance coefficient has been conducted, measuring the pressure drop through the valve with the flow rates given. Two results have been compared and show good agreement in the tendency. The flow patterns are visualised by pressure contour and streamlines. The main flow is shown to jet towards the centre from the holes and leads to a sharp pressure drop at small opening conditions. Counter-rotating vortex pairs are observed because part of the flow is stopped and reversed by the wall. Cavitation vapour emerges at the inlet of the holes at first, then fills the region adjoining to the hole-walls when the inlet pressure decreases. With a further pressure drop, the cavities expand and join together. The transient cavitation flow is calculated then, and shows the time evolution of cavitation shape. The cavities are scattered and distributed in several parts. Afterwards the large annular cavity near the holes extends and unites downstream, which tends to fall off the solid wall and move downward. Based on the analysis above, the structure of the valve can be further improved to reduce cavitation and its accompanying problems.

References

[1] Vaughan N D, Johnston D N and Edge K A 1992 *Proc. of the Institution of Mech. Engineers, Part C: J. of Mech. Eng. Sci.* **206**(2) 119-27
[2] Valdés J R, Rodríguez J M, Monge R, Peña J C and Pütz T 2014 *Energy Conversion and Management* **78** 776-86
[3] Tabrizi A S 2014 *J. Of Eng. Thermophysics* **23**(1) 27-38
[4] Chern M J, Hsu P H, Cheng Y J, Tseng P Y and Hu C M 2013 *J. of Energy Eng.* **139**(1) 25-34
[5] Gao H, Lin W and Tsukiji 2006 *Pro. of Institution of Mech. Engineers Part G-J. of Aerospace Eng.* **220**(G4) 253-65
[6] Jazi A M and Rahimzadeh H 2009 *Ultrasonics* **49**(6-7) 577-82

IOP Conf. Series: Materials Science and Engineering **72** (2015) 042036 doi:10.1088/1757-899X/72/4/042036

Numerical simulation of gas-solid two-phase flow in U-beam separator

X Y Zhou[1], H Z Zhang[2], X P Chen[1], J M Ruan[2] and H S Dou[1,3]
[1] Faculty of Mechanical Engineering & Automation, Zhejiang Sci-Tech University, Hangzhou 310018, China
[2] Dep. of Thermal Power Technology & Economic Operation, Huadian Electric Power Research Institute, Hangzhou 310018, China

Email: huashudou@yahoo.com

Abstract. Numerical simulation is carried out for gas-solid two-phase flow in a U-beam separator. In this study, the U-beam is altered with the inlet fins in order to improve the performance of the separator. The inlet fin angle of the separator are 30°, 35°, 40°, 45°, 50°, 55 ° and 60°. The governing equations are the Reynolds-Averaged Navier-Stokes equation with the standard k-ε model and the discrete phase model (DPM) describing the discrete two-phase flow as well as stochastic tracking model. Results show that the pressure drop deviation with fins is within 3% from those without fins. It is found that there is a maximum separation efficiency at the fin angle of 35°. Fin induces generation of a stagnation region which could collect particles and lead to change of vortical structures. The fin induced flow also causes the turbulent intensity inside the baffle to decrease to facilitate separation.

1. Introduction

The inertial separator is an apparatus to separate particles from gas, which relies on the difference of inertial force between gas and particles [1]. In the early years, a Swedish corporation introduced the U-beam separation technique for circulating fluidized bed boilers (CFBB) and later the U-beam separation was adopted by B &W Corporation [2]. In 1994, Ogawa put forward an inertial separation device manufactured with angle iron in his monograph [3].

A few research groups have also done a lot of works on it, for example: Xi'an Jiaotong University presented an impact type inertial separator with fluted tubes [4]; Zhejiang University developed a finned tubes impact gas-solid separator with heat transfer enhancement effect [5]; Tsinghua University has developed a combined multiple-channel inertial separator [6-7]. Besides, Cen et al. developed Fluctuation-Spectrum-Random-Trajectory Model [8]. Wang et al. used Reynolds Stress Model, taking into account the collision of particles and wall, to simulate inertial separator [9]. Li et al. investigated the influence of different structure parameters on the U-beam separator by using the discrete phase model, in which the solid phase is defined as discrete phase [10].

In this paper, the traditional U-beam separator is studied using numerical simulation method, and the separation efficiency is compared with experimental data to verify the reliability of computing. Then, the traditional U-beam separator is altered with inlet fins, and the effect of inlet fins on the performance of U-beam separator is analyzed.

[3] Corresponding author

2. Computing method and physical model

2.1. Computing method

In this research, since the volume fraction of solid particle is small, it is assumed that the gas is continuous phase and the solid particle is discrete phase. The governing equation for gas phase is the Reynolds-averaged Navier-Stokes equations, while the discrete phase model (DPM) is used for the solid phase. What's more, the discrete random walk (DRW) model is adopted to predict the dispersion of the particles due to turbulence, which determines the instantaneous gas velocity.

2.2. Physical model

For simulation of U-beam separation, the streamwise, wall-normal and spanwise directions are in x, y and z directions, respectively. The computational domain size ($L_x \times L_y \times L_z$) is $1.2\times1.0\times0.5$, whose unit is m. For single U-beam, its length and width are both 0.04 m, while the thickness is 0.003 m. Four rows of baffles are fixed 0.5 m apart from the entrance in a staggered formation, while the transverse and longitudinal spacing are both 0.04 m.

The modification of the traditional U-beam separator structure is as follow. Fins are added at the inlet of the baffle, whose length is 0.01m. An angle between the fin and the inner wall of the baffle is called fin angle α. To investigate the effect of fin, several cases are studied. The inlet gas velocity equals to 4 m/s and 10 m/s, while α is 30°, 35°, 40°, 45°, 50°, 55° and 60°. For the two-phase flow studied, the particle concentration is 0.025 kg/m^3.

The boundary conditions are as follow. The velocity at inlet is given, and gradient condition is given at the outlet. Symmetrical boundary condition is employed at centerline. On the wall boundary, non-slip conditions are used for the three velocity components. When calculating two-phase flow, the bottom wall is defined as "trap", inlet and outlet surface are defined as "escape", and other walls are defined as "reflect".

2.3. Model Validation

To verify the accuracy of simulation result, the simulation results are compared with experiment [10]. The solid particle is silica sand with a diameter of 220~1000 μm and density of 2650 kg/m^3. Figure 1 plots the separation efficiency. The numerical results produce a good agreement with experimental data, thus proving the reliability of numerical simulation.

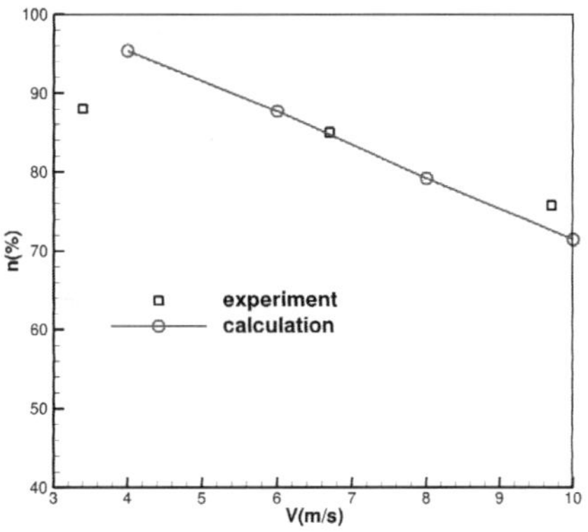

Figure 1. Separation efficiency of numerical computation and experiment data.

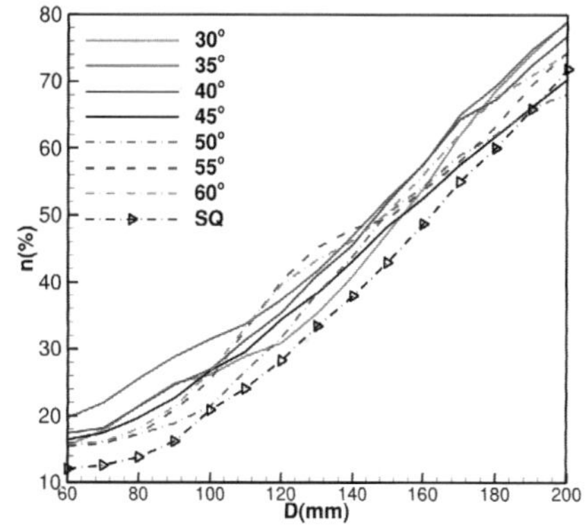

Figure 2. Separation efficiency under different fin angles for v=4 m/s. (SQ stands for square-shape traditional U-beam baffle without fins.)

As it is seen, when inlet velocity increases, the separation efficiency decreases rapidly because the drag force acting on the particles increases. As such, the particles are not easily deviated from the gas, and therefore are easier to be re-entrained because of colliding inside the baffle.

IOP Conf. Series: Materials Science and Engineering **72** (2015) 042036 doi:10.1088/1757-899X/72/4/042036

3. Results and analysis

3.1. Effect of fins on performance of U-beam separator

Table 1 shows the pressure drop of U-beam separator under different fin angles. As the fin angle α increases, the pressure drop follows a downward trend. The maximum deviation of pressure drop between the U-beam separator with fins and the traditional U-beam separators is within 3%. Furthermore, "SQ" stands for the square-shape baffle of traditional U-beam separator (without fins).

Table 1. Pressure drop of traditional U-beam separator and U-beam separator with fins.

α / v(m/s)	None	30°	35°	40°	45°	50°	55°	60°	Max deviation
4	198.4	201.8	200.2	200.1	200.5	200.1	198.6	197.5	1.7%
10	1236	1256.4	1267.3	1246.8	1249.3	1245.8	1237.4	1230.2	2.5%

Figure 2 shows the separation efficiency under different fin angles and the particle size range is 60~200 μm. For the velocity of 4 m/s, the separation efficiency of all seven fin angles has been enhanced, especially at 35°. Compared with the result of traditional type, an improvement of 8~13% is achieved for the case of 35° fin angle. However, for the velocity of 10 m/s, an improvement of 2~4% is achieved for the case of 35° fin angle.

To make further analysis of fin's effect on the separation efficiency, the bottom wall is divided into six zones signed as trap 0~5 as displayed in figure 3.

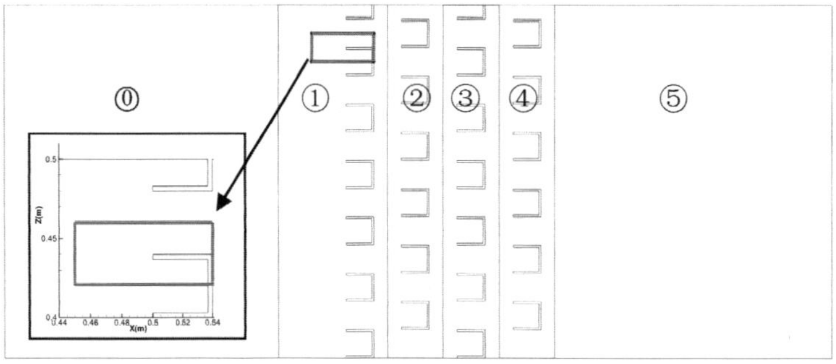

Figure 3. Partition of bottom wall and enlarged view of the cut piece.

Figure 4 shows the separation efficiency for two cases, which are U-beam separator with fins at α=35° and traditional U-beam separator. The separation efficiency of the former is 2~3 times of the latter in trap 1 with particle diameter of 60~120 μm. However, both the two separators show similar properties in other trap zones. When the particle size is larger than 120 μm, the separation efficiency of U-beam separator with fin angle of 35 improves significantly in trap 1 and trap 2, while in other trap regions their difference is not obvious.

3.2. Analysis of flow characteristics

A rectangular zone is cut from the cross section y=0.5 m as shown in large view in figure 3. It gives more details of the region whose length and width is 0.09×0.08 m, including half of the injection and collection area from top view. The distributions of the turbulence intensity, velocity and particle concentration are shown in figures 5 to 7 for three configurations, which are U-beam separator with fins at α=35°,α=50° and the traditional U-beam separator. In these figures, parameters are plotted at five locations of x at 0.45, 0.48, 0.5, 0.51 and 0.53 m.

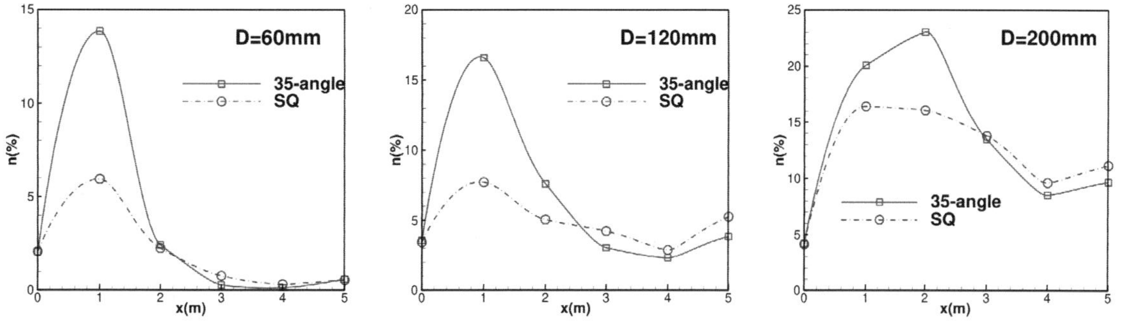

(a) D=60 mm (b) D=120 mm (c) D=200 mm

Figure 4. Separation quality in different trap zones between traditional U-beam separator and U-beam separator with fins when the particle size is of 60~200 μm.

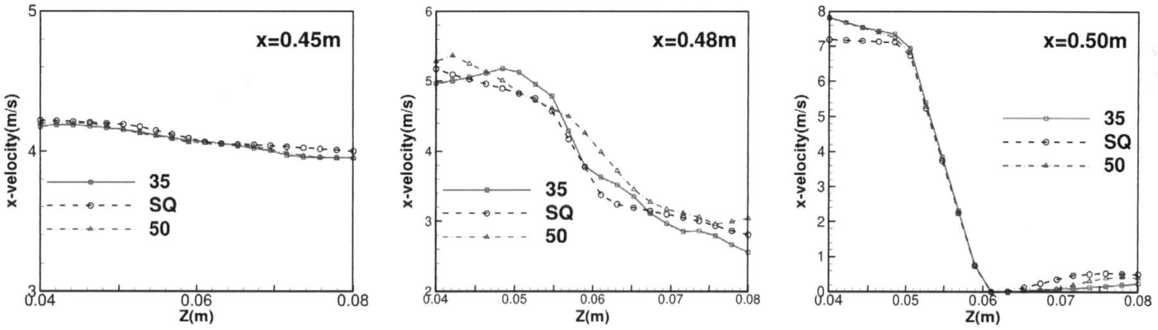

(a) x=0.45 m (b) x=0.48 m (c) x=0.50 m

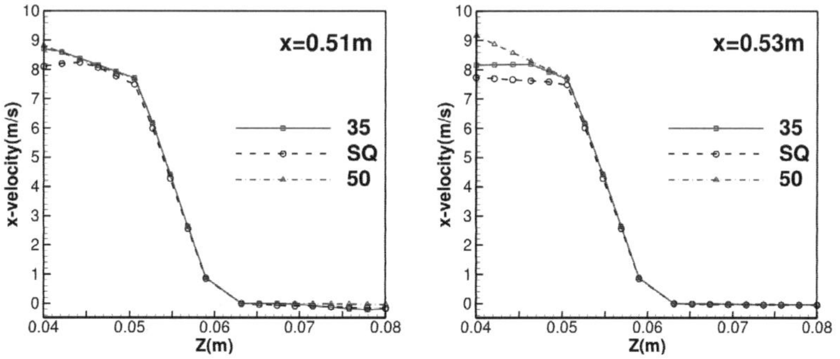

(d) x=0.51 m (e) x=0.53 m

Figure 5. Velocity distribution in the rectangular piece for three separator configurations: α=35°, α=50°and traditional type.

At five positions mentioned above, the velocity distribution has no significant difference for the three configurations as displayed in figure 5. However, the velocity at above mentioned five locations for fins at α=35° and 50° cases approach to zero earlier than the traditional type at x=0.5 m. Meanwhile, a negative streamwise velocity appears, indicating backflow inside the baffle.

Figure 6 plots the turbulent intensity, which is defined as $I=\left(2/3\cdot k\right)^{1/2}/\bar{u}$. It can be seen that the turbulent intensity of the three cases at the position x=0.45 m is almost the same. However, the turbulent intensity shows violent fluctuation with the increase of x value. The flow for cases at α=35° and 50° always tends towards to be stable earlier throughout the region, and inside the baffle the turbulent intensity for traditional U-beam separator is higher than those with fins.

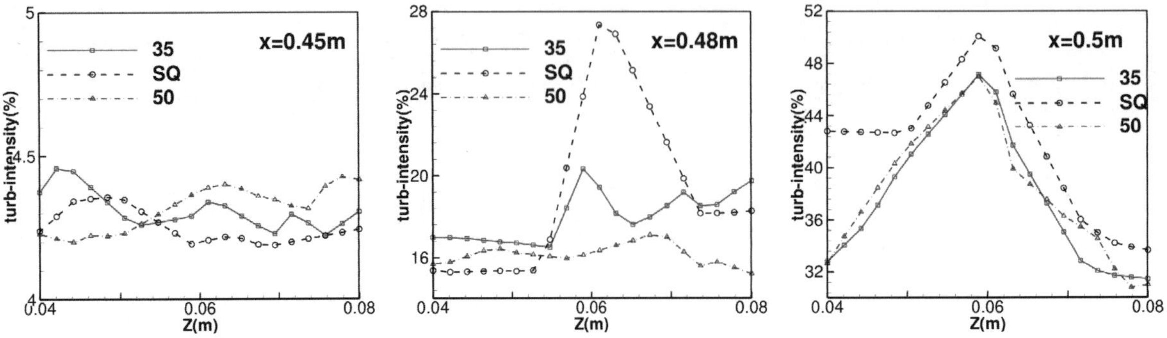

(a)x=0.45 m (b)x=0.48 m (c)x=0.50 m

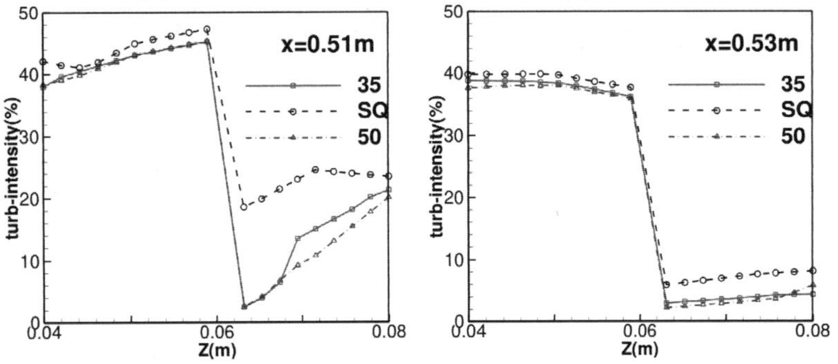

(d)x=0.51 m (e)x=0.53 m

Figure 6. Turbulence intensity in the rectangular piece for three kinds of U-beam separator: α=35°, α=50° and traditional type.

The researches on turbulent diffusion in [8] show that smaller particles diffuse faster than larger particles in gas, and diffusion enhances when the turbulence intensity of gas flow increases. For this reason, turbulence intensity can to some degree indicate particle's following ability with gas.

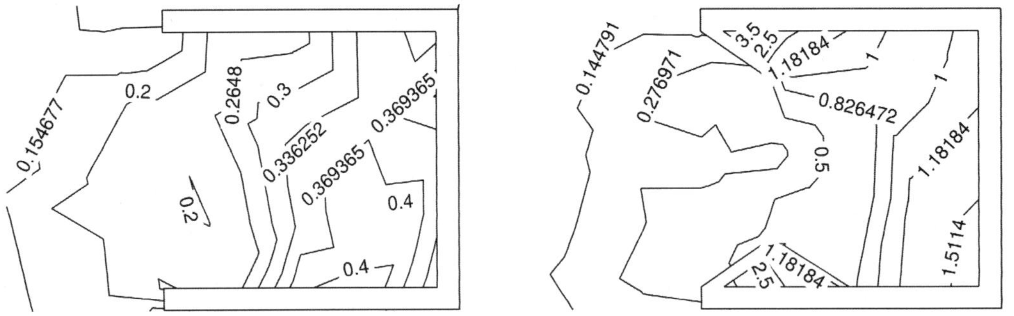

(a)traditional type (b)α=35°

Figure 7. Particle concentration in a single baffle.

Furthermore, there forms a stagnation region between the fin and the inner wall of baffle where the velocity is very close to zero as seen in figure 7. Meanwhile, turbulent intensity is very low in this region seen from figure 6(d). This is because when the gas flow is obstructed by fins, the particles, especially for small particles, are easier thrown into the stagnation region with the action of centrifugal force. Figure 7 shows that the particle concentration in stagnation region is significantly higher than other regions. Particles are difficult to escape from stagnation region and then separate from the gas eventually, thus enhance the separation efficiency of the U-beam separator.

4. Conclusions

Numerical simulation has been carried out to study the flow characteristics of U-beam separator. Results show that the separation efficiency of U-beam separator can be enhanced by fins added at the inlet of the baffle. The maximum separation efficiency is obtained at the fin angle of 35°. It is also found that the pressure drop of U-beam separator with fins has not been affected largely by fins, as the maximum pressure drop variation is within 3% from the traditional U-beam separator to the U-beam separator with fins. The separation efficiency could be increased by 13% at inlet velocity 4 m/s. At inlet velocity 10 m/s, there is still 4% increase of separation efficiency for fin angle of 35°. Simulation results indicate that fin induces generation of a stagnation region which could collect particles and lead to change of vortical structures. The fin also causes the turbulent intensity inside the baffle to decrease which helps separation. These influences play important roles in the separation efficiency improvement.

Acknowledgments

This work is supported by the Special Major Project of Science and Technology of Zhejiang province (No.2013C01139), the Fluid Engineering Innovation Team of Zhejiang Sci-Tech University (ZSTU) (No. 11132932611309), and the Science Foundation of Zhejiang Sci-Tech University (ZSTU) (No. 11130032241201).

References

[1] Che D F and Li H X 2007 *Multiphase Flow and its Application* (Xi'an: Xi'an Jiaotong University Press) 361 (in Chinese)
[2] Belin F and Flynn T J 1991 *Proc. of the 10th International Conference on FBC (San Francisco, USA ,30 April - 3 May, 1989)* **2** 287-94
[3] Ogawa A 1984 *Separation of Particles from Air and Gases* (CRC press)
[4] Zhang Y Z, Li Y L, Zhang X M, Jin D A, Duan J Z, Bao D W and Wang G H 1989 *Power Eng.* **9**(6) 9-13 (in Chinese)
[5] Chen K F, Li X D and Li Y X 1997 *Chemical Eng. J.* **66** 159-169 (in Chinese)
[6] Zhang X 1988 *Horizontal flow separator with two-stage inertia separation.* China CN87103655 (in Chinese)
[7] Zhang X 1987 *Fluidized-bed boiler.* China CN85109455 (in Chinese)
[8] Cen K F 1999 *The Theories and Technologies of Gas-Solid Separation* (Hangzhou: Zhejiang University Press) (in Chinese)
[9] Wang H G, Liu S and Jiang F 2004 *Journal of Engineering for Thermal Energy and Power.* **19**(4) 380-83 (in Chinese)
[10] Li H W, Liu J J, Liu J, Han H B and Li N 2012 *Int. Workshop on Information and Electronics Eng.(IWIEE), Procedia Eng.* **29** 3227-33
[11] Li H W, Han H B, Wang J and Zhang H 2011 *Acta Armamentarii* **32**(2) 225-29 (in Chinese)
[12] Tan X J, Chen L H and Li H J 2005 *Journal of Hydrodynamics. Ser A* **20** 910-15
[13] Baskakov A P, Mudrechenko A V, Bubenchikov A M, Starchenko A V, Gogolev A F and Markovich D M 2000 *Powder techno.* **107** 84-92

Analysis of vortex flow in a cyclone separators based on the energy gradient theory

L Liu[1], H S Dou[1,3], X P Chen[1] and H Z Zhang[2]
[1] Faculty of Mechanical Engineering & Automation, Zhejiang Sci-Tech University, Hangzhou 310018, China
[2] Dep. of Thermal Power Technology & Economic Operation, Huadian Electric Power Research Institute, Hangzhou 310018, China

Email: huashudou@yahoo.com

Abstract: Numerical simulation is carried out for the tangential inlet cyclone separator with different dimensionless parameter (the ratio of cone-tip diameter and cylinder diameter d/D is 0.342, 0.355, 0.368, 0.382 and 0.395). Then, the stability of the vortex flow in the separator is analyzed with the energy gradient theory. The governing equation is the Reynolds-averaged Navier-Stokes equations with Reynolds Stress Model (RSM). The finite volume method is employed to simulate the flow and the velocity, pressure, separation efficiency and the turbulent intensity are obtained. Results show that the location of the maximum tangential velocity is near the axis, and the highest outer vortex tangential velocity is achieved at the dimensionless rate of 0.355. When the d/D=0.355, the tangential velocity distribution in the outer vortex is the most close to velocity distribution characteristics of free vortex. The free vortex distribution has the best stability from the energy gradient theory which may be the reason why optimal performance can be achieved at d/D=0.355. To achieve high efficiency and low pressure drop for the device studied, d/D=0.355 is the optimal dimensionless diameter ratio.

1. Introduction

Cyclone separator is one of the most important industrial dust collectors with a lot of advantages, such as simple structure, low cost of operation and medium energy loss.

In order to improve the separation performance of cyclone separator, many researchers have put forward various feasible methods. Cao et al [1] carried out numerical investigation on the inner flow in cyclone separator, and found that reducing the inlet section area, the vortex finder diameter and the cone-tip diameter can make the tangential velocity increase, but simultaneously causing the pressure drop to rise. Qian et al [2] found that the cyclone separator attached a vertical tube can avoid the re-entrainment effectively and improve the separation efficiency. Khairy's [3] research showed that the pressure drop decreases with the increases of cylinder height, but its influence would reduce if the ratio between cylinder height and cylinder diameter is more than 1.8.

In this paper, using three-dimensional numerical simulation method, an analysis of the effects of cone-tip diameter on the flow field, pressure drop and separation efficiency of the cyclone separator, has been carried out to find out the optimal value of cone-tip diameter. According to the structure of combined

[3] Corresponding author

vortex, the energy gradient theory [4-10] is employed to study influencing mechanism of cone-tip diameter.

2. Numerical Simulation

The governing equations are based on the Reynolds-averaged Navier-Stokes (RANS) equation. For increasing the accuracy of numerical simulation, the Reynolds Stress Model (RSM) is used, which solves a transport equation for each component of Reynolds stress. What's more, the energy gradient method is applied to analyze stability of the structure of vortex.

2.1. Geometric model and boundary condition

Figure 1 is the geometric diagram of the cyclone separator, in which detail parameters can be seen from Table1. In this paper, numerical simulation is carried out for cyclone separators with tangential inlet for various different cone-tip diameters. Five sets of diameter are employed as shown in Table 2.

Table 1. Geometry parameters of cyclone separator.[11]

a/D	b/D	D_e/D	H/D	h/D	S/D
0.5	0.2	0.337	4	1.5	0.5

Table 2. Dimensionless ratio of d/D.

D/mm	d/mm	d/D
	65	0.342
	67.5	0.355
190	70	0.368
	72.5	0.382
	75	0.395

The cyclone separator has been divided into four regions: inlet, rotary, outlet and tube cone sections using Multi-Block grid technique. Unstructured grids are used for inlet region, but structured grids are employed for others regions. A series of numerical tests show that the total amount of grids used in this study is enough to accurate simulate the flow in the cyclone separator.

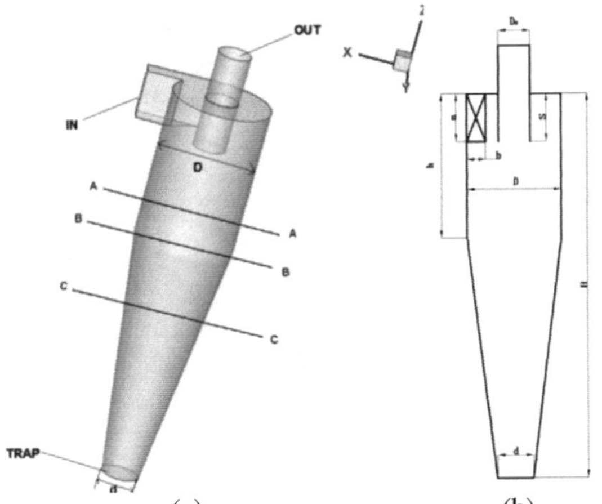

(a) (b)

Figure 1. Cyclone separator geometric diagram (In figure a, A-A, B-B, and C-C are the cross-section of the cyclone at the axial position Z=645mm, Z= 560mm,Z=360mm respectively. Figure b shows the geometric parameters of cyclone).

The boundary conditions are as follows. The velocity at inlet is given and that at outlet is extrapolated. On the wall boundaries, non-slip condition is used for the three velocity components. When considering two-phase flow, the inlet and outlet surface are defined as "escape" boundary condition. The bottom wall is defined as "trap", while others are defined as "reflect."

2.2. The energy gradient theory

In the past 10 years, Dou[4-10] proposed an energy gradient theory with the aim to clarify the phenomenon of transition from laminar flow to turbulence. According to the theory, for a given flow, a stability criterion can be written as follows for a half-period of disturbance.

$$F = \frac{\Delta E}{\Delta H} = \left(\frac{\partial E}{\partial n}\frac{2\bar{A}}{\pi}\right) \Big/ \left(\frac{\partial H}{\partial s}\frac{\pi}{\omega_d}u\right) = \frac{2}{\pi^2}K\frac{\bar{A}\omega_d}{u} = \frac{2}{\pi^2}K\frac{v'_m}{u} < Const \tag{1}$$

And

$$K = \frac{\partial E / \partial n}{\partial H / \partial s} \tag{2}$$

Here, F is a function of coordinates; K is a dimensionless field variable (function). E is the total mechanical energy per unit volumetric fluid, s is along the streamwise direction and n is along the transverse direction. H is the energy loss per unit volumetric fluid along the streamline for finite length. Furthermore, u is the streamwise velocity of the main flow. $V_m^{'} = \bar{A}\omega_d$ is the amplitude of the disturbance of velocity, here, $\bar{A}$ is the amplitude of the disturbance in the transverse direction, ω_d is the frequency of the disturbance.

When the vertical structure is free vortex, the value of energy gradient function K equals to 0 [4-10]. Thus, the flow is stable since the free vortex is able to restrain or absorb the disturbance.

3. Results and Discussions

3.1. Validation of the Numerical Model

Figure 2 plots the comparison of the simulation and experimental data [12] of tangential and axial velocities along the radius at the axial position z=560 mm for the case of d/D=0.382. The tendency of the tangential velocity distribution is consistent with the experimental data, especially the location of the maximum tangential velocity. The double vortex structure exactly shows a good conformity. On the contrary, the axial velocity distribution does not accord well with the experimental data. The possible source for this difference may come from that the influences of the wall boundary layer and the unstable vortex in the region.

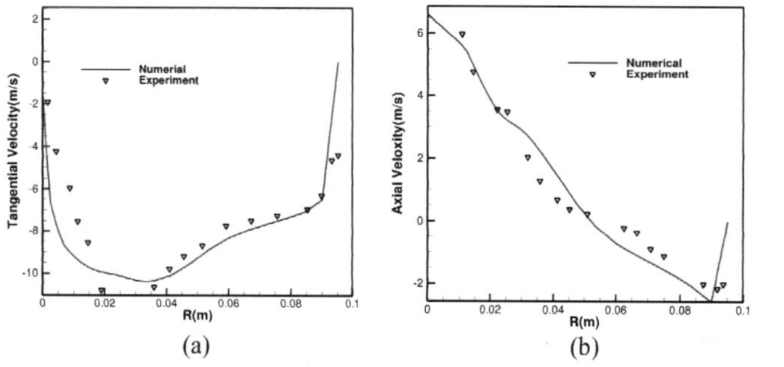

Figure 2. Comparison of the simulation and experimental velocity at the axial position Z=560mm. (figure (a) and (b) show the comparisons of the tangential velocity and axial velocity respectively.

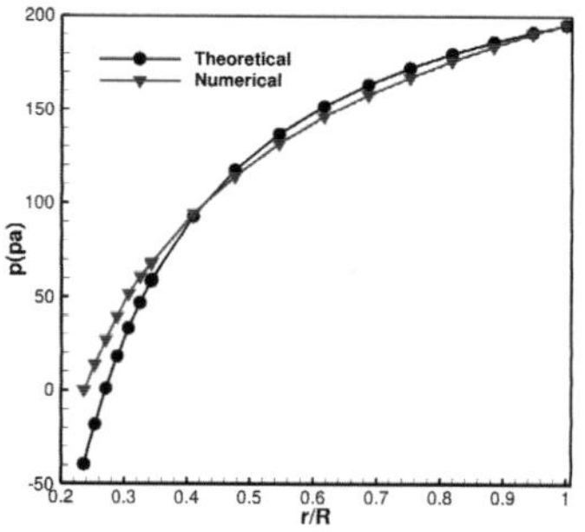

Figure 3. Comparison of static pressure between simulation and theory.

In order to check the accuracy of numerical simulation, the comparison of static pressure between numerical result and theoretical value is examined. A theoretical equation of static pressure distribution in the free vortex, based on the *Gas-solid Separation Theory and Technology* [13], is

$$p_k - p = \frac{\rho_g u_i^2}{2n}[(\frac{R}{r})^{2n} - 1] \quad (3)$$

where, ρ_k is the density of the fluid, p_k is the pressure near the wall, p is the pressure of optional position and n is the velocity distribution index which is calculated by using the empirical equation developed by Alanander, see the reference [13]. The comparison of the static pressure obtained from the simulation and theory expressed by equation 3 can be seen from figure 3.

The above comparison results show that the numerical simulation employed in this paper is reliable.

3.2. Tangential velocity analysis

In the cyclone separator, the working principle is based on that particles can be separated from the gas due to density difference between them under the action of the tangential velocity. Thus, the role of tangential velocity is very important.

Figure 4 shows the comparison of the tangential velocity at Z=360, Z=560 and Z=645 (mm) in the X=0 plane with different value of d/D. From the figure 4, the maximum tangential velocity causes the inner flow exhibiting a combined vortex structure, i.e., inner and outer flow (quasi-forced vortex and quasi-free vortex)[14]. In the inner vortex, the tangential velocity increases with the increasing of the radius near the axis, after reaching its peak the tangential velocity decreases gradually in outer vortex. It is noted that the outer vortex is the main separation zone, so wider range of outer vortex will help the separation. Higher tangential velocity in the outer vortex facilitates the particle motion to the wall of the cyclone separator. As is given in figure 4, for d/D=0.355, the location of the maximum tangential velocity is closest to the axis, and the outer vortex tangential velocity is higher than other d/D cases.

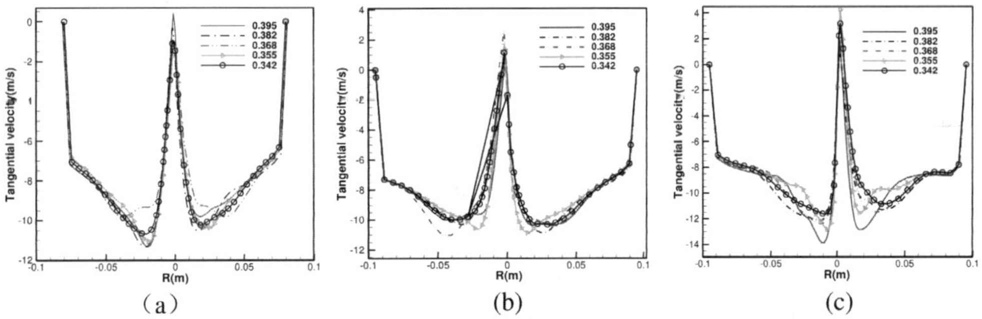

Figure 4. Tangential velocity distribution with different magnitude of the d/D at three axial stations (a) Z=360 mm, (b) Z=560 mm and (c) Z=645 mm.

3.3. Pressure drop and separation efficiency profiles

Pressure drop and separation efficiency are two of main performance parameters for the cyclone separators. Under the same volume fraction, the separation efficiency with various sizes of silica sand (5, 7, 9 and 11μm) are obtained, which is shown in figure 5. With the particle diameter increase, it is can be seen that the variation of separation efficiency has the same tendency for the different d/D. Figure 5 reveals that the higher separation efficiency is obtained at d/D=0.355. Meanwhile, the pressure drop increases with the increase of the magnitude of the d/D as shown in figure 5. The main reason for this phenomenon may be due to the variation of the contacting area of particles and the wall, which leads to the increase of the friction loss. In conclusion, at d/D=0.355, the cyclone separator achieves the highest separation efficiency and the relatively low pressure loss.

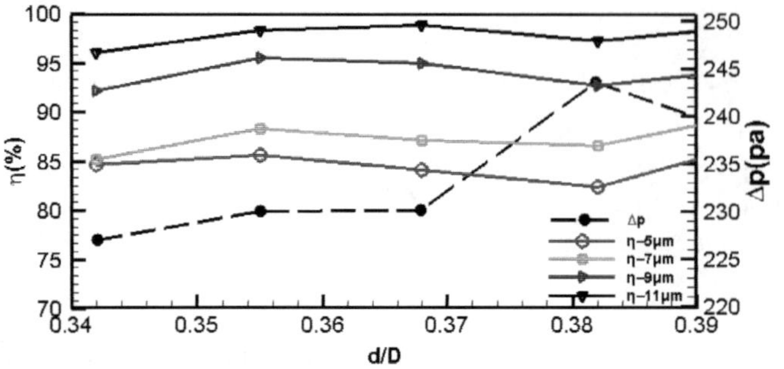

Figure 5. Pressure drop and separation efficiency with different magnitude of the d/D.

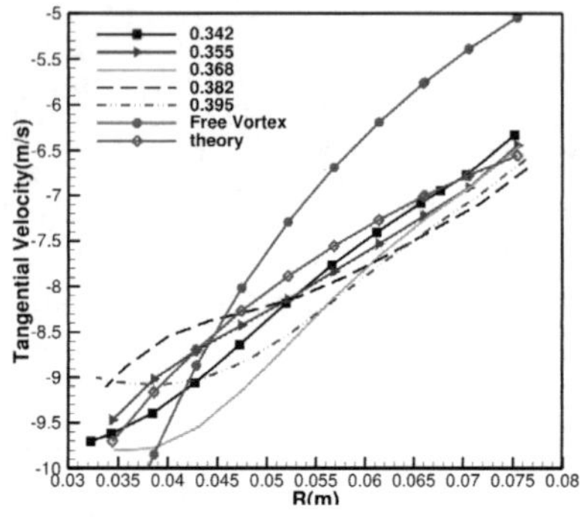

Figure 6. Tangential velocity distribution along radius.

3.4. Stability analysis

The vortex structure can mainly be divided into forced vortex and free vortex in inviscid fluid, and the moment of momentum is same in every position in free vortex [15]. However, due to the effect of the viscosity in cyclone separator, the outer vortex is a quasi-free vortex which is divided into half free vortex region and the boundary layer region. The velocity distributions equations of free vortex and half free vortex respectively as follows:

Free vortex

$$ur = c_1 \tag{4}$$

Half free vortex

$$ur^n = c_2 \tag{5}$$

where n is the velocity distribution index, generally n=0.5~0.9. In this paper, n is selected at 0.5; r is the radius; c_1 and c_2 are constants.

Figure 6 shows the tangential velocity distribution at Z=560 mm. From this figure, at the case of d/D=0.355, the tangential velocity distribution of real flow is the most close to the theoretical tangential velocity distribution in half free vortex and the distribution form in quasi-free vortex of real flow and free vortex is most similar.

Meanwhile, as shown in table 3, at d/D=0.355, the proportion of forced vortex in vortex structure is the least for various different structures.

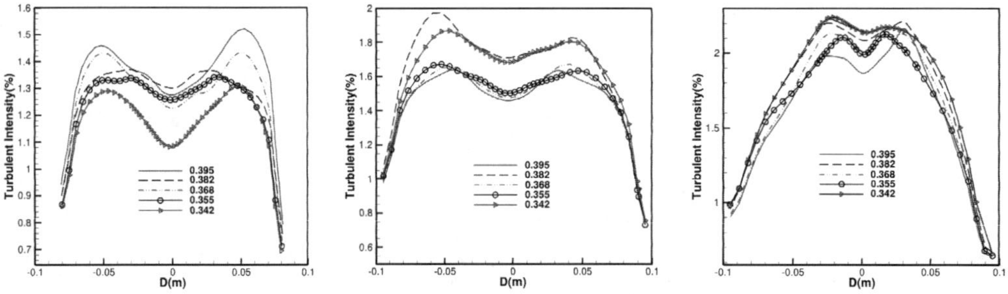

(a) (b) (c)

Figure 7. Comparison of turbulence intensity with different magnitude of the d/D at three axial station (a) Z=360mm,(b) Z=560mm and (c) Z=645mm.

Table 3. The proportion of forced vortex in combined vortex structure

d/D	Z=360	Z=560	Z=645
0.342	0.22	0.289	0.226
0.355	0.218	0.179	0.12
0.368	0.35	0.326	0.211
0.382	0.235	0.311	0.258
0.395	0.223	0.253	0.14

According to the energy gradient theory, the free vortex is stable and has the ability to restrain disturbances. Thus, when the quasi-free vortex area becomes bigger, the turbulence intensity is lower. The tangential velocity distribution in quasi-free vortex is close to free vortex, and thus the stability of the internal flow in cyclone separator is improved. Furthermore, figure 7 shows the comparison of the turbulence intensity distribution with different structures. For figure 7(a), turbulence intensity achieves the minimum and the maximum value at d/D=0.342 and 0.395, respectively. However, at d/D=0.355, the turbulence intensity is minimum as showed in figure 7(b) and figure7(c). Thus, the velocity distribution for *d/D*=0.355 is relatively stable. It is noted that there may also exists secondary flow (short-cut flow, vertical eddy and eccentric circulation vortex et al) besides the mainstream in cyclone separator. Moreover, Xue et al [16-17] found that whether secondary flow occurs or not has great relations to the turbulence intensity in a cyclone separator. The secondary flow will appears frequently when the turbulence intensity becomes larger, and then the flow will be of high instability.

In a word, we can obtain the conclusion that, at d/D=0.355, the half free vortex area is the largest and the turbulent intensity is relatively low, resulting in the flow to be most stable.

4. Conclusions

In this paper, considering the influence of dimensionless parameter d/D, the distributions of velocity and pressure, separation efficiency and vortex structure are studied by using numerical simulation method. Conclusions obtained are as follows:

1) At d/D=0.355, the location of the maximum tangential velocity is closest to the axis and tangential velocity in outer vortex is higher than others cases of geometric structures. Meanwhile, the cyclone separator achieves the highest separation efficiency and the relative small pressure drop at d/D=0.355. Thus, d/D=0.355 is the optimal dimensionless diameter ratio.

2) According to the energy gradient theory, the free vortex is stable, while the forced vortex is unstable. From the analysis for five different ratios (d/D), the quasi-free vortex occupies largest area at d/D=0.355 among the five cases. Thus, the velocity distribution for *d/D*=0.355 is most stable. This may be the reason why this case can get the optimal performance.

Acknowledgements

This work is supported by the Special Major Project of Science and Technology of Zhejiang province (No.2013C01139), the Fluid Engineering Innovation Team of Zhejiang Sci-Tech University (ZSTU) (No. 11132932611309), and the Science Foundation of Zhejiang Sci-Tech University (ZSTU) (No. 11130032241201).

References

[1] Cao Q Y, Ji G Q, Jin Y H and Wang J J 2008 *J. Fluid Machinery* **06** 34-38 (in Chinese)
[2] Qian F P and Zhang M Y 2006 *Sci. and Techno.* **02** 169-174 (in Chinese)
[3] Elsayed, K., & Lacor, C. (2014). Analysis and Optimisation of Cyclone Separators Geometry Using RANS and LES Methodologies. In *Turbulence and Interactions* pp 65-74 (Springer Berlin Heidelberg)
[4] Dou H S, Khoo B C and Yeo K S 2008 *J. of Therm. Sci.***47** 1422-35
[5] Dou H S 2011 *International J. of Physical Sci.* **6** 1411-25

[6] Dou H S, Khoo B C and Yeo K S 2008 *Int. J. of Therm. Sci.* **47** 1422-35

[7] Dou H S, Niu L and Cao S L 2014 *J. Eng. Thermophysics* **35**(6) (in Chinese)

[8] Dou H S and Khoo B C 2012 Stability of radial swirl flows *Proce. of 26th IAHR Symposiumon Hydraulic Machinery and Systems*, IOP Publishing, IOP Conf. Series: Earth and Environmental Science, 15 (2012) 012004 doi:10.1088/1755-1315/15/1/012004

[9] Dou H S and Xiao M N 2013 *J. Eng. Thermophysics* **34**(4) 658-61 (in Chinese)

[10] Dou H S Mechanism of Flow Instability and Transition to Turbulence 2006 *Int. J. of Non-Linear Mechanics* **5**(41) 512-17

[11] Wang H G and Liu S 2004 *J. of Eng. Thermophysics* S**1** 189-92 (in Chinese)

[12] Fraser S M, Abdel-Razek A M and Abdullah M Z 1997 *Proc. Instn. Mech. Engrs* **211** (E4) 247-57

[13] Cen K F, Ni M J and Yan J H 1999 *Gas-solid Separation Theory and Technology* (China: Zhejiang University) pp 327-29 (in Chinese)

[14] Qian F P and Zhang M Y 2007 *Chan. Eng. Technol.* **30(11)** 1564-70 (in Chinese)

[15] Jiang M H,Wang Z C,Zhao L X and He J 1999 *J. of Petroleum Machinery* **27**(1) 20-23(in Chinese)

[16] Xue X H, Wei Y D, Sun G G and Shi M X 2005 *J. of Eng. Thermophysics* **26**(2) 243-245(in Chinese)

[17] K Elsayed and Chris L 2011 *Computers & Fluids* **51** 48-59

IOP Conf. Series: Materials Science and Engineering **72** (2015) 042038 doi:10.1088/1757-899X/72/4/042038

Three-dimensional numerical simulation of air exhausted from submerged nozzles

J T Liu, S J Qin, T C Miao and D Z Wu[1]
Institute of Process Equipment, Zhejiang University, Road 38, Hangzhou, Zhejiang, 310008, China

E-mail: wudazhuan@zju.edu.cn

Abstract. Underwater exhaust produces an intricate unsteady two-phase flow field. For exploring the methods to predict the structure of air-water flow field and revealing the interaction of gas and water, three-dimensional underwater gas jet model with the VOF multiphase flow tracking method was adopted to simulate the transient flow field of gas jet into water. The air-water two-phase flow and its acoustic characteristic of turbulent gas exhausted from underwater nozzles were experimentally investigated in the early stages. Process of bubbles formation, detachment, fragmentation and coalescence were recorded clearly. The simulated results which were compared with the prior experimental results proved that the model almost accurately catches the behaviour of underwater bubbles. A few points were set in the two phase flow field to monitor pressure fluctuation. It had shown that higher air flow rate causes intense gas-column contraction and consequent bubble fragmentation, leading to higher amplitude and frequency of pressure fluctuation.

1. Introduction

Air discharging through an underwater nozzle generates powerful acoustic pressure [1, 2], which has been identified as one of the main noise sources in submerged equipment. The behavior of the two-phase flow field underwater is also responsibility for the efficiency of many industrial proceedings such as chemical engineering, metallurgy industry and energy industry, etc.

Since 1917, Rayleigh [3] was amongst the first to study the problem in cavitation and bubble dynamics. Many researchers show their interest in the same question, for a review, see Plesset and Prosperetti [4] and Leighton [5]. Sound generated by bubbles has been investigated in quantity [6] but most of those studies focused on a single bubble. Based on the theory of volumetric oscillations of the bubble, the sound spectrum generated by the formation of a bubble at low to moderate bubbling rates can be calculated theoretically and has been used for bubble sizing [7]. However, with the increase of airflow rate, multiple bubbles are involved in the flow field, resulting in the irregular acoustic signal, and no simple theories can predict the sound field. The bubble dynamics are affected by many factors, such orifice size, gas flow rate, and induced by liquid flow. A two-dimensional model using the VOF method was built to simulate the bubble formation process, at airflow rates from 1m/s to 20m/s, by Chen and Manasseh [8]. Manasseh and Chen et al [9] also carried out a lot of experimental researches on acoustic characteristic from underwater exhaust. They studied the effects of different gas flow rate and different diameter of nozzles. Meanwhile, the underwater two-phase flow field was observed. Experimental results indicated that gas velocity and the sound pressure level existed nonlinear

[1] Corresponding author.

relationship. Arghode et al [10] adopted five nozzles with different cross-sections to study the characteristics of submerged gas jet. The results showed that sound pressure level was affected by the orifice cross-section and pressure fluctuation was coupled with jet instabilities. At higher airflow rate, bubbling regime will transition to jetting regime. Zhao and Irons [11] applied combined Kelvin-Helmholtz and Rayleigh-Taylor instability analysis to obtain the critical velocity for jetting transitions. More recently, Shi Honghui [12], Chris Weiland [13] and Harby K. et al [14] investigated the flow structure of submerged high-speed gas jet experimentally. In numerical simulation of the complicated flows, Xu xiaoqiang [15] and Tang Jianing et al [16] successfully calculated the flow field with the VOF equations by two-dimensional model. The phenomena including expansion, bulge, necking and back-attack are highlighted in the jet process, which add the flow structure complexity. However, further studies are still necessary for the mechanism between sound generation and the flow structure of underwater exhaust.

In this paper, a 3D unsteady flow model was established to simulate air injected into opening quiescent water from nozzles which is located 760mm underwater with flow rates from 58m/s to 108m/s. The underwater two-phase flow fields were captured with RANS approaches and the VOF tracking method. Bubble formation, deformation, detachment, fragmentation and coalescence were simulated. The velocity and pressure fields caused by incompressible flow field fluctuation were analysis. The objective of this paper is to gain a good understanding of the relationship between the flow structure and the acoustic field.

2. Numerical methods

Consider air injected from a submerged nozzle into an open cylindrical container at room temperature. In this study, both air and water were assumed to be incompressible and immiscible. Bubbles were affected by inertial force, surface tension and buoyancy force, etc. Regardless of the phase transformation, there were obvious interfaces between air and water. The VOF (Volume of Fluid) methodology was used here to evolve and capture the interface.

2.1. Governing equations

The mean velocity $\overline{u}$ and pressure $\overline{p}$ of the underwater two-phase flow field could be described by the continuity and the Reynolds-averaged Navier-Stokes equations, which are written as:

$$\frac{\partial \overline{u_i}}{\partial x_i} = 0 \tag{1}$$

$$\frac{\partial \overline{u_i}}{\partial t} + \frac{\partial\left(\overline{u_i u_j}\right)}{\partial x_j} = -\frac{1}{\rho}\frac{\partial \overline{p}}{\partial x_i} + \frac{\partial}{\partial x_j}\left[\left(\nu+\nu_t\right)\left(\frac{\partial \overline{u_i}}{\partial x_j} + \frac{\partial \overline{u_j}}{\partial x_i}\right)\right] + g_i + \frac{1}{\rho}\overline{F_i^{ST}} \tag{2}$$

$$\rho = \varphi\rho_g + (1-\varphi)\rho_l \tag{3}$$

$$\nu = \varphi\nu_g + (1-\varphi)\nu_l \tag{4}$$

$$\overline{F_i^{ST}} = \sigma\kappa\frac{\nabla\phi}{\phi}, \quad \kappa = \nabla\cdot\frac{\vec{n}}{\left|\vec{n}\right|} \tag{5}$$

Here, ν is the kinematic viscosity; ν_t is the eddy viscosity; g_i is the gravitational acceleration; ρ is the density of one fluid. φ is the volume fraction in the local cell of one fluid. Its value may be 0 when the cell is full of liquid phase and 1 when the cell is full of gas phase. A value between 0 and 1 indicates that the cell contains the interface between gas and liquid if a gas-liquid two-phase system is involved. The subscript g and l represent the gas phase and the liquid phase, respectively. The surface tension $\overline{F_i^{ST}}$, which is the last term of equation (2), modelled by the Continuum Surface tension Force (CSF) [17], could be computed via equation (5), where σ is the surface tension coefficient; κ is the interface curvature; $\vec{n}$ is the surface normal which is defined as the gradient of ϕ.

2.2. Initial and boundary conditions

In this paper, the initial liquid level height in the cylindrical container is 760mm, the remaining part is air phase which is opened to the atmosphere (see figure 1). Pressure-outlet where the gauge pressure is zero is adapted at the outlet. The inlet is defined as velocity-inlet, where the volume fraction of air phase is unity. The boundary conditions are u=0 at the walls.

2.3. Construction of numerical model

The simulation domain co-ordinates are [-200,200], [-200,200], [-60, 1000] mm in the x, y, z directions respectively. Gas ejection direction lies in the z positive direction, and the Cartesian co-ordinate origin at the central axis of the nozzle. The simulation model is used with a structural hexahedron grid of a total number of 700,000 cells. In order to provide good resolution of the two phase interface and the jet boundary layer, these relevant domains are refined to properly capture the air-water two-phase interface. Outside these domains, for reducing the computational cost, the size of mesh increases in proportion up to reach the boundary. The mesh structure of the 3D submerged gas exhaust model is shown in figure 2.

Figure 1. Air volume fraction of initial conditions.

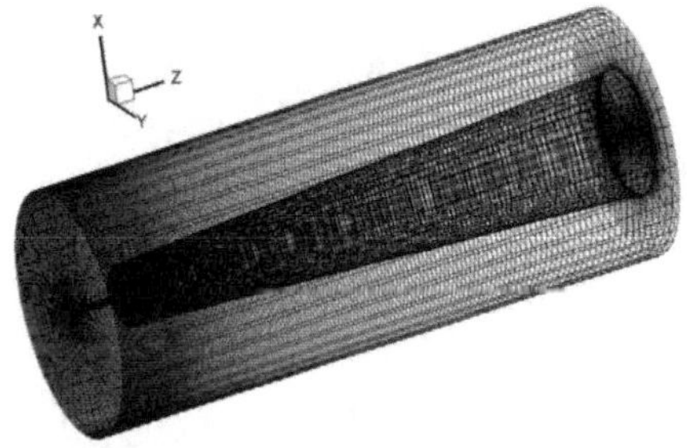

Figure 2. Diagram of 3D non-uniform cells.

3. Results and discussions

3.1. Underwater two-phase flow field

Gas discharge through a submerged orifice into quiescent water is an unsteady process. The air-water two-phase flow field changes with the change of time and space. The initial stage of underwater two-phase flow fields with air injection at two different gas velocities of 58m/s and 108m/s are shown in figure 3. In the beginning, there is a gas-column formed with the effect of inertial force from injected air. Under the combined effect of surface tension, buoyancy and viscous shear force, an obvious "necking" phenomenon exists at the end of the gas-column away from nozzle, resulting in the first small bubble separated from the gas-column. Air continuous supplied from the upstream nozzle feed the gas-column becoming fatter and fatter. Meanwhile, the first bubble divided into lots of small bubbles downstream. In course of time, bubble formation and detachment occurred around the orifice periodically. After the bubble detached from the nozzle, the successor bubbles form. The subsequent bubbles may be absorbed into the dominant bubble and stretched by the dominant bubble (shown in figure 5).

As the gas injection velocity increased, the dominant forces are fluid inertia and buoyancy, instead of surface tension force [18]. It can be seen from figure 3 that the periodic time of bubbles formation becomes shorter with the increasing of underwater gas flow rate. Turbulent kinetic energy increase brings more intense disturbance of the flow field. Under the strong disturbance, gas bubble could be more easily to break into lots of small bubbles. In the experimental video of underwater flow field, this phenomenon also exists in another form that there are more small bubbles in the flow field than lower gas injection velocity. In order to study the effect of bubble motion on flow field structure, bubble-bubble fragmentation and the resulting motion of the surrounding water under the influence are shown in figure 4. It can be found that the incoming gas puts the surrounding liquid in motion, and spiral movements exit around each bubble. Bubble-bubble fragmentation causes changes in direction of movement of partial flow field, and resulting in chaos in the surrounding flow field, which may just explain the reason of pressure fluctuation of the flow field around them. It is interesting to note that the surrounding liquid will enter into the large gas bubble when abruption occurs.

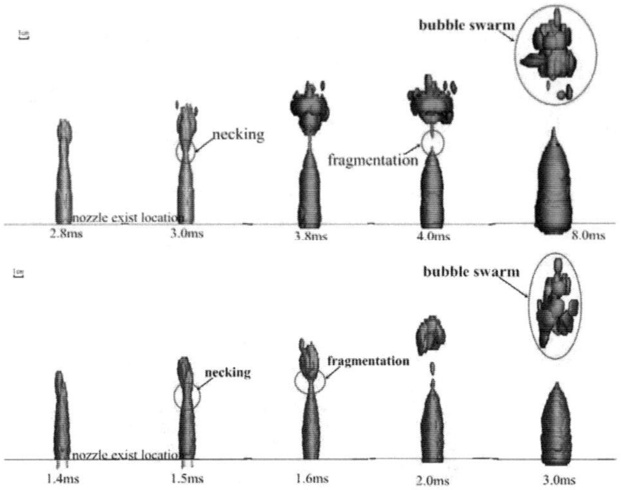

Figure 3. Initial stage of two-phase flow field with air injection velocity of 58m/s (top) and 108m/s (bottom).

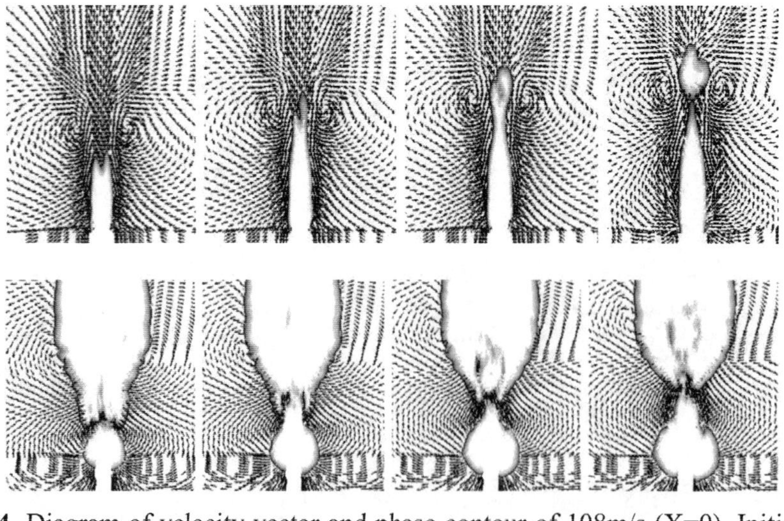

Figure 4. Diagram of velocity vector and phase contour of 108m/s (X=0). Initial stage: 1.2ms, 1.4ms, 1.5ms, 1.6ms (from left to right, top). Successor bubbles: 40ms, 42ms, 44ms, 46ms (bottom).

3.2. Comparison with the experimental results of underwater two-phase flow field

Either the experimental video or the simulated results had shown that the process of submerged bubbles generation is periodical. However, due to the complexity of the structure of flow field and the

IOP Conf. Series: Materials Science and Engineering **72** (2015) 042038 doi:10.1088/1757-899X/72/4/042038

intricacy of forces that bubbles taken, the form of submerged bubbles is not repeatable. Figure 5 displays the comparison of the underwater two-phase flow field between three-dimensional simulation and experiment results. It can be seen that the three-dimensional model which is used in this paper almost presents the underwater air-water flow field accurately.

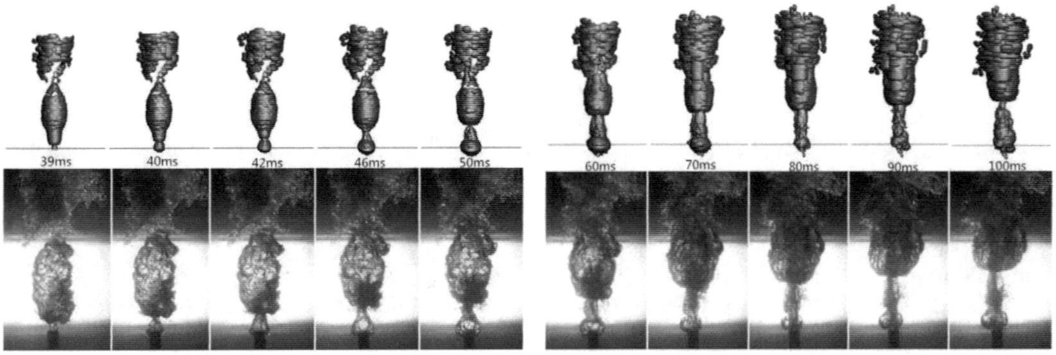

Figure 5. Comparisons at 108m/s of gas injection velocity.

3.3. *Pressure fluctuation and velocity fields*

Five points (seen in figure 6) were set in the underwater two-phase flow field to monitor the variation of pressure. On account of the consistence of the pressure fluctuation trend, here, only data of point A with the gas flow rate of 108m/s is taken as an example to analyse.

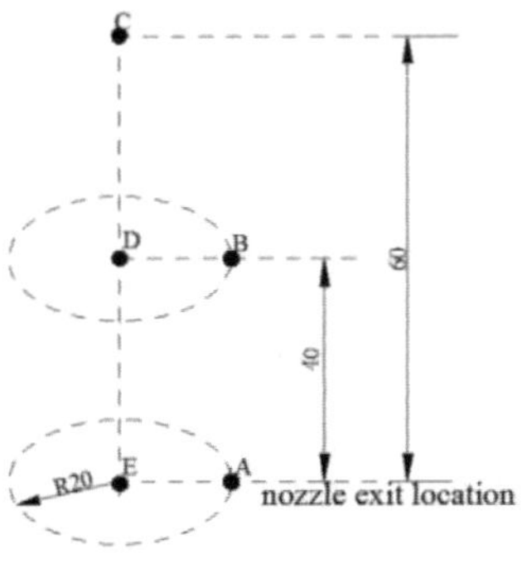

Figure 6. Schematic diagram of pressure monitor points.

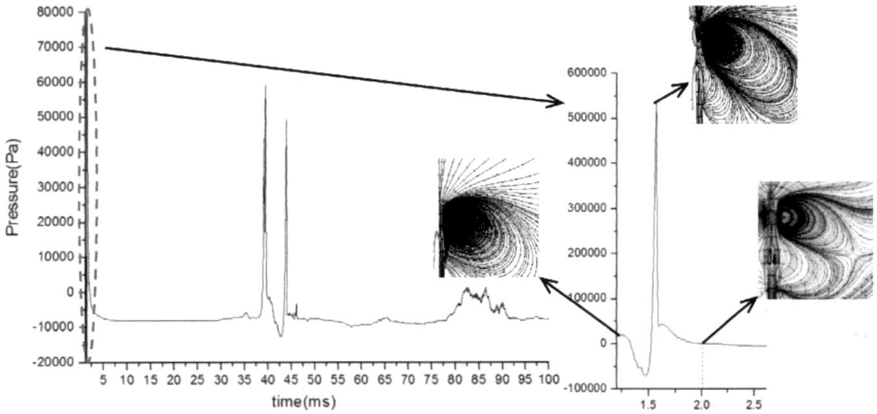

Figure 7. Pressure fluctuation at point A with gas velocity of 108m/s.

The history of hydrodynamic pressure with time at point A is demonstrated in Figure 7. It can be seen that the first bubble separated from the gas-column gives rise to a large pressure fluctuation. From the velocity vector diagram, it can be found that separation between the air bubbles causes vortex and reverse motion of the surrounding liquid, which should be responsibility for the pressure fluctuation during the time of 1.2ms to 1.6ms, 40ms to 44ms. Submerged bubble detached from the nozzle is also accompanied by a significant liquid pressure variation (see t=38ms to t=40ms in figure 5 and figure 7). These significant changes of the pressure could be the main contribution of the low frequency acoustic signal. As time progress, the air-water flow field becomes more complexity. When bubble detachment, fragmentation, collapse, coalescence and natural oscillation exited simultaneously, the pressure fluctuation presents irregular performance (see t=80ms to t=100ms in figure 5 and figure 7). Furthermore, with the increase of the gas flow rate, the peaks of pressure fluctuation are greater than those of lower gas flow rate.

4. Conclusions

Gas exhausted through an underwater nozzle into quiescent water with high gas flow rates have been successfully simulated in three-dimensional form. The simulation results show bubble-bubble fragmentation, interaction and coalescence, and bubble-nozzle detachment been supposed to from the early experimental observation. With the increase in air flow rate, the frequency of bubble detached from the submerged nozzle also increases. Moreover, the hydrodynamic pressure fluctuation in the air-water flow field was monitored. Bubble-bubble fragmentation and bubble-nozzle detachment both give rise to a significant pressure fluctuation, which should be the major contributor in the strong acoustic signatures. Simultaneously, the velocity vector fields were analysed to explain the phenomenon of pressure fluctuation. It is helpful to understand the formation and evolution of the submerged two-phase flow fields and pressure fields. This paper creates a foundation for further acoustic analysis.

Acknowledgements

The authors are grateful to Weiping Dai and Zongrui Hao for building the experimental equipment.

Reference

[1] Minnaert M 1933 *Philos. Mag.* **16** 235-48
[2] Strasberg M 1956 *J. Acoust. Soc. Am.***28** 20-26
[3] Rayleigh L 1917 *Philos. Mag.* **34** 94-98
[4] Plesset M S and Prosperetti A 1977 *Ann. Rev. Fluid Mech.* **9** 145-85
[5] Leighton T G 1994 *The acoustic bubble* (London: Academic Press)
[6] Longuet-higgins M S 1989 *J. Fluid Mech.* **201** 525-41
Longuet-higgins M S 1989 *J. Fluid Mech.* **201** 543-65
[7] Manasseh R, Lafontaine R F, Davy J, Shepherd I and Zhu Y G 2001 *Exp. Fluids* **30** 672-82
[8] Chen L, Wood S, Moore S and Nguyen B 2012 *Proc. of Acoustic(Fremantle, Australia, 21-23, November, 2012)*
[9] Chen L and Manasseh R 2001 *4th Int. Congress on Multiphase Flow(New Orleans, USA, 27,May-1 June, 2001)*
[10] Chen L, Manasseh R, Nikolovska A and Norwood C 2003 *8th Western Pacific Acoustic Conf.(Melbourne, Australia, 1 January 2003)*
[11] Arghode V K, Gupta A K and Yu K H 2008 *46th AIAA Aerospace Sciences meeting and exhibit(Reno, Nevada, USA, 7-10 January 2008)*
[12] Zhao Y F and Irons G A 1990 *Metallurgical Transaction* B **21** 997-1003
[13] Shi H H, Wang B Y and Dai Z Q 2010 *SCI China Phys. Mech.* **53** 527-35
[14] Chris W and Pavlos P V 2013 *Int. J. Multiphase Flow* **48** 46-57
[15] Harby K, Chiva S and Munoz-Cobo J L 2014 *Exp. Therm. Fluid Sci.* **53** 26-39
[16] Xu X Q, Deng J and Ren A L 2005 *J. Hydrodynamics* B **17** 204-08

IOP Conf. Series: Materials Science and Engineering **72** (2015) 042038 doi:10.1088/1757-899X/72/4/042038

[17] Tang J N, Wang N F and Wei Shyy 2011 *Acta. Mech. Sin.* **27** 461-72
[18] Brackbill J U, Kothe D B and Zemach C 1992 *J. Comput. Phys.* **100** 335-54
[19] Gavigan J J, Waston E E and King W F 1974 *J. Acoust. Soc. Am.* **56** 1094-99

IOP Conf. Series: Materials Science and Engineering **72** (2015) 042039 doi:10.1088/1757-899X/72/4/042039

Numerical investigation on cavitation in pressure relief valve for coal liquefaction

G F Ou, W Z Li, D H Xiao, Z J Zheng, H S Dou and C Wang[1]

The Flow Included Corrosion Institute, Faculty of Mechanical Engineering & Automation, Zhejiang Sci-Tech University, Hangzhou 310018, China

E-mail: wangchao@zstu.edu.cn

Abstract. The pressure relief valve for regulating the level of the high-pressure separator works under a pressure difference up to 15 MPa in the temperature of 415 ℃. Severe cavitation erosion and particle impact lead to the valve disc's mass loss. In this paper, three-dimensional turbulent cavitating flows in the pressure relief valve are numerically simulated to reveal the mechanism of mass loss at valve disc. The RNG k-ε turbulence model and the mixture model with a mass transfer for cavitation are employed to simulate the cavitating flow in the pressure relief valve. The result shows that there is phase change in the pressure relief process and cavitation bubbles would be transported by high-velocity backflow to the head of valve disc. For the local pressure higher than the saturated vapor pressure, the bubbles collapse at the head of disc and cavitation erosion is formed at the head of the disc. By comparing the cases of opening of 40%, 50%, and 60%, backflow velocity and cavitation region in front of the disc decrease with the opening increase. Therefore, during the actual operation, the pressure relief valve should be kept to a relatively large opening.

1. Introduction

Cavitation, a general fluid mechanics phenomenon, would occur when the local pressure in liquid is below the saturated vapor pressure at the operating temperature [1]. The cavitation bubbles would collapse if they are transported to regions of higher pressure, generating extremely high temperatures, pressures and velocities [2-3]. If a bubble collapses near a wall, the high speed micro-jet, shock wave and pulse pressure produced may cause serious local damage [4-7].

The pressure relief valve studied is located at the bottom of high-pressure separator which is used for regulating liquid level in high-pressure separator. This type of valve is required to work under the most severe service conditions in Direct Coal Liquefaction Project. The pressure relief valve developed by the USA, which is tested on a 200-t/d H-Coal pilot unit, has a maximum service life of 693 hours. The pressure relief valve developed by Japan sintered diamond spool, which is tested on a 150-t/d pilot unit, has a maximum service life of 1008 hours. During the first commissioning of Direct Coal Liquefaction Demonstration Project, the service life of the pressure relief valve does not exceed 1 day at minimum and is not longer than 200 hours at maximum [8].

The pressure relief valve works under a pressure difference up to 15 MPa in the temperature of 415 ℃ in Direct Coal Liquefaction Demonstration Project which is the first megaton industrial scale

[1] Corresponding author

IOP Conf. Series: Materials Science and Engineering **72** (2015) 042039 doi:10.1088/1757-899X/72/4/042039

demonstration project in the world [9]. Severe cavitation erosion and particle impact lead to the valve disc' mass loss [10]. The surface of valve disc is treated by WC (tungsten carbide) which has a good performance resisting particle erosion, but the disc can't bear severe cavitation erosion. The study on cavitation in pressure relief valve is of great engineering significance. Numerical simulation is a feasible choice to understand the flow characteristic compared to experiment for the severe operating condition.

In this paper, we simulate the three-dimensional turbulent cavitating flows in the pressure relief valve to reveal the mechanism of mass loss at valve disc and compare the difference of flow characteristic in the valve under different openings. Then, we analyze the mechanisms of cavitation formation and bubble collapse for various operation parameters.

2. Governing equations and numerical simulation

Numerical method is applied to study the cavitation flow in the pressure relief valve. The simulation model, governing equations, cavitation model, boundary conditions and discrete method are shown in the following part.

2.1. Geometrical model

The pressure relief valve is an angle type valve. Its flow channel and coordinate system are shown in Figure 1. The origin of coordinates is on the vertex of the valve disc when the pressure relief valve is closed (opening 0%). The media flows horizontally into the valve and gravity is in –y direction. The structure of valve disc is shown in Figure 2. The head of valve disc is a part of spherical surface with an elliptical surface after.

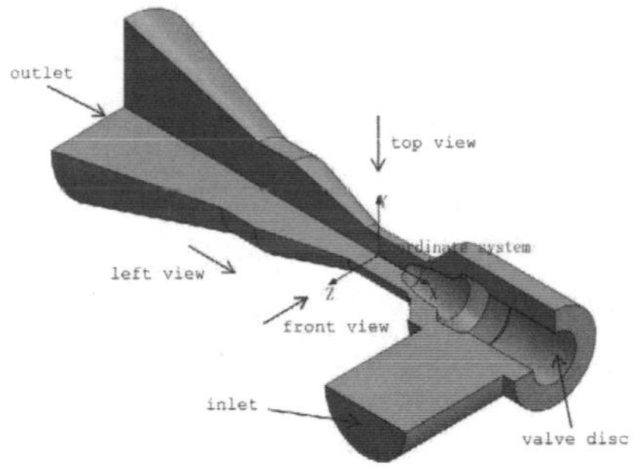

Figure 1. Flowing road structure.

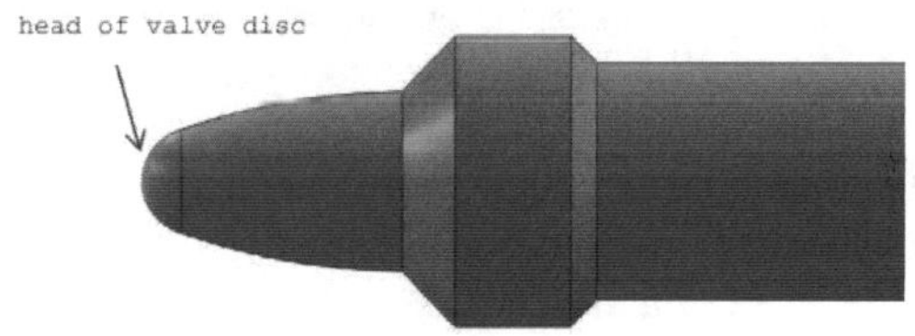

Figure 2. Valve disc structure.

2.2. Governing equations

When cavitation occurs, the phase change from liquid to vapor happens. A multiphase model has to be employed to describe the flow. The governing equations for a multiphase flow are based on a single-fluid approach, regarding the mixture as one liquid. The flow field is solved for the mixture continuity and momentum equations:

$$\frac{\partial}{\partial t}(\rho_m)+\nabla\cdot(\rho_m\vec{v}_m)=0 \tag{1}$$

$$\frac{\partial}{\partial t}(\rho_m\vec{v}_m)+\nabla\cdot(\rho_m\vec{v}_m\vec{v}_m)=-\nabla p+\nabla\cdot\left[\mu_m\left(\nabla\vec{v}_m+\nabla\vec{v}_m^T\right)\right]-\rho_m\vec{g} \tag{2}$$

Here the mixture density ρ_m and μ_m are defined as:

$$\rho_m=\alpha\rho_v+(1-\alpha)\rho_l \tag{3}$$

$$\mu_m=\alpha\mu_v+(1-\alpha)\mu_l \tag{4}$$

where the subscripts *m*, *v*, and *l* represent mixture, vapour and liquid phase respectively.

2.3. Cavitation model

The cavitation model employed here is developed by Schnerr & Sauer[11]. The liquid-vapor mass transfer is governed by the vapor transport equation:

$$\frac{\partial}{\partial t}(\alpha_v \rho_v) + \nabla \cdot (\alpha_v \rho_v \vec{v}_v) = R_e - R_c \tag{5}$$

where the mass transfer source terms R_e and R_c are connected to the growth and collapse of the vapor bubbles respectively.

The source terms are derived from the Rayleigh-Plesset equation and are defined as:

$$R_e = \frac{\rho_v \rho_l}{\rho} \alpha_v (1-\alpha_v) \frac{3}{R_B} \sqrt{\frac{2(p_v - p)}{3\rho_l}}, \text{ when } p_v \geq p \tag{6}$$

$$R_c = \frac{\rho_v \rho_l}{\rho} \alpha_v (1-\alpha_v) \frac{3}{R_B} \sqrt{\frac{2(p - p_v)}{3\rho_l}}, \text{ when } p_v \leq p \tag{7}$$

2.4. Discrete method

The numerical simulations are performed by using general CFD code Fluent. The RNG k-ε turbulence model and standard wall function are used in this study. The convective term is discretized by second-order upwind method and the diffusion term is approximated with central difference scheme. The PISO method is adopted to couple the pressure and velocity. The volume fraction is discretized by QUICK method.

2.5. Boundary conditions

The pressure boundary is chosen as the inlet and outlet condition. The inlet pressure is 18.7 MPa and outlet pressure is 2.9 MPa. Pipes and valve disc are set as wall boundary condition with no-slip. The inlet temperature is 415 ℃ and outlet temperature is 411.5 ℃ with variation of less than 1%. We consider the temperature in the valve is constant and ignore the effect of temperature variation. The rated travel of the pressure relief valve is 100 mm and normal opening is 40%~60%. In the numerical simulation, we compute the case of opening 40%, 50% and 60% to study its flow characteristic. The operating fluid property of the pressure relief valve is shown in Table 1.

Table 1. Operating fluid property of the pressure relief valve.

	Vapor	Liquid
Saturated vapor pressure (MPa)		2.0
Density (kg·m^{-3})	35.5	680.9
Dynamic viscosity (kg·m^{-1}·s^{-1})	2.08×10^{-5}	1.4×10^{-4}

3. Results and analysis

3.1. Flow field under the opening 40%

The pressure energy is converted into kinetic energy when the high pressure media flow through the throttle. Phase change would occur where the pressure is lower than saturated vapor pressure. The throttle area is regulated by changing the opening of the valve and then the flow rate is adjusted correspondingly to maintain the liquid level of the high-pressure separator to be stable.

Figure 3 shows the contours of the mixture pressure from different views and from which we can obtain a three-dimensional impression of the flow field in the pressure relief valve. The pressure at the inlet is 18.7 MPa and it drops significantly in the throttle section. The minimum pressure on the surface near the head of the disc and downstream is under the saturated vapor pressure where the phase change takes place first. There is a region of high pressure in front of the disc and maximum pressure reaches 6 MPa. Thus, there exists a pressure gradient toward the head of valve. The pressure on the head of valve disc is about 5 MPa which is much higher than the saturated vapor pressure, so the cavitation bubbles would collapse and cavitation erosion forms if they are transported to the head of valve disc.

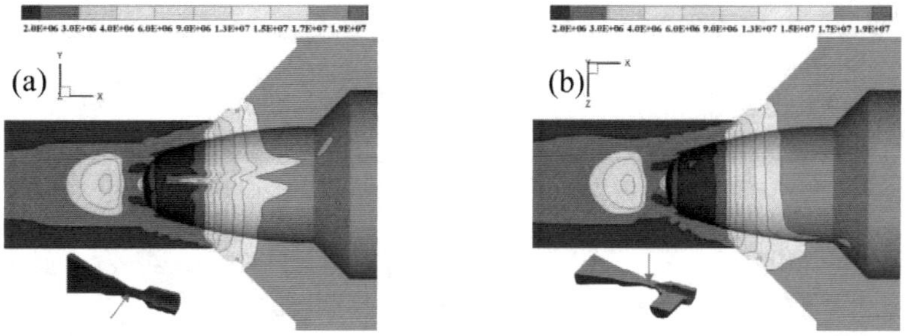

Figure 3. Contours of mixture pressure (Pa): (a) z=0 plane from front view; (b) y=0 plane from top view.

Figure 4 shows the contours and streamline of mixture velocity on z=0 plane from front view and y=0 plane from top view. The results show that there is an obvious increase of velocity when the media flows through the throttle and the maximum velocity reaches 220 m/s. There exist vortices in front of disc and the backflow velocity ups to 75 m/s. The pressure gradient mentioned before speeds up the backflow velocity. The high-velocity backflow would not only transport the bubbles to the head of disc leading to cavitation erosion, but also promote the solid particles to impact the surface of disc.

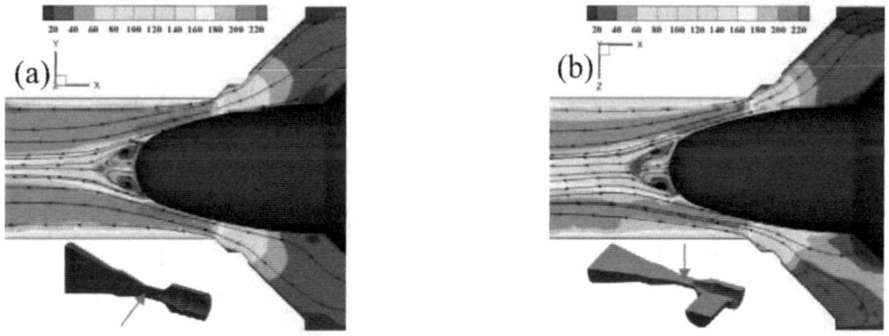

Figure 4. Contours of mixture velocity (m/s): (a) z=0 plane from front view; (b) y=0 plane from top view.

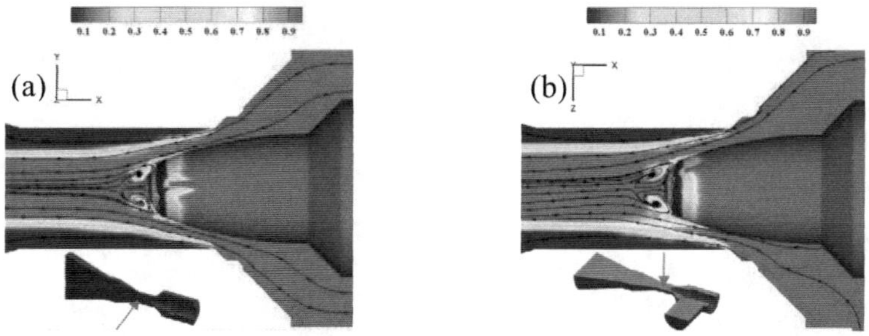

Figure 5. Contours of liquid volume fraction: (a) z=0 plane from front view; (b) y=0 plane from top view.

Figure 5 shows the liquid volume fraction contours and streamline of mixture on z=0 plane from front view and y=0 plane from top view. As the sum of liquid volume fraction and vapour volume fraction is 1, we can also get information of vapour volume fraction in Figure 5. There exist cavitation regions at the head of disc and downstream of the valve. The throttle area decreases due to the cavitation which leads to the degradation of valve flow performance. The cavitation bubbles drown by the high-velocity backflow would impact the head of disc. As the pressure on the head of valve disc is much higher than the saturated vapor pressure, cavitation bubbles transported to the disc would severely collapse and cavitation erosion is formed.

3.2. Comparison of different opening

The velocity in front of the disc is nearly axisymmetric, so we use the x-velocity component on the centreline to analysis the backflow region and the variation of backflow velocity. Figure 6 shows the x-velocity component of mixture on the centreline for opening 40%, 50% and 60%. In order to clearly compare the backflow in front of disc, we move the coordinate system along x-axis to the vertex of disc during the post processing of the data. The sign of velocity means the flow direction and positive represents that the flow is along the x-axis. The starting point of backflow under opening 40%, 50% and 60% are about 14 mm, 8 mm and 3 mm from disc vertex respectively. The maximum backflow velocities are 75 m/s, 37 m/s and 21 m/s respectively. It is found that not only the backflow region, but also the magnitude of the velocity decreases with the opening increase. The velocity downstream becomes larger, because the increasing of opening leads to a larger flow rate.

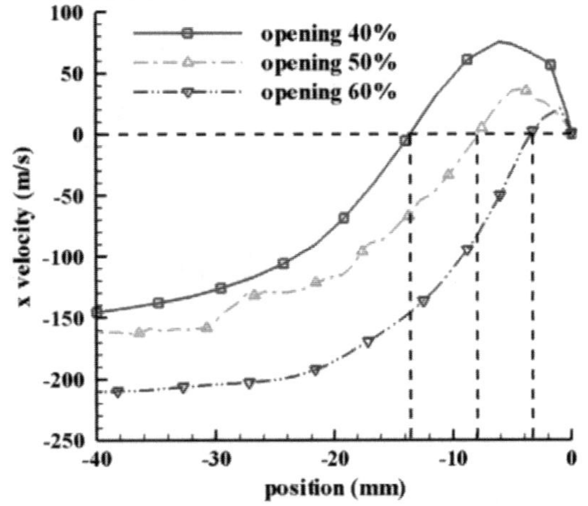

Figure 6. X-velocity component of mixture on the centreline

Figure 7 shows the liquid volume fraction on the head of valve disc from left view. The vapor forms an annular on the head of valve disc. The volume fraction of vapor increases firstly and then decreases from outside to inside. It results from that cavitation initiates on the outside and then bubbles collapse on the inside. It is seen that cavitation on the disc is gradually reduced with the opening increase.

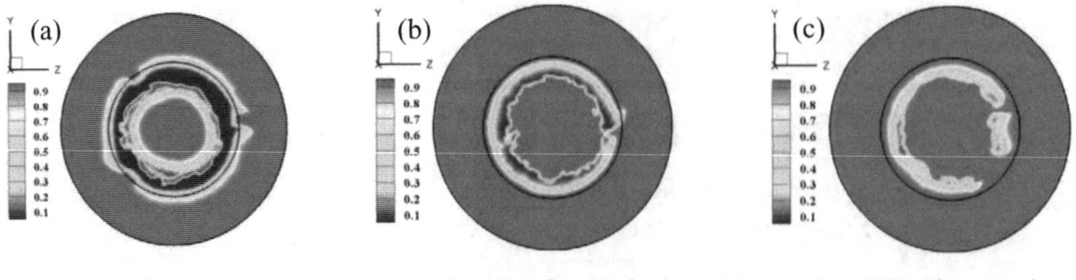

Figure 7. Liquid volume fraction on valve disc from left view: (a) opening 40%; (b) opening 50%; (c) opening 60%.

3.3. Damaged feature of the disc

Figure 8 shows the pictures of the damaged disc of high pressure relief valve from different views. Compared with the initial structure shown in Figure 2, the valve disc has a serious mass lose in the forepart and forms a concave shape surface where there are small pits on the concaved surface. The concave shape surface would not appear in the forepart without the high-velocity backflow. The high-velocity back flow that transports bubbles and particles to the head of disc is an important factor leading to cavitation erosion and particle erosion on the valve disc. The computation results have captured the cavitation erosion to the valve disc. According to these results, during the actual operation, the pressure relief valve should be kept to a relatively large opening to reduce the backflow and cavitation.

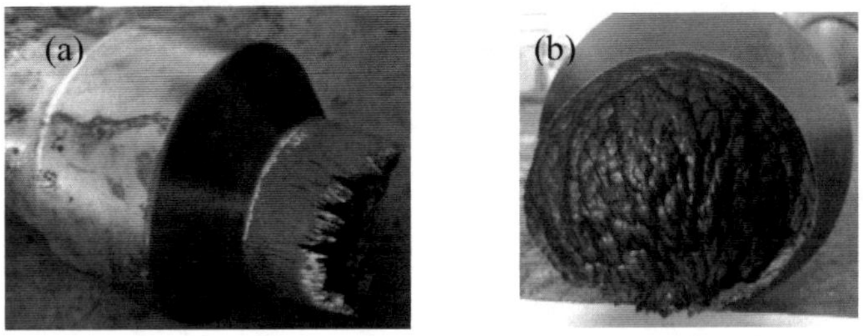

Figure 8. The pictures of damaged disc.

4. Conclusion

In this paper, we numerically investigate the three-dimensional turbulent cavitating flow in the pressure relief valve used in Direct Coal Liquefaction Demonstration Project and reveal the mechanism of cavitation erosion at valve disc. It is found that there exist cavitation regions at the head of disc and downstream of the valve disc. The high-velocity backflow in front of the disc drives the cavitation bubbles impacting the head of disc. For the local pressure higher than the saturated vapor pressure, the bubbles collapse at the head of disc and cavitation erosion is formed at the head of the disc. The computation results and analyses are accordance with the damaged feature of the disc found in actual operation. By comparing the cases of opening of 40%, 50%, and 60%, backflow velocity and cavitation region in front of the disc decrease with the increase of opening. Therefore, the pressure relief valve should be kept to a relatively large opening during the actual operation.

Acknowledgements

This work was supported by National Natural Science Foundation of China and ShenHua Group Corporation Limited United Program （U1261124), Specialized Research Fund for the Doctoral Program of Higher Education (20133318120004) and Graduate Student Innovation Research Project of Zhejiang Sci-Tech University (YCX13030).

References

[1] Franc J P and Michel J M 2004 *Fundamentals of Cavitation* (Berlin: Springer)
[2] Osterman A, Dular M and Sirok B 2009 *J. of Fluid Sci. and Techno.* **4**(1) 210-221
[3] Chen H S 2010 *Wear* **269** 602-606
[4] Pecha R and Compf B 2000 *Phys. Rev. Lett.* **84**(6) 1328-30
[5] Ding Z and Gracewski SM 1996 *J. Fluid Mech.* **309** 183-209
[6] Brujan E A, Ikeda T, Yoshinaka K, and Matsumoto Y 2011 *Ultrasonics Sonochemistry* **18**(1) 59-61
[7] Shima A 1997 *Shock Waves* **7** 33-42
[8] Shu G P, Shi S D and Li K J 2003 *The coal liquefaction technology* (Beijing: China coal industry publishing house)
[9] Zhang Y Z 2011 *J. of the China Coal Society* **36**(2) 179-184

[10] Ou G F, Rao J, Zhang L T, Zheng Z J and Ye J 2013 *Tribology* **33**(2) 155-161
[11] Schnerr G H and Sauer J 2001 Physical and numerical modeling of unsteady cavitation dynamics In 4th *Int. Conf. on Multiphase Flow(New Orleans, Louisiana, USA, 27 May- 1 June, 2001)*

IOP Conf. Series: Materials Science and Engineering **72** (2015) 042040 doi:10.1088/1757-899X/72/4/042040

Three-dimensional simulation of bubble dynamics in a narrow pipe using lattice Boltzmann method

D Y Shi[1], Z K Wang[1] and A M Zhang[2]
[1] College of Mechanical and Electrical Engineering, Harbin Engineering University, Harbin 150001, China
[2] College of Shipbuilding Engineering, Harbin Engineering University, Harbin 150001, China

E-mail: shidongyan@hrbeu.edu.cn;

Abstract. In the paper, a three-dimensional model of a gravity-driven bubble rising in a narrow pipe filled with viscous liquid is built using the lattice Boltzmann method. On the Cartesian grid, the free-energy multiphase lattice Boltzmann model and the no-slip bounce-back scheme are combined together to implement the bubble interface and the solid boundary treatment, respectively. To start with, the Laplace law for bubble interface is verified with the newly built model in this paper. Then the cases where the pipe with the radius 1.2 to 2.5 times the bubble radius are carried out to investigate the effects of pipe dimension on the bubble motion, including rising velocity, deformation and jet formation. Moreover, the asymmetric characteristics of bubble biases the centre axle are explored further. The results show that the boundary condition effect consisting of the pipe dimension and the offset of bubble biasing the centre axle is of great significance to the bubble dynamics in a narrow pipe. The former factor mostly affect the velocity characteristics of the bubble, while the latter one mostly focuses on the bubble deformation and trajectory.

1. Introduction

The motion of bubble in a pipe is a normal phenomenon that widely exists in engineering fields, such as hydraulic system, pipeline valve and heat collector, etc. Due to the difficulty in dealing with the multiphase interface and the complex fluid properties, quite a lot of researchers have investigated the process for a long time by using many numerical methods [1-3]. Still, many unknown phenomenon need people to uncover.

In researching bubble dynamics, quite a lot of work has been carried out in terms of experiments, theory, and numerical simulation. A common view is that solid walls have a significant effect on bubble dynamics. Zhang A M and Ni B Y et al. studied the bubble dynamics in the vortex field induced by solid boundaries [4], Popinet S et al. researched the effects of viscosity on bubble collapse near a solid wall [5], Zhang A M studied the characteristics of jet and impact combing the BEM and SPH [6]. Therefore, the wall effect has become an important branch in the research area. For the bubble dynamics in a narrow pipe, the wall effect is more obvious.

Generally, for a traditional method, the question can be described by solving a macroscopic control equation, such as the Navier-Stokes equation. However, the lattice Boltzmann method (LBM) based on the molecular theory can recover the macroscopic equation by adding up the particles distribution on the mesoscopic level. Due to its natural microcosmic property, it can easily deal with complex solid

IOP Conf. Series: Materials Science and Engineering **72** (2015) 042040 doi:10.1088/1757-899X/72/4/042040

boundary conditions and the microphysical quantity acting on the interface. Till now, a series of models treating gas-liquid interface, such as colour model, phase field model, and the free-energy model exist.

Combining the free-energy multiphase lattice Boltzmann model with the no-slip solid boundary condition, a buoyancy-driven bubble rising in narrow pipe is simulated. Both the relative radius size and the offset of bubble biasing the centre axle effecting the bubble motion and topological deformation are researched.

2. Numerical method and implementation

2.1. Lattice Boltzmann equations

Based on the free-energy model, Zheng H C et al. [7] introduced another distribution function g_i to describe the multiphase flow interface dynamics which can be written as

$$g_i(\mathbf{x}+\Delta\mathbf{x},t+\Delta t)-g_i(\mathbf{x},t)=-\frac{1}{\tau_{\mathrm{g}}}(g_i(\mathbf{x},t)-g_i^{\mathrm{eq}}(\mathbf{x},t))+Q\,, \tag{1}$$

where i is the discrete velocity direction. In the D3Q19 model, $i=0\sim18$, and in the D3Q7 model, $i=0\sim6$. For the distribution function g_i, the D3Q7 model is used. τ_{g} is the relaxation parameter, and $\tau_{\mathrm{g}}=0.6$ in this paper. $g_i^{\mathrm{eq}}(\mathbf{x},t)$ is the equilibrium distribution function satisfying the Max-well distribution [7,8]. The last term on the right hand of equation (1) is a modified term related to the distribution condition of particles at the two adjoining node positions. Combining the step by step operation [9] and introducing the controls on transfer efficiency between lattice nodes [10], equation (1) can be divided into two steps. The first step is

$$g_i'(\mathbf{x},t)=g_i(\mathbf{x},t)+\frac{1}{\tau_{\mathrm{g}}}(g_i^{\mathrm{eq}}(\mathbf{x},t)-g_i(\mathbf{x},t))\,, \tag{2}$$

and the second step is

$$g_i(\mathbf{x}+\Delta\mathbf{x},t+\Delta t)=g_i'(\mathbf{x}+\Delta\mathbf{x},t)+q(g_i'(\mathbf{x},t)-g_i'(\mathbf{x}+\Delta\mathbf{x},t))\,. \tag{3}$$

The first step only occurs on the local node position, so it can be easily implemented parallel process in three dimensional calculations. The second step is the streaming operation where q is the controlled parameter, $q=\dfrac{2}{1+2\tau_{\mathrm{g}}}$. Using the Chapman-Enskog expansion, equation (1) can recover the Cahn-Hilliard equation which is widely used to trace interface movement with the second-order accuracy. The Cahn-Hilliard equation can be written as

$$\partial\phi/\partial t+\nabla\cdot(\phi\boldsymbol{u})=\varpi\nabla^2\theta\,. \tag{4}$$

where ϕ is the order parameter, $\phi=\Sigma g_i$. ϖ is the mobility and θ is the chemical potential parameter. The evolution of the original distribution function which can recover the Navier-Stokes equations by Chapman-Enskog expansion is

$$f_i(\mathbf{x}+\Delta\mathbf{x},t+\Delta t)-f_i(\mathbf{x},t)=-\frac{1}{\tau_{\mathrm{f}}}(f_i(\mathbf{x},t)-f_i^{\mathrm{eq}}(\mathbf{x},t))+F_i\,. \tag{5}$$

where τ_{f} is the relaxation parameter and $f_i^{\mathrm{eq}}(\mathbf{x},t)$ is the equilibrium distribution function [11] corresponding to the distribution function f_i. In this paper, $\tau_{\mathrm{f}}=0.75$. F_i is discrete body force based on the lattice structure which plays an important role in the numerical simulation.

2.2. Body force implementation

Different from the traditional discrete operation of a body force, the discrete lattice effects must be considered into the lattice Boltzmann model especially the multiphase flow model. In 2002, Guo Z L et al. [12] examined several popular existing force treatments, and proposed an effective representation of the forcing term in which the discrete lattice effects are introduced. The representation can be written as

$$F_i = 3w_i(1-\frac{1}{2\tau_f})[(\mathbf{e}_i-\mathbf{u})+3(\mathbf{e}_i\cdot\mathbf{u})\cdot\mathbf{e}_i]\cdot F\,. \tag{6}$$

where w_i is the weighting factor, for the D3Q19 model, $w_0=1/3$, $w_{1\sim6}=1/18$, $w_{7\sim18}=1/36$. e_i is the discrete velocity whose value is determined by the lattice structure. For a single-phase flow simulation, the body force F is traditionally composed of the gravity, however the interface force must be considered in a multiphase flow. F can be written as $F=\theta\nabla\phi+G$, where G is the gravity acceleration. θ and ϕ are same with those defined in the previous article.

2.3. Solid boundary implementation

Another important issue that needs much more attention is the solid boundary implement. In the model, a no-slip bounce-back scheme is applied to deal with the curved walls. For a fluid node near to the solid wall, the distribution function coming from the solid node after the stream process can be calculated as [13]

$$f_{\tilde{i}}(\mathbf{x}_f,t+\Delta t)=f_i^{eq}(\mathbf{x}_f,t)-6w_i\rho(\mathbf{e}_i\cdot\mathbf{u}')\,. \tag{7}$$

$\tilde{i}$ is the negative direction with the discrete direction i. $\mathbf{x}_f$ stands for the fluid node position adjacent to the solid wall. $\mathbf{u}'$ is the intermediate velocity value on the virtual boundary composed of the midpoints between the solid node adjacent to the solid wall with $\mathbf{x}_f$. $\mathbf{u}'$ can be calculated through interpolation and extrapolation [13].

3. Results and discussion

3.1. Verification

In the simulation, twenty thousand iterations are used at the beginning to eliminate the spurious velocity caused by the initial conditions. Related parameters are defined as $\tau_f=0.6$, $\tau_g=0.75$, $\sigma=2.0$. The convergence is detected according to the Laplace Law, $\Delta P=2\sigma/R$. The results of bubbles with different scales are shown in figure 1.

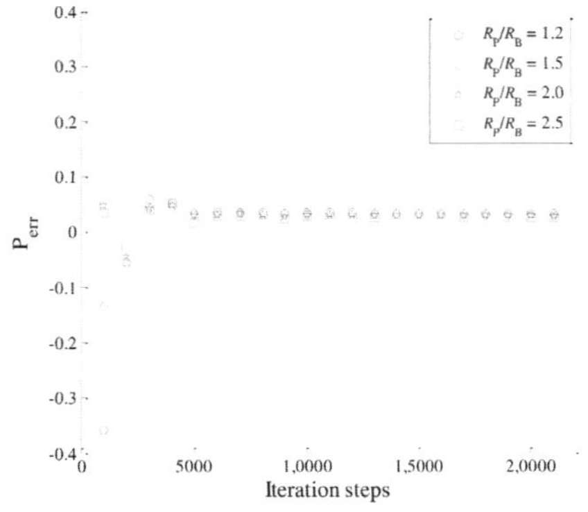

Figure 1. Duration curve of pressure difference inside and outside bubble.

Figure 1 shows the pressure difference P_{err} which is defined as

$$P_{err} = (2\sigma / R_B - \Delta P) \cdot R_B / \sigma, \quad (8)$$

with $R_P / R_B = 1.2 \sim 2.5$. R_P stands for the inner radius of the pipe, and R_B stands for the bubble radius. The curves show that the results converge well and match well with the theoretical value.

3.2. Effects of size ratio

Keeping the pipe radius stable, several cases where the radius of bubble is different are carried out. The detailed information is show in Table 1.

Table 1. Cases of simulating bubble rising in a pipe with different size ratio

Pipe Radius(R_P)	Relative Size(R_P / R_B)	Bubble Radius(R_B)
50	1.2	41.67
	1.5	33.33
	2.0	25
	2.5	20

Figures 2-3 show the rising process for bubbles with radius different times the pipe radius, $R_P / R_B = 1.2$ in figure 2, and $R_P / R_B = 2.5$ in figure 3. In each figure, time increases from left to right, and bubble terminal shape gradually formed. The bubbles tend to have a quite different deformation behaviour as the size ratio changes. This result indicates that the solid wall plays an important role in the evolution of the bubble shape as it rises under gravity. For the size ratio $R_P / R_B = 1.2$, the bubble experiences a large amount of deformation. By the joint action of fluid viscosity and solid wall effect, the interface adjacent to the wall moves slower than the central part, and a bubble with long strip shape is formed in the end. The phenomenon is consistent with the summary provided by Lee T [14] that if the size ratio is smaller than 1.25, the bubble deformation will not be as consistent. For the case, $R_P / R_B = 2.5$, the bubble rises as an isolated bubble in a free field. The results show that when the size ratio is larger than 1.5, bubbles can turn to be an approximately oblate spheroid as the wall effect decreases with large size ratio.

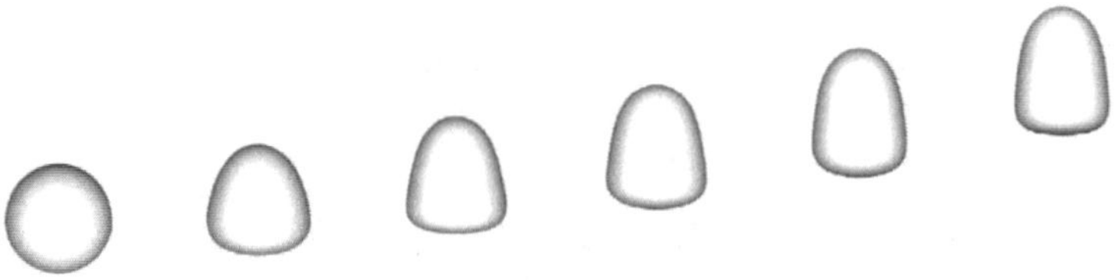

Figure 2. Bubble rising process with the size ratio $R_P / R_B = 1.2$.

Figure 3. Bubble rising process with the size ratio $R_P / R_B = 2.5$.

Figure 4 shows the vertical position of the bubble with different size ratio, the vertical position H is defined as the distance between the top node position of the bubble and the pipe bottom divided by R_P. The initial position of the four lines is determined by the bubble position and radius size. It can be found that the bubble rises faster as the size ratio increases which means that the wall can restraint bubble moving. Whatever, the bubble will come to a stable rising velocity just as it will reach a terminal shape. The results also show that the jet deformation becomes inapparent, as the size ratio decreases, and a negative jet velocity appears when $R_P / R_B = 1.2$.

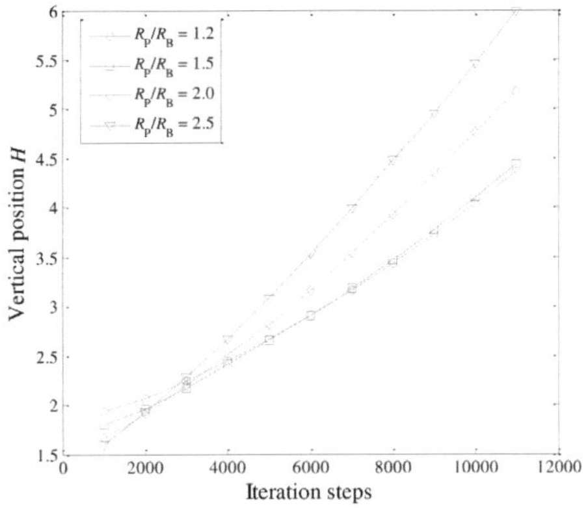

Figure 4. Vertical position of bubble coaxial with the pipe centre axle defined by the distance between its top node and the pipe bottom divided by R_P.

3.3. *Effects of offsets*

In the earlier paragraphs, bubble dynamics in a narrow pipe where the bubble is coaxial with the pipe centre axle is studied. Next, effect of bubbles biasing the centre axle on bubble motion is researched. Keeping the pipe and bubble radius stable, offset values range from 10 to 25 as shown in Table 2.

Table 2. Cases of simulating bubble rising in a pipe with different offsets biasing the centre axle

Pipe Radius	Bubble Radius	Relative Size	Offset value(ΔS)
50	20	2.5	10
			15
			20
			25

Figure 5. Bubble rising process with the offset $\Delta S = 25$.

Figure 5 shows the bubble rising process in a pipe where the bubble biases the pipe centre axle with the offset $\Delta S = 25$ .Due to the phenomenon becomes apparent, as the offset increases, the case with the offset $\Delta S = 25$ is chosen to be shown in figure 5. Three main phenomena can be found. Firstly, under the condition of solid wall, a jet towards the wall is induced. Further, one can see the researches by Zhang A M [6] and Popinet S [5] to know solid walls affecting bubble jet motion. Secondly, the wall will attract the bubble at the initial stage, and then repel it. The bubble will produce an oscillate-rising trajectory. The phenomenon is consistent the results presented by Samaroo R [15]. The third one is the bubble will approach to an equilibrium position coaxial with the pipe.

Figures 6-7 indicate the position information of the bubble. Figure 6 show the vertical position of bubble defined as the distance between its top node and the pipe bottom divided by R_P. Figure 7 show the horizontal position of bubble defined as the distance between its rightmost node and the pipe left side divided by R_P. The figures show that the offset has worse effects on the horizontal position of bubble than on the vertical position. Comparing with figure 4, the results also indicate that size ratio mostly effects on the velocity characteristics of the bubble, while offset focuses on the bubble deformation and trajectory when the size ratio is not too small. It can also be seen in figure 7 that bubbles with different offsets biasing the pipe centre axle will finally come close to the equilibrium position on the pipe centre axle.

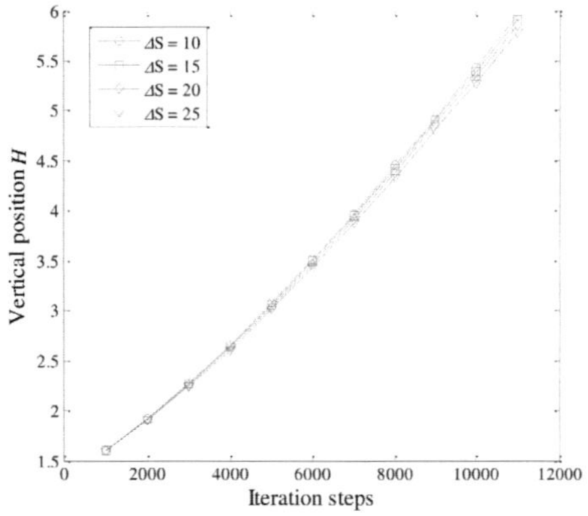

Figure 6. Vertical position of bubble biasing the pipe centre axle defined as the distance between its top node and the pipe bottom divided by R_P.

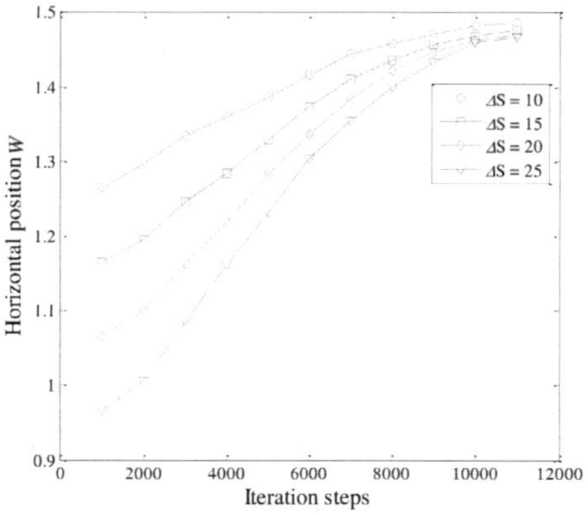

Figure 7. Horizontal position of bubble biasing the pipe centre axle defined as the distance between its right node and the pipe left side divided by R_P.

4. Conclusion

By using the lattice Boltzmann method, a three dimensional numerical model of bubbles moving in a narrow pipe is established. Effects of size ratio (R_P / R_B) and offsets that bubbles bias the pipe centre axle (ΔS) on bubble dynamics are researched. The results can be summarized as:

1) Relative to size ratio which is not too small, offsets can induce a sensible horizontal movement and asymmetrical jet towards the solid wall which mainly influence the bubble deformation and trajectory.

2) When the size ratio is small, for example, smaller than 1.5, the wall effect becomes significant for coaxial cases. Bubble rising velocity goes down, and the interface adjacent to the wall moves slower than the central part, and a bubble with long strip shape is formed in the end.

3) Bubbles with a offset biasing the pipe axle will produce an oscillate-rising trajectory, and approach to an equilibrium position coaxial with the pipe.

Reference

[1] Chiu P H and Lin Y T 2011 *J. Comput. Phys.* **230** 185-204
[2] Yu Z, Yang H and Fan L S 2011 *Chem. Eng. Sci.* **66** 3441-51
[3] Simiano M, Lakehal D, Lance M and Yadigaroglu G 2009 *J. Fluid Mech.* **633** 191-231
[4] Zhang A M, Ni B Y, Song B Y and Yao X L 2010 *Appl. Math. Mech-Engl.* **31** 449-60
[5] Popinet S and Zaleski S 2002 *J. Fluid Mech.* **464** 137-63
[6] Zhang A M, Yang WS, Huang C and Ming F R 2013 *Comput. Fluids* **71** 169-78
[7] Zheng H W, Shu C and Chew Y T 2006 *J. Comput. Phys.* **218** 353-71
[8] Guo Z, Shi B and Wang N 2000 *J. Comput. Phys.* **165** 288-306
[9] Lee T and Lin C L 2005 *J. Comput. Phys.* **206** 16-47

[10] Shi D Y, Wang Z K and Zhang A M 2014 *Chinese J. of Theoretical and Applied Mechanics* **46** 224-33
[11] Huang H, Zheng H, Lu X and Shu C 2010 *Int. J. Numer. Meth. Fl.* **63** 1193-207
[12] Guo Z L, Zheng C and Shi B 2002 *Phys. Rev. E* **65** 046308
[13] Ladd A J C 1994 *J. Fluid Mech.* **271** 285-309
[14] Amaya-Bower L and Lee T 2011 *Chem. Eng. Sci.* **66** 935-52
[15] Samaroo R, Kaur N, Itoh K, Lee T, Banerjee S and Kawaji M 2014 *Nucl. Eng. Des.* **268** 203-14

Numerical investigation of performance of vane-type propellant management device by VOF methods

J T Liu[1], C Zhou[2], Y L Wu[2], B T Zhuang[1] and Y Li[1]
[1] Beijing Institute of Control Engineering, China Aerospace and Technology Corporation, Beijing 100190, China
[2] State Key Laboratory of Hydroscience and Engineering, Tsinghua University, Beijing 100084, China

E-mail: liujintao86@hotmail.com

Abstract. The orbital propellant management performance of the vane-type tank is so important for the propellant system and it determines the lifetime of the satellite. The propellant in the tank can be extruded by helium gas. To study the two phase distribution in the vane-type surface tension tank and the capability of the vane-type propellant management device (PMD), a large volume vane-type surface tension tank is analysed using 3-D unsteady numerical simulations. VOF methods are used to analyse the location of the interface of the two phase. Performances of the propellant acquisition vanes and propellant refillable reservoir in the tank are investigated. The flow conductivity of the propellant acquisition vanes and the liquid storage capacity of propellant refillable reservoir can be affected by the value of the gravity and the volume of the propellant in the tank. To avoid the large resistance causing by surface tension in an outflow of a small hole, the design of the vanes in a propellant refillable reservoir should have suitable space.

1. Introduction

Propellant management device (PMD) has been used as a kind of surface tension devices in liquid propellant tanks for over forty years. Surface tension forces are negligible in most engineering problems. However, surface tension forces are significant and often dictate the location and the orientation of liquid within tanks in the low gravity environment of orbit vehicles.

A PMD component is defined as a structure which creates as internal, or closed, flow path along which propellant can flow. Using PMD can provide a lightweight, straightforward, easily manufactured, and highly reliable propellant tank for spacecraft.

Jaekle analyzed the capability of the PMD by analyzing the influence of vanes[1], sponges[2], galleries[3], traps and troughs[4] to the propellant distribution in the tank with micro gravity. Tam[5,6] proposed a new PMD, which was capable of transferring both gas-free propellant and liquid-free pressurant upon demand. The PMD performance analysis utilized the same design methodology and conservative approaches as all previous PMD design efforts.

The study of propellant distribution in the tank with micro gravity was studied by Chen[8] using experimental test. It is found that increases in gap size, fluid viscosity, contact angle, and vane edge roundness decrease capillary advance rate while increases of the vane thickness and the vane obliquity angle enhance the capillary advance. Estes[9] analyzed the development and implementation of a

IOP Conf. Series: Materials Science and Engineering **72** (2015) 042041 doi:10.1088/1757-899X/72/4/042041

process for producing a highly wettable Aluminum PMD for the GPM hydrazine tank. Lenahen[10] investigated the natural frequencies and damping effects of Diaphragm-Implemented spacecraft propellant tanks using computational methods.

In order to study the flow distribution in the propellant tank with micro gravity, two kinds of conditions of a tank with PMD was chosen for the analysis of the performance of the PMD.

2. Propellant Tank Geometry

The study was conducted with a model propellant tank. The structure of propellant tank is shown in Figure 1. The propellant tank has four vanes which are close to the wall of the tank. The PMD consists of four long blades and 8 short blades.

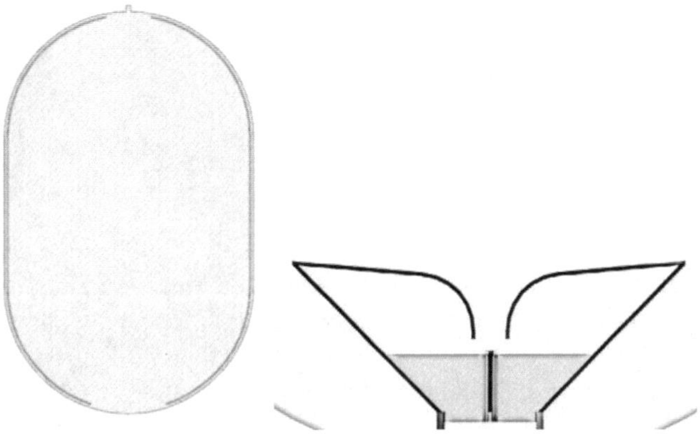

Figure 1. Propellant tank and PMD

3. Computational Method

The study was conducted with a model propellant tank. The structure of propellant tank is shown in Figure 1. The propellant tank has four vanes which are close to the wall of the tank. The PMD consists of four long blades and 8 short blades.

3.1. Volume of fluid model

The volume of fluid (VOF) method determines the shape and location of free surface based on the concept of a fractional volume of fluid.

3.1.1. Continuity Equation. The continuity equation for the mixture is

$$\frac{\partial}{\partial t}(\rho) + \frac{\partial}{\partial x_i}(\rho u_i) = R \tag{1}$$

R is the source term. ρ is the mixture density. u_i is the mass-averaged velocity.

The properties appearing in the transport equations are determined by the presence of the component phases in each control volume. In a two-phase system, for example, if the phases are represented by the subscripts 1 and 2, and if the volume fraction of the second of these is being tracked, the density in each cell is given by

$$\rho = \alpha_1 \rho_1 + \alpha_2 \rho_2 = (1 - \alpha_2)\rho_1 + \alpha_2 \rho_2 \tag{2}$$

α is the volume fraction of the phase.

The volume fraction equation will not be solved for the primary phase; the primary-phase volume fraction will be computed based on the following constraint:

$$\alpha_1 + \alpha_2 = 1 \tag{3}$$

3.1.2. Momentum equation. The momentum equation is

$$\frac{\partial}{\partial t}\rho u_j + \frac{\partial}{\partial x_i}\rho u_i u_j = -\frac{\partial p}{\partial x_j} + \mu \frac{\partial}{\partial x_i}\left(\frac{\partial u_i}{\partial x_j} + \frac{\partial u_j}{\partial x_i}\right) + \rho g_j \tag{4}$$

t is time. g is the acceleration of gravity. μ is the coefficient of viscosity.

For the effect of surface tension,

$$\frac{\partial p}{\partial x_j} = -\sigma \frac{\partial}{\partial x_j}\left(\frac{1}{R}\right) \tag{5}$$

σ is the coefficient of surface tension force, *R* is radius, so the momentum equation can be transformed into,

$$\frac{\partial}{\partial t}\rho u_j + \frac{\partial}{\partial x_i}\rho u_i u_j = \sigma \frac{\partial}{\partial x_j}\left(\frac{1}{R}\right) + \mu \frac{\partial}{\partial x_i}\left(\frac{\partial u_i}{\partial x_j} + \frac{\partial u_j}{\partial x_i}\right) + \rho g_j \tag{7}$$

3.2. Simulation conditions

The commercial CFD code FLUENT was used to perform the simulations. The SIMPLEC algorithm was used to enforce mass conservation. No slip boundary condition was using to solve the flow on the wall. Two phases in the tank were He and N_2H_4, and the coefficient of surface tension force was 0.03 N/m. For present unsteady flow calculation, the time step was 0.0001 s. For all calculations, simulations were run until convergence, which was determined by a reduction in the residual error to less than 0.0001.

Three dimensional calculations were performed to investigate the influence of flow distribution in the tank. The model's grids, which were composed of an unstructured hexahedron and tetrahedron, were developed using ICEM, which is a commercial software package used for CFD discretization. Hexahedral grids were used for each component of the propellant tank. The mesh of the tank is shown in Figure 2.

4. Result and Discussion

4.1. Propellant distributions with small vane in PMD

4.1.1. Propellant for 10% volume. Propellant distributions in the tank with zero gravity are shown in Figure 2. The red color is the propellant, the blue color is gas in the tank. For zero gravity condition, uniform distribution can be seen of the propellant in the tank due to the force of surface tension. The propellant in the tank flow along the vanes and the height of the propellant along the right vane reaches half of the height of the tank for 6.0s. Due to the scale of the PMD is small and the height of the vanes in the PMD is in accordance with the cylinder in the PMD, the surface tension force becomes a resistance at the top of the vane in the PMD. The propellant will not fill the PMD. Figure 3 shows propellant distributions in the tank with 0.001 gravity at -X coordinate. The flow in the PMD also has the same result as shown in Figure 2, but the propellant tends to flow to the X coordinate.

4.1.2. Propellant for 5% volume. Propellant distributions in the tank with zero gravity are shown in Figure 4. Propellant distributions in the tank with 0.001g gravity are shown in Figure 5. When the residual propellant is 5% volume of the tank, small gravity will not affect the distribution of the propellant in the tank. The propellant firstly flows into the PMD and then a little back to the tank. So it can be seen the addtion of the volume of propellant in the PMD at first and the volume gradually decreases when t>2.0 s.

It can be comcluded that the store of propellant by the vane is stronger than the PMD with lower height vanes. The surface tension force becomes resistance force at the top of the vanes in the PMD. The PMD will not be filled up by the propellant.

Figure 6 shows the flow rate between the tank and the PMD with 10% volume of prpellant in the tank and the volume of prpellant in the PMD is shown in Figure 7. The flow rate has a linear relationship with time when t<1.8s, and the flow rate has a maximum value at t=1.8s. The flow rate gradually decreases when t>1.8s and the slope of the flow rate line reduces. When t=7.0s, the flow rate decrease to 2.0ml/s. The propellant will not maintain the thrust and the PMD needs to be improved.

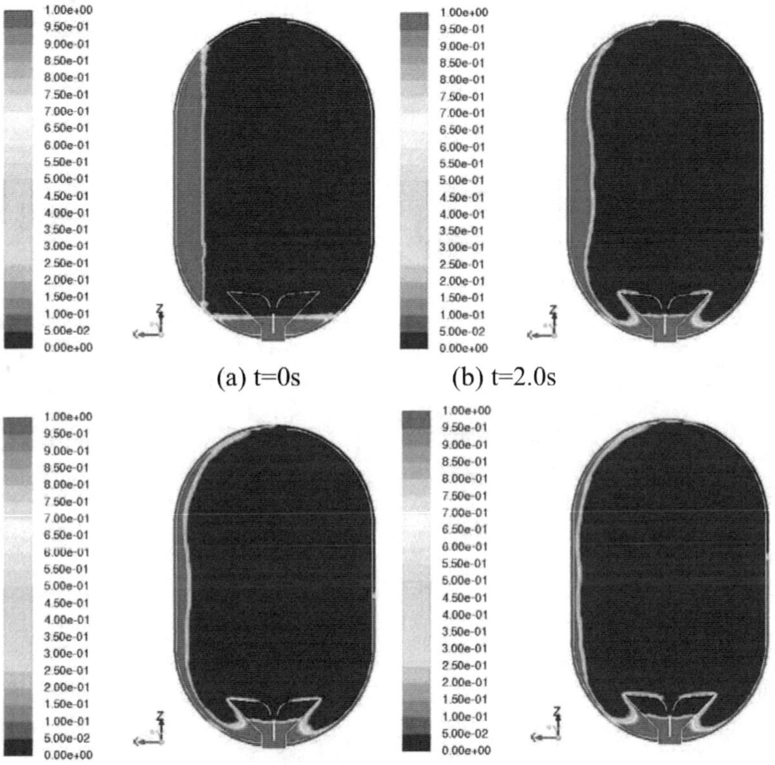

(a) t=0s (b) t=2.0s

(c) t=4.0s (d) t=6.0s

Figure 2. Zero gravity condition

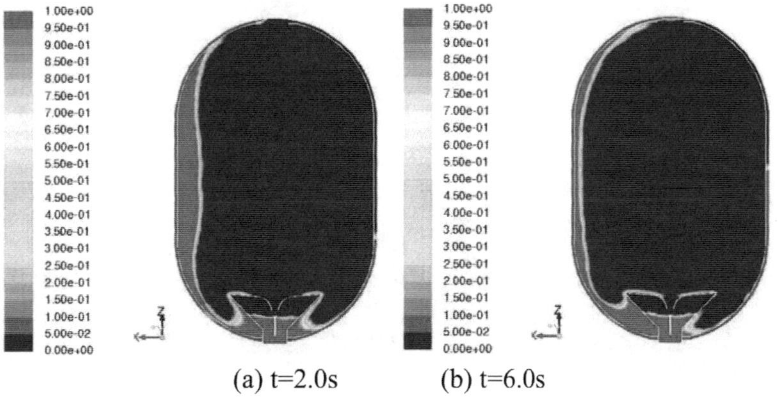

(a) t=2.0s (b) t=6.0s

Figure 3. 0.001 gravity in –X coordinate

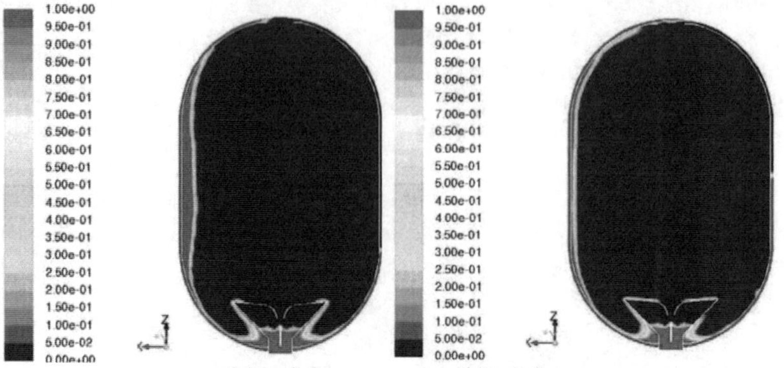

(a) t=2.0s (b)t=6.0s

Figure 4. Zero gravity condition

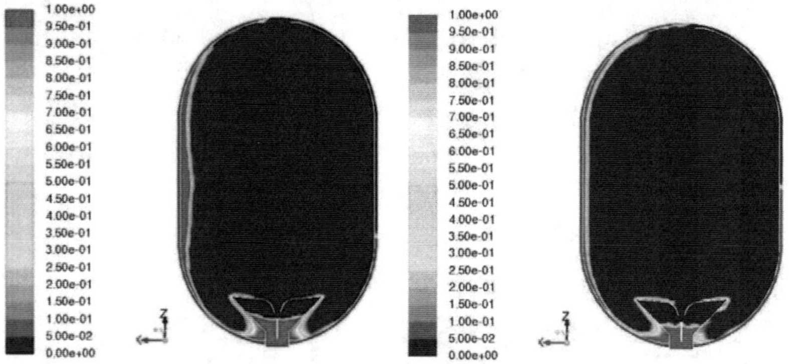

(a) t=2.0s (b) t=6.0s

Figure 5. 0.001 gravity in –X coordinate

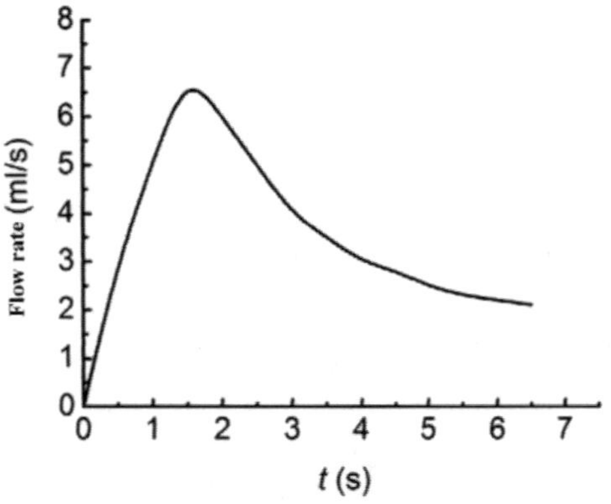

Figure 6. Flow rate between the tank and PMD

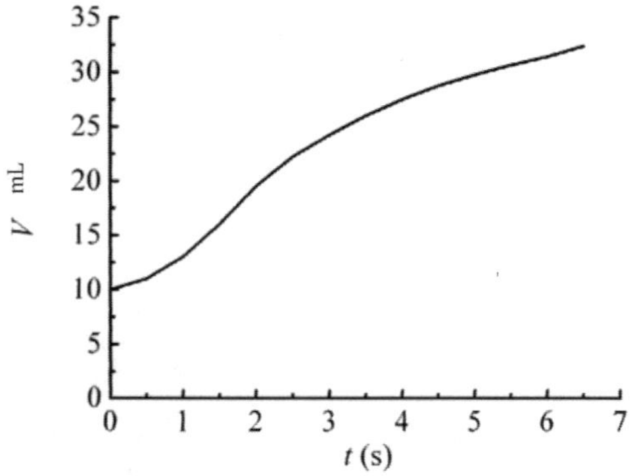

Figure 7. Volume of propellant in the PMD

4.2. Optimization of propellant management device

To improve the resistance force at the top of the vanes, the short vanes in the PMD are lengthened and the structure of the PMD are shown in Figure 8. The distence between the top of the vane and the up casing of PMD is *l*.

The result of the propellant distribution in the tank for the improved PMD is shown in Figure 9. The result is calculated with 0.001 gravity in –X coordinate. The propellant will reanch the top of the

PMD when t=1.5s. Because the distence between the top of vane in the PMD and the up casing is small, the propellant will flow to the up hole of the PMD, and the hole for exhausting gas in the PMD is plugged up. Lot of gas will be left in the PMD. It will affect the stable working of the thrust. Figure 10 shows the flow rate between the tank and the PMD. The lenghten of the vane in the PMD increases the flow rate.

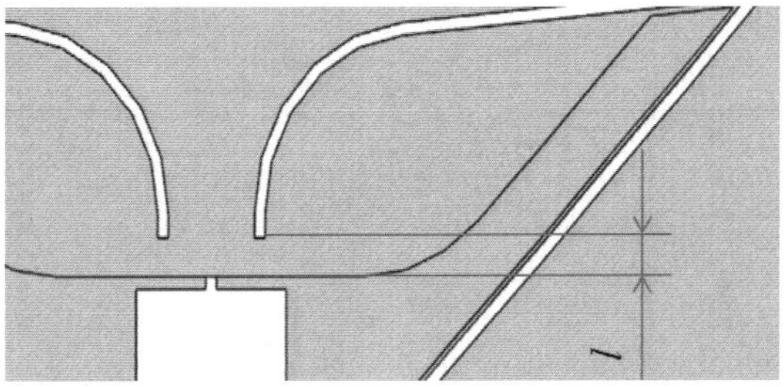

Figure 8. Structure of the improved PMD

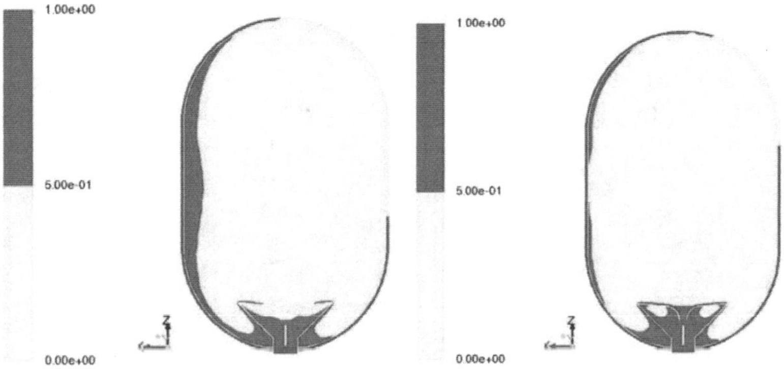

(a) 1.5s (b) 6.0s

Figure 9. Propellant distribution for the improved PMD

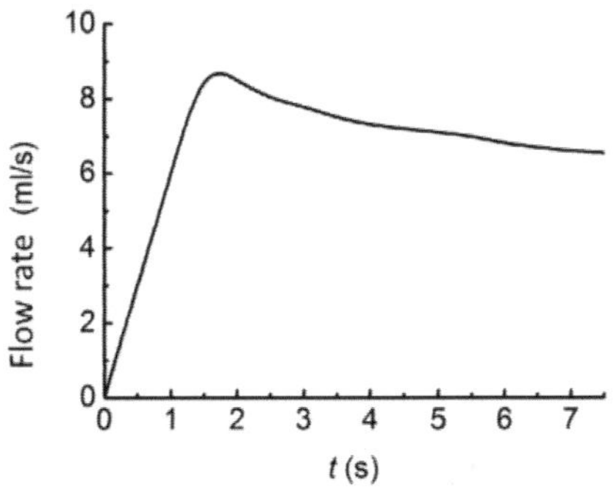

Figure 10. Flow rate between the tank and the PMD

In order to let the gas in the PMD flow into the tank through the up hole, the distance between the top of vane in the PMD and the up casing is increased and the improved PMD-2 is shown in Figure 11. The result of the propellant distribution for the improved PMD-2 is shown in Figure 12. The addition of the distance will improve the propellant distribution in the PMD and the hole in the PMD will not be plugged up. The flow rate between the tank and the PMD-2 is almost 9 ml/s. The performance of the improved PMD-2 can maintain the working of the thrust.

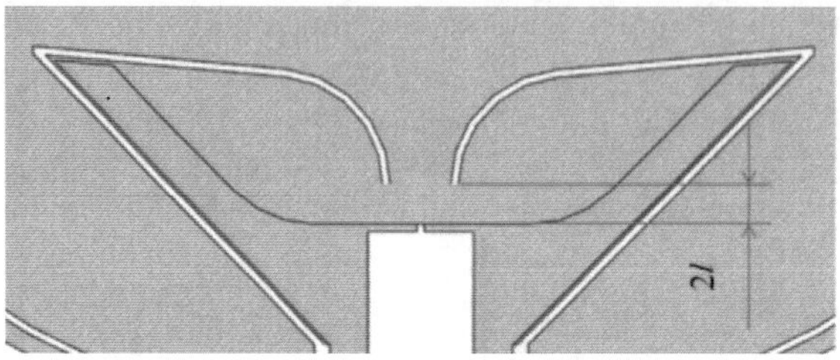

Figure 11. Structure of the improved PMD-2

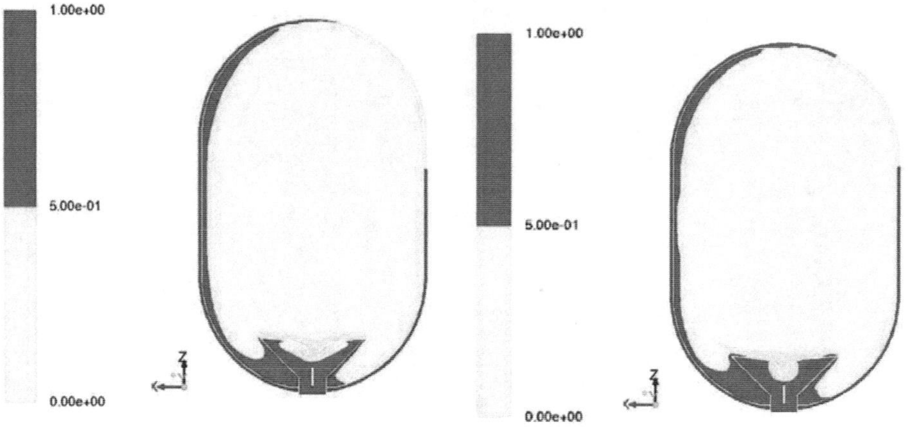

(a) 2.5s (b) 6.0s

Figure 12. Propellant distribution for the improved PMD-2

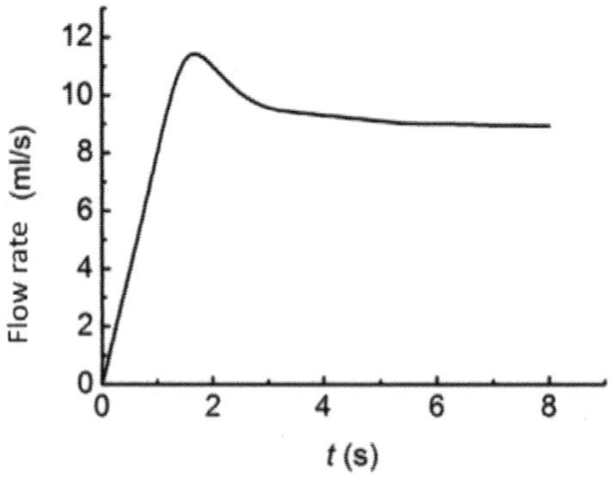

Figure 13. Flow rate between the tank and the PMD-2

5. Conclusion

This paper involves the propellant distribution and optimization of a PMD in a tank. The VOF model could be used to simulate the flow in a PMD with micro gravity. Flow distribution can be influenced by the change of the value of gravity when the gravity is large than 0.001g. For the scale of the PMD is small, the surface tension force may become a resistance at the top of the vane in the PMD. The avoidance of the resistance is to modify the shape of PMD and choose a suitable distance for the PMD structure. The PMD is one of the key components in the tank, it determines the flow rate between the PMD and the tank.

Acknowledgments
The authors would like to thank the National Natural Science Foundation of China (No. 51406010) and project supported by Open Research Fund Program of State key Laboratory of Hydroscience and Engineering, NO: sklhse-2014-E-02.

References

[1] Jaekle D E 1991 Propellant management device conceptual design and analysis: Vanes *AIAA/SAE/ASME/ASEE 27 Joint Propulsion Conf. (Sacramento, CA, 24-26 June 1991)*

[2] Jaekle D E 1993 Propellant management device conceptual design and analysis: Sponges *AIAA/SAE/ASME/ASEE 29 Joint Propulsion Conf. (Montery, CA, 28-30 June 1993)*

[3] Jaekle D E 1997 Propellant management device conceptual design and analysis: Galleries *AIAA/SAE/ASME/ASEE 33 Joint Propulsion Conf. (Montery, CA, June 28-30, 1997)*

[4] Jaekle D E 1995 Propellant management device conceptual design and analysis: Traps and troughs *AIAA/SAE/ASME/ASEE 31 Joint Propulsion Conf. (San Diego, CA, 10-12 June 1995)*

[5] Tam W D and Jaekle J 2005 Design and Manufacture of an Oxidizer Tank with Surface Tension PMD *AIAA* **2005** 3734

[6] Tam W H, Ballinger I, Jaekle Jr D E 2006 Conceptual Design of Space Efficient Tanks *42nd AIAA Joint Propulsion Conf. & Exhibit (Sacramento, CA, 9-12 July 2006)*

[7] Tam W and Ballinger I 2008 Surface Tension PMD Tank for On Orbit Fluid Transfer *44th AIAA/ASME/SAE/ASEE Joint Propulsion Conf. & Exhibit (Hartford, CT, 21-23 July 2008)*

[8] Chen Y K and Collicott S H 2004 Experimental Study on the Capillary Flow in the Vane-Wall Gap Geometry *42nd AIAA Aerospace Sciences Meeting and Exhibit (Reno, Nevada, 5-8 Jan 2004)*

[9] Estes R H 2012 Development and Implementation of a Process for Producing a Highly Wettable Aluminum PMD for the GPM Hydrazine Tank *48th AIAA/ASME/SAE/ASEE Joint Propulsion Conference & Exhibit (Atlanta, Georgia, 30 July-01 Aug 2012)*

[10] Lenahen B, Bernier A and Gangadharan S 2012 A Computational Investigation for Determining the Natural Frequencies and Damping Effects of Diaphragm Implemented Spacecraft Propellant Tanks *48th AIAA/ASME/SAE/ASEE Joint Propulsion Conference & Exhibit (Atlanta, Georgia, 30 July-01 Aug 2012)*

IOP Conf. Series: Materials Science and Engineering **72** (2015) 042042 doi:10.1088/1757-899X/72/4/042042

Numerical analysis and experiment research on fluid orbital performance of vane type propellant management device

Q Hu, Y Li, H L Pan, J T Liu and B T Zhuang
Beijing Institute of Control Engineering, Haidian District, Beijing 100084, China

E-mail: hqbuaa03@sohu.com

Abstract. Vane type propellant management device (PMD) is one of the key components of the vane-type surface tension tank (STT), and its fluid orbital performance directly determines the STT's success or failure. In present paper, numerical analysis and microgravity experiment study on fluid orbital performance of a vane type PMD were carried out. By using two-phase flow model of volume of fluid (VOF), fluid flow characteristics in the tank with the vane type PMD were numerically calculated, and the rules of fluid transfer and distribution were gotten. A abbreviate model test system of the vane type PMD is established and microgravity drop tower tests were performed, then fluid management and transmission rules of the vane type PMD were obtained under microgravity environment. The analysis and tests results show that the vane type PMD has good and initiative fluid orbital management ability and meets the demands of fluid orbital extrusion in the vane type STT. The results offer valuable guidance for the design and optimization of the new generation of vane type PMD, and also provide a new approach for fluid management and control in space environment.

1. Introduction

Vane type PMD which is the core part of the vane-type STT mainly comprises of vane type structure parts, such as propellant acquisition vanes (PAV) and propellant refillable reservoir (PRR) [1-2]. Basing on surface tension, the PMD is a new device which implements initiative and complete fluid management and control in space environment, and its operational principle is that by vane type parts it can achieve the separation between liquid and gas, fluid transmission, storage, orientation and discharge, which overcomes the shortcomings of net type PMD, such as complicated structure, being polluted easily, heavy weight, poor performance of volume expanding and incapable filling and discharge fluid repetitively. The PMD is currently the most advanced fluid management and control in space environment, and it can satisfy kinds of flux requirements and be applicable to different microgravity environment, especially large satellite platform with relatively more small microgravity level. So far, main geostationary orbit satellites internationally have applied the vane-type STTs, such as A2100, SB4000, and ETS-Ⅷ [3-5].

Fluid orbital management performance of PMD will directly determine the orbital STT's success or failure. However, because of restrictions on structure function of vane-type PMD, it is unable to carry out long-term verification test on fluid management performance of the PMD, so numerical analysis on fluid orbital management performance of the PMD becomes especially important and is required to complete first, which can offer valuable guidance for the design and optimization of the PMD. Then verification test and research on fluid orbital management performance of PMD and results of numerical analysis must be carried out by microgravity drop tower tests. Researchers from abroad have studied mechanism of fluid orbital performance of vane type PMD in the microgravity from the early 70's, and achieve a great deal of findings through tremendous manpower and resources. While others in our country began researching that in the twenty-first century, and gain some valuable achievements [6-11]. In present paper, aiming at a vane type PMD, rules of its fluid management, transmission and distribution are revealed by numerical analysis and microgravity drop tower tests, which provide valuable guidance and datum support for design and optimization of vane-type STT.

2. Numerical model and test system of vane type PMD

In the paper, the vane type PMD shown in figure 1 mainly comprises of a PRR and four PAV, which can implement initiative and complete propellant orbital management and control. Four PAV arranged in four directions +X, -X, -Y, and +Y are used for transferring propellant. The PRR wrapped up by the inside and outside cones is located at liquid outlet of STT's bottom, and a small PRR is arrange in the PRR, which is a radial wedge structure comprised of some blades. The PRR acts on storing and supplying propellant for STT.

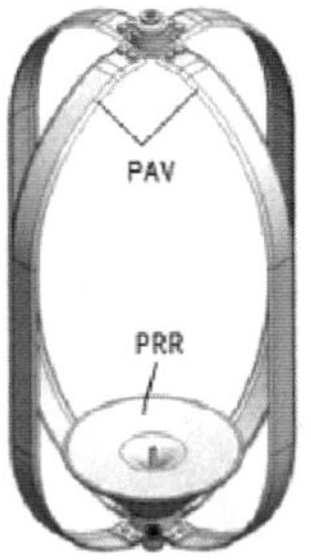

Figure 1. Sketch map of vane type PMD.

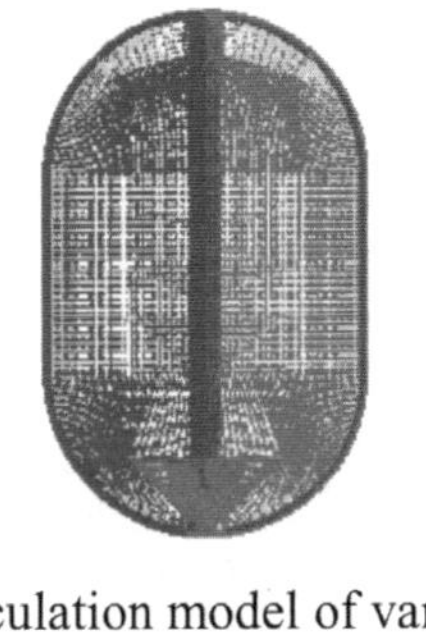

Figure 2. Calculation model of vane type PMD.

2.1. Numerical model

By using two-phase flow model of volume of fluid (VOF), rules of fluid distribution in vane-type STT under microgravity environment are gotten by numerically calculating.

VOF model is an algorithm which is to solve two-phase flow with gas and liquid and calculate flow of fluid free surface. By leading in fluid volume fraction function and its control equation, the algorithm can show the density of mixed fluid and make tracks for positions of fluid free surface. Due to laminar flow generally in vane-type STT, basic equations of VOF model comprise of the physical equation, the continuity equation and the momentum equation.

The physical property of fluid is determined by volume fraction of different phases in mixed fluid, and the physical equation expresses physical property of different volume fraction. Owing to only two-phase mixed flow in the STT, density properties equation of mixed fluid is given below.

$$\rho = \alpha_1 \rho_1 + \alpha_2 \rho_2, \qquad \alpha_1 + \alpha_2 = 1 \tag{1}$$

Where ρ is the density of mixed fluid, where α_1 and α_2 are volume fractions of the first phase and the second phase respectively, and where ρ_1 and ρ_2 are densities of the first phase and the second phase respectively, which are given values.

Basic expression of the fluid continuity equation is given below.

$$\frac{\partial \rho}{\partial t} + \frac{\partial}{\partial x}(\rho u) = R \tag{2}$$

Where u is velocity of mixed fluid, and where R is the source term.

According to equation (1) and equation (2), the transport equation of different phase volume fraction is gotten and given below.

$$\frac{\partial \rho}{\partial t}+\frac{\partial}{\partial x}(\alpha_q u)=R \tag{3}$$

Where α_q is volume fraction of the q-th phase, and where R is caused by net phase transition. Because the value of R is very small in this model, it is negligible while solving the equation (3).

The momentum equation of mixed fluid is given below.

$$\frac{\partial}{\partial t}(\rho u_j)+\frac{\partial}{\partial x_i}(\rho u_i u_j)=-\frac{\partial p}{\partial x_j}+\mu\frac{\partial}{\partial x_i}\left(\frac{\partial u_i}{\partial x_j}+\frac{\partial u_j}{\partial x_i}\right)+\rho g_j \tag{4}$$

Where p is inner pressure of the STT, where u_j and u_i are liquid phase velocity and gas phase velocity respectively, where x_j and x_i are liquid phase position and gas phase position respectively, where t is the time, where g_j is microgravity acceleration, and where μ is coefficient of viscosity. Due to the action of surface tension, the equation in the gas-liquid interface is given below.

$$\frac{\partial p}{\partial x_j}=-\sigma\frac{\partial}{\partial x_j}\left(\frac{1}{r}\right) \tag{5}$$

Where σ is coefficient of surface tension, and where r is surface curvature radius. According to equation (4) and equation (5), the final momentum equation of mixed fluid is given below.

$$\frac{\partial}{\partial t}(\rho u_j)+\frac{\partial}{\partial x_i}(\rho u_i u_j)=\sigma\frac{\partial}{\partial x_j}\left(\frac{1}{r}\right)+\mu\frac{\partial}{\partial x_i}\left(\frac{\partial u_i}{\partial x_j}+\frac{\partial u_j}{\partial x_i}\right)+\rho g_j \tag{6}$$

According to the structure of the PMD shown in figure 1, calculation model of performance analysis is established. The computing grids are generated by using the form of structured grids, and the grids near walls of structure are further thinner and denser. The grid number of the model is approximately one million, and figure 2 shows the numerical model of grids.

While solving the model, all the walls are set to boundary condition of solid walls, whose boundary condition is without penetration and slither. The flow near walls of structure is simulated by using logarithmic function.

The computational domain includes liquid and gas, and the distributions of liquid and gas in the domain are confirmed via setting volume fraction of each phase in advance. Now, the first phase is nitrogen, and the second phase is the propellant named ADN.

2.2. Test system and abbreviate model

Microgravity drop tower tests which are most usual in the means of microgravity tests can provide short microgravity time, but can supply relatively low microgravity level. In the paper, these microgravity tests have been carried out by the hundred meters drop tower of National Microgravity Laboratory (NML) [12]. The free fall experiment facility of the drop tower provides microgravity time which is about 3.5s, and the facility comprises of the falling module, the deceleration and recovery system, the release system, the control system, the measurement system, and auxiliary equipments and so on. The falling module is the core part of the facility, and is the important and special equipment used for loading test models in the drop tower test system. The module is divided into two groups: the internal double module and the internal single module. The microgravity level provided by the internal double module grades between 10^{-4} g_0 and 10^{-5} g_0, and The microgravity level provided by the internal single module grades between 10^{-2} g_0 and 10^{-3} g_0.

According to the requirement of microgravity drop tower experiment research, the internal single module is used for the tests in the paper, and the test system with abbreviate models is built, which is shown in figure 3 and figure 4 below. The test system consists of abbreviate models, the test bracket, the lighting device and the picture acquisition device and so on, and can be used for microgravity test and verify flow behaviors of vane type PMD. In the system, the high resolution camera with CCD is using for recording fluid climbing process, distribution and reorientation process in abbreviate models.

Figure 5 below shows the abbreviate model of the vane-type STT used in the microgravity tests, which consists of a vane type PMD and shells. The PMD uses titanium alloy material, and these shells use plexiglass material whose light transmittance is approximately 90%, so it is easy to observe test results. Anhydrous ethanol is used for the test medium of propellant, its static contacting angles with titanium alloy and plexiglass are approximately 0°.

Figure 3. Sketch map of test system.

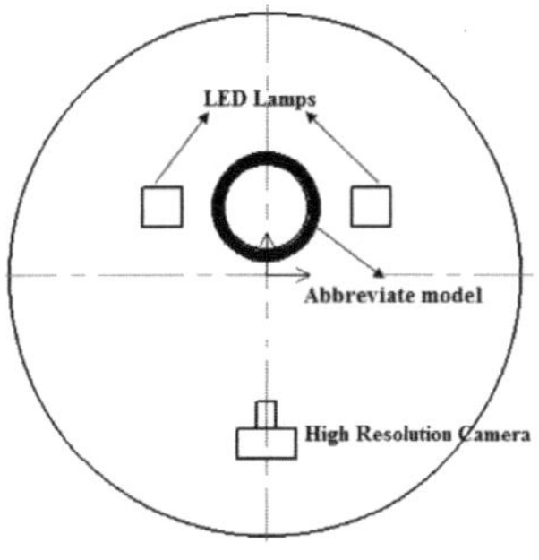

Figure 4. Sketch map of assembly placement.

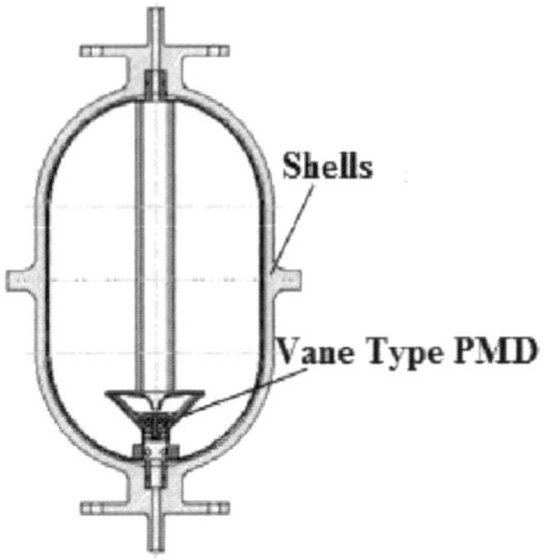

Figure 5. Sketch map of abbreviate model structure.

3. Results and discussions

3.1. Results and discussions of numerical analysis

When the microgravity acceleration under fluid-sinking mode is 5×10^{-5} g_0, the reorientation process with filling ratio 10% is shown in figure 6 below, and in the figure the darker park is fluid and the lighter part is gas. Seeing from figure 6, owing to the driving of surface tension, the fluid is transfered to the PRR by the PAV and the filling phase of PRR is beginning. The fast filling fluid into PRR has been achieved under the small PRR and the fluid can't block the through vent on the inside cone of PRR, it makes sure that the gas in PRR discharges smoothly during the filling phase, and there are none phenomena or trend of fluid cutoff on PAV during the filling phase. After 13s of the reorientation process, the PRR has been nearly filled with fluid. The results of numerical analysis coincide with the desired effects of design.

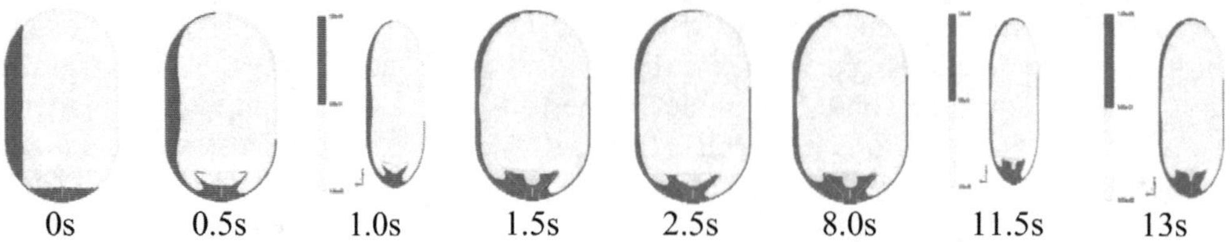

Figure 6. Reorientation process under fluid-sinking mode.

Then the discharging process under microgravity lateral acceleration with 1×10-3g0 is shown in figure 7 below, and in the figure the darker park is fuid and the lighter part is gas. The fluid surface distribution of the 8th second during the above-mentioned reorientation process is selected as the initinal condition of the discharging process, and the discharging rate of flow is 8ml/s. According to figure 7, during the discharging process, not only the fluid in PRR is discharged providing some flux, but also the fluid outside PRR is discharged for providing the other flux via wickets under PRR. There are none fluid cutoff on PAV or in PRR during the discharging process, so the vane type PMD can satisfy the flow requirement of 8ml/s. The results of numerical analysis achieve the desired effects of design.

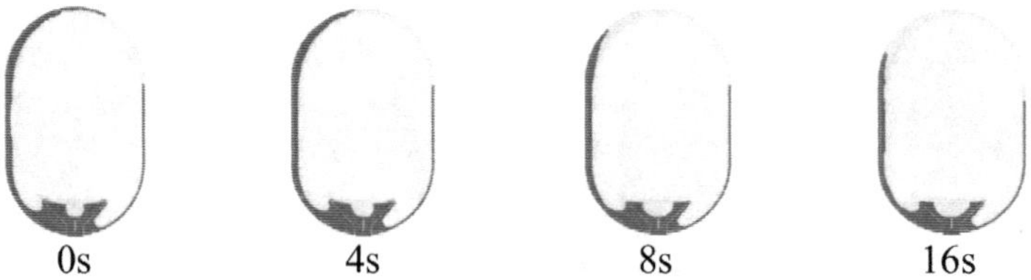

Figure 7. Discharging process under microgravity lateral acceleration.

When the microgravity lateral acceleration is 1×10^{-3} g_0, the reorientation process with filling ratio 2 % is shown in figure 8 below, and in the figure the darker park is fuid and the lighter part is gas. If the driving of fluid surface tension isn't enough during the end phase of STT's life, then it is possible to appear some bad phenomenons, such as inadequate flow and liquid discharging with some gas. But seeing from figure 8, owing to the driving of surface tension, fluid mainly stores near PRR and the fast filling fluid into PRR has been achieved still under the small PRR, and fluid covers the liquid outlet of STT's bottom throughout. So it makes sure that some fluid without gas can be discharged out from STT, and it infers that the orbital extrusion efficiency of STT is more than 98%, which satisfies assignment requirement of fluid orbital extrusion. The results of numerical analysis also achieve the desired effects of design.

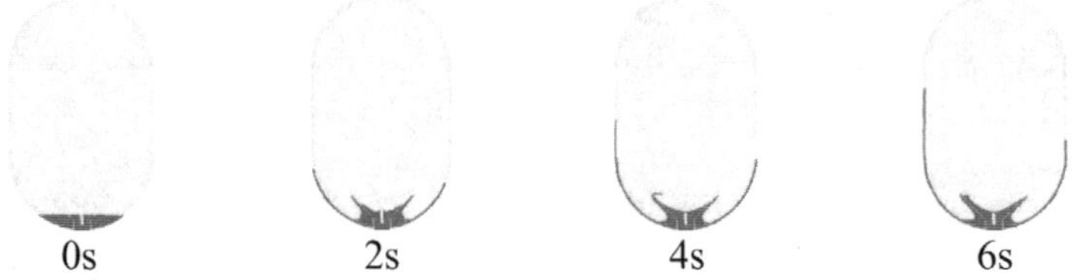

Figure 8. Reorientation process under microgravity lateral acceleration.

3.2. Results and discussions of microgravity drop tower tests

Figure 9 below shows the fliud reorientation process with filling ratio 10% in the abbreviate model under fluid-sinking mode. Before releasing the falling module, fluid surface mainly maintain level surface, and all the microgravity time is 3.5s. As shown in figure 9, by using the driving of surface tension, the fluid rapid climbs along the gap region between PAV and inner walls of STT after releasing the falling module, which is due to the smaller curvature radius of the gap region. When the microgravity time is 1.5s, the fluid has climbed along PAV nearly to STT's top, and a concave liquid surface between PRR and inner walls is shaped slowly during the climbing process. After 1.5s, liquid suerface and radius along PAV are mainly invariant, but the concave menniscus dents quickly, because the fluid between PRR and inner walls is fast absorbed into PRR, so it indicates that the PRR has good fluid storage ability. Meanwhile fluid covers the liquid outlet of STT's bottom throughout. The experiment results indicate that the vane type PMD has good and initiative fluid orbital management ability and can availably achieve the separation between fluid and gas interface and providing propallent without gas for thrusters.

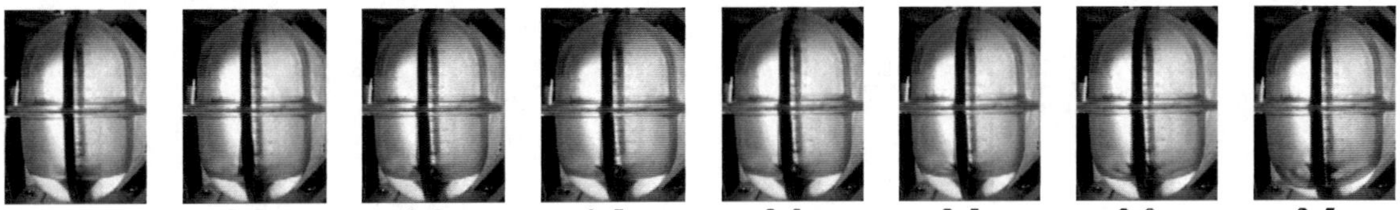

0s 0.5s 1.0s 1.5s 2.0s 2.5s 3.0s 3.5s

Figure 9. Reorientation process with filling ratio 10% in the abbreviate model.

Figure 10 below shows the fliud reorientation process with filling ratio 2% in the abbreviate model under fluid-sinking mode. According to figure 9, by using the driving of surface tension, the fluid rapid climbs along the gap region between PAV and inner walls of STT and a concave liquid surface between PRR and inner walls is shaped under microgravity environment. After 1.0 s, the fluid between PRR and inner walls is fast absorbed into PRR, which indicates that the PRR has good fluid storage ability. Finally the fluid mainly reserves in PRR and the region between PRR and inner walls and covers the liquid outlet of STT's bottom whole, so it makes sure that some fluid without gas can be discharged out from the abbreviate model, and it infers that the orbital extrusion efficiency of the abbreviate model is more than 98%. The experiment results indicate that the goal of microgravity experiment verification in principle on fluid distribution condition at the end of STT's life and extrusion performance of STT has been achieved.

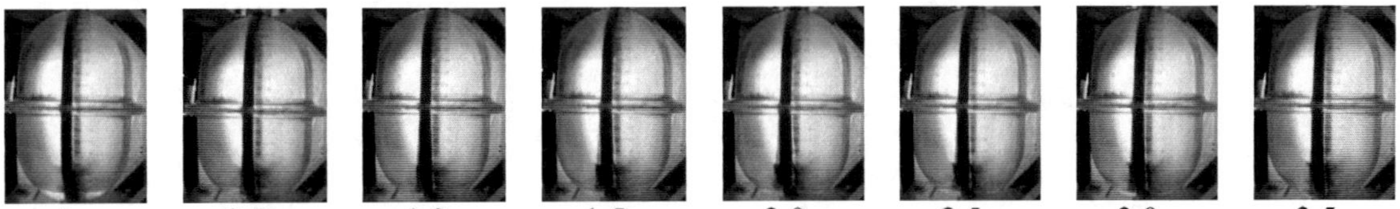

0s 0.5s 1.0s 1.5s 2.0s 2.5s 3.0s 3.5s

Figure 10. Reorientation process with filling ratio 2% in the abbreviate model.

Comparing the results mentioned above, the researchers find the results of numerical analysis and the results of microgravity experiment study are relatively identical and attain the effect of verifying analysis and tests each other, and the numerical analysis are correct and effective.

4. Conclusions

Numerical analysis and microgravity experiment study on fluid orbital management performance of a vane type PMD have been carried out, and fluid management characteristic and fluid transmission and distribution rules of the vane type PMD have been obtained under microgravity environment. The results of numerical analysis and microgravity experiment study are relatively identical, and the results offer valuable guidance for the design and optimization of new generation vane type PMD and have been used for the development of vane-type STT, which promote the update of domestic STT products.

The analysis and tests results show that the vane type PMD has good and initiative fluid orbital management ability and meets the demands of fluid orbital extrusion in the vane type STT. Meanwhile the vane type PMD provides a new approach for fluid management and control and is favor to enhance the technical level of fluid management and control in space environment.

Acknowledgments

The authors would like to thank National Natural Science Foundation of China (No. 51306018).

References

[1] Jaekle Jr D E 1991 Propellant management device conceptual design and analysis: vanes *The 27th Joint Propulsion Conf.* (Sacramento, CA, USA, 24-26 June 1991) pp 2172-91
[2] Jaekle Jr D E 1993 Propellant management device conceptual design and analysis: sponges *The 29th Joint Propulsion Conf.* (Monterey, CA, USA, 28-30 June 1993) pp 1970-93
[3] Michael J, Debreceni, Kuo T K and Jaekle Jr D E 2004 *Development of a composite wrapped propellant tank* AIAA 3505
[4] Netter G, Renner U and Dreyer M 1999 *Design and verification of a standard surface tension propellant* AIAA 2178
[5] Debreceni M J, Lay W D, Kuo T K and Jaekle Jr D E 2001 *Propellant tank for an advanced communications satellite* AIAA 3826
[6] Hu Q, Li Y, Pan H L et al. 2013 *Aero. Control and Application* **39** (3) 58-62
[7] Zhuang B T, Li Y, Pan H L et al. 2012 *Aero. Control and Application* **38** (6) 1-5
[8] Hu Q, Li Y, Geng Y B et al. 2010 *Aero. Control and Application* **36** (3) 55-8
[9] Hu Q, Chen J and Li Y 2009 *Aero. Control and Application* **35** (6) 54-7
[10] Zhuang B T, Li Y, Pan H L et al. 2014 *Aero. Control and Application* **40** (1) 27-30
[11] Li Y, Zhao C Z, Pan H L et al. 2008 *J. Astronautics* **29** (1) 24-8
[12] Zhang X Q, Yuan L G, Wu W D et al. 2005 *Sci. in China Ser. Eng.* **35** (5) 523-34

IOP Conf. Series: Materials Science and Engineering **72** (2015) 042043 doi:10.1088/1757-899X/72/4/042043

Numerical simulation of cavitation flow characteristic on Pelton turbine bucket surface

C J Zeng, Y X Xiao, W Zhu, Y Y Yao and Z W Wang[1]
Department of Thermal *Engineering*& State Key Laboratory of Hydroscience and Engineering, Tsinghua University, Beijing 100084, China

E-mail: wzw@mail.tsinghua.edu.cn

Abstract. The internal flow in the rotating bucket of Pelton turbine is free water sheet flow with moving boundary. The runner operates under atmospheric and the cavitation in the bucket is still a controversial problem. While more and more field practice proved that there exists cavitation in the Pelton turbine bucket and the cavitation erosion may occur at the worst which will damage the bucket. So a well prediction about the cavitation flow on the bucket surface of Pelton turbine and the followed cavitation erosion characteristic can effectively guide the optimization of Pelton runner bucket and the stable operation of unit. This paper will investigate the appropriate numerical model and method for the unsteady 3D water-air-vapour multiphase cavitation flow which may occur on the Pelton bucket surface. The computational domain will include the nozzle pipe flow, semi-free surface jet and runner domain. Via comparing the numerical results of different turbulence, cavity and multiphase models, this paper will determine the suitable numerical model and method for the simulation of cavitation on the Pelton bucket surface. In order to investigate the conditions corresponding to the cavitation phenomena on the bucket surface, this paper will adopt the suitable model to simulate the various operational conditions of different water head and needle travel. Then, the characteristics of cavitation flow the development process of cavitation will be analysed in in great detail.

1. Introduction

The Pelton turbine can be applied to a wide range water head from 30 to 3000 m. The unit working under great water head has a high flow velocity and the cavitation may occur on the bucket surface. The Pelton runner is the most important component of the unit for energy transformation. During the operation process the bucket cut into the high velocity jet, thus the buckets suffer from a pulsed stress. The operation condition is so bad that the occurrence of cavitation will damage the bucket surface and even destroy the runner structure.

The internal flow is quite complicated in the Pelton turbine and there exists controversy in the academic circle that whether the cavitation occurs in the Pelton turbine. By applying the Bernoulli equation, Zhou et al hold the opinion that the Pelton runner operates under atmospheric pressure and there is no low pressure area on the working surface of bucket. The damage of bucket rear surface and side surface is a kind of fatigue failure caused by the repeated impact [1]. However according to the practice runner operation condition, Li pointed out that for a certain runner used in high water head, the cavitation easily occurs near the splitter tip due to the poor manufacture process [2]. Rossetti et al

[1] Corresponding author.

simulated the cavitating flow phenomenal in the Pelton turbine. The occurrence place and the development process of water vaporization and condensation is analyzed during the bucket cut-in process to the jet flow. The qualitative analysis shown that the pitting cavitation occurred at the rear surface of splitter tip and the position is coincidence with the damage part of the test case runner [3].

With the development of the computational fluid dynamics, the unsteady multiphase internal flow of the Pelton turbine can be simulated. This paper adopted Eulerian-Eulerian multiphase flow model and analyzed the water-air flow characteristic in the Pelton turbine without cavitation at first stage. Then with the identification of low pressure area on the bucket surface, the numerical simulation method for cavitation flow in the Pelton turbine was investigated and the cavitation flow characteristic was analyzed.

2. Simulation model

2.1. Governing equations

This simulation adopted the Eulerian-Eulerian multiphase flow model. As the flow in the Pelton turbine is a free surface flow where the interface is well defined, the homogeneous multiphase model is employed to simplify the simulation. Thus the common flow field is regarded as shared by all fluids. Before the numerical analysis of water-air-vapour multiphase flow, the unsteady water-air flow simulation is conducted to get a reasonable initial condition for the cavitation flow analysis.

The homogeneous model assumes all fluids share the common flow field. And the flow field can be obtained by the continuity equation and momentum equations, which is dependent on the volume fractions of all phases through the properties ρ and μ as shown below.

$$\rho = \sum_{\alpha=1}^{N_p} r_\alpha \rho_\alpha \tag{1}$$

$$\mu = \sum_{\alpha=1}^{N_p} r_\alpha \mu_\alpha \tag{2}$$

$$\mathbf{U}_\alpha = \mathbf{U}, 1 \le \alpha \le N_p \tag{3}$$

$$\frac{\partial \rho}{\partial t} + \frac{\partial}{\partial x_j}\left(\rho U_j\right) = 0 \tag{4}$$

$$\frac{\partial \rho U_i}{\partial t} + \frac{\partial}{\partial x_j}\left(\rho U_i U_j\right) = -\frac{\partial p}{\partial x_i} + \frac{\partial}{\partial x_j}\left(\tau_{ij} - \rho \overline{u_i u_j}\right) + S_M \tag{5}$$

The Reynolds stresses arise from the nonlinear convective term in the un-averaged equations. They reflect the fact that the convective transport due to turbulent velocity fluctuations will act to enhance mixing over and above that caused by thermal fluctuations at the molecular level. To close the equation, the eddy viscosity turbulence models are introduced to model the additional terms of Reynolds stresses. And the Shear Stress Transport model is chosen for achieving highly accurate predictions of the onset and the amount of flow separation under adverse pressure gradients. The equation for the turbulence viscosity, μ_t, turbulence kinetic energy, k, and turbulent frequency, ω, are as shown below.

$$\mu_t = \frac{\rho a_1 k}{\max\left(a_1 \omega, S F_2\right)} \tag{6}$$

$$\frac{\partial(\rho k)}{\partial t} + \frac{\partial}{\partial x_j}\left(\rho U_j k\right) = \frac{\partial}{\partial x_j}\left[\left(\mu + \frac{\mu_t}{\sigma_{k3}}\right)\frac{\partial k}{\partial x_j}\right] + P_k - \beta' \rho k \omega + P_{kb} \tag{7}$$

$$\frac{\partial(\rho\omega)}{\partial t}+\frac{\partial}{\partial x_j}(\rho U_j\omega)=\frac{\partial}{\partial x_j}\left[\left(\mu+\frac{\mu_t}{\sigma_{\omega3}}\right)\frac{\partial\omega}{\partial x_j}\right]+\frac{2\rho(1-F_1)}{\sigma_{\omega2}\omega}\frac{\partial k}{\partial x_j}\frac{\partial\omega}{\partial x_j}+\frac{a_3P_k\omega}{k}-\beta_3\rho\omega^2+P_{\omega b} \quad (8)$$

The Rayleigh Plesset model is adopted to simulate the interphase mass transfer between the water and water vapor. The Rayleigh-Plesset equation provides the basis for the rate equation controlling vapor generation and condensation. And the equation describing the growth of a gas bubble in a liquid is given by:

$$R_B\frac{d^2R_B}{dt^2}+\frac{3}{2}\left(\frac{dR_B}{dt}\right)^2+\frac{2\sigma}{\rho_f R_B}=\frac{p_v-p}{\rho_f} \quad (9)$$

Neglecting the second order terms (which is appropriate for low oscillation frequencies) and the surface tension, and assuming there are N_B bubbles per unit volume, the total interphase mass transfer rate per unit volume of vaporization and condensation is:

$$\dot{m}_{fg}=F\frac{3r_g\rho_g}{R_B}\sqrt{\frac{2}{3}\frac{|p_v-p|}{\rho_f}}\operatorname{sgn}(p_v-p) \quad (10)$$

2.2. *Physical model*

A vertical axis 19-bucket Pelton turbine is investigated. The ration of the bucket inner breadth B and the reference runner diameter D_{ref} is B/D_{ref}=0.25. The turbine optimum unit speed n_{11}=41 rpm with one injector, a rated speed of 500rpm with the design head of 456 m. In the case of high water head, gravity has little influence on the turbine performance. Thereby, as the incoming jet is assumed to be ideal, the computational domain of Pelton turbine can be simplified to the half because of the structure symmetrical characteristic as shown in Figure 1.

Figure 1. Computational domain

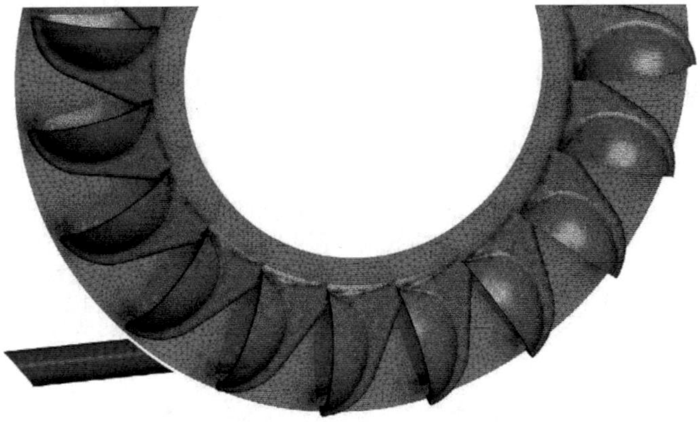

Figure 2. Grid distribution

Because of the complex shape of bucket and a better capture of cavitation flow, an unstructured mesh is used to refine the local grid of critical area in bucket surface. The total mesh has about 1,160,000 elements and 260,000 nodes. The computational domain grid and the refined grid of the bucket surface are shown in Figure 2.

The numerical simulation is carried out with the ANSYS CFX 13 code. The continuity and momentum equations are discretized using an upwind scheme with the physical advection terms weighted by a gradient-dependent blend factor, providing a good trade-off between diffusion and dispersion. A first-order backward Euler scheme is used for the transient terms. According to the flow condition, the unsteady simulation time step was set as 1.58×10^{-4} s in which the runner rotates 1/40 of the angle between two adjacent buckets. For obtaining a higher calculation precision to satisfy the practical requests, the convergence residual was set as 10^{-4}.

3. Results and discussion

3.1. Numerical results of no cavitation flow

Firstly, this article investigated the water-air two phase flow in the Pelton turbine. The internal unsteady flow results of Pelton turbine have been obtained without the consideration of cavitation. As the runner buckets cut through the free jet periodically, the working process of a single bucket is periodically. Figure 3 shows the torque variation curves of the cutting procedure. The detailed analysis of torque variation process can be seen in the reference [4].

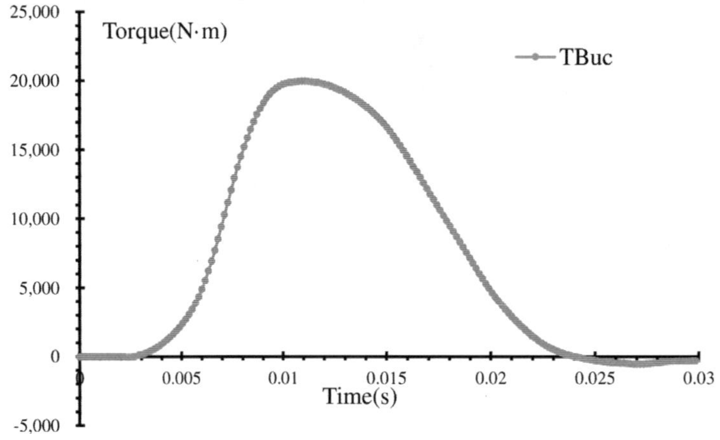

Figure 3. Torque variation curves of a single bucket

The pressure distribution on the bucket front surface during its working process is analysed in Figure 4. As the bucket cutting through the free jet, the free jet began to impact on the front surface and became into free water sheet. The light blue isosurface of water volume fraction indicates the free jet boundary and the white lines indicates the boundary of the free water sheet flow region on the bucket surface. As can be seen from the Figure 4(a), the low pressure area began to appears near the cut-out among the water sheet flow region. With the development of flow, the low pressure expanded to the most at 3.52e-3 s and disappeared at 5.17e-3 s as shown in Figure 4(b) and Figure 4(c). The low pressure area among the water sheet flow region indicated the cavitation may occur on the bucket front surface.

3.2. Numerical results of cavitation flow

Based on the water-air two phase numerical results, this article conducted water-vapour-air three phase simulation on the internal flow of Pelton turbine to study the cavitation flow. The mass transfer model between water and water vapour adopted the Rayleigh Plesset model and the saturation pressure was set as 3574 Pa of 25 ℃ water. The numerical result is shown in Figure 5. As the water sheet flow spreads on the bucket front surface, cavity began to grow near the splitter. In Figure 5(a), the red vapour surface shows the initial cavity on the bucket front surface. The cavitation area shows good accordance with the low pressure area in Figure 4(a). Also the inflated cavity in Figure 5(b) and the shrunk cavity in Figure 5(c) are consistent with the pressure distributions in Figure 4.

3.3. Effect of cavitation on turbine performance

The torque variation curves of single bucket are compared in Figure 6. It can be seen from the difference between two cases that the maximum torque diminished and the working process is shorten because of the cavitation. It can be distinguished that the average torque of runner decrease because of the cavitation. Because the efficiency of the injection mechanism is of the runner efficiency, the cavitation would decrease the output capacity of whole unit.

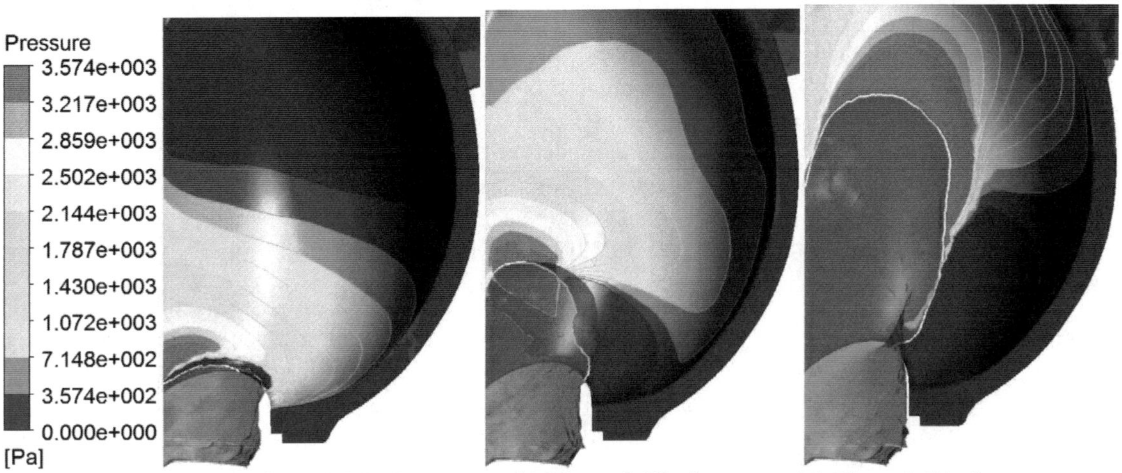

(a) Time=1.86e-3s (b) Time=3.52e-3s (c) Time=5.17e-3s

Figure 4. Low pressure area on the Bucket front surface

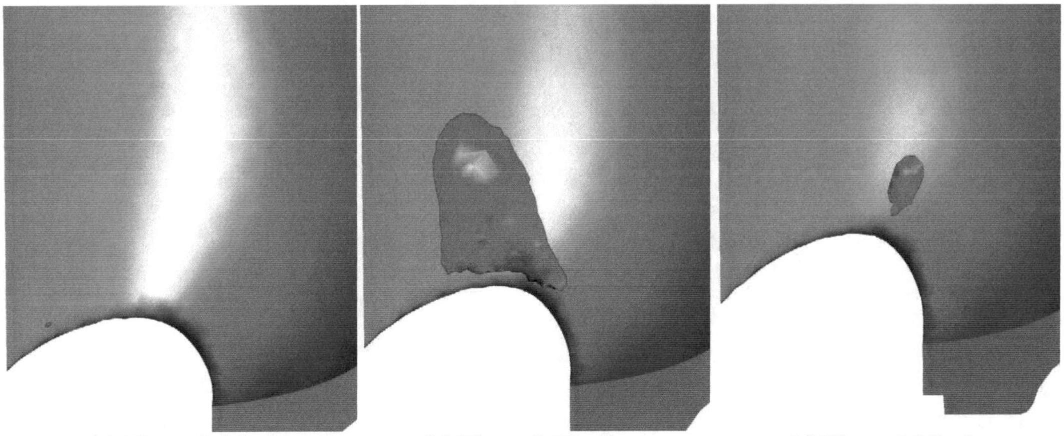

(a) Time=1.86e-3s (b) Time=3.52e-3s (c) Time=5.17e-3s

Figure 5. Cavity on the Bucket front surface

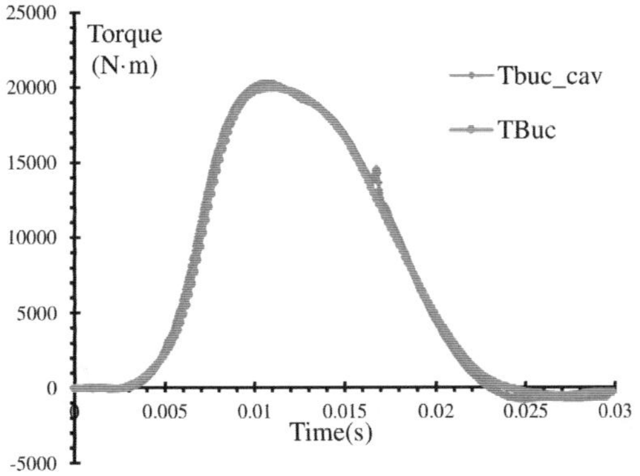

Figure 6. Bucket Torque curves of two cases

4. Conclusions

This article simulated the three dimensional unsteady flow in Pelton turbine. With the consideration of cavitation model, the water-vapour-air three phases numerical method is investigated. The main conclusion can be draw as follows:

(1) The water-vapor-air three phase model can be applied to simulation of cavitation flow in Pelton turbine. The predicted cavitation area is in good accordance with the low pressure area in the two phase simulation.

(2) Cavity may occur on the front surface of Pelton bucket. The three phases numerical results indicated cavity would initial near the cut-out on the front surface at the beginning phase of bucket cutting through the free jet.

(3) Cavitation can influence the unit performance. The numerical results indicated cavitation would decrease the output of Pelton runner.

Acknowledgments

Special thanks are due to the National Natural Science Foundation of China under Contract (No. 51479093), State Key Program of National Science of China (Grant No. 51439002), State Key Laboratory of Hydroscience and Engineering (Grant No. 2014-KY-05) for supporting the present work.

References

[1] Zhou W T 2008 *Large Electric machine and Hydraulic Turbine* **2** 44-53.
[2] Li L 2014 Design and Application of Pelton Runner *Transform and Upgrade Traditional Industries to Develop Strategic Emerging Industries Conference Proc.*
[3] Rossetti A, Pavesi G, Ardizzon G and Santolin A 2014 *J. Fluid Eng.* **136**(8) 081304
[4] Xiao Y X, Wang Z W, Zhang J, Zeng C and Yan Z 2014 *Proc. of the Institution of Mech. Engineers, Part A: J. of Power and Energy* **228**(1) 46-55
[5] ANSYS *ANSYS CFX User's Guide* (ANSYS Inc.).
[6] Xiao Y X, Cui T, Wang Z W and Yan Z G 2012 *26th IAHR Symposium on Hydraulic Machinery and Systems(Beijing, China, 19-23, August, 2012*)
[7] Xiao Y X, Han F Q, Zhou J L and Kubota T 2007 *J. of Hydrodynamics, Ser. B* **19**(3) 356-64
[8] Xiao Y X 2013 *IOP Conference Series: Materials Sci. and Eng.* **52**(5) 052032

IOP Conf. Series: Materials Science and Engineering **72** (2015) 042044 doi:10.1088/1757-899X/72/4/042044

Experimental and CFD analysis for prediction of vortex and swirl angle in the pump sump station model

C G Kim[1], B H Kim[1], B H Bang[2] and Y H Lee[3,4]

[1] Department of Mechanical Engineering, Korea Maritime and Ocean University, 727 Taejong-ro, Youngdo-gu, Busan 606-791, Republic of Korea
[2] Daewoo Institute of Construction Technology, 60, Songjuk-dong, Jangan-gu, Suwon-si, Gyeonggi-do 440-800, Republic of Korea
[3] Department of Mechanical and Energy system Engineering, Korea Maritime and Ocean University, 727 Taejong-ro, Youngdo-gu, Busan 606-791, Republic of Korea

E-mail: lyh@kmou.ac.kr

Abstract. Sump model testing is mainly used to check flow conditions around the intake structure. In present paper, numerical simulation with SST turbulence model for a scaled sump model was carried out with air entrainment and two phases for prediction of locations of vortex generation. The sump model used for the CFD and experimental analysis was scaled down by a ratio of 1:10. The experiment was performed in Korea Maritime and Ocean University (KMOU) and the flow conditions around pump's intake structure were investigated. In this study, uniformity of flow distribution in the pump intake channel was examined to find out the specific causes of vortex occurrence. Furthermore, the effectiveness of an Anti Vortex Device (AVD) to suppress the vortex occurrence in a single intake pump sump model was examined. CFD and experimental analysis carried out with and without AVDs produced very similar results. Without the AVDs, the maximum swirl angle obtained for experimental and CFD analysis were 10.9 and 11.3 degree respectively. Similarly, with AVDs, the maximum swirl angle obtained for experimental and CFD analysis was 2.7 and 0.2 degree respectively. So, with reference to the ANSI/HI 98 standard that permits a maximum swirl angle of 5 degree, the use of AVDs in experimental and CFD analysis produced very desirable results which is well within the limit.

1. Introduction

It is understood that quite a number of problems faced in a pumping station are related to the design of sump or intake rather than pump design. As an important part of the pumping station, sump is designed to provide uniform, swirl and entrained air free flow to the pump intake. Undesirable flow conditions (entrained air, non-uniform flow distribution, vortices, etc.) can reduce the pump efficiency; induce vibration, noise, cavitation, lead to excessive bearing loads of impeller and structure damage. Due to the quite complex nature of the flow, although the flow conditions causing pump problems are well established, there are no specific solutions to eliminate them. There are some design guidelines for the specific geometrical and hydraulic constraints of the pump sump, for any project, the model study is the only tool for solving potential problems in new designs and rectifying problems observed in existing installations.

[4] Corresponding author.

IOP Conf. Series: Materials Science and Engineering **72** (2015) 042044 doi:10.1088/1757-899X/72/4/042044

Rajendran et al.[1] have made numerical analysis for the flow characteristics of a sump model with pump intake and good agreements were achieved by comparing the numerical results with the experiments. Iwano et al.[2] have introduced a numerical method for the submerged vortex by analyzing the flow in the pump sump with and without baffle plates. Lee et al.[3] have conducted the CFD analysis of a multi-intake pump sump model to check the flow uniformity by predicting the location, number and vorticity of the vortex. The cavitation phenomena has been studied by various investigators. Van et al.[4] have investigated the cavitation inception behavior of a mixed-flow pump impellers model and the calculated results matched well with the experiments. Okamura et al.[5] have evaluated two numerical cavitation prediction methods used in the pump industry manufacturers, and attempts were made to improve the cavitation performance. Li et al.[6] have conducted the numerical analysis on the basis of the development liquid/vapor interface tracking method to predict the cavitation characteristics with a centrifugal pump impeller model.

Experiments were conducted in using a sump station model scaled down by a factor of 10. An Anti-vortex device (AVD) was designed and the experiments were conducted with and without AVDs. Numerical simulations were done on the same model and the experimental results and the numerical results were compared on the basis of swirl angle and vortex.

2. Sump model

Table 1 shows the various dimensions of the model parameters both real scale model and the model used for the test. The design was adopted from the guidelines dictated by ANSI/HI 9.8[8] standard. It consists of 2 drum screens, 2 inlets, 3 pumps and one dummy channel. The actual sump station length and width are 65,800 mm and 34,000 mm respectively whereas the model test size is 6,580 mm long and 1,700 mm wide.

Table 1. Model parameters.

Model Parameters	Real scale model	Model test
Bell throat diameter(d)	900 mm	90 mm
Bell mouth diameter(D)	2,000 mm	200 mm
Bell mouth ground Clearance	1,300 mm	130 mm
Rated flow rate	18,000 m^3/hr	56.9 m^3/hr
Rated flow vel.(bell mouth)	1.203 m/s	0.381 m/s
Sump station length	65,800 mm	6,580 mm
Sump station width	34,000 mm	1,700 mm

3. Experimental setup

Figure 1(a) shows the overall experimental setup in which the water is circulated through 3 channels with the help of 55 kW pump. An inverter is implied to control the rpm of the pump and hence the mass flow rate. The pump operates at 8 m head and the maximum flow rate attainable is 900 m^3/h. Electro-magnetic flow meters are installed in each channel. Flow meters measurement range varies from 0.3 m/s- 10 m/s with accuracy of F.S±0.5%. Figure 1(b) shows the front channel and bell mouth made from acrylic resin such that the flow pattern and vortex formation can be easily observed.

A four blade zero-pitch swirl meter, which is supported by a low friction bearing, is installed at about four suction bell throat pipe diameters downstream from the pump suction to measure swirl angle of pump approaching flow. For the swirl meter, the tip to tip blade diameter is 0.75d and the length in flow direction is 0.6d and one of the four blades is painted yellow as a reference to count revolutions. Measuring tape is attached in front of the acrylic to determine the water level. Wire mesh used inside the sump with drum screen has same porosity as the real case.

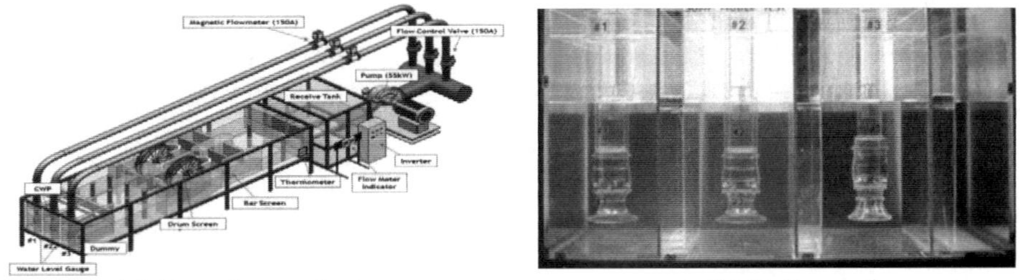

(a) Experimental setup (b) Bell mouth

Figure 1. Experimental Model setup.

Several attempts were made to finalise the distibutors and the bottom AVD for this design. Figure 2 depicts the AVD design and the distibutors. Various distibutors combination were tried to ensure the uniform flow inside the channel. Similarly curtain walls were implied to reduce the free surface vortices and bottom AVD to remove the submerged vortices. A through investigation was made to ensure the optimum flow and swirl angle. As ANSI/HI 9.8 standard states the maximum swirl angle allowed is 5°, several combinations were tried to obtain the same. 3 distributors, 1 bottom AVD and 1 curtain wall is implied in each channel for the experimentation. Figure 2(b), (c) shows the bottom AVD and the distributors location.

(a) (b) (c)

Figure 2. AVD design. (a) Bottom AVD; (b) Distributor at the front side of Common Bay; (c) Distributor at the backside of Common Bay

3.1. Test Condition

Table 2 summaries the overall operating conditions of pump station. 3 cases of experiments are listed in the table. Each pump operates with the rated flow rate of 56.9 m^3/hr. For the LR-1 case, all the three pumps are in operation with both stop logs and both inlet gates open. For the LR-5 case, pump 1 does not operate (pumps 2 and 3 operating) with stop log 1 and inlet gate 1 in closed position. For the LR-6 case, pumps 1 and 2 are in operation (pump 3 does not operate) with stop log 2 and inlet gate 2 in closed position. Each x shown in the table shows the location of closed positions/pumps.

3.2. Swirl angle Calculation

Swirl angle predicts the intensity of flow rotation. Swirl angle is calculated by experimental method according to HI standard and this is used to check the flow rotation in the suction pipe. The observation time for the swirl meter rotation was above 10 minutes. The HI recommended criterion for swirl angle is within 5 degrees. Equation (1) is experiment calculation method with swirl meter. The swirl angle θ is defined as follows:

$$\theta = \tan^{-1}\frac{V_\theta}{V_z} \tag{1}$$

where V_z = axial mean velocity = $\frac{4Q}{\pi d^2}$, V_t = tangential mean velocity at the swirl meter = $\frac{\pi dn}{60}$, Q = flow rate (m^3/s), d = bell throat diameter (m), n = revolution of swirl meter (rpm).

For the swirl angle calculating method with CFD, the key point is to obtain the average tangential velocity, so the swirl check circles were created. But for numerical method, the swirl check circle is located at the section of 4d height with diameter of 0.25d, 0.5d and 0.75d respectively. Swirl angle θ is defined as follows**:**

$$V_\theta = \frac{\sum_1^N V_{0.25i}/N + \sum_1^N V_{0.5i}/N + \sum_1^N V_{0.75i}/N}{3} \times 2 \quad (2)$$

where V = circumferential velocity of check point in check circle, respectively (m/s), N = check point total numbers in single check circle.

Table 2. Operating conditions of pump station.

Case	Pump operation	Stop log 1	Stop log 2	Inlet gate 1	Inlet gate 2	Operating condition
LR-1	#1 #2 #3	o	o	o	o	
LR-5	x #2 #3	x	o	x	o	
LR-6	#1 #2 x	o	x	o	x	

4. Numerical method

A structured hexahedral mesh was generated using ICEM CFD. Initially, the difference between a single phase calculation and a two phase calculation was done. In the one phase numerical analysis, there is no noticeable effect in pressure at the region behind the distributor when it is inserted. But in two phase analysis, there is a pressure drop behind distributor and its effect has to be considered. Therefore, for all numerical cases, a two phase steady state calculation was selected. Therefore, the numerical analysis was carried using a finite volume solver ANSYS CFX v.13.0. The Shear Stress Transport (SST) turbulence model was selected to model the turbulence in the numerical model with a total 9000000 after mesh sensitivity testing.

5. Results

5.1. Experimental and CFD results comparison

Table 3 shows the comparison between the experimental and CFD results without the AVD installed for each of the cases. Currently, numerical simulations have only been able to predict the location of the formation of vortices but are unable to determine the strength of the vortex. The CFD result shows good prediction of the location of the vortex. In addition, the swirl angles are also compared and show that the CFD result predicts high swirl angles and in some cases almost accurately.

Table 4 shows the results when the model has the AVD installed. Both results show that no vortices form and the swirl angle have significantly reduced.

IOP Conf. Series: Materials Science and Engineering **72** (2015) 042044 doi:10.1088/1757-899X/72/4/042044

The results from both tables are plotted in figure 4 with the HI standard guide line highlighted at a swirl angle of 5 degrees. The differences in the results can be due to some unavoidable errors in experiments, such as the water flow causing the swirl meter to fluctuate and not completely turn, and the innate problem that CFD cannot exactly replicate experimental conditions. In spite of this, the CFD results follow the trend of the experimental results closely. Predicting swirl angles above the standard when the AVD is not installed and the reduction in swirl angle when the AVD installed. The CFD results overall predict are more stable flow due to the fact that the calculations are done in steady state.

Table 3. Comparison of vortex region and swirl angle without AVD by CFD and experiment.

Case	Scale	Pump	CFD vortex region	Test vortex region	Swirl angle	
					Test degree(°)	CFD degree(°)
LR-1	1:10	#1	LRSW [a], BW [b], B [c]	RSW, BW, B	10.7	5.5
		#2	LRSW, BW, B	LRSW, BW, B	7.6	4.6
		#3	LRSW, BW, B	N/A	N/A	3.9
LR-5		#2	LSW, BW, B	LRSW, BW, B	7.2	7.3
		#3	LSW, B	N/A	N/A	11.3
LR-6		#1	LSW, B	RSW, B	3.3	3.9
		#2	LRSW, B	LRSW, B	6.7	6.4

[a] Left right side wall vortex; [b] Back wall vortex; [c] Bottom vortex; [d] Free surface vortex

Table 4. Comparison of vortex region and swirl angle with AVD by CFD and experiment.

Case	Scale	Pump	CFD vortex type	Test vortex type	Swirl Angle	
					Test degree(°)	CFD degree(°)
LR-1	1:10	#1	NV	NV	2.7	0.07
		#2	NV	NV	0.9	0.19
		#3	NV	N/A	N/A	0.04
LR-5		#2	NV	NV	1.9	0.23
		#3	NV	N/A	N/A	0.08
LR-6		#1	NV	NV	1.4	0.1
		#2	NV	NV	1.1	0.08

Figure 5 depicts the velocity contours and vortex core region. Figure 5(a) shows the velocity contours at the plane 1 meter above the bottom without an AVD and Figure 5(b) shows the same with an AVD. While comparing the two figures, it is observed that there is a strong jet flow in the common bay in figure 5(a) and so a high swirl angles develop. But in figure 5(b), the distributors help to make the flow uniform, reduce the axial flow and hence results in a very small swirl angle. Figure 5(c) shows the vortex core region obtained by submerged bottom vortex method. The side walls (left and right), back wall and bottom vortex are show in the figure with red circles.

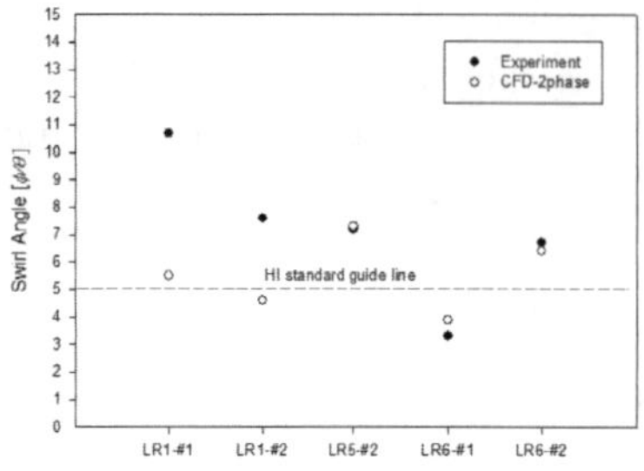

(a) without AVD

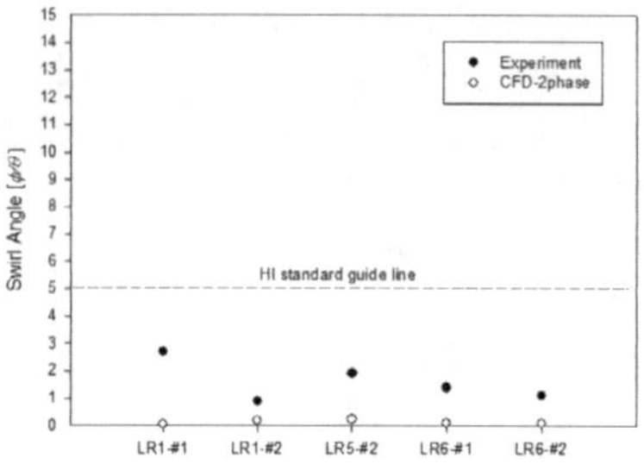

(b) with AVD

Figure 4. Experimental setup.

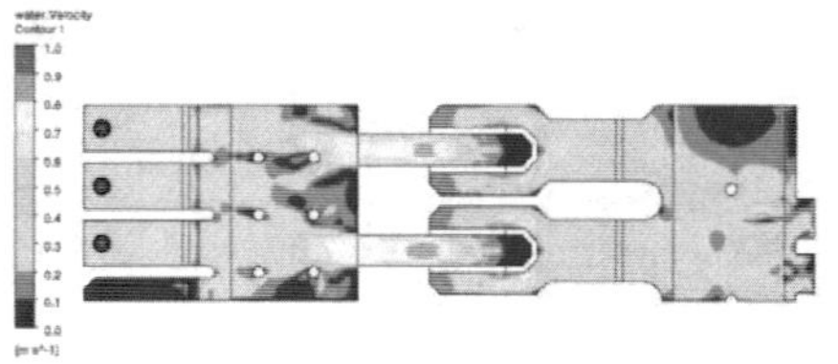

(a) Velocity Contours (without AVDs)

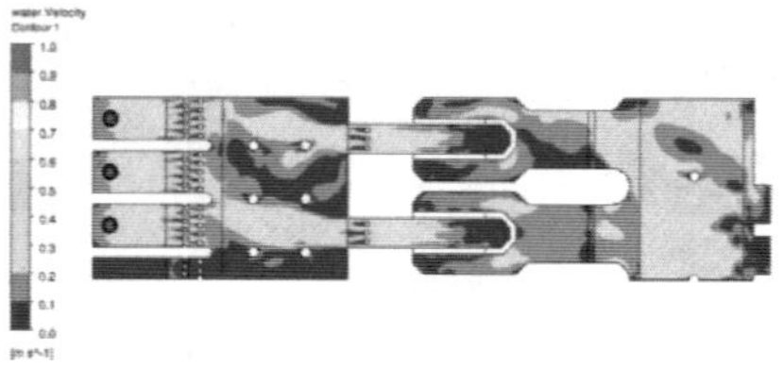

(b) Velocity Contours (with AVDs)

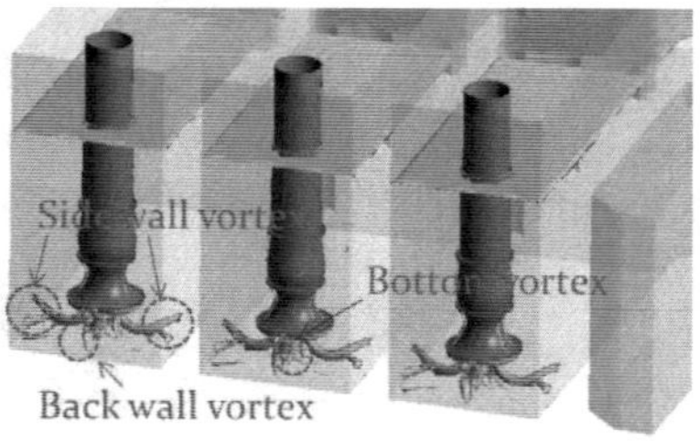

(c) Vortex Core Region of without AVDs (Vorticity : $5S^{-1}$)

Figure 5. Velocity contours and vortex core region.

6. Conclusions

Experiments were conducted in dedicated sump model installed in the laboratory of KMOU and numerical simulations were performed in ANSYS CFX 13.0. Using these two methods the prediction of vortex and swirl angle were performed from the obtained results with and without AVD. This paper can be concluded with the following points.

1) In an initial investigation, the effect of the distributor is correctly predicted and the pressure drop across it can be seen in two- phase calculation whereas one phase calculation can lead to incorrect predictions near the bell mouth almost negating the effect of AVD.

2) CFD results show good prediction of the location of vortex and the swirl angles comparison find good agreement with one another. Side walls (left and right), back wall and bottom vortices i.e. all submerged vortices are also characterised.

3) Strong jet flow and high swirl angle is observed in common bay without the use of an AVD. This can be controlled by using an AVD, thus making the flow uniform and reducing the axial flow. The vortex is suppressed and eliminated and the swirl angle is reduced with the use of an AVD.

4) The differences in the results can be due to some unavoidable errors in experiments, such as the water flow causing the swirl meter to fluctuate and not completely turn, and the innate problem that CFD cannot exactly replicate experimental conditions. In spite of this, the CFD results follow the trend of the experimental results closely.

Acknowledgement
This Research was supported by the Daewoo Institute of Construction Technology for Optimization of Sump Station Project, South Korea.

References

[1] Rajendran V P, Constantinescu S G and Patel V C 1998 Experiments on Flow in Model Water-pump Intake Sump to Validate a Numerical Model *Proc.of ASME Fluids Eng. Division Summer Meeting* (Wahington, DC, USA, 21-25 June 1998) 98-5098
[2] Iwano R, Shibata T, Nagahara T and Okamura T 2002 Numerical Prediction Method of a Submerged Vortex and Its Application to the Flow in Pump Sumps with and without a Baffle Plate *Proc. of the 9th Int. Symp. on Transport Phenomena and Dynamics of Rotating Machinery (Honolulu, Hawaii, USA, 10-14 Feb 2002)* pp 1-6
[3] Choi J W, Choi Y D, Kim C G and Lee Y H 2010 *J. Mech. Sci. and Tech.* **24**(7) 1389-400
[4] van Os M J, Op de woerd J G H and Jonker J B 1997 A parametric study of the cavitation inception behavior of a mixed-flow pump impeller using a three-dimensional potential flow model *Proc. of ASME Fluids Eng. Division Summer Meeting (Vancouver, Canada, 22-26 June 1997)* 97-3374
[5] Dupont P and Okamura T 2003 *Int. J. Rotating Machinery* **9**(3) 163-70
[6] Li J, Liu L J, Li G J and Feng Z P 2007 *J. Eng. Thermophysics* **28**(6) 948-50
[7] Kim C G, Choi Y D and Lee Y H 2012 *Earth and Environment Science* IOP IAHR XXVI-083
[8] Hydraulic Institute 2012 *American National Standard for Pump Intake Design* ANSI/HI 9.8

IOP Conf. Series: Materials Science and Engineering **72** (2015) 042045 doi:10.1088/1757-899X/72/4/042045

Numerical simulation of cavitation for a horizontal axis marine current turbine

Q Guo[1], L J Zhou[1,3] and Z W Wang[2]
[1] College of Water Resources and Civil Engineering, China Agricultural University, Beijing 100083, China
[2] Department of Thermal Engineering, Tsinghua University, Beijing 100084, China

E-mail: zlj09@263.net

Abstract. Marine current turbines, compared with the analogous wind turbines, have the potential to suffer cavitation. This paper focuses on the effect of cavitation on a marine current turbine and implements the two-phase flow simulations based on the Rayleigh-Plesset cavitation model. It can be found that under the influence of cavitation, the power and thrust coefficients of turbine decrease and especially near the blade tips. Due to the non-uniform hydrostatic pressure along the depth of water, the present work takes this effect into account and finds a larger cavitation area appears on the blade tips at a shallow submergence. Meanwhile the power and thrust performance of one blade change during its rotating period. In order to analyze the cavitation performance of turbine locates at a certain water depth, the variations of inflow velocity and rotor rotating speed are implemented. It indicates that, with the inflow velocity increasing, the C_P and C_T of turbine with higher rotation speed will significantly reduce due to the occurrence of cavitation, and the influence on the turbine with lower speed is small. These predicting results can provide implications for the safe and stable operation of marine current turbines.

1. Introduction

Cavitation is a general fluid mechanics phenomenon and usually has a potentially undesirable effect on the hydraulic machinery. Marine current turbines show many similarities in design and operating characteristics to the wind turbines with high-aspect ratio blades, but an important difference is the marine current turbines has the potential to suffer cavitation [1]. Cavitation may damage the blade surface material and affect the operations of marine current turbines.

An investigation of cavitation performance on two-dimensional sections can be used as a starting point in the preliminary design stage for marine current turbines, as the cavitation tunnel experiments and the numerical simulations for several suitable sections have demonstrated that how the section shape and some parameters of the foil affect the cavitation performance [1-2]. The three-dimensional cavitation flow conditions has been shown by a test in a cavitation tunnel with a model of the marine current turbine, and the investigation has indicated that careful design of blade angles and tip speeds should be possible to prevent cavitation on full scale marine current turbines [3]. Wang D et al [4] found the turbine can experience strong unstable sheet and cloud cavitation as well as strong tip vortex

[3] Corresponding author.

cavitation at a shallow depth by experimental results. Batten W M J et al. [5] predicted that the cavitation could occur on both sides of the blade and both for the design case or when the blade is heavily pitched for power regulation, so all areas of operation would need to be considered when designing a rotor. According to the definition of the cavitation number, which is a function of pressure and velocity, it can be inferred that the likelihood of cavitation is related to the locations upon blades, specifically the greatest chance of cavitation damage will occur at the blade tips when at the highest point of rotation [6]. Due to the variation of the hydrostatic pressure in different water depth, the cavitation number in one point on the blade will change with the rotor rotating, especially on the blade tip of a large turbine [7]. Based on the interest of the cavitation characteristics in the process of rotating, this paper simulates a marine current turbine model with the effect of the vertical hydrostatic pressure. The tests of cavitation are always carried out in a fully enclosed tunnel by reducing the pressure, where the pressure field differs from the undersea environment for the prototype turbine, thus the present work studies the effect of incident flow velocity on the cavitation.

2. Methodology

2.1. Computational domain and mesh generation

The model turbine was from the cavitation tunnel tests of Bahaj A S et al [3] with a 0.8 m diameter (D) rotor arranged to the fully enclosed tunnel with the size of 5 m×2.4 m×1.2 m. The experimental results with 20° hub pitch angle and U_T=1.4 m/s at zero yaw were used as the reference for the simulations. The computational domain included two parts as the rotating inner domain with the rotor located at the coordinate system's origin and the stationary outer domain with symmetrical sizes in both height and width directions. The unstructured tetrahedral mesh was generated in the inner domain and the structured hexahedral mesh was for the outer domain. Mesh sensitivity was guaranteed by comparison with the variation of mesh nodes number. Figure 1 shows the schematic computational domain and the mesh structure in a cross section.

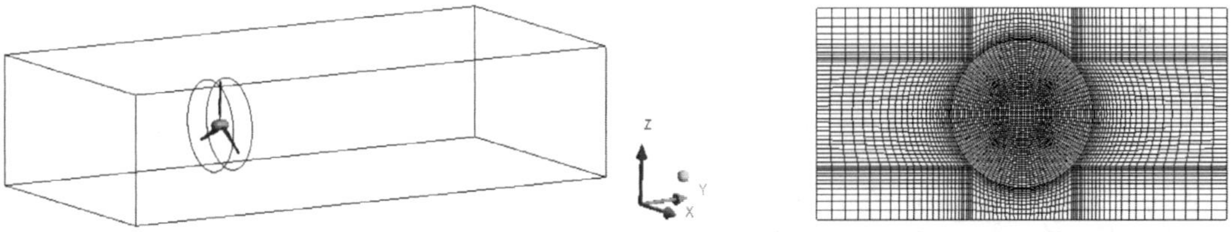

(a) The schematic computational domain. (b) The mesh structure in a cross section.

Figure 1. Computational domain and mesh generation.

2.2. Numerical simulation

A constant inflow condition of U_T was applied to the inlet with $2D$ upstream distance from rotor. Opening condition with different pressures based on the cavitation numbers (σ) defined as equation (1) [3] was set in downstream.

$$\sigma=\frac{P_T - P_V}{0.5\rho V^2} \tag{1}$$

where P_T is the tunnel pressure, P_V is vapour pressure with the value of 2000 N/m^2, ρ is the fluid density and V is referred to the velocity at the blade tip as $V=\sqrt{U_T^2+(\omega R)^2}$ relating to the angular velocity ω and rotor radius R.

Numerical simulations were performed in ANSYS CFX with RNG k-ε turbulence model and Rayleigh-Plesset cavitation model. The high resolution scheme was used for the advection scheme.

3. Results and discussion

3.1. Cavitation features and effects

The simulation type is transient with the consideration the effect of gravity. According to the experimental results that cavitation appeared until the cavitation number was reduced to about 0.9 and there was strong cavitation with cavitation number was below about 0.4, so the experimental condition of σ =0.44 at TSR=8.7 was compared with a non-cavitation condition with σ =1.34 (P_T=101325 N/m^2) at the same TSR. For marine current turbines, the non-dimensional power and thrust coefficients (C_P and C_T) were defined as $C_P = T\omega/(0.5\rho U_T^3 A)$ and $C_T = F/(0.5\rho U_T^2 A)$, where T and F is the torque and thrust, and A is rotor swept area. Considering the blockage corrections [3], figure 2 shows the azimuthal variation of corrected C_P and C_T for the single blade (dash line) and for the whole rotor (solid line) on both σ =0.44 and σ =1.34 conditions. The azimuthal angle θ was defined as the locations of the coloured blade in its period of rotating clockwise shown in figure 2.

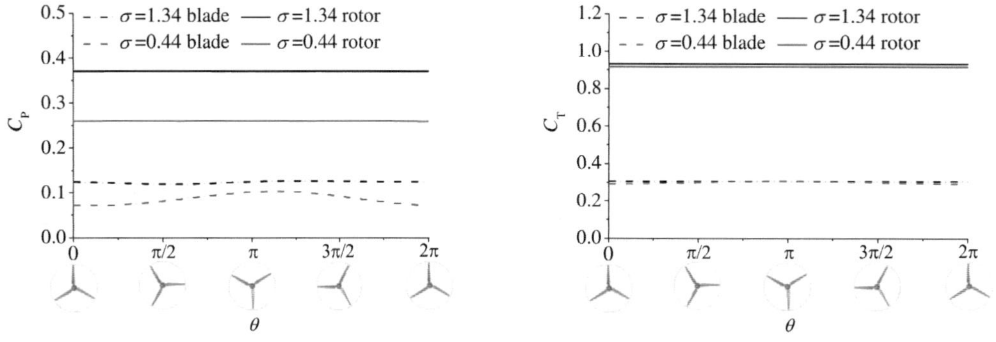

Figure 2. Azimuthal variation of C_P and C_T for the single blade and rotor.

Compared with the available data from the towing tank experiment with 20° hub pitch at 1.4m/s towed speed in [3], the deviation for simulation results of σ =1.34 were less than 5% for C_P and 1% for C_T. This deviation is mainly due to the differences between the test tank and cavitation tunnel results compared in the experiment and the simplified rotor geometry without support structures used in the simulation. The calculations of σ =1.34 can be used as a validated reference for the cavitation condition (σ =0.44) within the error range. Figure 2 reveals the power performance was negatively affected by the cavitation with the declined C_P values, but the effect on the thrust performance was small. It is interesting that, for the condition of σ =0.44, the single blade has the maximum value for C_P with its tip near the lowest point of rotation, and the C_P values decline with the higher location of the single blade . The variation of C_T values has the same trend within small scope.

In order to further analyse the effect from cavitation on the power and thrust performance for the single blade, figure 3 shows the radial variation of C_P and C_T for three azimuthal locations on both non-cavitation and cavitation conditions. Here the single blade except the blade root is divided into 16 radial slices whose δC_P and δC_T can be obtained.

It can be seen that the effect on C_P and C_T from cavitation mainly appears near the blade tip region within r/R >0.7 and it is more serious when the blade locates at the higher point of rotation with θ =0. Considering the hydrostatic pressure varied with the water depth, it can be speculated that the pressure is lower with a shallow depth of submergence and where the cavitation will be more serious. According to the relationship between the negative pressure coefficient (-$C_{pressure}$) and the cavitation number (σ) [8], figure 4(a) shows the distribution of -$C_{pressure}$ for a foil near blade tip (at r/R=0.9) at three azimuthal locations. It indicates that the cavitation mainly appears near the trailing edge on the suction side of blades, and the radial length of cavities on blade ① is longer than that on the other blades slightly shown in figure 4(b) by the iso-surface of the vapour volume fraction. On the pressure side, there is even small region with cavitation near the leading edge for the blade with shallow depth of submergence, shown in figure 4(a, c).

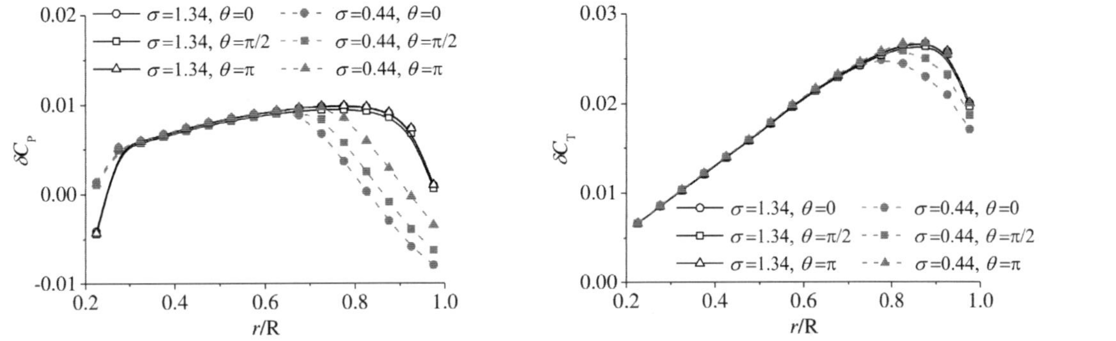

Figure 3. Radial variation of δC_P and δC_T for the single blade at different azimuthal angle.

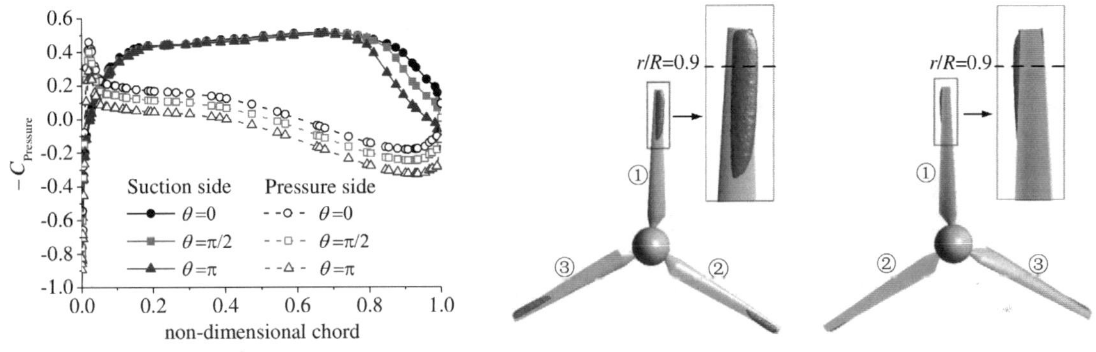

(a) Pressure coefficient for a foil at r/R=0.9. (b) Suction side. (c) Pressure side.

Figure 4. Cavitation position and the distribution of pressure coefficient on a foil section.

Based on the variation of cavitation region and the power and thrust performance mentioned above, one blade operates in the rotating period will undergo the alternating effect and may lead to the fatigue damage or noise. The main factor is the hydrostatic pressure in different water depth and its effect will be more serious with the size of turbines increasing [7]. Experimental study on cavitation always use the closed tunnel with pressure decreased, but the ambient pressure is relative higher when the prototype turbines operate in river or ocean. In addition to the effect of pressure, the cavitation number is related to the velocity as shown in equation (1), thus the next section focus on the effect of velocity on the cavitation.

3.2. The effect of velocity on cavitation

The marine current turbines always locate at a relative deep depth of water to avoid interference of waves [9], and then the ambient pressure will be higher than an atmosphere generally. In order to simplify the calculation model, the free surface is neglected and the ambient pressure is set as an atmosphere which represents a shallow submergence for the turbine. The velocity V in equation (1) can be transformed into $V=U_T\sqrt{1+\mathrm{TSR}^2}$ which is related to the inflow velocity U_T and the non-dimensional TSR. The mean peak tidal currents are generally over 2m/s in the world [5]. Figure 5 shows the power and thrust performance with blockage corrections of present turbine under different TSR conditions with the various inflow velocities. At the same time, the cavitation numbers (σ) at each condition are shown near the marks. The cavitation regions are shown with the distribution of pressure in figure 6 on some typical conditions. Due to the limit time, these results are from the steady simulations.

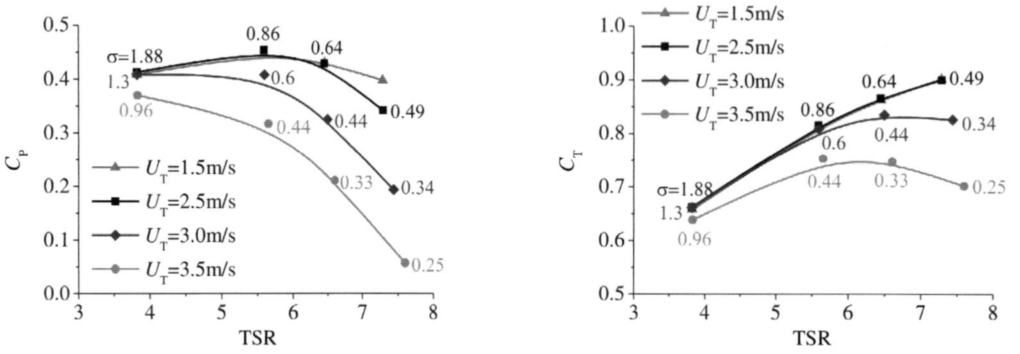

Figure 5. Power and thrust performance of the turbine under different inflow velocity.

There is no distinct cavitation in the simulations with U_T=1.5 m/s and the most suitable condition is near TSR=6 which was demonstrated in the experiment [3]. Thus the performance can be used as non-cavitation results in the subsequent analysis. For the calculations with U_T=2.5 m/s, the C_P has a decreasing trend at higher TSR with a lower value of σ, as shown in Figure 6(c) with cavities near blade tip, while the C_T still agrees well with the non-cavitation result. With the increasing of U_T, the optimal condition with maximum C_P value moves to the lower TSR due to the occurrence of serious cavitation at higher TSR conditions where the C_T decreases significantly. It is worth noting that Figure 6(d) shows the cavitation appears on the leading edge in a long and narrow region, but its power and thrust performance has almost not been affected. Until U_T increases to 3.5 m/s, there are also some reduction on C_P and C_T. In general, the cavitation comes from the variation of inflow velocity has more serious effect on the conditions at higher TSR. Therefore, properly reducing the rotational speed of rotor may be beneficial to prevent the cavitation. However for a specific power output, the rotor operates at lower speed need to support a larger torque [7], which requires a stronger structure. In figure 5 and 6 the model turbine is located at a shallow submergence with an atmosphere as the ambient pressure approximately. With the submerged depth increasing, the ambient pressure will rise and lead to a high cavitation number, thus the cavitation performance will be improved with the cavitation area reducing. According to the distribution of pressure on the suction side in figure 6, it can be speculated that the variation trend for the cavitation region. Considering the blockage effect in the confined tunnel, the cavitation characteristics cannot be corrected thus a further research in a wider computational domain is needed.

4. Conclusions

The two-phase flow calculations using CFX's homogenous two-phase model together with the Rayleigh-Plesset cavitation model was performed on a marine current turbine model. Some observations made as results are as follows.

(1) Based on the unsteady simulation including the hydrostatic pressure varied with the water depth, it validates that the more serious cavitation occurs on the blade tips when at the shallow water depth of submergence. The turbine shows a significantly reduction on C_P but very little change on C_T, and these effects from cavitation are embodied in the region near blade tips especially. Meanwhile, the power and thrust performance of one blade will change but the performance of the whole rotor is almost invariable in its rotating period.

(2) Assuming the turbine locates at a certain water depth, the variation of inflow velocity has effects on the cavitation characteristics. With the incident velocity increasing, the C_P and C_T of turbine with higher rotation speed will significantly reduce due to the occurrence of cavitation. In contrast, there is

small influence on the turbine operating at lower TSR conditions although some cavities appearing on the leading edge of blade.

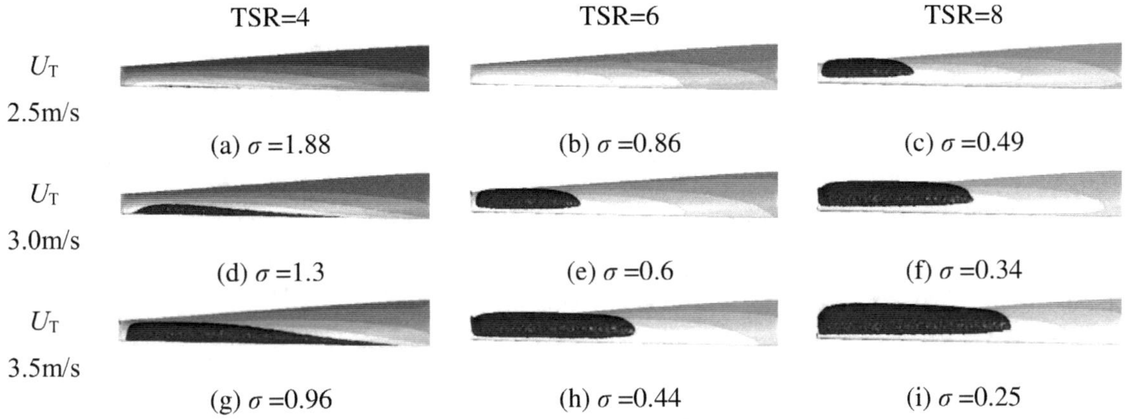

Figure 6. Cavitation positions and pressure contours on the suction side under different inflow velocity.

Acknowledgements
The authors thank the Open Research Fund Program of State Key Laboratory of Hydroscience and Engineering (No. sklhse-2012-E-01) and the Specialized Research Fund for the Doctoral Program of Higher Education (No. 20130002110072) for supporting the present work.

References
[1] Batten W M J, Bahaj A S, Molland A F and Chaplin J R 2006 *Renew. Energy* **31** 249-56
[2] Molland A F, Bahaj A S, Chaplin J R and Batten W M J 2004 Measurements and prediction of forces, pressures and cavitation on 2-D sections suitable for marine current turbines *Proc. Inst. Mech. Eng. Part M: J. Eng. Marit. Environ.* **218** (2) 127-38
[3] Bahaj A S, Molland A F, Chaplin J R and Batten W M J 2007 *Renew. Energy* **32** 407-426
[4] Wang D, Atlar M and Sampson R 2007 *Proc. Inst. Mech. Eng. Part A: J. Power Eng.* **221** 219-31
[5] Batten W M J, Bahaj A S, Molland A F and Chaplin J R 2008 *Renew. Energy* **33** 1085-96
[6] Bahaj A S and Myers L E 2001 Cavitation prediction in operating marine current turbines *Proc. Conference C76 of the Solar Energy Society (Belfast, UK, 6-14 Sept 2001)*
[7] Ying Y, Li W, Liu H W and Lin Y G 2008 *Compressor Blower & Fan Technology* **4** 8-11
[8] Zhao W G, Zhang L X, Xin X P, Shao X M and Li W 2011 *Jixie Gongcheng Xuebao* **47** 171-6
[9] Bahaj A S, Myers L E, Rawlinson-Smith R I and Thomson M 2011 *J. Offshore Mech. Arct. Eng.* **134** 021104

IOP Conf. Series: Materials Science and Engineering **72** (2015) 042046 doi:10.1088/1757-899X/72/4/042046

Study about the influence of cavitation on the dynamic characteristics for the sliding bearing

L M Zhai, Y Y Luo and Z W Wang[1]
State Key Laboratory of Hydroscience and Engineering & Department of Thermal Engineering, Tsinghua University, Beijing 100084, China

E-mail: wzw@tsinghua.edu.cn

Abstract. Sliding bearings are employed to support the rotor system and limit the vibration amplitude. In high speed rotor system, cavitation often occurs in the oil film and affects the dynamic characteristics of the sliding bearing greatly. In this paper, numerical method is adopted to simulate the cavitation in the oil film with homogeneous two-phase mixture flow using Singhal-et-al cavitation model in the commercial code FLUENT-solver. Cases without cavitation model were also calculated at the same time. Many computations with different frequency ratios were conducted. Then the rotor dynamic characteristics of the sliding bearing were retrieved. The results show that the cavitation has great influences on the pressure distribution in the oil film. As the rotational speed or whirling speed of the journal increases, the cavitation will become prominent. The dynamic coefficients of the bearing such as stiffness and damping with cavitation model considered are quite different from that without cavitation. So it is worth to pay attention to and do further study about the cavitation in the sliding bearing in the high speed rotor system.

1. Introduction

Sliding bearings has been widely used in rotating machinery which enables modern rotating machinery more stable and efficient. Oil is usually used in the bearing for lubricating the journal, supporting the shaft weight and taking away the viscosity friction heat. During the operation, the displacement perturbation will induce the rotor vibration which includes two kinds of motion, rotational speed around the shaft center and whirling speed around the balanced location. In this situation, the dynamic characteristics of the bearing such as direct and coupled stiffness and damping have great effects on the vibration amplitude. However, when the rotational speed or the whirling speed is too high, the local pressure in the oil film will fall below the specific value and induce the cavitation phenomenon, which will change the dynamic properties of the bearing and affect the vibration characteristics.

Cavitation around a hydrofoil has been thoroughly researched [1, 2], but the cavitation in the sliding bearing is been studied in recent years. Many experiments show the oil film doesn't rupture completely, but presents thin strips [3]. Sun DC [4] did the experiment to show that the negative zone includes both air bubbles and oil vapors. Some numerical studies for the cavitation in the sliding bearing are also conducted. By solving the Reynolds equations, Li Qiang [5] put forwards a new method which can introduce the JFO boundary condition more easily, and compared the numerical

[1] Corresponding author.

results with the experimental results. Yang Ying [6] also solved the Reynolds equations with cavitation model using finite difference method. Furthermore, she united the energy equation to get the temperature distribution in the film. Guo Zenglin [7] first built the 3D CFD model for the sliding bearing. The results with CFD method matched the results by solving the Reynolds equation. However, the cavitation effects in the bearing was not considered. Song Yin [8] put forwards a new cavitation model based on air solubility. With this model, Computation for a journal bearing using CFD method was conducted. The numerical results matches well with the experiments. The new method shows better accuracy than the Half-Sommerfeld cavitation boundary condition. However, there are few papers that discuss the influence of the cavitation on the dynamic characteristics of the sliding bearing. It is meaningful for the rotor-bearing coupled system.

This paper focus on the cavitation in the sliding bearing. A 3D model of the oil film flow field was built. Homogeneous two-phase mixture flow with Singhal et al cavitation model was used in the commercial code FLUENT-solver. Cases without cavitation model were also calculated at the same time. Many computations with different frequency ratios were conducted. Then the rotor dynamic characteristics of the sliding bearing were retrieved. The influence of the cavitation on the dynamics properties of the sliding bearing were discussed in detail.

2. Numerical method

1.1. Governing equations

Mixture model is chosen for the cavitation computation in this paper. All fluids are in one flow field share one velocity field. The mixture model assumes that each quantity transported for all phases are same. The continuity and momentum equations are as follows:

1.1.1. Continuity equation. Mixture phase:

$$\frac{\partial \rho}{\partial t} + \nabla \cdot (\rho \boldsymbol{u}) = 0 \tag{1}$$

Bubble phase:

$$\frac{\partial(\alpha_v \rho_v)}{\partial t} + \nabla \cdot \alpha_v \rho_v \boldsymbol{u} = R_e - R_c \tag{2}$$

1.1.2. Momentum equation:

$$\frac{\partial}{\partial t}(\rho \boldsymbol{u}) + \nabla \cdot (\rho \boldsymbol{u}\boldsymbol{u}) = \rho \boldsymbol{g} - \nabla p + \frac{1}{3}\nabla[(\mu + \mu_t)\nabla \cdot \boldsymbol{u}] + \nabla \cdot [(\mu + \mu_t)\nabla \boldsymbol{u}] \tag{3}$$

Where $\rho = \alpha_v \rho_v + \alpha_l \rho_l$ is the density of mixture phase, ρ_v is bubble phase density, ρ_l is the liquid phase density, α_v is the volume fraction of the bubble phase, α_l is the volume fraction of the liquid phase, $\boldsymbol{u}$ is the time mean velocity, μ is the dynamic viscosity coefficient of the mixture phase, μ_t is the eddy viscosity coefficient of the mixture phase, R_e is the bubble formation rate, R_c is the steam condensation rate.

1.2. Cavitation model

There are three cavitation model in ANSYS FLUENT including Singhal et al. model, Zwart-Gerber-Belamri model and Schnerr and Sauer model. Singhal-et-al cavitation model is used here which is based on the "full cavitation model", developed by Singhal et al. [9]. It has the capability to account for multiphase flows or flows with multiphase species transport, the effects of slip velocities between the liquid and gaseous phases, and the thermal effects and compressibility of both liquid and gas phases.

Using the two-phase continuity equations, Singhal et al. derives an expression of the net phase change rate (R) as follows:

IOP Conf. Series: Materials Science and Engineering **72** (2015) 042046 doi:10.1088/1757-899X/72/4/042046

$$R = (n4\pi)^{\frac{1}{3}}(3\alpha)^{\frac{2}{3}}\frac{\rho_v\rho_l}{\rho}\left[\frac{2}{3}\left(\frac{P_B - P}{\rho_l}\right)\right]^{\frac{1}{2}} \tag{4}$$

Here R represents the vapour generation or evaporation rate, that is, the source term R_e. n is the bubble number density. α is the vapour volume fraction. l is liquid phase, v is vapor phase, ρ is mixture density All terms, except n, are either known constants or dependent variables. In the absence of a general model for estimation of the bubble number density, the phase change rate expression is rewritten in terms of bubble radius (R_B), as follows:

$$R = \frac{3\alpha}{R_B}\frac{\rho_v\rho_l}{\rho}\sqrt{\frac{2}{3}\left(\frac{P_B - P}{\rho_l}\right)} \tag{5}$$

As for bubble collapse or the condensation process, though it is expected to be different from that of bubble growth, as a first approximation, also often used to model the bubble collapse by using the absolute value of the pressure difference and treating the right side as a sink term.

It may be noted that in practical cavitation model, the local far-field pressure P is usually taken to be the same as the cell center pressure. The bubble pressure P_B is equal to the saturation vapour pressure in the absence of dissolved gases, mass transport and viscous damping, that is, $P_B = P_v$. Where, P_B is bubble pressure, P_v is saturation vapor pressure.

Based on equation (5), Singhal et al. proposed a model where the vapor mass fraction is the dependent variable in the transport equation. This model accommodates also a single phase formulation where the governing equations is given by:

$$\frac{\partial}{\partial t}(f_v\rho) + \nabla\cdot\left(f_v\rho\vec{V_v}\right) = \nabla\cdot(\Gamma\nabla f_v) + R_e - R_c \tag{6}$$

Where, f_v is vapor mass fraction, f_g is noncondensable gases, Γ is diffusion coefficient.

The rates of mass exchange are given by the following equations:

If $P \le P_v$

$$R_e = F_{vap}\frac{max\left(1.0,\sqrt{k}\right)\left(1 - f_v - f_g\right)}{\sigma}\rho_v\rho_l\sqrt{\frac{2}{3}\frac{(P_v - P)}{\rho_{ell}}} \tag{7}$$

If $P > P_v$

$$R_c = F_{cond}\frac{max\left(1.0,\sqrt{k}\right)f_v}{\sigma}\rho_v\rho_l\sqrt{\frac{2}{3}\frac{(P_v - P)}{\rho_l}} \tag{8}$$

The saturation pressure is corrected by an estimation of the values of the turbulent pressure fluctuations:

$$P_v = P_{sat} + \frac{1}{2}(0.39\rho k) \tag{9}$$

The constants have the values $F_{vap} = 0.02$ and $F_{cond} = 0.01$. In this model, the liquid-vapor mixture is assumed to compressible. Also, the effects of turbulence and the noncondensable gases have been taken into account.

1.3. Approach for the dynamic characteristics of the journal bearing

Due to some factors such as unbalanced mass or axial misalignment, the journal usually rotates offset the bearing center (eccentricity, e) which can be resolved into two kinds of motion, rotation around the journal center O_1 with speed Ω, and the rotation around the bearing center O with speed ω, as shown in Figure 1. The forces that the oil film impacts to the rotor in a plane perpendicular to the axis of rotation are depicted in Figure 1, and decomposed into components in the direction x and y ,where this coordinate system is fixed in the framework of the rotor. The instantaneous forces are denoted by $F_x(t)$, $F_y(t)$, and the time-averaged values of these forces in the stationary frame are denoted by F_{0x}, F_{0y} which can be described using the following function:

$$F = f(x, y, \dot{x}, \dot{y}, \ddot{x}, \ddot{y}) \tag{10}$$

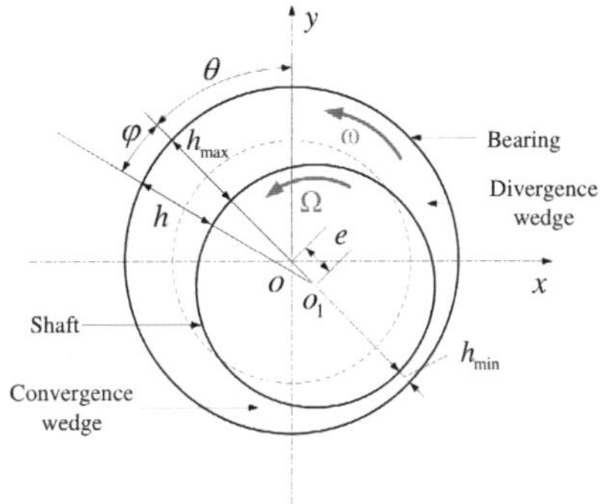

Figure 1. Schematic diagram of the journal bearing.

It is usually assumed that this displacement is sufficiently small so that linear perturbation model is accurate. The fluid force can be expanded in linear Taylor's series as following equation (11):

$$\begin{bmatrix} F_x \\ F_y \end{bmatrix} = \begin{bmatrix} F_{x0} \\ F_{y0} \end{bmatrix} + \begin{bmatrix} K_{xx} & K_{xy} \\ K_{yx} & K_{yy} \end{bmatrix} \begin{bmatrix} x(t) - x_0 \\ y(t) - y_0 \end{bmatrix} + \begin{bmatrix} C_{xx} & C_{xy} \\ C_{yx} & C_{yy} \end{bmatrix} \begin{bmatrix} x(\dot{t}) - \dot{x_0} \\ y(\dot{t}) - \dot{y_0} \end{bmatrix} + \begin{bmatrix} M_{xx} & M_{xy} \\ M_{yx} & M_{yy} \end{bmatrix} \begin{bmatrix} x(\ddot{t}) - \ddot{x_0} \\ y(\ddot{t}) - \ddot{y_0} \end{bmatrix} \tag{11}$$

Where $[K]$ is the stiffness matrix, $[C]$ is the damping matrix, $[M]$ is the mass matrix, $x(t)$ and $y(t)$ are the displacement of the journal along direction x and y. x_0 and y_0 are the displacement at the equilibrium position.

One particular feature of the rotor dynamic matrices, $[K]$, $[C]$ and $[M]$, deserves special note. There are many geometries in which the rotor dynamic forces should be invariant to a rotation of the x, y axes [10]. Such will be the case only if

$$\begin{gathered} K_{xx} = K_{yy} = K; \ K_{xy} = -K_{yx} = k \\ C_{xx} = C_{yy} = C; \ C_{xy} = -C_{yx} = c \\ M_{xx} = M_{yy} = M; \ M_{xy} = -M_{yx} = m \end{gathered} \tag{12}$$

Assumed that $\boldsymbol{x(t)}$ reaches to the maximum displacement when t is 0. So

$$\begin{gathered} x(0) = e, y(0) = 0, x(\dot{0}) = 0, \\ y(\dot{0}) = e\omega, x(\ddot{0}) = -e\omega^2, y(\ddot{0}) = 0 \end{gathered} \tag{13}$$

Plug the equations (12) and (13) into formula (11), the following formulas can be obtained.

$$\frac{F_x}{e}|_{t=0} = M\omega^2 - c\omega - K$$
$$\frac{F_y}{e}|_{t=0} = -m\omega^2 - C\omega + k \quad (14)$$

Therefore, all six rotor dynamic coefficients can be directly evaluated from quadratic curve fits to the graphs of F_x and F_y against ω.

3. Numerical model and boundary conditions

To provide a better grid, hex grid was used to mesh the 3D flow field of the oil film in figure 2. The structure parameters of the sliding bearing is list in the Table. 1. Here the eccentricity ratio is 0.2 long the direction X, and it is assumed that it is linear with the eccentricity. After grid independence check, the flow field of the water film was meshed with the grid number of 162,690 and 10 layers in the clearance. Because both the rotational effect and squeeze effect should be considered, rotating coordinate system was used in this study. In this method, a rotating coordinate system with the speed ω same as the whirling speed is set for the whole flow region. The rotating speeds of the journal and bearing wall are set as $\Omega - \omega$ and $-\omega$ relative to the rotating coordinate system, respectively. The clearance is so small relative to the radius that the Reynolds number is very small, so the laminar model was chosen. Computation cases with difference combination of rotational speed and whirling speed are conducted. Here, 2000 Pa total pressure is used as the orifice inlet boundary. Static pressure (0 Pa) is set as the outlet boundary condition. The rotational speed changes from 1000 rpm to 5000 rpm. At each rotational speed, 7 whirling speeds were chosen as the frequency ratios (ω/Ω) are 0, 0.25, 0.50, 0.75, 1.00, 1.25, 1.50.

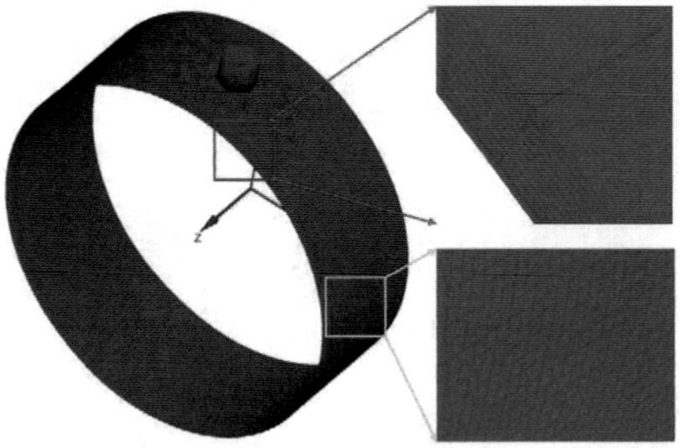

Figure 2. Computational grids for the sliding bearing.

Table 1. Some parameters of the bearing and oil.

Item	Symbol	value
Pad radius [mm]	R_1	20.00
Journal radius [mm]	R_2	19.98
Average gap [mm]	c	0.02
Bearing length [mm]	L	15.00
Oil orifice radius [mm]	r	4.00
Eccentricity ratio[mm]	e	0.2
Oil density [kg/m3]	ρ	890
Oil viscosity [kg/(m·s)]	ν	0.048
Vaporization pressure [Pa]	P_{vapor}	7550
Surface Tension Coefficient [N/m]	ξ	0.035

4. Results and analysis

4.1. Influence of the cavitation on the pressure in the oil film

Figure 3 shows the numerical result comparison between with cavitation and without cavitation model when rotational speed $\Omega = 3000$ rpm. In these cases the journal has an eccentricity ratio 0.2 along the direction X. Both the rotation speed and whirling speed rotate around the direction Z using the right-hand rule. Figure 3(a~c) show the results without whirl speed, while (e~f) show that of synchronous whirl. Comparing (a) with (b), the high pressure are formed in the convergence wedge. However, there is obvious zone of negative pressure in the divergence wedge in figure 3(a), while no negative pressure zone is observed in figure 3(b) considering the cavitation model. It can also be seen that the air volume fraction distribution concentrates in the divergence wedge where the local pressure usually falls down along the rotational direction, which explains why no negative pressure zone is observed in the figure 3(b). In figure 3(d), the high pressure zone is located in the divergence wedge while the negative pressure zone is in the convergence wedge because the squeezing effect of the whirl is greater than the rotational effect of the spinning in Synchronous whirl motion. Without cavitation model, high pressure is still located in the divergence wedge and no negative pressure exists as shown in figure 3(e) which can be explained by the air volume fraction distribution in figure 3(f).

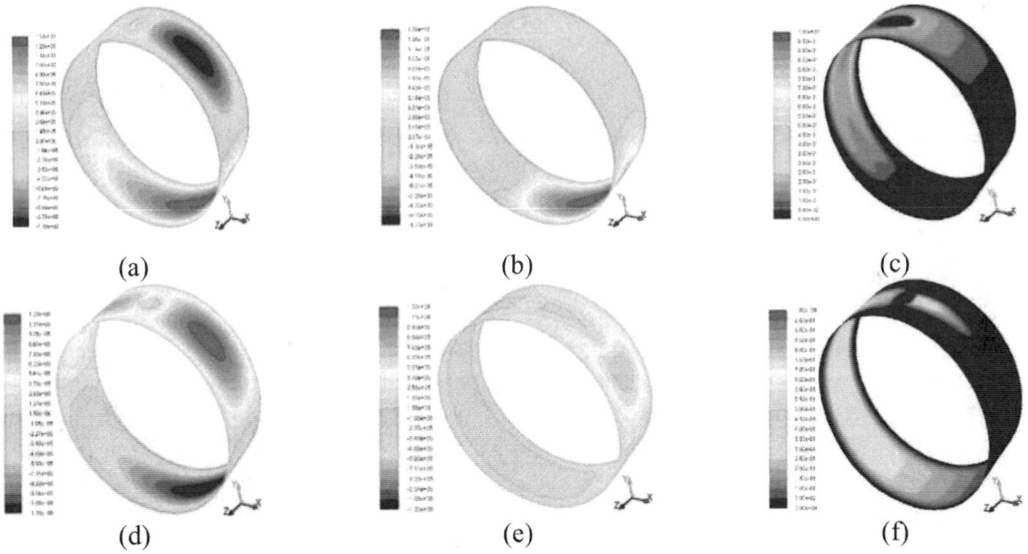

Figure 3. Computational result comparison with cavitation and without cavitation when $\Omega = 3000$ rpm: (a~c) $\omega = 0$ rpm, (e~f) $\omega = 3000$ rpm, (a, d) without cavitation, absolute pressure, (b, e) with cavitation, absolute pressure, (c, f) *with* cavitation, air volume fraction

Figure 4 shows the influence of the whirl speed on the oil film force in X direction at different rotational speeds. Figure 4(a) shows the results without cavitation while figure 4(b) considers cavitation. Five cases with rotational speeds of 1000~5000 rpm are calculated where 7 frequency ratio (ω/Ω) of 0.0~1.5. Fx increases almost linearly as the frequency ratio and rotational speed increase when no cavitation model is considered as shown in figure 4(a). In contrast, considering cavitation, Fx shows obvious 2 order nonlinearity as the frequency ratio increases in figure 4(b). Figure 5 shows the influence of the whirl speed on the oil film force in Y direction at different rotational speeds. Quite difference can also be observed between no cavitation model and cavitation. Fy decreases almost linearly as the rotational speed increases in figure 5(a), while the Fy shows 2 order nonlinearity in figure 5(b).

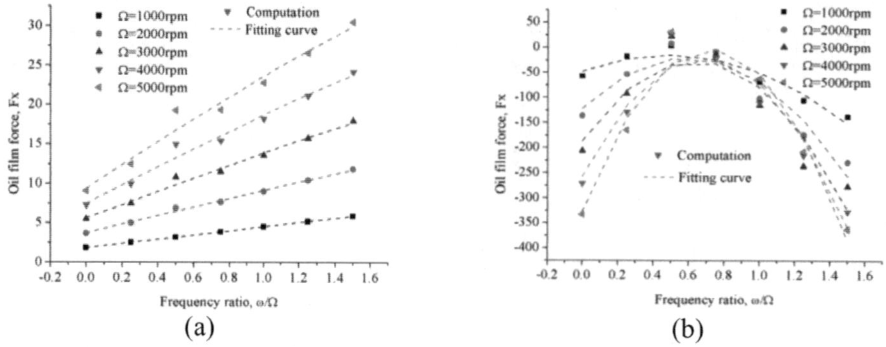

Figure 4. Oil film force on the journal in x direction with difference frequency ratio: (a) with cavitation (b) no cavitation

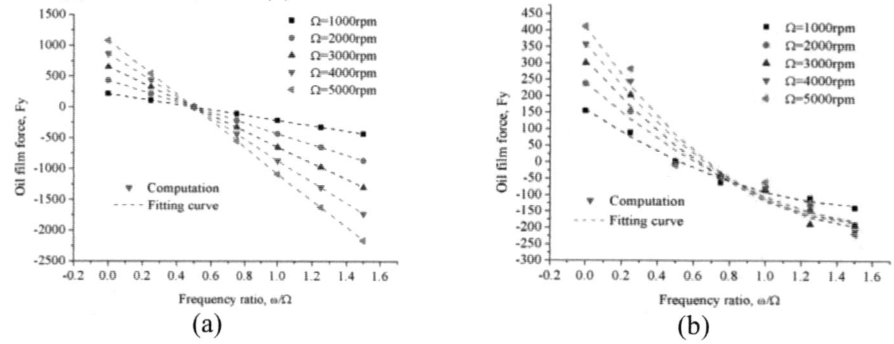

Figure 5. Oil film force on the journal in y direction with difference frequency ratio: (a) with cavitation (b) no cavitation

4.2. Influence of the rotational speed on the pressure properties in the oil film

Figure 6 shows the influence of the rotational speed on the pressure distribution in the oil film with no whirl speed, ω = 0 rpm. The high pressure is formed in the convergence wedge. As the rotational speed increases from 1000 to 5000 rpm, the pressure increases are shown in the figure 6(a)~(b). Correspondingly, the cavitation area and air volume fraction increase in the divergence wedge with the rotational speed as shown in figure 7. Figure 8 shows the cavitation rate of the oil film. It can be seen that the cavitation rate increase as the rotational speed increases at each frequency rate except the rate 0.75. At each rotational speed, the cavitation rate first decreases, reaches the lowest point at frequency rate 0.5, and then increases as the frequency rate increases. Therefore, half-speed whirl can weaken the cavitation in the oil film.

4.3. Approach for the dynamic characteristics of the journal bearing

According to the equation (14), there are six unknown variables including direct and coupled stiffness, damping and added mass. One whirl speed can provide two equations for Force X and Force Y. So at least three whirl speed (ω) need to be calculated for closing the equation sets. Here seven whirl speeds are used at each rotational speed to improve the calculation accuracy. The Fx and Fy at each set of rotational speed and whirl speed with cavitation and no cavitation are shown in figure 4 and 5.

Based on the equations (14) and the results in figure 4 and 5, the dynamics characteristics of the sliding bearing including stiffness, damping are calculated as shown in figure 9. In figure 9(a), the direct stiffness (K) and coupled stiffness (k) are nonlinear with the rotational speed. Direct stiffness with cavitation is larger than that without cavitation, while coupled stiffness with cavitation is smaller than that without cavitation. Figure 9(b) shows the damping coefficients of the bearing. The direct and coupled damping without considering cavitation keep almost constant with the rotational speed, while the damping considering cavitation decreases as the rotational speed increases. The direct damping with cavitation is smaller than that without cavitation, similar to the coupled damping. Therefore the cavitation has great effects on the stiffness and damping.

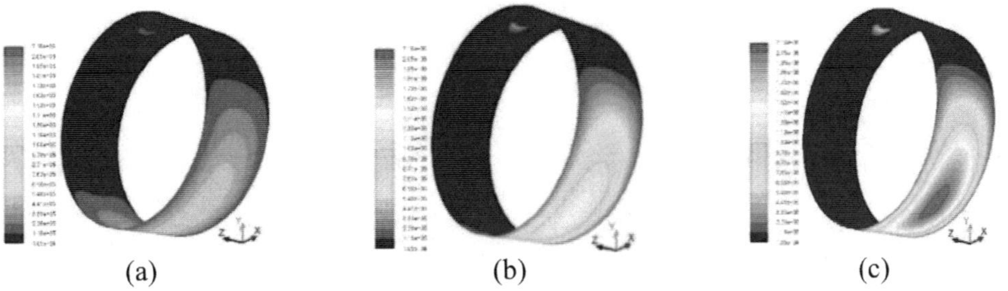

(a) (b) (c)

Figure 6. Influence of rotational speed on the pressure distribution when $\omega = 0$ rpm
(a) $\Omega = 1000$ rpm (b) $\Omega = 3000$ rpm (c) $\Omega = 5000$ rpm

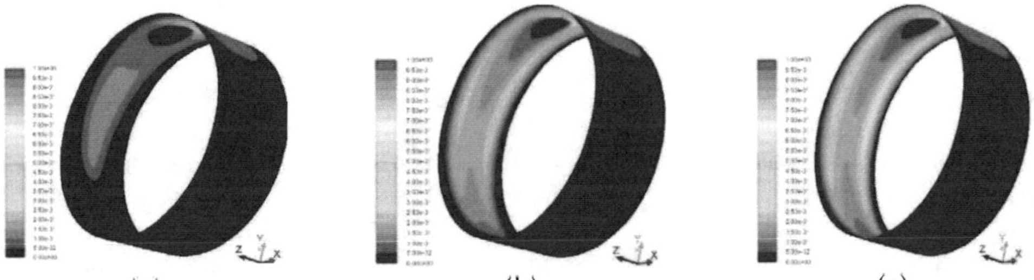

(a) (b) (c)

Figure 7. Influence of rotational speed on the air volume fraction when $\omega = 0$ rpm
(b) $\Omega = 1000$ rpm (b) $\Omega = 3000$ rpm (c) $\Omega = 5000$ rpm

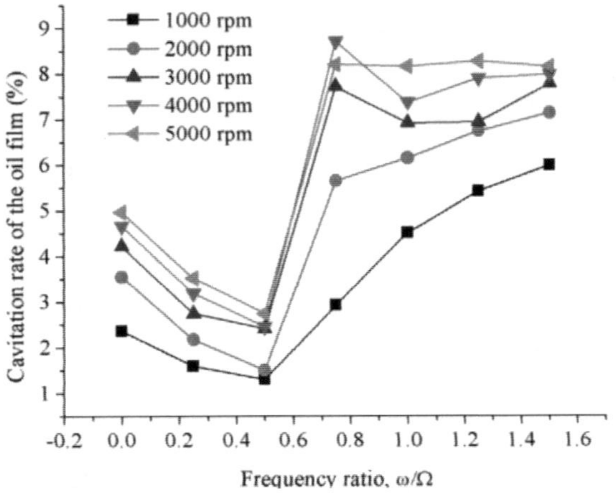

Figure 8. Cavitation rate of the oil film.

According to the bearing-rotor coupled dynamic theory, the coupled stiffness and direct damping have great influences on the stability of the rotor system. Negative coupled stiffness and positive direct damping contribute to the stability of the rotor system, while positive coupled stiffness and negative direct damping have the opposite impact. However, if both the coupled stiffness and damping are positive or negative, another method need. There is a stability criterion called whirl ratio (γ) combining both stiffness and damping effect described as the following formulation:

$$\gamma = \frac{k}{\Omega C} \tag{15}$$

Where γ is the whirl ratio, k is the coupled stiffness, Ω is the rotational speed, and C is the direct damping.

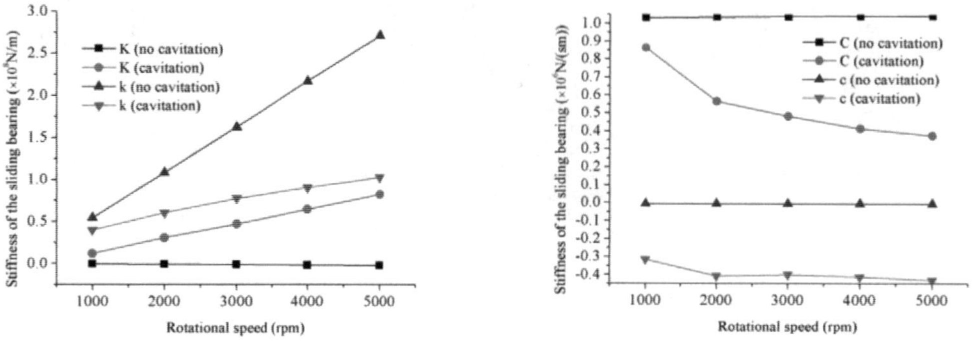

(a) (b)

Figure 9. Dynamic characteristics of the sliding bearing with cavitation and no cavitation. (a) Direct and coupled stiffness (b) Direct and coupled damping.

If both k and C are positive, the rotor will be stable when $\gamma < 1$, and the smaller the γ is , the more stable the rotor will be. If both k and C are negative, the rotor will be stable when $\gamma > 1$, and the larger γ is , the more stably the rotor will be.

Because the coupled stiffness and direct damping are positive, the stability criterion (15) is used here. As listed in Table 2, the whirl ratio γ without cavitation keeps almost constant as the rotational speed increases from 1000 rpm to 5000 rpm, while γ with cavitation increases from 0.440 to 0.522. The whirl ratio is less than 1. So the bearing is stable when the rotational speed is less than 5000 rpm. And decreasing the rotational speed contributes to strengthen the stability of the rotor system.

Table 2. Stability of the sliding bearing.

Rotational speed (rpm)	**Stability (γ)**	
	No cavitation	**Cavitation**
1000	0.499	0.440
2000	0.498	0.507
3000	0.498	0.508
4000	0.497	0.520
5000	0.497	0.522

5. Conclusions

In this paper, numerical method is adopted to simulate the cavitation in the oil film with homogeneous two-phase mixture flow using Singhal-et-al cavitation model in the commercial code FLUENT-solver. Cases without cavitation model were also calculated at the same time. Some conclusions can be obtained.

1) Comparing with the cases without considering cavitation, the cavitation changes the pressure distribution in the oil film. No negative pressure zone exits in the film because of the film rupture.

2) As the rotational speed or whirling speed of the journal increases, the cavitation will be more obvious. The dynamic coefficients of the bearing such as stiffness and damping with cavitation model

are quite different from that without cavitation. So it is worth to pay attention to and do further study about the cavitation in the sliding bearing in high speed rotor system.

In the presented analysis, there are still some deficiencies in the computation. Because this paper focus on influence of the cavitation on the dynamics of the sliding bearing, so some computation processes are simplified. The rotating coordinate system and relative rotating speed are used to simulate the transient process approximately which is not the real transient computation and loses some transient terms. It is assumed that the oil force is linear with the eccentricity of the journal which is usually nonlinear. In addition, the selection of the cavitation model can also affect the numerical results. So future work should be done to improve the numerical method to obtain more accurate results.

Acknowledgement
The authors thank the Specialized Research Fund for the Doctoral Program of Higher Education of China (Grant No. 20120002110011) and National Natural Science Foundation of China (Grant No. 51409148) for supporting the present work.

References
[1] Ji B, Luo X W and Arndt R E A 2014 *Ocean Engineering* **87** 64-77
[2] Ji B, Luo X W and Wu Y L 2013 *International Journal of Multiphase Flow* **51** 33-43
[3] Sun M L and Zhang Z M 1997 *Journal Bearing Journal of Shanghai University* **5** 500-7
[4] Sun D C and Brewe D E 1991 *Journal of Tribology* **133** 287-94
[5] Li Q, Zheng S Y and Liu Shulian 2008 *Bearing* **12** 1-4
[6] Yang Y and Lu C H 2010 *Transactions of the Chinese Society for Agricultural Machinery* **41** 215-6
[7] Guo Z L, Hirano T and Gordon K R 2005 *Journal of Engineering for Gas Turbines and Power* **127** 445-51
[8] Song Y, Li X S and Gu C W 2010 *J Tsinghua Univ (Sci & Tech)* **50** 1047-52
[9] Singhal Ashok K, Athavale M M, Li H Y and Jiang Y 2002 *J. Fluids Eng.* **124** 617-24
[10] Christopher E B 1994 Hydrodynamics of Pumps *Oxford Science Publications 81*

IOP Conf. Series: Materials Science and Engineering **72** (2015) 042047 doi:10.1088/1757-899X/72/4/042047

Numerical simulation and analysis of the internal flow in a Francis turbine with air admission

A Yu[1], X W Luo[1, 2] and B Ji[1, 3]
[1] State Key Laboratory of Hydroscience and Engineering, Tsinghua University, Beijing 100084, China
[2] Beijing Key Laboratory of CO_2 Utilization and Reduction Technology, Tsinghua University, Beijing 100084, China

Email: jibin@mail.tsinghua.edu.cn

Abstract. In case of hydro turbines operated at part-load condition, vortex ropes usually occur in the draft tube, and consequently generate violent pressure fluctuation. This unsteady flow phenomenon is believed harmful to hydropower stations. This paper mainly treats the internal flow simulation in the draft tube of a Francis turbine. In order to alleviate the pressure fluctuation induced by the vortex rope, air admission from the main shaft center is applied, and the water-air two phase flow in the entire flow passage of a model turbine is simulated based on a homogeneous flow assumption and SST k-ω turbulence model. It is noted that the numerical simulation reasonably predicts the pressure fluctuations in the draft tube, which agrees fairly well with experimental data. The analysis based on the vorticity transport equation shows that the vortex dilation plays a major role in the vortex evolution with air admission in the turbine draft tube, and there is large value of vortex dilation along the vortex rope. The results show that the aeration with suitable air volume fraction can depress the vortical flow, and alleviate the pressure fluctuation in the draft tube.

1. Introduction

To meet the variable demands of power system and take the task of power-frequency control in an electrical grid, hydro turbines are often operated under a wide range of loads. Thus it is necessary for a hydro turbine to be frequently operated under off-design conditions, especially part load conditions. People have acknowledged that the internal flow in a hydro turbine becomes rather complex at part load operations, and a kind of special swirling flow in the turbine draft tube is very vital for the safe operation of a hydropower station. This swirling flow started from the runner exit results in the development of a processing helical vortex called "vortex rope". The vortex rope is usually associated with large pressure fluctuations in the draft tube and can be rather harmful for the stable operation of a Francis turbine, especially at part load condition and often causes structural troubles. The vortex rope phenomenon has attracted a widespread attention and the intensive investigations on draft tube pressure pulsation on various specific speed turbines were performed in 1990's [1-3]. Due to the high costs of experiment investigation, and experiments can hardly reflect the detailed characteristics of the internal flow and sometimes the results of model test are not valid for the prototype, numerical simulations are needed for better understanding of the flow mechanism of the vortex rope. Computation methods for capturing the vortex rope have been studied for a long time, and some

[3] Corresponding author.

achievements have been published [4-9]. Turbulent model plays an extremely important role during the simulation for the rather complex unsteady flow in a draft tube. Studies show that Reynolds Stress Models (RSM) and Large Eddy Simulation (LES) have a more accurate result than $k-\varepsilon$ models [4-7]. Three turbulent models for vortex simulation have been compared in Reference [10].

On the other hand, researchers and engineers have presented many methods to alleviate the pressure fluctuation in a draft tube theoretically and technically, such as geometrical optimization, water injection and aeration. To aerate nature air into the draft tube is one of the most efficient way to depress the pressure fluctuations. Koichi Nakanishi and Tsuneo Ueda [11] made a systematic study by model tests. They point out the measurement of the most reasonable amount of air flow and the determination of the position of the air pipe installation. When air admission is considered, the flow in a turbine becomes a two-phase turbulent flow. This makes the simulation hard to get convergence. Qian [12] and Liao [13] simulated the pressure fluctuations with different ventilation volume based on a water-air two phase flow model, and analyzed the relationship between pressure pulsation in the whole passage and air admission. These researches focus on aeration's effectiveness of fluctuation alleviating, and little attention have been spend on internal flow characteristic and the interaction between air admission and vortex movement.

In this paper, three-dimensional unsteady turbulent flow in the whole passage of a model Francis turbine is simulated based on the homogeneous flow assumption. The pressure fluctuation caused bay vortex rope is analyzed via Fast Fourier Transform (FFT) and compared to the experiment. The relationship of air admission and pressure fluctuation in the turbine draft tube is discussed based on the internal flow analysis and vorticity transport equation.

2. Model turbine geometry and numerical methods

This paper focuses on the aeration influence on the internal flow in a model Francis turbine, whose structure is showed in figure 1. The Francis turbine consists of a spiral casing, stay vanes and guide vanes, a runner, and a draft tube. An aeration hole is made along the turbine shaft center. The parameters of the model turbine are as follows: runner diameter i.e. D_1=420 mm, runner blade number Z_b=17, relatively height of the guide vane b_0=0.18257.The turbine operates at the head H=30.07 m, rotation speed 907.407 r/min, and guide vane opening $\alpha_0 = 12.175°$.

For the purpose of achieving better convergence and accurate results, structured mesh is generated for the computation domain. The localized refinement is applied in region close to the vane leading and trailing edges. The whole domain consists of 1,600,000 nodes and 1,490,000 elements.

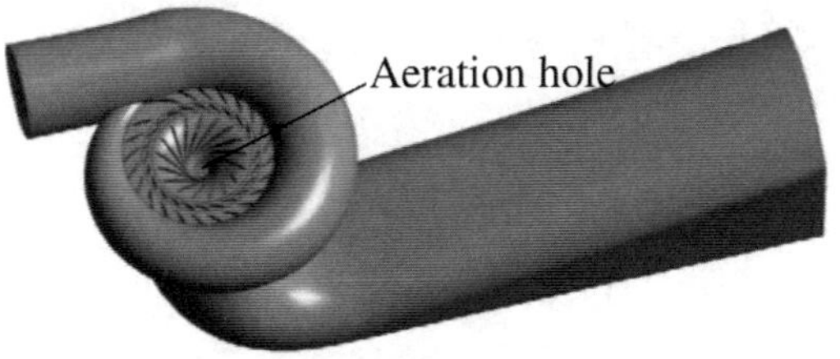

Figure 1. Computation domain for the turbine.

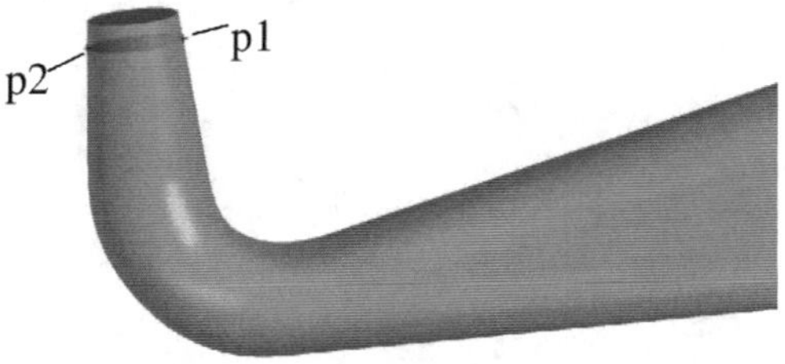

Figure 2. Monitoring points in computation domain.

In the simulation, the RANS equations as well as SST k-ω turbulence model are used. Total pressure and static pressure are assigned for inlet and outlet conditions respectively. For the case of aeration condition, the flow in the turbine is assumed as a kind of homogeneous mixture.

The time step is set as 0.001102 s at first, corresponding to a rotating angle of 6 degree per time step. Then 3 degree per time step is adopted for the subsequent simulation.

3. Results and analysis

For the convenience of the analysis, one monitoring section near the inlet of the draft tube is chosen. In order to investigate the effect of pressure fluctuations, two monitoring points i.e. p1 and p2 are set in the draft tube wall, as shown in figure 2. Note that two monitoring points are located on the monitoring section, and at the opposite position.

3.1. Pressure fluctuation without air admission

To verify the effectiveness of the numerical methods, the pressure fluctuation without air admission at the monitoring pointp1was measured by experiment (The experimental test of the model Francis turbine was conducted on the test rig at Harbin Electric Company Ltd, China).Fast Fourier Transform (FFT) is applied to process both calculation and experiment data to obtain the pressure fluctuation frequency and the corresponding amplitude.

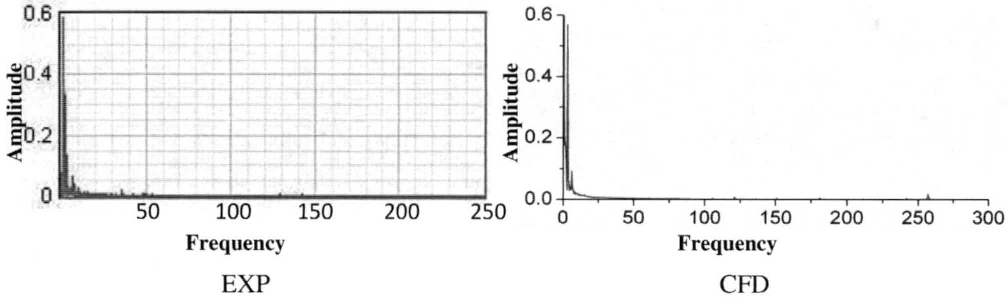

Figure 3. Comparison between calculated and measured pressure fluctuation.

Figure 3 shows the comparison between the experiment and CFD result for dimensionless pressure fluctuation. The CFD curve meets the experiments well both in frequency and amplitude. The dominant frequency and the secondary frequency predicted by CFD are3.11Hz with the amplitude of 0.57, and 6.19Hz with the amplitude of 0.11, while that recorded by the experiments are 3.09 Hz with the amplitude of 0.59, and 6.09 Hz with the amplitude of 0.14. It is clear that the pressure fluctuation frequency is smaller than the runner rotation frequency of 15.12 Hz, and the pressure fluctuation is induced by the vortex rope in the draft tube shown in figure 4, where iso-pressure surface is used to illustrate the rope.

The shapes of vortex rope at several instants are shown in figure 4.The rope rotates in the draft tube and the frequency corresponds to the dominant frequency observed at monitoring point p1.

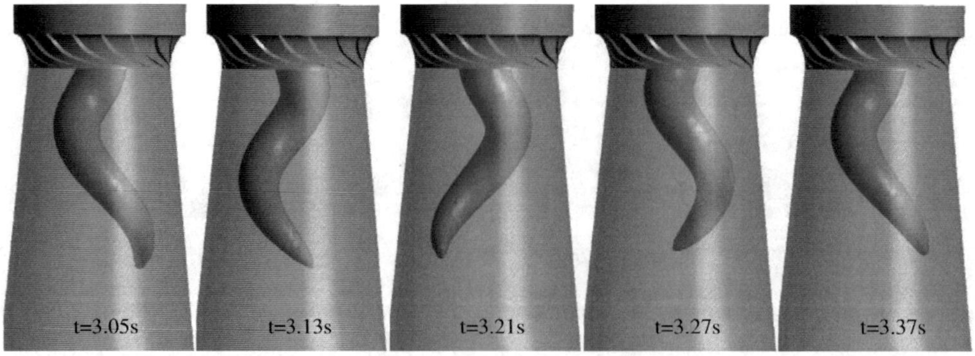

Figure 4. Vortex rope shapes in one cycle.

3.2. Aeration effect on pressure fluctuation

In order to alleviate the pressure fluctuation, the air is admitted into the draft tube from the aeration hole. The size of the hole is decided using the same way as Reference [13].

According to Li [14], the amplitude of pressure fluctuation may increase when the volume of the air admission is small. Only when the supplied air volume is large enough to change the eccentric distance of vortex rope obviously, the pressure fluctuation will be alleviated. In this paper, the air volume is simulated in a wide range from Q_{air}=0.5% to Q_{air}=4% of water flow-rate at the inlet plane. Simulation results with various air admissions are shown in figure 5.

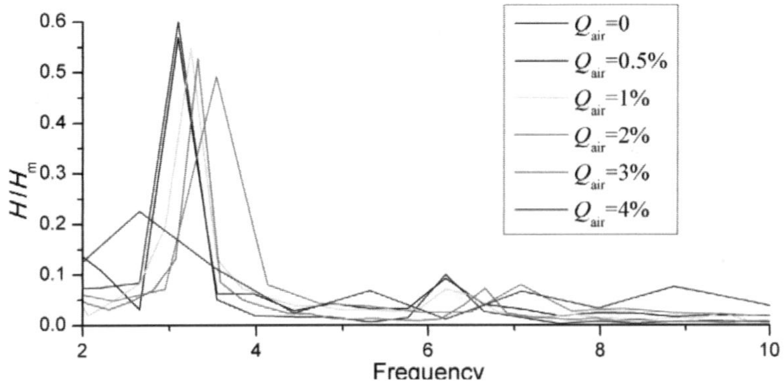

Figure 5. Pressure fluctuation sat monitoring point p1 at different $Q_{air.}$

It is noted that the pressure fluctuation characteristics predicted by the simulation agree well with the phenomena mentioned in Reference [14]. When Q_{air} is very small e.g. Q_{air} =0.5%, the pressure fluctuation amplitude increase a little and the frequency changes little. According to Li [14], the pressure fluctuation occurs because of the uneven distribution of static pressure. Figure 6 shows the static pressure distribution on the monitoring section at an instant, and the eccentric vortex rope makes an inhomogeneous pressure distribution. With the rotation of the rope, pressure transducer at p1 can measure a pulsation. The larger the eccentric distance and rope diameter are, the larger the amplitude could be. For small amount of air admission, the rope diameter increases, and the pressure fluctuation amplitude increases a little.

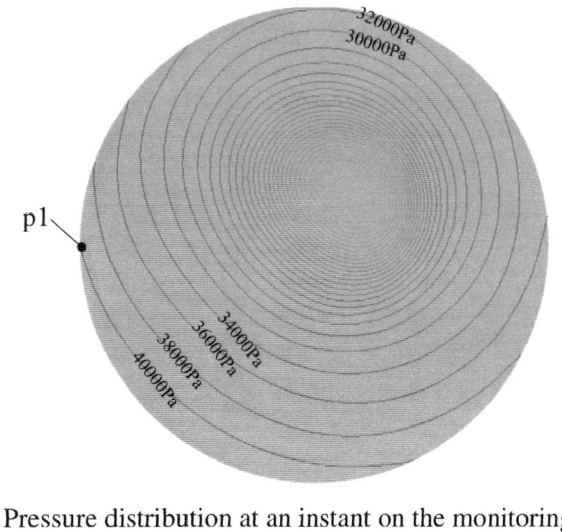

Figure 6. Pressure distribution at an instant on the monitoring section.

When Q_{air} increase from 1% to 3%, the pressure fluctuation amplitude decreases a little, and the effects of the vortex rope are alleviated. Figure 5 also indicates that the pressure fluctuation frequency increases with the air admission. When Q_{air} increases to 4%, the amplitude decreases substantially. There seems no dominant component in this operation condition. Based on figure 7 where the static pressure distribution are shown, the low pressure core on the monitoring section becomes smaller with the increase of air admission and the pressure distribution for Q_{air}=4% seems much more homogeneous than that without air admission.

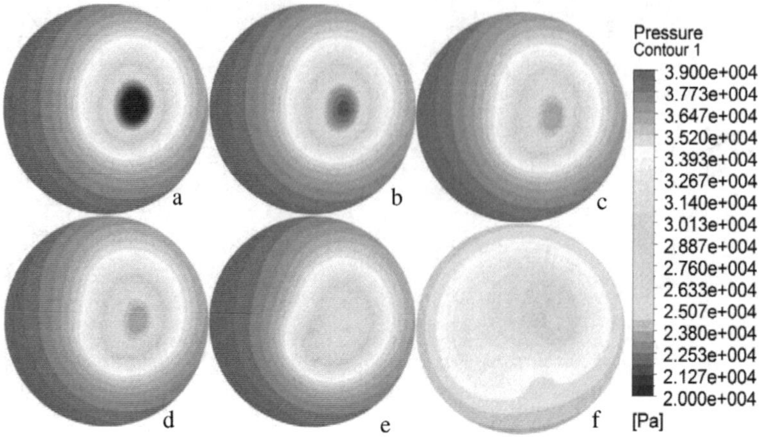

Figure 7. Pressure distributions with different air admission: (a) Q_{air}=0, (b) Q_{air}=0.5%, (c) Q_{air}=1%, (d) Q_{air}=2%, (e) Q_{air}=3%, (f) Q_{air}=4%.

Figure 8 shows the iso-surface with the air volume fraction of 10% with different air admission. It is noted that the rope diameter increases, but the orbit of the rope has a larger eccentric with the increase of the air admission.

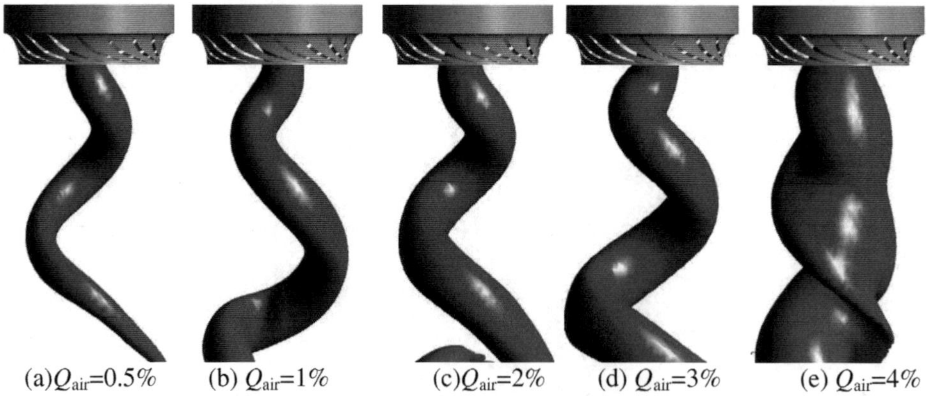

(a)Q_{air}=0.5% (b) Q_{air}=1% (c)Q_{air}=2% (d) Q_{air}=3% (e) Q_{air}=4%

Figure 8. Iso-surface shaving 10% air volume fraction with different air admissions

3.3. Vorticity Distributions

The vorticity transport equation is employed to better understand the vortex-air interactions. The equation is shown as

$$\frac{D\boldsymbol{\omega}}{Dt} = (\boldsymbol{\omega} \cdot \nabla)\boldsymbol{V} - \boldsymbol{\omega}(\nabla \cdot \boldsymbol{V}) + \frac{\nabla \rho_{\mathrm{m}} \times \nabla p}{\rho_{\mathrm{m}}^2} + \frac{1}{R_{\mathrm{e}}}(\nabla^2 \boldsymbol{\omega}) \quad (1)$$

where ω: vorticity, $\boldsymbol{V}$: velocity, ρ_{m}: mixed density, R_{e}: Reynolds number.

In this equation, the vorticity production of a fluid includes four parts: the first term on the right-hand-side i.e. $(\omega\cdot\nabla)V$ represents the vortex stretching; the second term i.e. $\omega(\nabla\cdot V)$ means the vortex dilation caused by volumetric expansion/contraction; the third term i.e. $(\nabla\rho m\times\nabla p)/\rho_m^2$ is called the baroclinic torque, which is nonzero only at the area with the air, and is caused by the misaligned pressure and density gradients; and the last term named as viscous diffusion, can be ignored as its effect on the vorticity transport is much smaller than the other terms. Thus, only vortex dilation and baroclinic torque terms are analyzed.

Figure 9 shows the typical instantaneous distributions of the vortex-dilatation and baroclinic torque terms in vortex transport equation projected on the monitoring section. The result shows that there are vorticity variations near the vortex rope, and the vortex dilation term plays a major role in the vertical flow in the draft tube. At the case of small air admission and without air admission, a clear vortex pair appears along the vortex rope. This indicates that the vertical flow is strong at those operation conditions. With Q_{air} =2% and 3%, there are two vortex pairs with less intensity. This phenomenon is visible from both the vortex dilation and baroclinic torque terms. At the case of Q_{air} =4%, the effect of the baroclinic torque term almost disappears, and that of vortex dilation becomes much depressed. These results can explain that the vertical flow is still strong at small air admission, and the pressure fluctuation is not alleviated yet. Consequently, a suitable air admission e.g. Q_{air} =4% for this work is necessary to depress the vortex movement and alleviate the pressure fluctuation. Thus, the analysis of the vortical flow evolution is helpful to understand the pressure fluctuation behavior in the turbine draft tube.

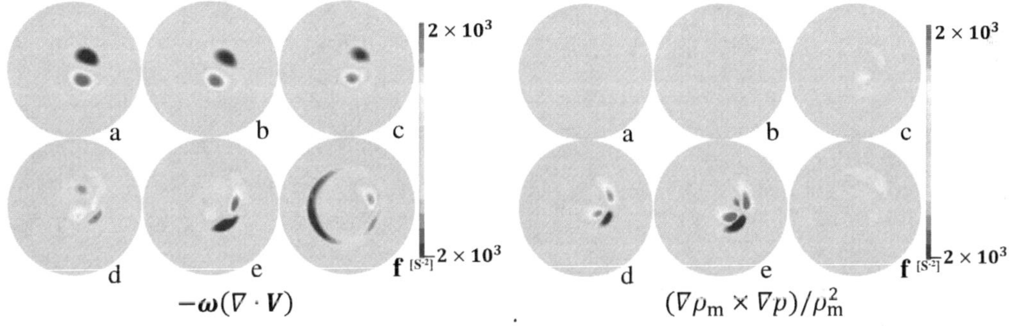

Figure 9. Contours of vortex dilation and baroclinic torque terms at the monitoring section: (a) Q_{air}=0, (b) Q_{air}=0.5%, (c)Q_{air}=1%, (d) Q_{air}=2%, (e) Q_{air}=3%, (f) Q_{air}=4%

3. Conclusions

The internal flow in a model Francis turbine is studied with particular emphasis on the interaction between the flow field and the vortices. Based on the present study, the following conclusions can be drawn:

(1) The numerical simulation reasonably predicts the pressure fluctuations in the draft tube, which agrees fairly well with experimental data.

(2)The analysis based on the vorticity transport equation shows that the vortex dilation term plays a major role in the vortex evolution with air admission in the turbine draft tube. There is large value of vortex dilation along the vortex rope.

(3) The aeration with suitable air volume fraction can depress the vortical flow, and alleviate the pressure fluctuation in the turbine draft tube.

Acknowledgments

This work is financially supported by the National Natural Science Foundation of China (Project Nos. 51179091,51376100, and 51206087) and the National Key Technology Research and Development Program (Grant No. 2011BAF03B01).

Reference

[1] Kercan V, Bajd M, Djelić V, Lipej A and Jošt D 1994 Work group on the behavior of hydraulic machinery under steady oscillatory conditions *Proc. of the 17th IAHR symposium on hydraulic machinery and cavitation (Beijing, China, 15-19 Sept 1994)* A6
[2] Kercan V, Bajd M, Djelić V, Lipej A and Jošt D 1996 *Proc. of the 18th IAHR symposium on hydraulic machinery and cavitation* vol II *(Valencia, Spain, 16-19 Sept 1996)* pp 994-1003
[3] Kercan V 2001 Ph.D thesis (Rijeka Croatia: University in Rijeka)
[4] Ji B, Luo X W, Arndt R and Wu Y 2014 *Ocean Eng.* **87** 64-77
[5] Ji B, Luo X W, Arndt R Peng X and Wu Y 2015 *Int. J. Multiphase Flow* **68** 121-134
[6] Luo X W, Ji B, Peng X, Xu H and Nishi M 2012 *J. Fluid Eng.* **134**(4) 041202
[7] Ji B, LuoX W, Wu Y, Peng X and Duan Y 2013 *Int. J. Multiphase Flow* **51** 33-43.
[8] Miyagawa K, Tsuji K, Yahara J and Nomura Y 2002 *Proc. of the 21st IAHR symposium on hydraulic machinery and systems(Lausanne, Swizerland, 9-12 Sept 2002)*
[9] Zhou L J, Wang Z W and Tian Y 2006 *23rd IAHR symposium on hydraulic machinery and systems (Yokohama, Japan, 18-21 Oct 2006)*
[10] Jošt D and Lipej A 2011 *J. Mech. Eng.* 57 (6) 445-456
[11] Nukunishi K and Uedo T 1964 Air supply into draft tube of Francis turbine
[12] Qian Z, Yang J and Huai W 2007 *J Hydrodyn.* **19** (4) 467-472
[13] Liao W L, Ji J T, Lu P and Luo X Q 2008 *J. Hydraulic Eng.* **8** 18
[14] Li Q 1981 *Waterpower* **5** 8

IOP Conf. Series: Materials Science and Engineering **72** (2015) 042048 doi:10.1088/1757-899X/72/4/042048

Research on wear properties of centrifugal dredge pump based on liquid–solid two-phase fluid simulations

G J Peng[1,3], Y Y Luo[2] and Z W Wang[2]

[1] Research Center of Fluid Machinery Engineering and Technology, Jiangsu University, Zhenjiang 212013, China

[2] State Key Laboratory of Hydroscience and Engineering & Department of Thermal Engineering, Tsinghua University, Beijing 100084, China

E-mail:pgj@ujs.edu.cn

Abstract. The impeller and casing of dredge pump are worn by sediment in the flow. However, there are few studies about abrasion of the impeller and casing for normal pump operating conditions. This paper investigated the relationship between the wear rates on the surfaces of the impeller as well as casing and the sediment concentration, with the distribution of the wear rates for normal pump operating condition analyzed. An Eulerian-Lagrangian Computational Fluid Dynamics (CFD) procedure was used to simulate steady liquid–solid two-phase flow for various operating conditions. The Finnie model was then used to predict the abrasion. The results show that, the wear rate relative value of impeller and casing surface increase as the sediment concentration increases. The wear rate relative value of impeller and casing surface is larger when the pump is in low flow rate condition, and the value of casing surface is larger than that of the impeller. The wear rate relative value of pump is low when pump is in high efficiency condition. This paper shows the abrasion characteristics on the impeller and casing with sediment flow and provides reference data for predicting the abrasion conditions in the flow passage components for a dredge pump.

1. Introduction

With the development of China's coastal port infrastructure, dredge manufacturing industry is developing at an alarming speed. China has become a veritable dredger built big country. The dredge pump is the key equipment dredger; the annual demand for dredge pumps and dredge pump accessories is huge. Severe wear makes dredge pump itself damaged and affects reliability and stability of operation, resulting in dredge pump efficiency and head lowered, shortening the life of the flow components. Research on wear properties of centrifugal dredge pump within different condition is meaningful.

With the development of computational fluid dynamics (CFD) and the improvement in performance of computer, it is possible to use numerical simulation to study two-phase flow in fluid machinery.

Patankar N A [1] [2] used LNS method to simulate the flow of particles, solving fluid phase continuity and momentum equations in the Euler-type grid, using Lagrangian method to simulate

[3] Corresponding author. Tel.: +86 51188780284.

IOP Conf. Series: Materials Science and Engineering **72** (2015) 042048 doi:10.1088/1757-899X/72/4/042048

particle phase, using the fluid phase momentum equation to process momentum exchange between particles and fluid.

Wu [3] used k-ε-Ap two-phase turbulence model to calculate two-phase turbulent in turbine runner and according to the three-dimensional velocity field to estimate sediment abrasion within runner blade surface. Li [4] used k-ε two-equation model and SIMPLE algorithm to simulate the internal flow of slurry pump. Li [5] used k-ε-Ap model to simulate the solid-liquid two-phase flow in the desulfurization pump, and particle concentration distribution, velocity distribution and wear characteristics were studied. Liu [6] used Eulerian-Lagrangian turbulence model to study sediment abrasion in turbines. By calculating the wear rate of the guide vanes, the result is more consistent with the actual. Cao [7] thought that the factors that affect the wear included collision wall velocity, impact angle, particle concentration. Because solid particles flow laws in the fluid machinery are complex, the conclusions of many researchers are not exactly same.

In this paper, the solid-liquid two-phase flow in a centrifugal dredge pump was simulated, with the pump casing and impeller abrasion in the different concentrations and different flow conditions analyzed.

2. Numerical theory

Numerical simulations of the two-phase flow were based on the solution of single phase flow without sediment. The two methods used to analyze two-phase flows are Eulerian-Eulerian method and Eulerian-Lagrangian method. In this study the Eulerian-Lagrangian method was used.

2.1. Governing equations

2.1.1 Continuity equation.

$$\frac{\partial \rho}{\partial t}+\frac{\partial \rho u_i}{\partial x_i}=0 \qquad (i=1,2,3) \tag{1}$$

Variable i from 1 to 3 stands for the x, y, z axis, ρ is the density and u_i are the u, v and w velocity components in the x, y and z directions.

2.1.2 Fluid phase governing equations.

$$\frac{\partial(\rho u\Phi)}{\partial x}+\frac{\partial(\rho u\Phi)}{\partial y}+\frac{\partial(\rho u\Phi)}{\partial z}=\frac{\partial}{\partial x}(\Gamma_\Phi\frac{\partial\Phi}{\partial x})+\frac{\partial}{\partial y}(\Gamma_\Phi\frac{\partial\Phi}{\partial y})+\frac{\partial}{\partial z}(\Gamma_\Phi\frac{\partial\Phi}{\partial z})+S_\Phi+S_{p\Phi} \tag{2}$$

where Φ is a governing variable that represents u, v, or w for the momentum equation, and S is the general dissipation source term. For details see the reference [8].

2.2. Abrasion prediction model

The abrasion on the wall surface caused by the solid particles is usually modeled as a function of the particle motion, particle characteristics and wall surface characteristics. For most metallic surfaces, the abrasion is a function of the impact angle and the velocity of the particle. The Finnie model [9] is:

$$Er=kV_p^n f(\alpha) \tag{3}$$

$$f(\alpha)=\begin{cases}\frac{1}{3}\cos^2\alpha & \tan\alpha>\frac{1}{3}\\ \sin 2\alpha-3\sin^2\alpha & \tan\alpha\le\frac{1}{3}\end{cases} \tag{4}$$

Er is the abrasion rate defined as the ratio of the target material abrasion mass to the impact particles mass; k is an empirical coefficient; V_p is the particle velocity and α is the angle between the particle trajectory and the wall surface. For details see the reference [8].

3. Liquid–solid two-phase numerical simulations

3.1. Basic pump parameters
Flow rate Q= 9000 m^3/ h, head H =34 m, rotate speed n =258 r/min, the impeller diameter D =1940 mm, blade number z =3.

3.2. Three-dimensional numerical model
The dredge pump whole flow passage is composed of a casing, impeller and inlet, as shown in figure 1. An unstructured mesh was used in the simulation.

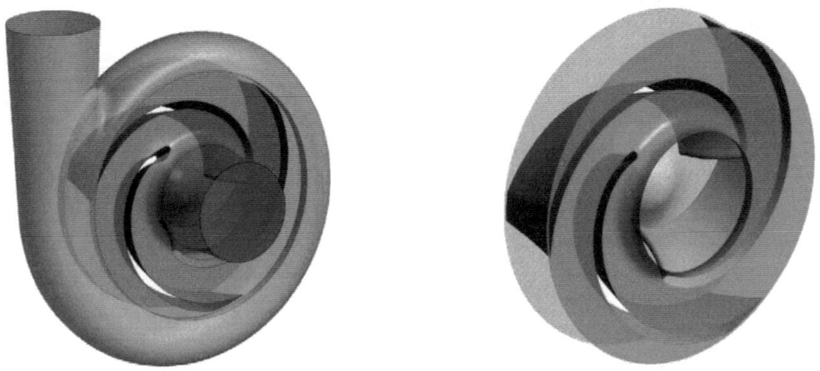

(a) Flow passage (b) Impeller

Figure 1. Dredge pump three-dimensional model.

3.3. Operating conditions and boundary conditions
Rotate Speed is 258 r/min, the flow rate are 3000, 5000, 7000, 9000, 11000, 13000, 15000 m^3/h, and the volume of sediment concentration is 5%, 10% for a total of 128 operating conditions.

The velocities were set at the inlet with fluid phase velocity set according to the inlet mass flow and the solid phase velocity set to the same velocity as the fluid phase. The pressure at the pump outlet was specified the value of 600000 Pa. The wall surface was a rough surface with a roughness of 0.2 mm. The discrete particles were solid with the shape assumed to be spherical. The side wall considered to have perfectly elastic collisions, and collisions between particles were neglected as well as the influence of particles on the flow field. The particles were assumed to have a minimum diameter of 0.005 mm, a maximum diameter of 0.25 mm, a median diameter of 0.20 mm, a standard deviation of 0.15 mm and a particle density of 1950 kg/m^3.

4. Numerical results and analysis

4.1. Wear rate relative value
The liquid–solid two-phase flow simulations were used to model the wear rate distribution on the pump casing and impeller surfaces. The results gave the total and maximum wear rates on the casing and impeller surfaces. The relative wear rate was then calculated relative to the maximum. Figure 3 shows the pump casing, the impeller surface of the wear rate of the relative value with the flow rate variation. C05 represents sediment volume concentration of 5%, and C10 represents a volume concentration of 10% sediment. Figure 2 shows the dredge pump performance curve, effective range occurs between the flow rate of 11000 m^3/h and 15000 m^3/h.

4.2. Abrasion distribution characteristics for various operating conditions

Figure 3 and figure 4 show that C10 curves are higher than C05 curve, which indicates that the wear rate relative value increased with the sediment concentration to exacerbate abrasion of dredge pump flow parts.

Figure 3 shows that the wear rate relative value of pump casing surfaces relative value increases first then decreases with the increasing of flow rate. The maximum of wear rate relative value occurs between flow rate of 5000 m^3/h and 7000 m^3/h. The reason is that the backflow is severe during the flow district and the dredge pump efficiency is relatively low. With the increase of flow rate the backflow from pump tongue decreased, the efficiency of the pump improved and the wear reduced. Figure 4 shows the maximum of wear rate relative value of impeller occurs at flow rate of 5000 m^3/h and keeps unchanged when the flow rate is from 9000 m^3/h to 15000 m^3/h. Generally, the wear rate value of the pump casing surface is higher than that of the impeller.

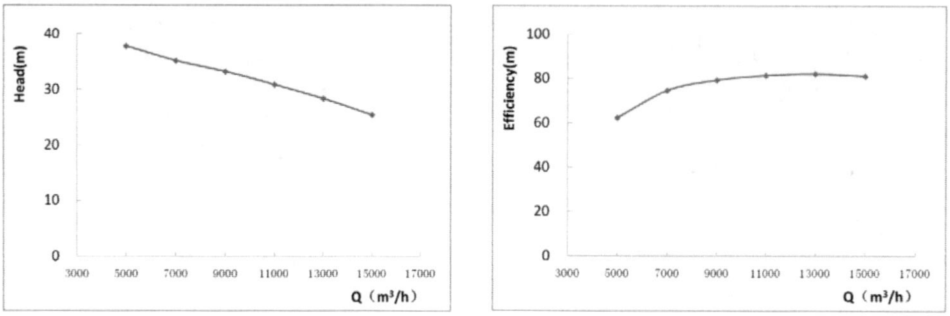

Figure 2. Dredge pump performance curve.

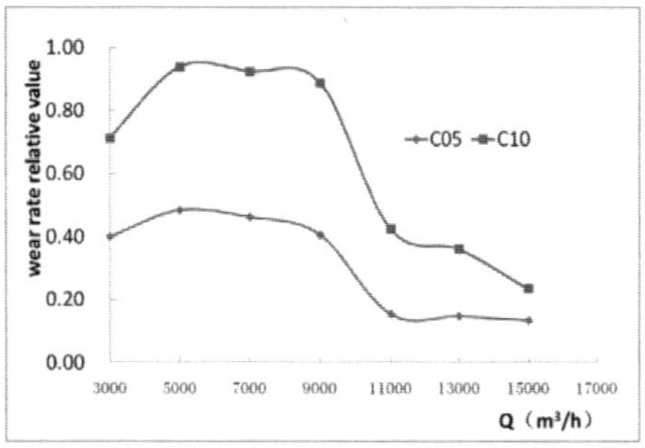

Figure 3. Wear rate relative value of pump casing.

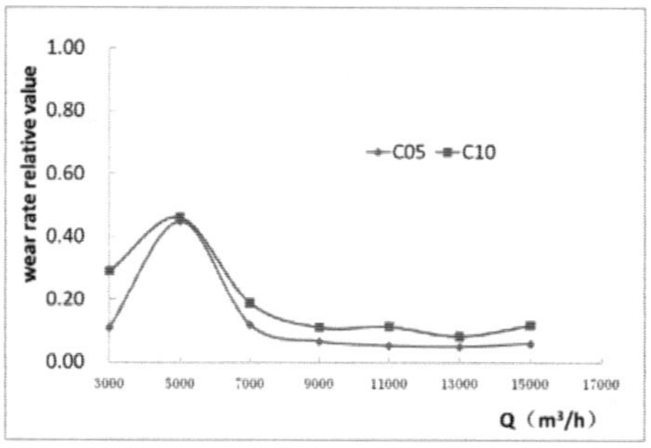

Figure 4. Wear rate relative value of impeller.

Figure 5 shows that the wear area is at the back of the impeller, the blades of the head and the intersection between surface of the front cover and the work face of blade. Figure 6 shows that the wear area of pump casing is between 3 to 8 section, and it is relatively large at flow rate of 7000 m^3/h.

5. Conclusions

The main conclusions are as follows:

1) With the increase of sediment concentration, the wear rate relative value increased accordingly.

2) The wear rate relative value of impeller and casing surface is larger when pump is in low flow rate condition and the value of casing surface is larger than that of the impeller.

3) The wear rate relative value of pump is low when pump is in high efficiency condition. So the dredge pump should be avoided to run in a small flow rate condition.

4) This paper shows the abrasion characteristics on the impeller and casing with sediment flow and provides reference data for predicting the abrasion conditions in the flow passage components of a dredge pump.

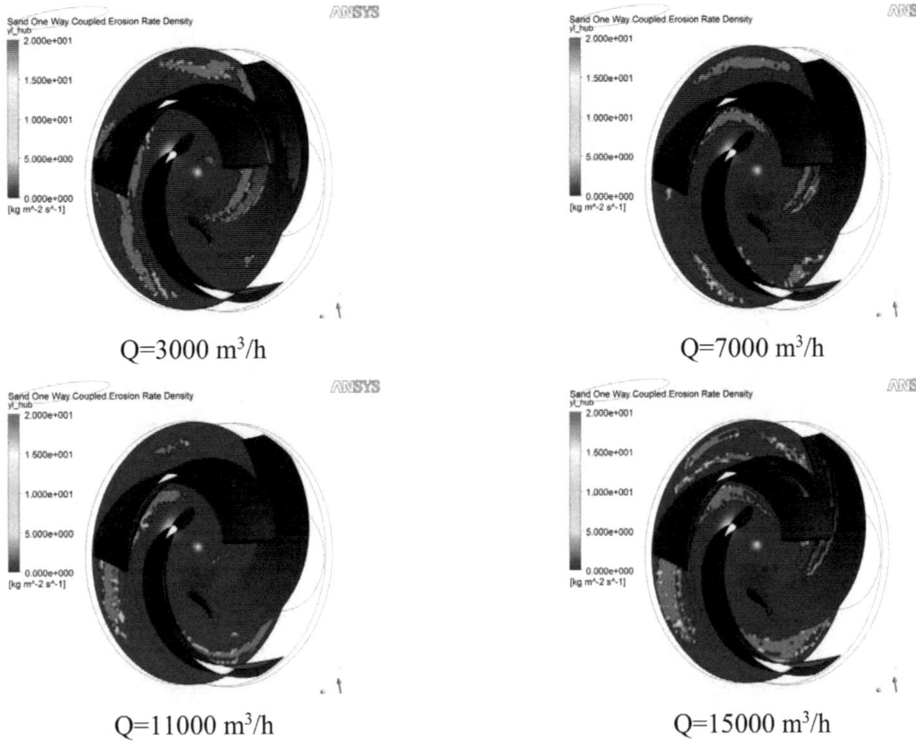

Figure 5. Wear rate distribution in pump impeller-C05

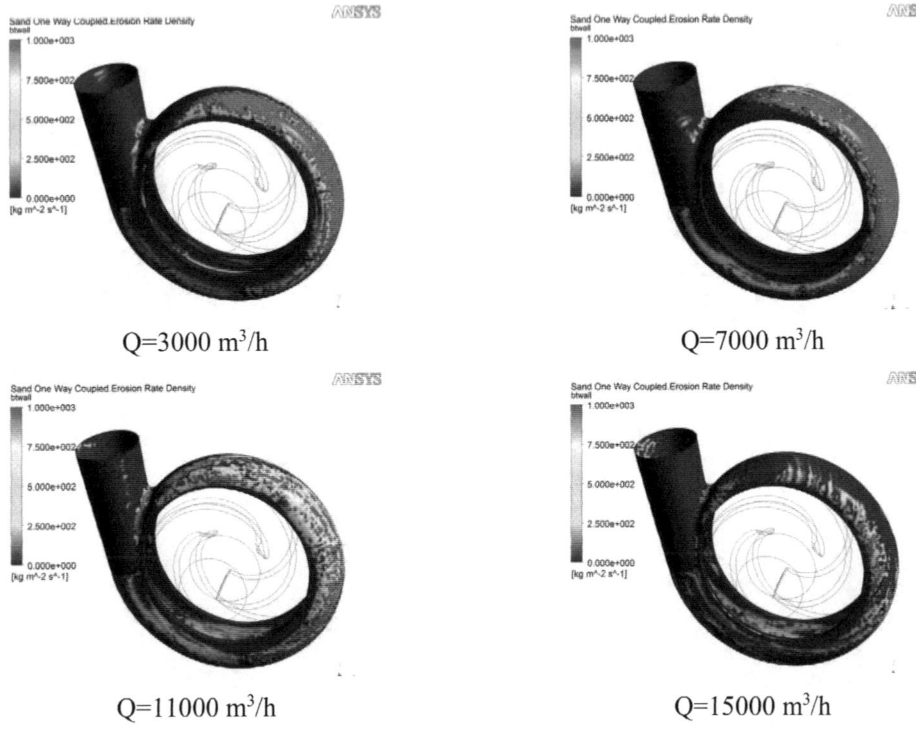

Figure 6. Wear rate distribution in pump casing-C05.

Acknowledgments
Special thanks are due to the Jiangsu University Senior Personnel Funded Project (12JDG097) and the Open Research Fund Program of State Key Laboratory of Hydroscience and Engineering (No. sklhse-2014-E-01) for supporting the present work.

References

[1] Patankar N A and Joseph D D 2001 *Int. J. Multiphase Flow* **27** (10) 1659-84
[2] Patankar N A, Joseph D D 2001 *Int. J. Multiphase Flow* **27** (10) 1685-706
[3] Wu W Z, Wu Y L, Ren J et al. 2000 *J. Eng. Thermophysics* **21** (6) 709-12
[4] Li R N, Xu Z F, Han W et al. 2007 *Drainage and Irrigation Machinery* **25** (2) 5-8
[5] Li Y, He W Q, Zhu Z C et al. 2009 *Drainage and Irrigation Machinery* **27** (2) 124-28
[6] Liu X B 2000 *J. Sichuan University of Science and Technology* **19** (2) 79-84
[7] Cao H M, Zhang S F, Gao C et al. 2008 *Pipeline Technique and Equipment* **1**(1) 20-9
[8] Peng G J, Wang Z W, Xiao Y X et al. 2013 *Eng. Failure Analysis* **33** 327-35
[9] Finnie I 1972 Some Observation on the Erosion of Ductile Materials *Wear* **19** 81-90

IOP Conf. Series: Materials Science and Engineering **72** (2015) 042049 doi:10.1088/1757-899X/72/4/042049

Prediction of pressure fluctuation of a hydraulic turbine at no-load condition

T J Chen[1], X J Wu[2], J T Liu[3] and Y L Wu[1]
[1] State Key Laboratory of Hydroscience and Engineering, Tsinghua University, Beijing 100084
[2] Hunan M&W Energy Saving Technology LTD, Hunan, China, 410000
[3] Beijing Institute of Control Engineering, Beijing 100190

Email: ctj@mail.tsinghua.edu.cn

Abstract. In order to study characteristics of pressure fluctuation of a turbine during the starting period, a turbine with guide vanes device at no-load condition was investigated using RNG k-ε turbulence model. The inner flow distribution and pressure fluctuation characteristics were analyzed. Results show that the pressure fluctuations in the region between the runner and guide vanes are different around the runner inlet. The dominant frequency of pressure fluctuation in the vaneless space close to the casing outlet is the blade passing frequency, while the dominant frequency at the rest region is the twice of the blade passing frequency. The increase of amplitude of pressure fluctuation close to the casing outlet can be attribute to the large scale stall at suction side of the runner inlet.

1. Introduction

Hydraulic turbine is commonly used as a residual pressure energy conversion device in petrochemical industry. It plays an important role in decreasing costs and improving economic efficiency. Hydraulic turbine will run through runaway operating point during start-up and shutdown process. When a hydraulic turbine runs at partial conditions, there will be a complex internal separation vortex and reverse flow, which may cause hydraulic turbine unstable and even cause the entire system to a larger pressure pulsation. The research on pressure fluctuation characteristics at runaway point of a hydraulic turbine has an important significance for understanding the stability of a hydraulic turbine.

At present, the selection and design of hydraulic turbine was done empirically by analyzing the operating characteristics of reverse pump. Most of hydraulic turbines in engineering application were designed by reverse pump [1-3] and optimizing the pump impeller to change the performance of a hydraulic turbine [4-6]. Studies on the stability of the hydraulic turbine are still needed. Yang Junhu et al [9] analyze effects on the external characteristics of the hydraulic turbine based on numerical simulation by changing the blade shape. Yang Sun Saint et al [10] investigated external characteristics of a hydraulic turbine by changing the length of the blade using experiment and numerical simulation method. Ferandez [11] and Yang Junhu [12] studied the influence of vanes on the performance of a hydraulic turbine, and they found that there is a certain intrinsic link between efficiency and flow rate coefficient of hydraulic turbine. The constant hydraulic turbine characteristic is more obvious and the pressure will not change with the changes in flow rate by the use of guide vanes. Inhama [7] analyzed the force on runner and the runaway speed of a hydraulic turbine which design from a reverse pump.

IOP Conf. Series: Materials Science and Engineering **72** (2015) 042049 doi:10.1088/1757-899X/72/4/042049

Shahram [8-9] analyzed detailed external characteristics of hydraulic turbine under different speed by experiment and numerical simulations, and it proposed a reasonable range of rotating speed of a reverse pump when it was used as a hydraulic turbine.

In this paper, the external characteristics and internal flow field of a hydraulic turbine at runaway operating point are analyzed by numerical simulation. The internal flow characteristics as well as rotor-stator interaction of the hydraulic turbine are investigated. The stability of the hydraulic turbine at runaway operating point is also studied.

2. Model of hydraulic turbine

A hydraulic turbine with guide vanes is chosen and modeled and the parameters of the hydraulic turbine are shown in table 1. Three-dimensional hydraulic model of the hydraulic turbine is shown in figure 1. D is the diameter of the runner. B is the number of the blades. V is the number of the guide vanes. n is the rotating speed. The hydraulic turbine is mainly composed of guide vanes, runner and double suction outlet.

ICEM is used to mesh the four components. The mesh on blades and vanes is used 12 layers for more accurate description near the wall region. Each hydraulic turbine components are divided by hexahedral mesh. The mesh of the runner is shown in figure 2. The total number of mesh of the entire hydraulic turbine is 6.5 million.

Table 1. Parameters of pump-turbine.

D (m)	B	V	n (r/min)
0.45	9	20	3000

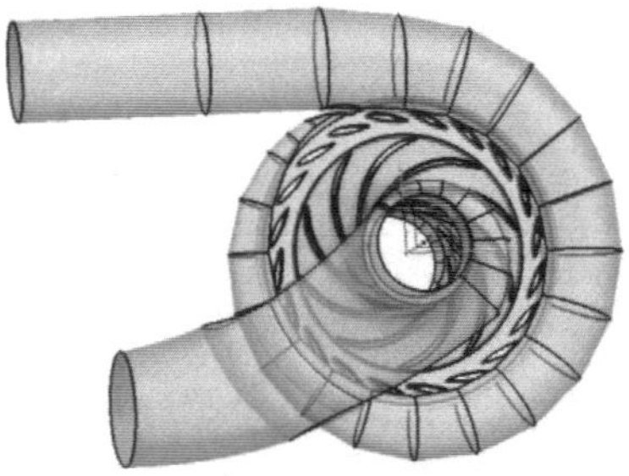

Figure 1. Profile of the turbine.

Figure 2. Mesh of the runner.

In order to analyze the stability of the hydraulic turbine during start-up and shutdown process at the run runaway point, unsteady calculations are performed.

3. Turbulence model and boundary conditions

When a hydraulic turbine runs at partial conditions, the internal flow is complex and flow separation will appear. In order to capture the complex flow in hydraulic turbine and obtain more accurate results of pressure fluctuation, RNG k-ε model is used to calculate the flow in the hydraulic turbine. Equations of RNG k-ε turbulence model are shown as follows:

$$\frac{\partial}{\partial t}(\rho k)+\frac{\partial}{\partial x_i}(\rho k u_i)=\frac{\partial}{\partial x_j}\left[\left(\mu+\frac{\mu_t}{\sigma_k}\right)\frac{\partial k}{\partial x_j}\right]+P_k-\rho\varepsilon \quad (1)$$

$$\frac{\partial}{\partial t}(\rho\varepsilon)+\frac{\partial}{\partial x_i}(\rho\varepsilon u_i)=\frac{\partial}{\partial x_j}\left[\left(\mu+\frac{\mu_t}{\sigma_\varepsilon}\right)\frac{\partial\varepsilon}{\partial x_j}\right]+C_{1\varepsilon}\frac{\varepsilon}{k}P_k-C_{2\varepsilon}^*\rho\frac{\varepsilon^2}{k} \tag{2}$$

where P_k is the production of turbulence kinetic energy; $C_{1\varepsilon}$ is a constant; σ_k and σ_ε is the Prandtl number of k equation and ω equations, respectively.

Total pressure inlet at the casing inlet and pressure outlet at double suction outlet of the hydraulic turbine is set. Interface is used for the data switching between turbine runner inlet and guide vanes as well as runner outlet and double suction. The runner rotates for 1° is used as the time step. SIMPLEC algorithm is used to couple pressure and velocity. The second-order upwind scheme is used to discrete equations. No-slip boundary conditions are used on the walls. Computing time sustains the runner rotating five cycles and the residual of 0.0001 is used as the convergence conditions within each time step.

4. Results and conclusions

3.1 Time domains of pressure fluctuations

For the hydraulic turbine, the rated head is 230 m and the rated flow is 0.8 m^3/h. Pressure fluctuation measuring points are shown in figure 1. Measuring points between the vanes and the runner which are A, B, C and D has an equally distribution along the circumferential direction. Time domain of pressure fluctuations on measuring points A, B, C and D are shown in figure 3. The relative amplitude of the pressure fluctuation is shown in table 2.

Pressure fluctuation at point A has maximum relative amplitude. Point C has minimum amplitude of pressure fluctuation. The amplitude of pressure fluctuation circumferentially increases after a small decrease at clockwise direction from point A. Point A which is near the volute outlet. There have different sections on the left and right sides of the volute, causing different flow rate, resulting in the increase of amplitude of pressure fluctuation. Point C is far away from the dramatic changes in the volute section, so the amplitude of pressure fluctuation is lower than the other points.

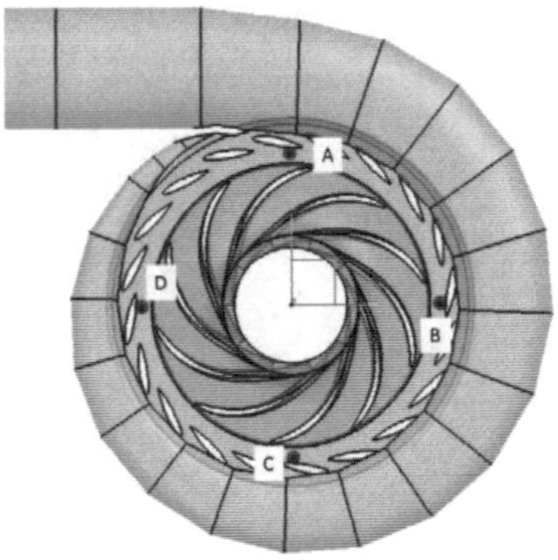

Figure 3. Locations of monitor points.

3.2 Frequency domains of pressure fluctuations

The Fourier transform results of pressure fluctuation at different points are shown in figure 5. Analysis of the amplitude of pressure fluctuation corresponding to the characteristic frequency is shown in table 2.

Table 2. Relative amplitude of pressure fluctuation (△*H*/*H*)

A	B	C	D

IOP Conf. Series: Materials Science and Engineering **72** (2015) 042049 doi:10.1088/1757-899X/72/4/042049

ΔH/H)	8.2%	7.2%	5.6%	6.8%

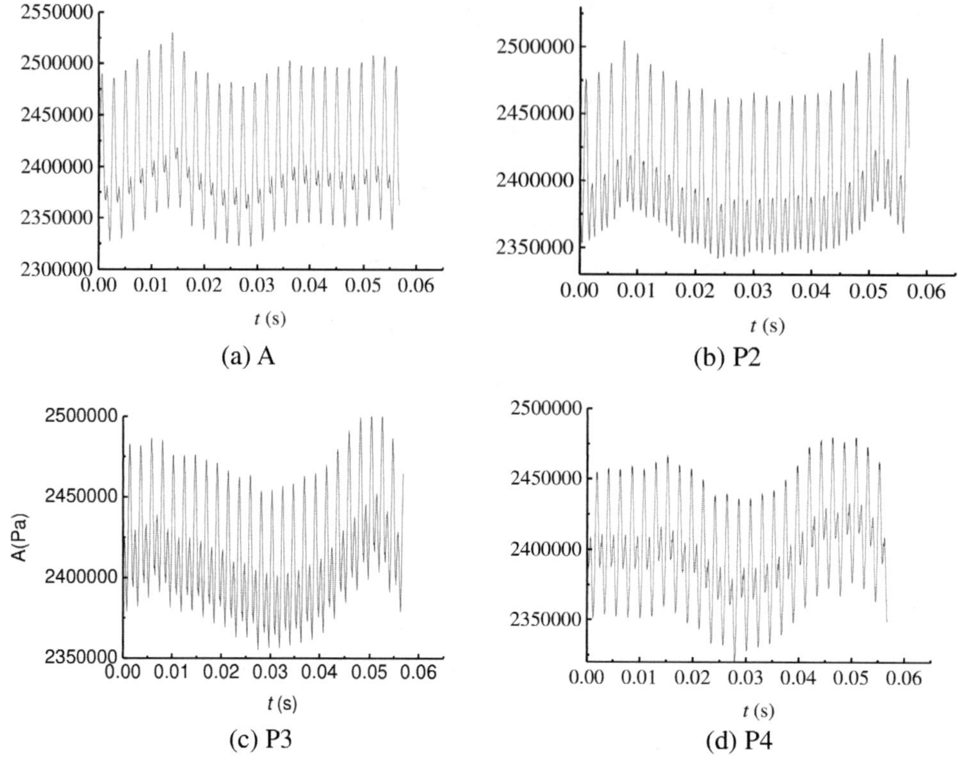

Figure 4. Pressure of each point at time domain.

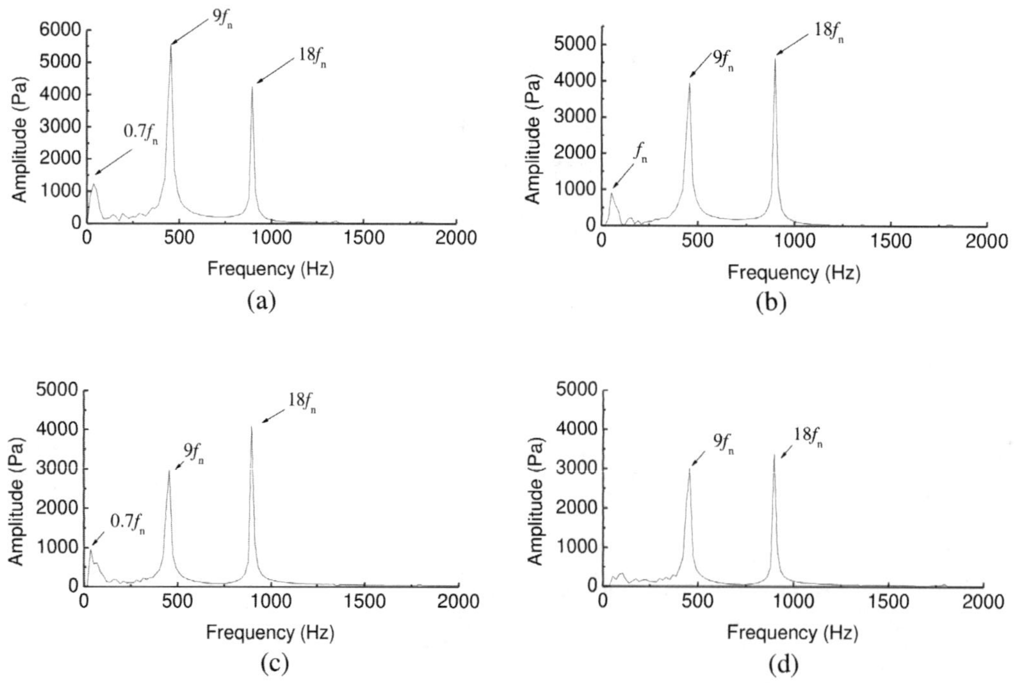

Figure 5. Pressure at frequency domain.

Dominant frequencies of the pressure fluctuations between the vanes and the runner inlet are different along the circumferential position. The dominant frequency is the blade passing frequency at point A corresponding to the position of volute outlet. The pressure fluctuation is mainly affected by the number of runner blades at point A, and there is a low frequency which is 0.7 times of the rotating frequency. The lower frequency component may be related to the stall vortex near the vaneless space. The second frequency at point A is twice of blade passing frequency, and the amplitude of the second frequency component is 90% of the dominant frequency component. The dominant frequency is twice of the blade passing frequency at point B, and the second frequency is the blade passing frequency. The amplitude of the rotating frequency component has a higher proportion in the spectrum of pressure fluctuation. The dominant frequency and second frequency at point C are the same as point B. The amplitude corresponding to the second frequency is approximately 70% of the dominant frequency component. Point C also has a low-frequency which is 0.7 times of rotating frequency. The dominant frequency is twice of the blade passing frequency at point D, and the second frequency is the blade passing frequency. The amplitude of the second frequency component at point D is 95% of the dominant frequency component. The amplitude of blade passing frequency decreases at clockwise beginning from point A. The amplitude of twice of the blade pass frequency begins to decrease after the first increase in the clockwise direction from the point A, and it has a maximum value at the point B.

3.3 Flow analysis

Streamlines in the runner at runaway condition of the hydraulic turbine are shown in Figure 6. Large scale vortex appears at the suction side of the blade in the one of the passage of the runner. The large scale vortex locates just adjacent to the volute outlet section, which may be the reason for causing the increase of the amplitude of pressure fluctuation.

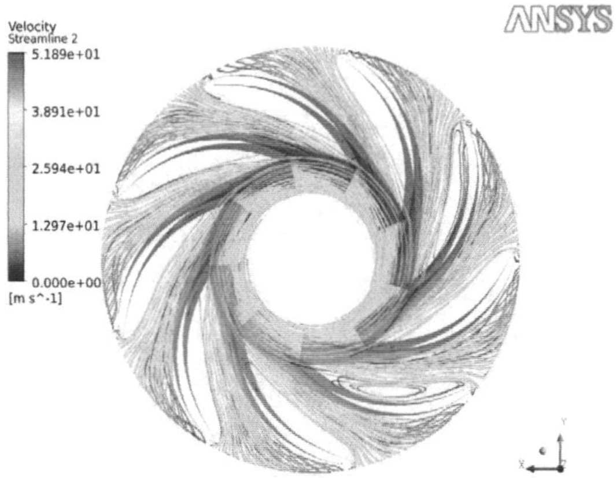

Figure 6. Streamlines in the runner.

4. Conclusions

When a hydraulic turbine operates at runaway point, the position of the maximum amplitude of the pressure pulsation locates near the tongue of the volute. The dominant frequency changes along the circumferential position at the runaway condition in the vaneless space. The dominant frequency of the pressure fluctuation in the vanelss space near the volute tongue is the blade passing frequency, and the dominant frequency at other point far away from the volute tongue is the twice of the blade passing frequency. The increase of the amplitude of pressure fluctuation near the volute tongue can be attributed to the large-scale vortex in the runner.

Acknowledgments
The authors would like to thank project 51406010 supported by National Natural Science Foundation of China and project supported by open Research Fund Program of State key Laboratory of Hydroscience and Engineering, NO. sklhse-2014-E-02.

References

[1] Derakhshan S and Nourbakhsh A 2008 *Exp. Thermal and Fluid Science* **32** (3) 800-7
[2] Williams A A 1994 *J. Power and Energy* **208** (1) 59-66
[3] Yang J H and Wang X H 2011 *J. Drainage and Irrigation Machinery Eng* **29** (4) 287-29 (in Chinese)
[4] Yang S S, Kong F Y, Xue L et al. 2012 *Transactions of the Chinese Society for Agricultural Machinery* **43** (7) 104-7 (in Chinese)
[5] Fernandez J, Blance E, Parrondo J, et al. 2004 *J. Power and Energy* **218** (4) 265-71
[6] Yang J H, Gong C H, Xia S Q, Luo K K and Li H L 2014 *J. Drainage and Irrigation Machinery Eng* **32** (2) 113-118 (in Chinese)
[7] Hitoshi I, Junichiro F and Yoshiyuki N 1999 Study of reverse running pump turbine *Proc. 3rd ASME/JSME Joint Fluids Eng. Conference (San-Francisco, California, USA, 18-23 July 1999)* pp1-6
[8] Shahram D and Ahmad N 2008 *Exp Thermal and Fluid Science* **32** 800-7
[9] Shahram D, Bijal M and Ahmad N 2009 *J. Fluid Eng.* **131** (2) 8-34

IOP Conf. Series: Materials Science and Engineering **72** (2015) 052050 doi:10.1088/1757-899X/72/5/052050

Application of the "Full Cavitation Model" to the fundamental study of cavitation in liquid metal processing

G S B Lebon[1], K Pericleous[1], I Tzanakis[2] and D Eskin[2]

[1] Centre for Numerical Modelling and Process Analysis, The University of Greenwich, London SE10 9LS, United Kingdom
[2] Brunel Centre for Advanced Solidification Technology (BCAST), Brunel University, Uxbridge, Middlesex UB8 3PH, United Kingdom

Email: G.S.B.Lebon@greenwich.ac.uk

Abstract. Ultrasonic cavitation treatment of melt significantly improves the downstream properties and quality of conventional and advanced metallic materials. However, the transfer of this technology has been hindered by difficulties in treating large volumes of liquid metal. To improve the understanding of cavitation processing efficiency, the Full Cavitation Model, which is derived from a reduced form of the Rayleigh-Plesset equation, is modified and applied to the two-phase problem of bubble propagation in liquid melt. Numerical simulations of the sound propagation are performed in the microsecond time scale to predict the maximum and minimum acoustic pressure amplitude fields in the domain. This field is applied to the source term of the bubble transport equation to predict the generation and destruction of cavitation bubbles in a time scale relevant to the fluid flow. The use of baffles to limit flow speed in a launder conduit is studied numerically, to determine the optimum configuration that maximizes the residence time of the liquid in high cavitation activity regions. With this configuration, it is then possible to convert the batch processing of liquid metal into a continuous process. The numerical simulations will be validated against water and aluminium alloy experiments, carried out at Brunel University.

1. Introduction

Significant improvement of quality and properties in metallic materials is observed when melt is treated with ultrasound [1][2]. These improvements are primarily due to ultrasonic cavitation, with the creation, growth, pulsation, and collapse of bubbles in the liquid. However, this technology has not been successfully transferred to the industry due to the difficulty in treating large volumes of liquid metal, as is required by processes such as continuous casting. A fundamental study of the ultrasonic treatment of melt is thus required to circumvent these difficulties.

The full cavitation model was developed by Athavale et al [3][4] to provide the capability for multidimensional simulation of cavitating flows, the modelling of which is crucial to the design of many engineering devices [5]. In their approach, the authors derived source terms for the bubble mass fraction transport equation from the Rayleigh-Plesset equation [6], which governs the evolution of a spherical bubble [7], to predict the formation and collapse of bubbles in cavitating flows. This model has been used in the modelling of the solidification structure evolution by Nastac [8].

In this study, the full cavitation model is modified to compute the bubble concentration in a launder. The novelty here is to understand the effect of ultrasonic treatment on flowing melt, paving the way to the ultrasonic treatment of liquid metal in batch.

IOP Conf. Series: Materials Science and Engineering **72** (2015) 052050 doi:10.1088/1757-899X/72/5/052050

2. Numerical method

The approach consists of solving the wave equation and using the acoustic pressure solution in the source term of a non-condensable gas mass fraction, f_g, equation. The Reynolds-Averaged Navier-Stokes (RANS) equations are solved along with the mass fraction transport equation. The density of the fluid is calculated as a function of the non-condensable gas mass fraction.

2.1. Governing equations

2.1.1. Fluid equations. Fluid flow is governed by the RANS equations:

$$\frac{\partial \rho}{\partial t} + \nabla \cdot (\rho \boldsymbol{u}) = 0$$

$$\frac{\partial(\rho u_i)}{\partial t} + \nabla \cdot (\rho \boldsymbol{u} u_i) = \nabla \cdot [(\mu + \mu_t)\nabla u_i] + S_{u_i} \quad (2)$$

where ρ is the fluid density, $\boldsymbol{u}$ is the fluid velocity, μ is the dynamic viscosity, μ_t is the eddy viscosity, and S_{u_i} are the momentum sources. The fluid density is related to the vapour and non-condensable gas mass fractions, f_v and f_g, according to:

$$\frac{1}{\rho} = \frac{f_v}{\rho_v} + \frac{f_g}{\rho_g} + \frac{1-f_v-f_g}{\rho_l} \quad (3)$$

where ρ_v is the density of the vapour in the bubbles, ρ_g is the density of the non-condensable gas–hydrogen for aluminium melt–, and ρ_l is the liquid density [4]. The standard $k - \varepsilon$ model [9] is used for closure in the RANS formulation.

2.1.2. Wave equation. The wave equation is:

$$\frac{\partial^2 p}{\partial t^2} - c^2 \frac{\partial}{\partial x_i}\left(\frac{\partial p}{\partial x_i}\right) = c^2 S_p \quad (4)$$

where p is acoustic pressure, c is the speed of sound, and S_p are the wave source terms.

2.1.3. Full cavitation model. The liquid-bubble mass transfer is governed by the cavity transport equation [4]:

$$\frac{\partial(\rho_v f_v)}{\partial t} + \nabla \cdot (\rho_v \boldsymbol{u} f_v) = \nabla \cdot (\Gamma \nabla f_v) + S_G - S_C \quad (5)$$

where $\Gamma = \mu + \mu_t$ is the effective exchange coefficient, and S_G and S_C are the mass transfer source terms related to the growth and collapse of the cavitation bubbles respectively. S_G and S_C are derived from the Rayleigh-Plesset equation [6] and are given by [4]:
when $p < p_v$:

$$S_G = C_G \frac{\sqrt{k}}{S} \rho_l \rho_v \left[\frac{2}{3}\frac{p_v - p}{\rho_l}\right]^{\frac{1}{2}} (1 - f_v - f_g) \quad (6)$$

when $p > p_v$:

$$S_C = C_C \frac{\sqrt{k}}{S} \rho_l \rho_v \left[\frac{2}{3}\frac{p - p_v}{\rho_l}\right]^{\frac{1}{2}} (f_v) \quad (7)$$

where $C_G = 0.02$ and $C_C = 0.01$. p_v is the sum of the vapor pressure and an estimation of the local values of turbulent pressure fluctuations [4]:

$$p_v = p_{sat} + 0.39\rho k/2 \quad (8)$$

2.1.4. Modification of the full cavitation model for melt modelling. In liquid aluminium, cavitation due to vapour pressure is hard to attain and cavitation is mainly due to non-condensable gases, mainly hydrogen. The full cavitation model is therefore modified to account for this difference.

The vapour mass fraction is ignored and the density and mass fraction transport equations are re-written as:

$$\frac{1}{\rho} = \frac{f_g}{\rho_g} + \frac{1-f_g}{\rho_l} \tag{9}$$

$$\frac{\partial(\rho_g f_g)}{\partial t} + \nabla\cdot(\rho_g \boldsymbol{u} f_g) = \nabla\cdot(\Gamma\nabla f_g) + S_G - S_C \tag{10}$$

where the sources are now given by:
when $p_{min} < (p_g + 0.39\rho k/2)$:

$$S_G = C_G\frac{\sqrt{v_{ch}}}{S}\rho_l\rho_g\left[\frac{2}{3}\frac{p_g+0.39\rho k/2-p}{\rho_l}\right]^{\frac{1}{2}}(1-f_g) \tag{11}$$

when $p_{max} > (p_g + 0.39\rho k/2)$:

$$S_C = C_C\frac{\sqrt{v_{ch}}}{S}\rho_l\rho_g\left[\frac{2}{3}\frac{p-p_g-0.39\rho k/2}{\rho_l}\right]^{\frac{1}{2}}(f_g) \tag{12}$$

The coefficients of Athavale et al [3] are used in the source terms, although their application to this modified model has to be validated from experiment. v_{ch} is estimated as 1 % of the mean flow speed.

2.1.5. Modelling flow and bubble generation in the macro- time scale. In order to run simulations in a time scale more appropriate for the flow, the minimum and maximum acoustic pressures p_{min} and p_{max} are first computed from a microsecond time scale simulation, with the time-step chosen so as to ensure that a minimum of 20 points are available per ultrasonic cycle. These stored pressures are then respectively applied to the generation and collapse source terms given by equations (11) and (12).

2.2. Discretization

2.2.1. Leap-frog scheme for the wave equation. Wave equation (4) is discretized using a second-order scheme. Acoustic pressures are stored at cell centres at each time step. Velocity components are stored on faces half time step apart from the pressures.

$$p^k = p^{k-1} - \frac{c\Delta t}{x_i}\left(u_{i,dwnd}^{k-\frac{1}{2}} - u_{i,upwnd}^{k-\frac{1}{2}}\right) + S\Delta t$$

$$u_i^{k+\frac{1}{2}} = u_i^{k-\frac{1}{2}} - c\Delta t\left(p_{dwnd}^k - p_{upwnd}^k\right) + S_i\Delta t$$

where u_i are the wave velocities.

2.2.2. Finite volume method. The finite volume method is used to discretize the RANS and transport equations.

3. Problem description

In order to model continuous treatment of melt, the launder, shown in Figure 1, is used for the computational domain. The sonotrode is at the center of the domain and the tip is immersed 1 cm below the liquid surface. The inlet and outlet are at the lowest and highest y values respectively. Clean liquid, that is without bubble, enters the domain at a velocity of 0.01 m s^{-1} in the y direction. The launder problem is solved for both water and aluminium. The launder boundaries are full reflective to ultrasound.

The distance between the baffles is varied as a function of the sound wavelength. In water, the wavelength λ of ultrasound at 20 kHz is 7.4 cm, and in aluminium, the wavelength is 23 cm. The case is run for each liquid for the following distances L between baffles: 0.5 λ, 1 λ, and 1.5 λ. The material properties for the melt – aluminium –, and water are shown in Table 1.

Table 1. Material properties of liquid aluminium and water [1].

Material Property	Aluminium (700 °C)	Water (20 °C)
Sound speed $\boldsymbol{c}$ (m s^{-1})	4600	1481
Density $\boldsymbol{\rho_l}$ (kg m^{-3})	2350	1000
Dynamic viscosity $\boldsymbol{\mu}$ (mPa s)	1.3	0.798
Surface tension (hydrogen interface) $\boldsymbol{S}$ (N m^{-1})	0.87	0.072
Vapour pressure $\boldsymbol{p_{sat}}$ (kPa)	0	4.24

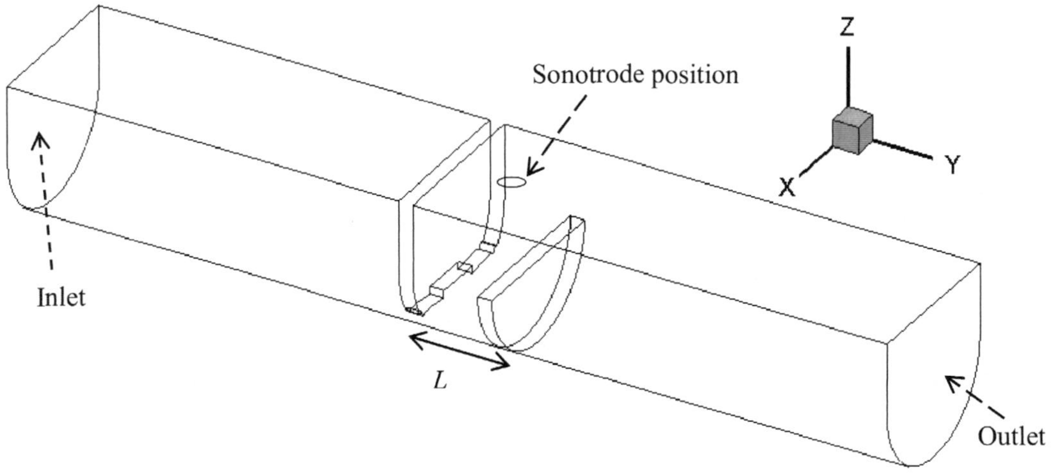

Figure 1. 50 cm x 9 cm x 8 cm launder. L denotes the length between baffles, of thickness 8 mm each. The sonotrode is immersed 1 cm into the free surface at the center of the domain.

4. Results

4.1. Acoustic run

The wave equation (4) is solved with a time step of 1 μs, and the minimum and acoustic pressure in the domain is obtained, as shown in Figure 2 and Figure 3 for the baffles separated at 37 cm, half the wavelength of sound in water at 20 °C. As expected, the extreme values of pressure are at their highest just below the sonotrode, in the middle of the domain. The cavitation threshold of -1.0 x 10^5 Pa is achieved in the whole domain for this configuration.

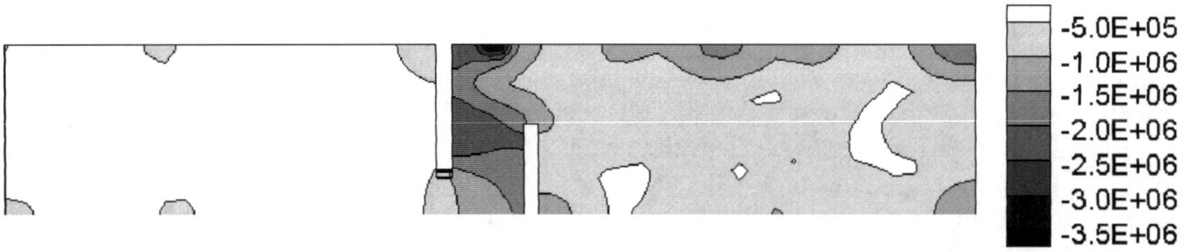

Figure 2. Predicted minimum acoustic pressure (Pa) in domain for a configuration with baffles separated at a distance of 0.5 $\boldsymbol{\lambda}$ for water.

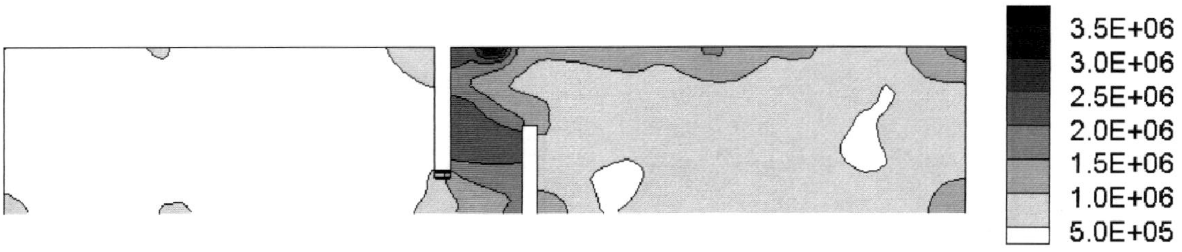

Figure 3. Predicted maximum acoustic pressure (Pa) in domain for a configuration with baffles separated at a distance of 0.5 λ for water.

The low values of maximum acoustic pressure at the left of the first (upstream) separation hinders the collapse of bubbles, and can yield large values of bubble concentration upstream, as shown in the next section.

4.2. Cavitation run

The full cavitation model is run for 20 s of simulation time with a time step of 1 ms for both water and aluminium. Figure 4 and Figure 5 show the bubble mass fraction below the sonotrode at the end of the run time for both the water and aluminium simulations. Figure 6 and Figure 7 describe the variation of the bubble concentration along the axis of the launder.

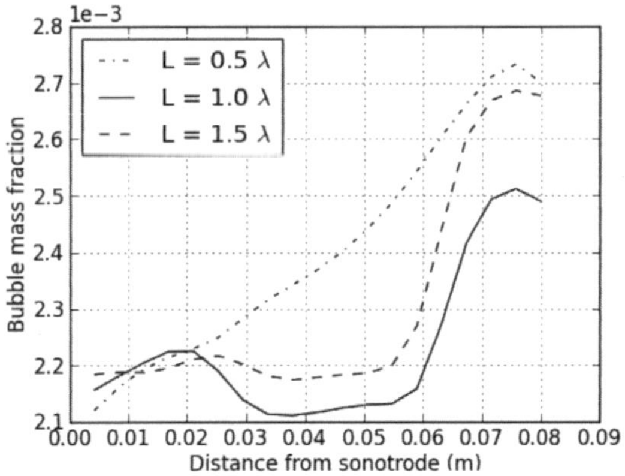

Figure 4. Bubble mass fraction along the axis of the sonotrode for water after a run time of 20 s. Mass fraction values are taken along the axis of the sonotrode.

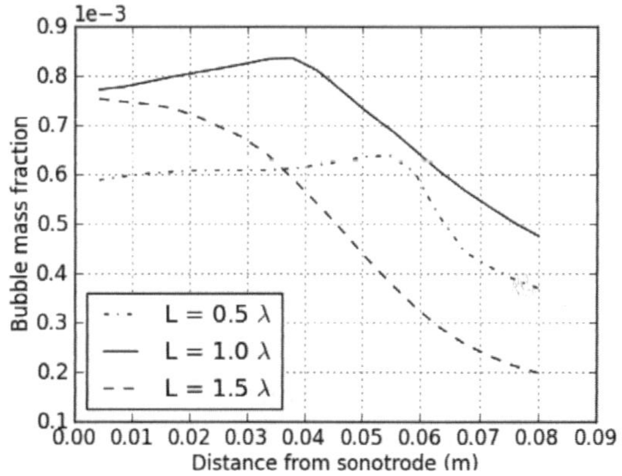

Figure 5. Bubble mass fraction along the axis of the sonotrode for aluminium after a run time of 20 s. Mass fraction values are taken along the axis of the sonotrode.

For water, the baffle configuration that maximises the bubble concentration in the domain is when the baffles are separated by a distance of 0.5 λ. The flow pattern generated with this optimum separation distance effectively convects the bubbles downstream, resulting in a large the bubble concentration downstream. For aluminium, a separation of 1.0 λ between the baffles is the optimum configuration. More bubbles are also convected downstream with this configuration.

The small opening of the upstream baffles forces a strong current at the bottom of the launder, resulting in a recirculation of the fluid beneath the sonotrode region, as shown in Figure 8 for aluminium and a baffle separation of 1.0 λ. The shortest path for exiting the domain is along the strong bottom current. Table 2 lists the shortest residence time for each configuration.

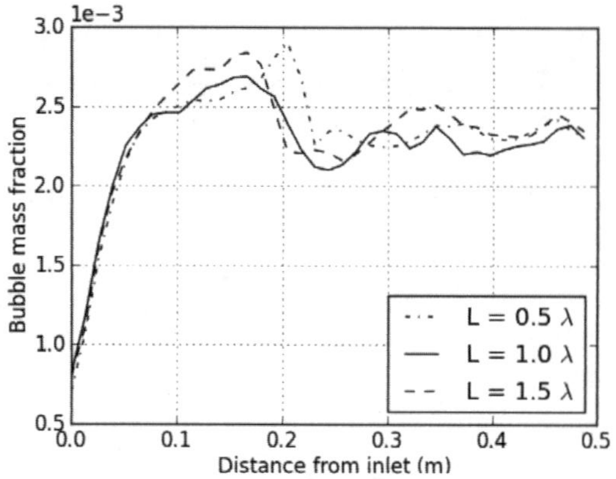

Figure 6. Bubble mass fraction across the launder for water after a run time of 20 s. Mass fraction values are taken along the axis of the launder.

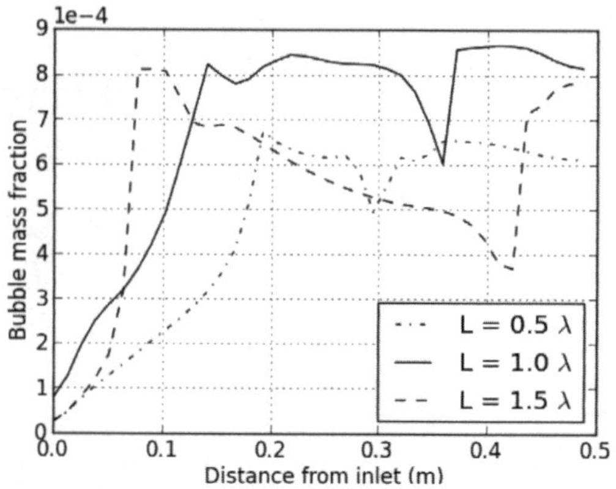

Figure 7. Bubble mass fraction across the launder for aluminium after a run time of 20 s. Mass fraction values are taken along the axis of the launder.

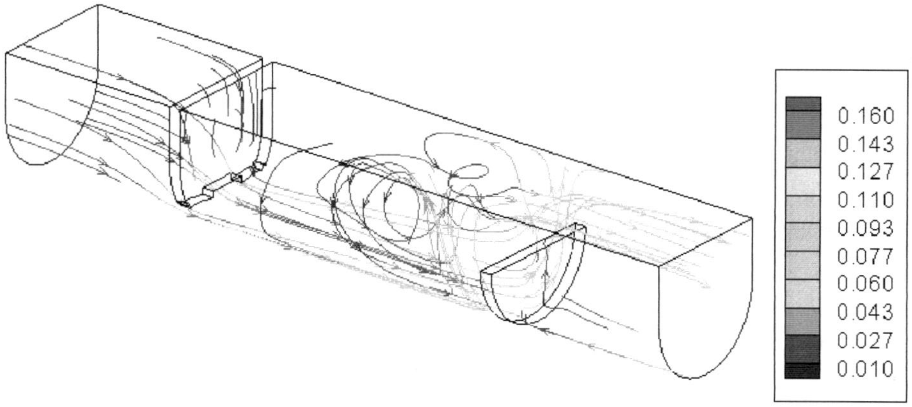

Figure 8. Recirculation in configuration with baffles separated at a distance of 1.0 λ for aluminium. Contour colours represent speed in m s^{-1}.

Table 2. Minimum residence time (s) for each configuration. The following values denotes the time taken for the liquid to leave the domain along the shortest path.

L (λ)	$L_{aluminium}$ (cm)	Aluminium at 700 °C (s)	L_{water} (cm)	Water at 20 °C (s)
0.5	11.5	16.0	3.7	21.2
1.0	23.0	12.5	7.4	15.5
1.5	34.5	9.8	11.1	13.8

5. Conclusion

In search of a complete model, multi-scale model for the ultrasonic treatment of liquid metals, this study computes the bubble concentration for the flow in a launder with baffles to create recirculations around the sonotrode. Cavitation threshold pressures are achieved in the large part of the domain, and bubble concentration as a function of baffle separation can be predicted in a timescale of seconds. The optimum configuration for the ultrasonic treatment of aluminium melt is found to be with a baffle separation of 1.0 λ.

Acknowledgement
The authors are grateful to the UK Engineering and Physical Sciences Research Council (EPSRC) for financial assistance for this research in contract numbers: EP/K00588X/1 and EP/K005804/1.

References
[1] Eskin G 1998 *Ultrasonic Treatment of Light Alloy Melts* (Amsterdam: Gordon and Breach)
[2] Campbell J 1981 *Int. Met. Rev.* **26** 71
[3] Athavale M, Jiang Y, Li H and Singhal A 2002 *Int. J. Rot. Mach.* **8** 45
[4] Singhal A, Athavale M, Li H and Jiang Y 2002 *J. Fluid Eng.* **124** 617
[5] Dular M, Bachert R, Stoffel B, and Širok B 2005 *Eur. J. Mech B.Fluid* **24** 522
[6] Plesset MS 1949 *J. Appl. Mech.* **16** 227
[7] Brennen C 1995 *Cavitation and Bubble Dynamics* (Oxford: Oxford University Press)
[8] Nastac L 2011 *Metall. Mater. Trans. B* **42** 1297
[9] Launder B and Spalding D 1974 *Comp. Meth. In Appl. Mech. & Eng.* **3** 269

IOP Conf. Series: Materials Science and Engineering **72** (2015) 052051 doi:10.1088/1757-899X/72/5/052051

Heat transfer performance of an oscillating heat pipe under ultrasonic field with dual frequency

B W Fu[1,2], N N Zhao[1, 2,4], H B Ma[3] and F M Su[1,2]
[1] Institute of Marine Engineering and Thermal Science, Marine Engineering College, Dalian Maritime University, Dalian, 116026, China
[2] Key Laboratory of Marine, Mechanical & Manufacturing Engineering of the Ministry of Transport, Dalian Maritime University, Dalian 116026, China
[3] Department of Mechanical & Aerospace Engineering, University of Missouri, Columbia MO 65211, USA

E-mail: znn@dlmu.edu.cn

Abstract. The oscillating motion and heat transfer capacity in an oscillating heat pipe (OHP) under the effect of ultrasound was investigated experimentally. Using the electrically-controlled piezoelectric ceramics, the ultrasonic sound was applied to the evaporating section of the OHP. The heat pipe was tested with or without the ultrasonic sound. The effect of ultrasound on the heat transfer performance was conducted with ultrasound of single frequency or dual frequency. The experimental results demonstrate that the OHP under the effect of the ultrasonic sound with dual frequency performs better than that one with single frequency.

1. Introduction

Extensive experimental investigations and theoretical analysis have been conducted, and show that the heat transfer performance of an OHP depends on many factors such as working fluid, tilt angle, dimension, filling ratio, turn number and heat flux level [1-10]. While these investigations have provided an insight into the mechanisms of oscillating motions and heat transfer occurring in the OHP, many factors affecting the heat transport capability have not been fully understood. Ma et al. [11] indicate that the temperature difference caused by the heat addition in the evaporator and heat rejection in the condenser is the primary reason for the oscillating motion. Pai et al. [12] presented a nonlinear thermomechanical finite-element model of U-shaped three-plug oscillating heat pipes (OHPs) and results show that an OHP is a parametrically excited nonlinear thermomechanical system. Ji et al. [13] and Hao et al. [14] improved the heat transfer capacity of the OHP by treating the inner surface. Most recently, Zhao et al. [15-16] demonstrated that ultrasonic sound can significantly reduce the startup power and enhance the heat transfer capacity of the OHP.

In this paper, the heat transfer performance of an oscillating heat pipe under the ultrasonic field with dual frequencies was investigated experimentally in order to find out whether ultrasonic sound with dual frequency can increase oscillating motion and enhance the heat transfer capacity of OHPs.

[4] Corresponding author.

2. Experimental setup

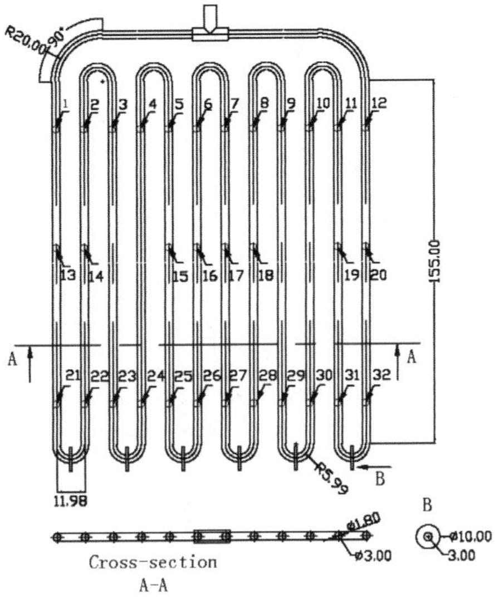

Figure 1. Schematic of the oscillating heat pipe with locations of the PZTs and thermocouples. (dimension unit: mm)

Figure 1 illustrates a 6-turn closed-loop oscillating heat pipe. The size of the OHP has the overall dimensions of 15.5 cm by 15.5 cm. The OHP consisting of the condenser, adiabatic section, and evaporator was made of copper tubing with an inner diameter of 1.8 mm and an outer diameter of 3.0 mm. Six piezoelectric ceramics made of Pb-based Lanthanumdoped Zirconate Titanates (PZT) were placed on the evaporating section. Each piezoelectric ceramic has an inner diameter of 3.0 mm and an outer diameter of 10.0 mm. The OHP was instrumented with 32 T-type calibrated thermocouples. The maximum error of the calibrated thermocouples was ±0.1 °C.

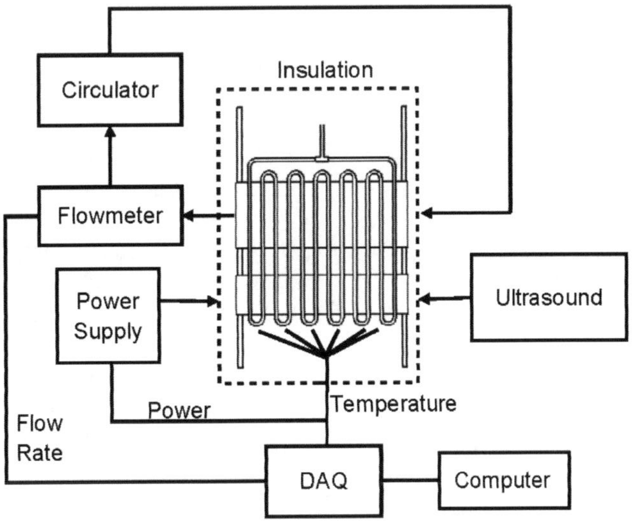

Figure 2. Schematic of the experimental system.

The schematic of the experimental system is shown in figure 2. In order to reduce the thermal contact resistance, the copper tubing were laid into 1.5 mm deep semi-circular grooves machined into the 6.4 mm thick copper heat spreading plates for the evaporator sections, and 20.0 mm thick copper plate in the condenser section. The evaporator plate was sized to match the strip heater's dimensions of 38.0 mm by 155.0 mm while the condenser plate was sized to 64.0mm by 155.0 mm. The adiabatic region was 53 mm long. The heat pipe was charged with high performance liquid chromatography (HPLC) grade water at the filling ratio of 50%. The OHP was encased in thermal insulation material and wrapped in aluminium foil to prevent the heat losses. The condenser section was cooled by a Julabo F34 temperature controlled circulator. Temperature data was collected via a NI SCXI-1600 DAQ at 100 Hz controlled by a computer. The ultrasonic sound was generated by the piezoelectric ceramics. A control circuit was used to adjust the ultrasound by varying voltage and frequency respectively. The operating temperature for all the tests were 40 °C.

3. Results and discussion

The same OHP was tested with and without the ultrasound to demonstrate the ultrasonic effect on the heat transfer performance of the OHP. The heat load tested was 25W, 40W, 60W, 80W, 100W, 150W and 200W respectively. The total power on the six PZTs was regulated to 0.1W no matter what frequencies were put on the PZTs. Figure 3 illustrates the ultrasonic effect on the thermal resistance of the OHP under different frequencies. As shown, the ultrasonic sound can effectively reduce the thermal resistance of the OHP no matter the power frequency on all the six PZTs were 14kHz or 474kHz. It also indicates that the thermal resistance was further reduced when the power frequency on one half of the PZTs was 14kHz and the other half of the PZTs was 474kHz.

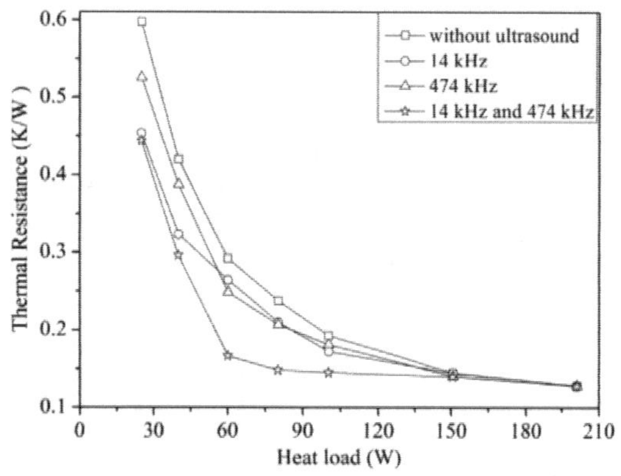

Figure 3. Ultrasonic effect on the thermal resistance.

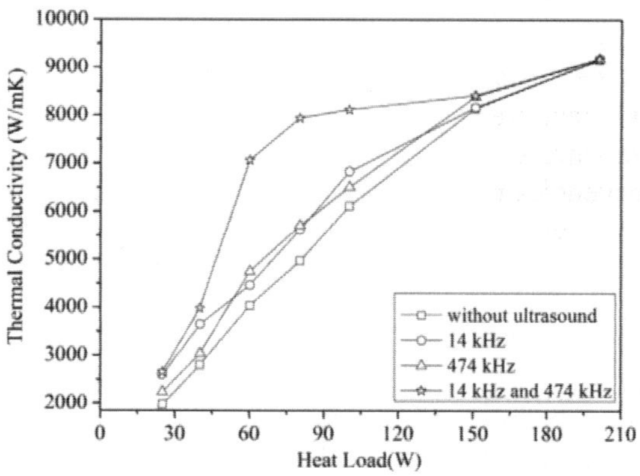

Figure 4. Ultrasonic effect on the effective thermal conductivity.

Figure 4 illustrates the ultrasonic effect on the effective thermal conductivity of the OHP under different frequencies. Same results were achieved that the effective thermal conductivity of the OHP performed best under the ultrasound with dual frequencies. Moreover, the ultrasonic effect works better when the heat load was less than 150W.

In order to better determine the ultrasonic sound effect on the heat transfer enhancement of an OHP, a relative enhancement, which is called the enhancement percentage, E, is defined as,

$$E = (A - B) / A \times 100\% \tag{1}$$

where A and B are the thermal resistance of the OHP without and with the ultrasonic effect respectively. Figure 5 shows the ultrasonic effect on the enhancement percentage of the OHP. As shown, when ultrasound with dual frequency were added in the OHP, the heat transfer enhancement was the highest. In addition, it can be found that the enhancement percentage depends on the heat load. The highest enhancement percentage of the thermal resistance of the OHP was up to 42.8% when the heat load was 60 W with the effect of dual frequency ultrasound.

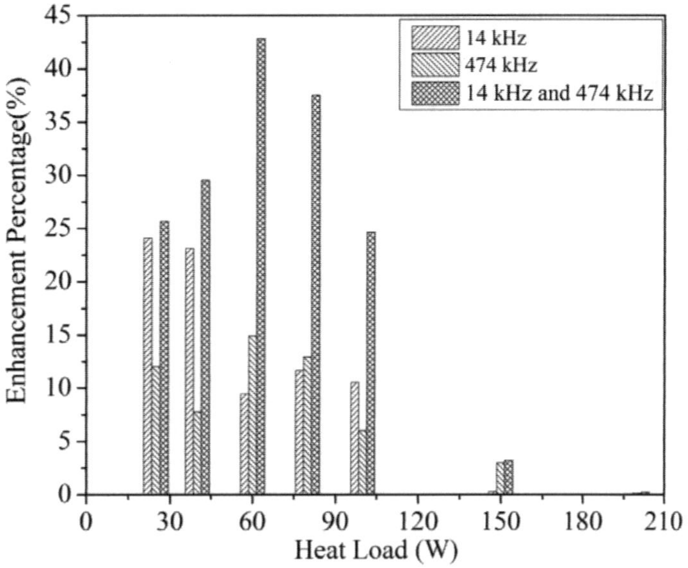

Figure 5. Enhancement percentage of the ultrasonic effect on the OHP.

4. Conclusions

The heat transfer performance of an oscillating heat pipe under the ultrasonic field with dual frequencies was investigated experimentally. Results show that ultrasonic sound with dual frequencies performs better improvement on the heat transfer capacity of the OHP comparing to one frequency. In addition, the heat transfer enhancement of an OHP under the effects of ultrasonic sound depends on the heat load.

Acknowledgments

This research work was supported by the Fundamental Research Funds for the Central Universities of China under Grant No. 3132014044 and 3132014320, the National Natural Science Foundation of China under Grant No. 51076019 and 51376027 and the Office of Naval Research under Grant No. N00014-11-01-0334 directed by Dr. Mark Spector.

References

[1] Zhang Y and Faghri A 2008 *Heat Transf. Eng.* **29** 20–44
[2] Shafii M B, Faghri A and Zhang Y W 2001 *J. Heat Transf.-Trans. ASME* **123** 1159–72
[3] Qu W and Ma H B 2007 *Int. J. Heat Mass Transf.* **50** 2309–16
[4] Zhang Y W and Faghri A 2002 *Int. J. Heat Mass Transf.* **45** 755–64
[5] Zhao N, Zhao D and Ma H 2013 *J. Therm. Sci. Eng. Appl.* **5** 011005
[6] Ma H B, Wilson C, Borgmeyer B, Park K, Yu Q, Choi S U S and Tirumala M 2006 *Appl. Phys. Lett.* **88** 143116
[7] Ma H B, Wilson C, Yu Q, Park K, Choi U S and Tirumala M 2006 *J. Heat Transf.-Trans. ASME* **128** 1213–6
[8] Khandekar S, Schneider M, Schafer P, Kulenovic R and Groll M 2002 *Microscale Thermophys. Eng.* **6** 303–17
[9] Thompson S M, Ma H B, Winholtz R A and Wilson C 2009 *J. Heat Transf.-Trans. ASME* **131** 043210
[10] Chien K H, Lin Y T, Chen Y R, Yang K S and Wang C C 2012 *Int. J. Heat Mass Transf.* **55** 5722–8
[11] Ma H B, Hanlon M A and Chen C L 2006 *Microfluid. Nanofluidics* **2** 171–9

[12] Pai P F, Peng H and Ma H 2013 *Int. J. Heat Mass Transf.* **64** 623–35
[13] Ji Y, Chen H, Kim Y J, Yu Q, Ma X and Ma H B 2012 *J. Heat Transf.* **134** 074502
[14] Hao T, Ma X, Lan Z, Li N, Zhao Y and Ma H 2014 *Int. J. Heat Mass Transf.* **72** 50–65
[15] Zhao N, Zhao D and Ma H B 2013 *J. Heat Transf.* **135** 074503
[16] Zhao N, Fu B and Ma H B 2014 *Appl. Phys. Lett.* **104** 263105

IOP Conf. Series: Materials Science and Engineering **72** (2015) 052052 doi:10.1088/1757-899X/72/5/052052

Simulation and experimental study of cavitation region caused by longitudinal and transverse vibration of casting ultrasonic radiator

X Q Li[1,2], R Q Li[1,2,3], F Dong[1,2], P H Chen[1,2] and R P Jiang[1,2]

[1] School of Mechanical and Electrical Engineering, Central South University, Changsha 410083, China

[2] State Key Laboratory of High Performance Complex Manufacturing, Central South University, Changsha 410083, China

E-mail:lll87430@126.com

Abstract. The dynamic simulation of casting ultrasonic vibration system were studied based on dynamic characteristics of the system and numerical simulation of ultrasonic sound pressure field under different vibration depths in water was performed. According to the simulation results, the cavitation region was estimated, and the experiment of cavitation erosion of aluminum foil in water and amplitude test were done to verify simulation results. The results showed that the longitudinal vibration at the end of face was the major vibration of ultrasonic radiation, while there was a strong vibration on cylindrical surface. Results of amplitude test are consistent with simulation results. Cavitation region caused by ultrasonic vibration mainly below the end face of radiation. The scope and intensity of cavitation were inversely related to the distance from the end face of radiation. With the vibrating depth increased, some small cavitation regions were found at the side of radiation, cavitation region distributed un-continuously below radiation. Cavitation field measurement and simulation results were basically consistent.

1. Introduction

In ultrasonic casting, ultrasonic cavitation generated in the molten metal is considered to be the main reason to promote nucleation and grain refinement. With the effect of the strong ultrasonic wave in melt, sparse acoustic phase and compression phase alternating, based on the above effect, the micro-bubbles would expand, compress, vibrate and eventually collapse, when they collapse, high-intensity shock wave in a very limited volume is produced instantly, prompting to break primary crystal in aluminum alloy melt and increase the number of heterogeneous crystallization nuclei, so that ultrasonic cavitation significantly refine solidification structure [1-4]. In the molten metal, due to high-temperature, internal invisible and high strength chemical changes, there is no effective, quantitative measurement to measure the intensity and range of cavitation. Most of the existing research to test cavitation is in the water [5-7], or observing aluminum alloy solidification treated by ultrasound to characterize the effect of ultrasonic cavitation indirectly.

In this paper, modeling and simulation of ultrasonic vibration system and sound field distribution in water were studied, the sound pressure field distribution in water and the amplitude of longitudinal and

[3] Corresponding author.

IOP Conf. Series: Materials Science and Engineering **72** (2015) 052052 doi:10.1088/1757-899X/72/5/052052

transverse vibration were got. Cavitation region was pre-estimated, amplitude and water cavitation tests verified the simulation results.

2. Simulation of ultrasonic vibration and cavitation region

2.1. Model and parameters

Figure 1 shows the structure diagram of ultrasonic casting system and the finite element model for sound-fluid-temperature coupled field. Ultrasonic system mainly contain sandwich transducer, the first amplitude transformer, the secondary amplitude transformer and radioactive rod. In this paper, the electrodes, connecting bolts dynamic between the amplitude amplifiers and flange in the nodal plane position was ignored because of tiny effects on dynamic characteristics. However, front and rear covers of piezoelectric ceramic transducer and pre-loaded bolts should be taken into account. Fluid field is analyzed at the infinity, and the fluid129 units are adopted to simulate the acoustic absorbing boundary which extends to infinity in the two-dimensional plane. The fluid field is simulated by the infinite semicircle region. As shown in figure 1(b).

Material of ultrasonic vibration system is shown in figure 1 mainly comprises titanium, aluminum, 45 steel and piezoelectric ceramics PZT-8. The fluid medium is water. As to fluid material, density of water is $1000 kg/m^3$, sound velocity is 1500 m/s in water.

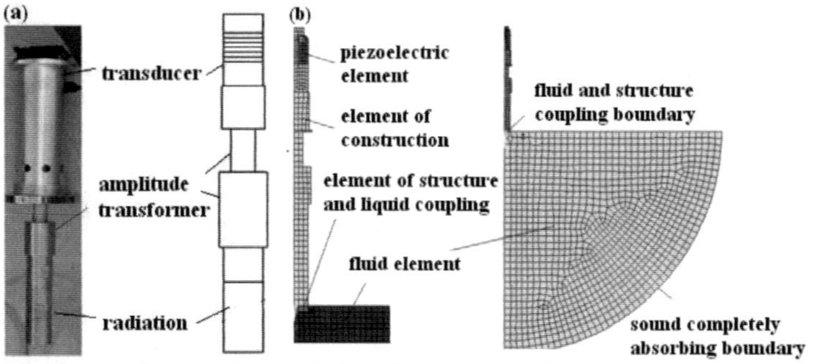

Figure 1. Schematic models of ultrasonic vibration system: (a) geometrical model, (b) computational model and FEM model.

2.2. Governing equations

2.2.1. Kinetic equation. Structural dynamics analysis based on the following basic equation:

$$M\{\ddot{u}\}+C\{\dot{u}\}+K\{u\}=\{F\} \tag{1}$$

M, C, K are mass matrix, damping matrix and stiffness matrix of the system, $\{F\}$ is load force vector, $\{u\}$ is displacement vector. In this paper, the above matrices and vectors are extended to a more generalized form, so physical problems can be transformed into the same equation form, finally the individual physics problems will be solved as pure mathematics problem.

2.2.2. Sound field control equation. When the ultrasonic wave propagation in the fluid, the sound pressure distribution obey Helmholtz equation[8]:

$$\omega^2 \bullet c^{-2} P+\nabla^2 P=0 \tag{2}$$

where P is sound pressure amplitude in fluid, N/m^2. Angular frequency $\omega=2\pi f$, f is the vibration frequency of the sound wave, Hz. c is propagation velocity of sound waves in the fluid, m/s. Solution of the partial differential equations is only connected with ω, c and boundary conditions, as long as those parameters are determined, it will determine the ultrasound field distribution in the region.

2.2.3. The fluid-structure coupling equations. When analyze fluid-structure coupling problems, the sound pressure of the fluid interface should be taken into account. Integrate Sound pressure at the interface S and take the sound pressure shape function matrix into equation (1), then combined the equation with interface energy waves of the discrete equation, fluid-structure coupling problems are got as equation (3),

$$\begin{bmatrix} [M_e][0] \\ [M^{fs}][M_e^P] \end{bmatrix} \begin{Bmatrix} \{\ddot{U}_e\} \\ \{\ddot{P}_e\} \end{Bmatrix} + \begin{bmatrix} [C_e][0] \\ [0][C_e^P] \end{bmatrix} \begin{Bmatrix} \dot{U} \\ \dot{P}_e \end{Bmatrix} + \begin{bmatrix} [K_e][K^{fs}] \\ [0][K_e^P] \end{bmatrix} \begin{Bmatrix} \{U_e\} \\ \{P_e\} \end{Bmatrix} = \begin{Bmatrix} \{F_e\} \\ \{0\} \end{Bmatrix} \quad (3)$$

where $[M^{fs}] = \rho_0 [R_e]^T, [K^{fs}] = -[R_e]$.This is the sound-fluid-structure coupling equation.

2.3. Boundary conditions and solving

1) Harmonic responding analysis, respectively coupling the positive and negative electrode node voltage freedom of the piezoelectric ceramic sheet, the negative electrode with a load Volt =0, the load applied to the positive electrode is 300 V. Set the frequency range is 15-30 kHz, the minimum frequency step size is 4 Hz.

2) Fluid-structure coupling analyze, applying fluid-structure coupling boundary condition at the junction of the ultrasonic radiation and fluid. Make each node in equipotential state. Frequency range is 17000 Hz~22000 Hz, sub-step is 1000 steps, each step interval is 5 Hz, define full structural damping ratio is 4.2%.

2.4 Simulation results and analysis

2.4.1 Amplitude distribution on end face and side face of ultrasonic radiation rod. Figure 2 indicates longitudinal amplitude distribution was sinusoidal on the end face, the vibration amplitude was the largest at the center, and then decreases monotonously from the center to the edge, but the overall distribution was relatively uniform, the vibration amplitude was more than 10 μm in each point of end face. Distribution of lateral transverse vibration was obvious inequality, the displacement peak was located in 50 mm from the end, the size was 3.15 μm. It's monotonically decreasing along with the peak to two sides, vibration amplitude attenuate to less than 1 μm near the location on the end face of radiation rod.

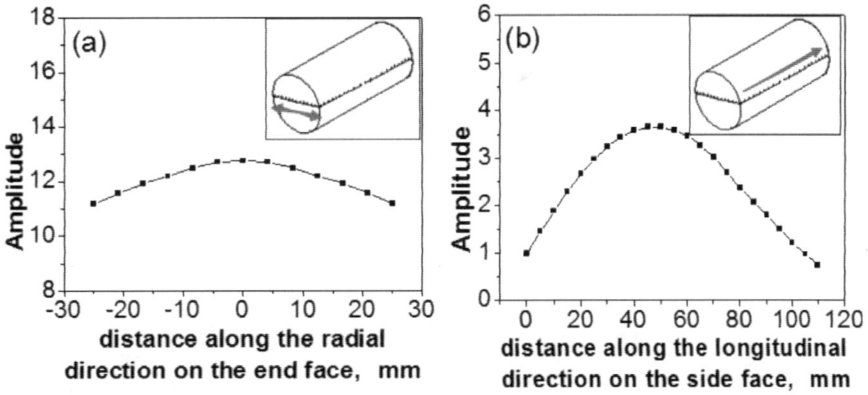

Figure 2. Simulation results of amplitude distribution: (a) on end face, (b) side face.

2.4.2 Sound pressure field distribution in water. The simulation results of sound pressure field distribution in water with different ultrasonic immersed depth shown in figure 3 indicated that sound pressure field distribution mainly below the end face of radiation. With the vibrating depth increased, some small regions were found in the side of radiation. The scope and intensity of sound pressure were inversely related to the distance from the end face of radiation.

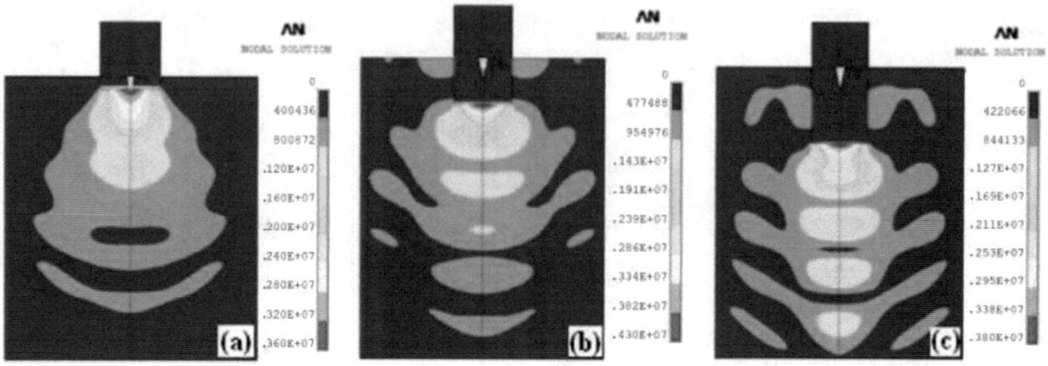

Figure 3. Sound pressure field distribution in water, ultrasonic immersed depth is: (a) 10mm, (b) 40mm and (c) 70mm.

To further analyze the sound field distribution, we defined ultrasonic cavitation range in water. The condition cavitation generated is that the sound pressure amplitude value exceeds the cavitation threshold value. Previous studies have shown that the cavitation threshold value is usually 0.1-1MPa in water, and it's 0.9 MPa when the water is in the air. Therefore, demarcate the area where amplitude value of sound pressure exceeds 0.9 MPa in figure 3, the ultrasonic cavitation range is shown in figure 4.

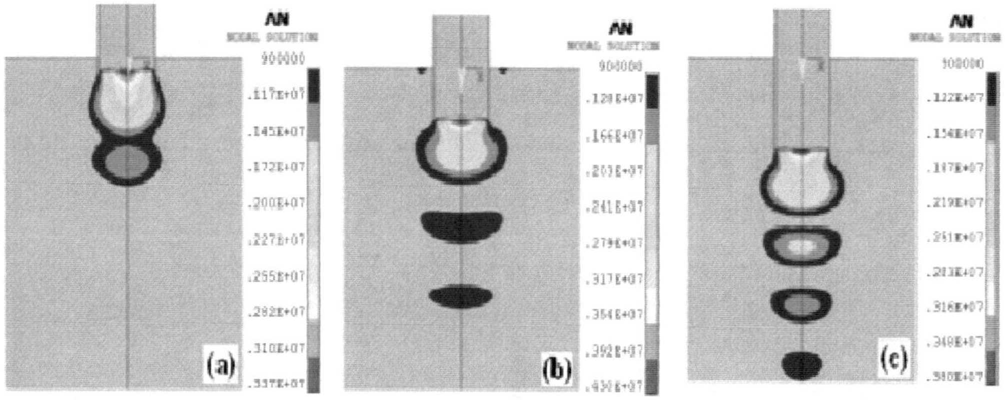

Figure 4. Ultrasonic cavitation region with different ultrasonic vibration depth in water: (a) 10mm, (b) 40mm and (c) 70mm.

Figure 4 indicates cavitation region caused by ultrasonic vibration mainly below the end face of radiation.The maximum value of cavitation appeared below the end face, the farther away from the end face, the smaller the range of cavitation region and the lower the cavitation intensity. When vibrating depth was 10 mm, there was only one region below the end face. When it was 40mm, there were three dispersive zones below the end face. Cavitation region also appeared on the side of radiation, but it was very small. When vibrating depth was 70 mm, there were four dispersive zones below the end face, but there was no region on the side of radiation.

3. Cavitation erosion of aluminum foil in water and amplitude test

3.1 Testing program

Amplitude test diagram is shown in figure 5. On the end face of the radiation along the diameter direction there are 13 measuring points. The origin is set in the center of the circle, the interval between each measurement point is 4 mm. On the side face of the radiation along the longitudinal direction there are 22 measuring points. The position of origin is 1mm away from the end face, the interval between each measurement point is 5 mm. To minimize the effects of external interference in the measurement process, all testing were completed in the vibration isolator.

Cavitation erosion of aluminum foil test is shown in figure 6. Cut aluminum foil appropriately according to the size of beaker and ultrasonic radiation, so that it can precisely fixed in the beaker. After ultrasonic vibration remove the foil from breaker, dry it and observe cavitation morphology and distribution of cavitation pore.

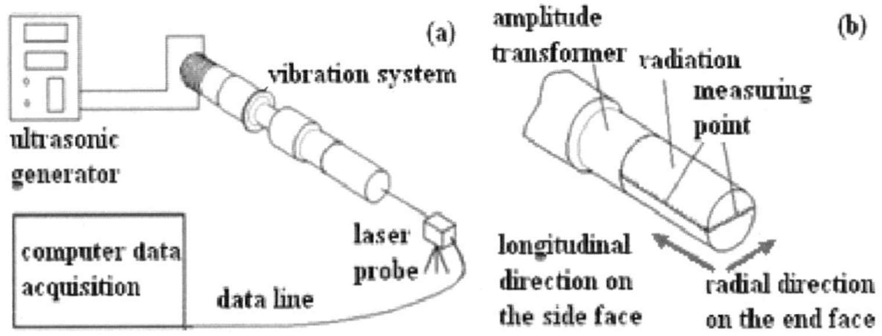

Figure 5. Amplitude test schematic: (a) schematic diagram of the test, (b) location map of test points.

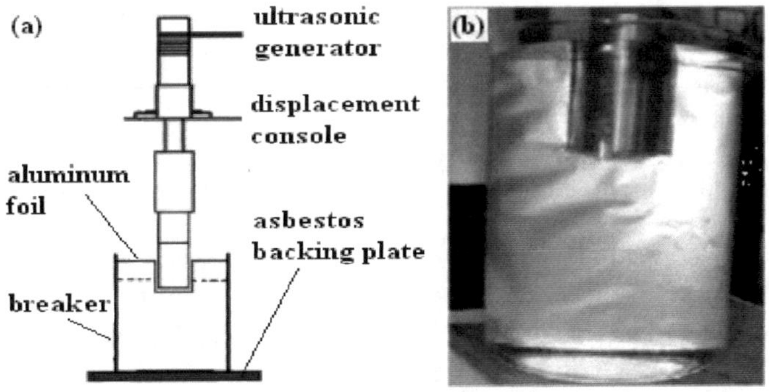

Figure 6. Cavitation erosion test of aluminum foil: (a) schematic of test, (b) photo of test.

3.2 Experimental results and discussion

Amplitude distribution test results of ultrasonic radiation compared with the simulation results as shown in figure 7 indicates that there was some deviation in longitudinal vibration distribution on side surface, the measured value were bigger than simulation results. And there was another small peak appear on lateral surface where is 90 mm away from end face. This was mainly because the connecting bolts between amplitude transformer were not considered in finite element modeling process. But on the whole test measurement and simulation results were basically consistent. This indicates that the radiation pole resonant vibration in the test is a single frequency, the vibration type determine there were a certain extent longitudinal vibration on the side surface.

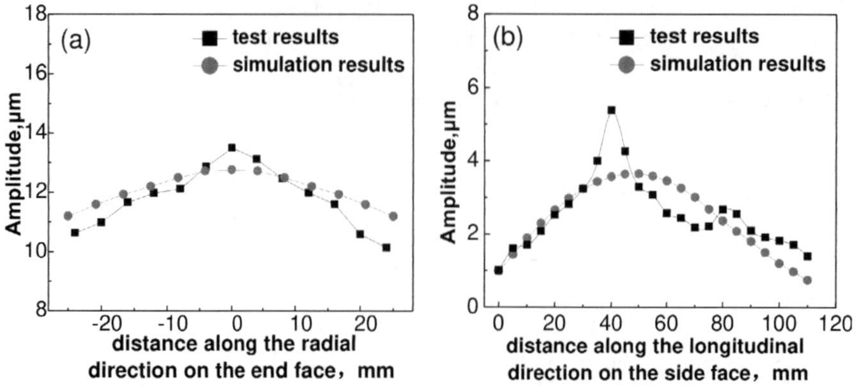

Figure 7. Amplitude distribution test results of ultrasonic radiation compared with the simulation results.

Aluminum foil cavitation erosion morphology is shown in figure 8. Figure 8(a) shows that the corrosion pits concentrated in the range of about 40mm below the end face of the tool head. The farther away from the end face, the number of corrosion pits gradually reduced and distributed gradually thinning. This indicates that the sound pressure amplitude decreased, cavitation intensity weakens. Figure 8 (a1, a2) which is a partially enlarged view shows that in cavitation region the corrosion pit size about 0.1-1mm and in the area of non-cavitation foil surface is smooth. Measured results agree well with the simulation results. The results shown in figure 8(b, c) suggest that the cavitation region was mainly distributed below the end face and nearby side surface of radiation, when ultrasonic immersed depth is 40 mm and 70 mm.There were three discrete cavitation region appeared below the end face, where the cavitation area below the rod end face was the most obvious area, its range was largest and cavitation erosion pore distribution is the most densely as shown in figure. 8(b2,c2), which indicates the cavitation effect was most significant in this place, with the place farther away from the end face of radiation, the cavitation zone range was greatly reduced and the pore distribution was increasingly sparse, but it could be clearly observed as shown in figures 8(b3, c3). This suggests that the cavitation also happened in this region, but its intensity decreased. Through measuring, the cavitation area farthest away from the end face is approximately 150mm. The range of cavitation is small nearby the radiation side surface, and pore distribution is sparse, this suggest the cavitation also happened in this region, but the strength is weak.

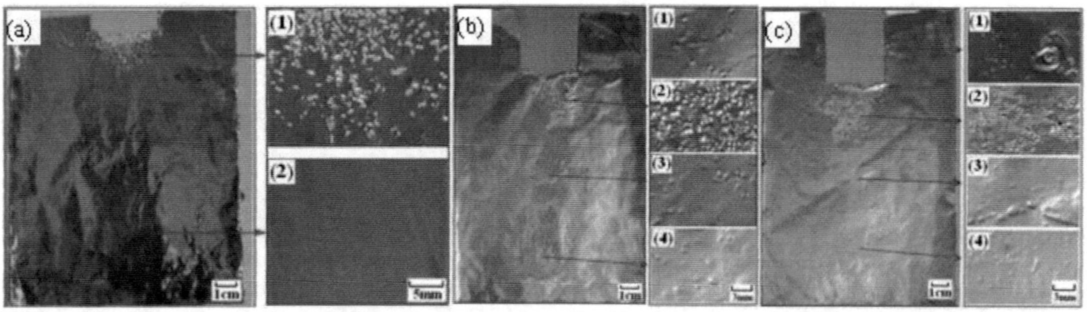

Figure 8. Aluminum foil cavitation erosion region distribution when ultrasonic immersed depth is: (a) 10mm, (b) 40 mm and (c) 70mm in water.

Compare simulation results with test results which is shown in figures 5 and 8, it can be found that the range of ultrasonic cavitation measured by test was bigger than that measured by simulation. The reasons for this phenomenon could be found from amplitude simulation and test, actual amplitude value of transverse vibration of radiation rod was bigger than simulation results. Overall, the measured cavitation field and simulation results were basically consistent, which indicated that the sound pressure field finite element simulation is accurate.

4. Conclusion

1) The amplitude simulation and test results show that the longitudinal vibration in the end face was the major vibration of ultrasonic radiation. Longitudinal amplitude distribution was sinusoidal on the end face, the vibration amplitude was the largest at the center, and then decreases from the center to the edge. Distribution of lateral transverse vibration is obvious inequality, It is monotonically decreasing along with the peak to two sides.

2) Cavitation region test and simulation results indicate that cavitation region caused by ultrasonic vibration mainly below the end face of radiation. The scope and intensity of cavitation were inversely related to the distance from the end face of radiation. With the vibrating depth increased, cavitation region distributed un-continuously below radiation, some small cavitation regions were found in the side of radiation. Cavitation field measurement and simulation results were basically consistent.

Acknowledgments

Authors are grateful to the National Basic Research Program of '973' (grant no. 2010CB731706 and 2012CB619504).

References

[1] Piao S Z 1999 *Ultrasonics Handbook* vol 1 ed Feng N (Nanjing: Nanjing University Press) p 103
[2] Singhal A K, Athavale A M, Li H Y and Jiang Y 2002 *J. Fluid Eng.* **124** 617-24
[3] Li Y L, Feng H K, Cao F R, Chen Y B and Gong L Y 2008 *Materials Sci. and Eng.* A **487** 518-23
[4] Liu X B, Osawa Y, Takamori S and Mukai T 2008 *Materials Sci. and Eng.* A **487** 120-3
[5] Liu L Y, Yang Y, Liu P H and Tan W 2014 *Ultrasonics Sonochemistry* **21** 566-71
[6] Shahram N, Seyed H H and Meysam M R 2014 *Chemical Eng. Research and Design* **92** 166-73
[7] Kyuichi Y, Toru T, Judy L Teruyuki K, Atsuya T and Yasuo I 2010 *Ultrasonics Sonochemistry* **17** 460-72
[8] Bao G, Wei G W and Zhao S 2004 *Int. J. Numer. Meth. Engng.* **59** 389

IOP Conf. Series: Materials Science and Engineering **72** (2015) 052053 doi:10.1088/1757-899X/72/5/052053

High speed observation of HIFU-induced cavitation cloud near curved rigid boundaries

Z G Zuo, F B Wang, S H Liu[1] and S J Wu

State Key Laboratory of Hydroscience and Engineering, Department of Thermal Engineering, Tsinghua University, Beijing 100084, China

E-mail: liushuhong@tsinghua.edu.cn

Abstract. This paper focuses on the experimental study of the influence of surface curvature to the behaviour of HIFU-induced cavitation cloud. A Q-switched ruby pulse laser is used to induce cavitation nuclei in deionized water. A piezoelectric ultrasonic transducer (1.7 MHz) provides a focused ultrasound field to inspire the nucleus to cavitation cloud. A PZT probe type hydrophone is applied for measuring the HIFU sound field. It was observed that the motion of cavitation cloud located near the boundary is significantly influenced by the distance between cloud and boundary, as well as the curvature of the boundary. The curvature was defined by parameters λ and ξ. Convex boundary, concave boundary, and flat boundary correspond to $\xi<1$, $\xi>1$ and $\xi=1$, respectively. Different behaviours of the cloud, including the migration of the cloud, the characteristics of oscillation, etc., were observed under different boundary curvatures by high-speed photography. Sonoluminescence of the acoustic cavitation bubble clouds were also studied to illustrate the characteristics of acoustic streaming.

1. Introduction

In recent years, precision damage with cavitation bubbles attracts more and more attentions, especially in the field of therapeutics, etc. It is recognized that the damaging power of cavitation bubbles is mainly caused by the high-speed microjet during the collapse of the bubble near a boundary wall. The behaviors of the bubble and the microjetare strongly influenced by the physical and geometrical conditions of the boundary.

Benjamin [1] found that a solid boundary induces asymmetric flow around a bubble, resulting in bubble migration toward the boundary due to the secondary Bjerknes force [2], and finally the formation of a liquid microjet. Tomita [3] recognized that the strength of jet is related to the parameter λ ($\lambda=L/R_{max}$), the distance between bubble center and target wall over the maximum radius of the migrate bubble. In recent research, Tomita's experiments [4] show that when a comparable sized bubble is located near a rigid boundary the bubble motion is significantly influenced by the surface curvature of the boundary. The curvature is characterized by a parameter ξ, giving convex walls for $\xi<1$, concave walls for $\xi>1$ and a flat wall when $\xi=1$. Uneven target walls lead to larger velocity of a liquid jet than the flat wall.

In the study of utilizing acoustic cavitation for calculus fragmentation, the cavitation clouds are induced by a HIFU (High Intensity Focused Ultrasound) device near the surface of the stones. The collective behavior of the bubble clouds, resulting from the interactions between clouds and the targets'

[1] Corresponding author.

surface [5], will have significant influence on the fragmentation efficiency. It has been shown that some similarities exist between the ensemble oscillations of the cavitation clouds induced by HIFU, and a single bubble oscillation model, in terms of the oscillating frequency spectrum [6]. It is thus postulated that the behavior of the cavitation clouds, and hence the fragmentation efficiency, might be influenced by the curvatures of the boundary walls.

A series of experiments has been designed and carried out to study this relationship between surface curvature and the behavior of cavitation clouds induce by HIFU, and preliminary results are presented below.

2. Experimental arrangements

In order to experimentally investigate the influence of surface curvature on the behavior of the HIFU-induced cavitation clouds, the following procedure was established:

1) Cavitation clouds were generated near the surfaces with different curvatures of artificial kidney stones made from a ceramic Begostone powders, by exposure under the acoustic field of a HIFU transducer in deionized water. The volumetric loss of the stones against time were then recorded, by using a 3D scanning technique, in order to evaluate the fragmentation efficiency.

2) In order to observe the behavior of the cavitation clouds, another set of experiments was carried out in degassed NaCl solutions. Surfaces with different curvatures were realized with machining of stainless steel samples. A Q-switched ruby laser was used to induce cavitation nucleiat prescribed locations. Cavitation clouds then grew from these nuclei under the radiation of HIFU acoustic field. A high-speed video camera was utilized to observe the oscillation and movement of the clouds.

3) In order to identify the cavitation intensities and the characteristics of acoustic streaming, sonoluminescence tests were carried out in luminolsolutions, by utilizing a photomultiplier and a digital SLR camera.

An overall schematic of the experimental setup is shown in figure 1. Detailed experimental devicesare illustrated below.

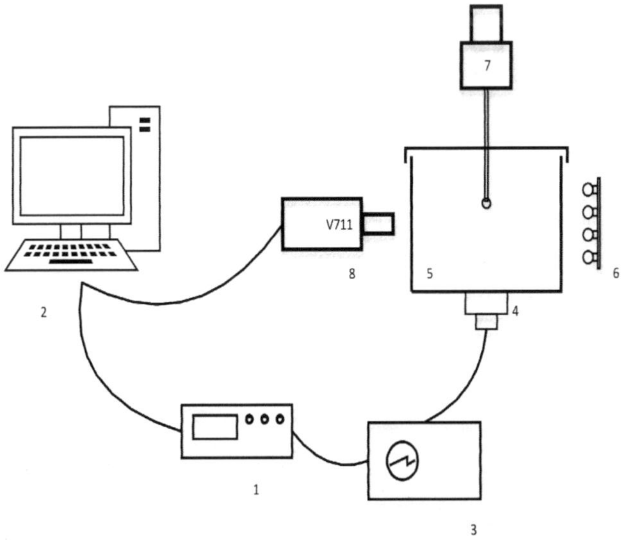

Figure 1. Schematic of the experimental setup: 1. Waveform generator; 2. Computer; 3. Power amplifier; 4. HIFU transducer; 5.Water chamber; 6. Illumination light source; 7. Sample holder and 3D positioning system; 8. High-speed video camera.

2.1. Water chamber

Since the focused sound field produced by HIFU transducer possesses a conical shape, and the cavitation clouds appear at the spindle-shaped focal region, the chamber was specially designed around the focus of the transducer. The main body of the chamber was fabricated with plexiglass, with a quartz glass window on each of the four side walls, forming a 60×60×80 mm viewing zone. This will

IOP Conf. Series: Materials Science and Engineering **72** (2015) 052053 doi:10.1088/1757-899X/72/5/052053

benefit the observation of the behavior of cavitation clouds. A PZT HIFU transducer was secured vertically at the chamber bottom, as shown in figure 2.

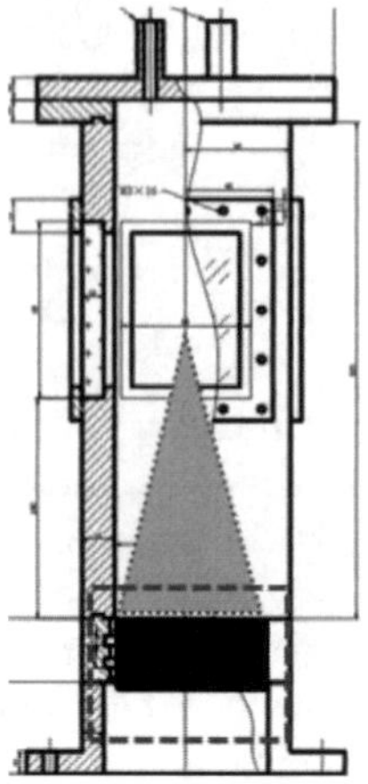

Figure 2. Water chamber.

2.2. HIFU system

Two PZT HIFU transducers have been used in the research, with natural frequency of 1.7 MHz and 1.1 MHz, and diameters of 70 mm, a focal length of 150mm and 100 mm, respectively. The transducers were driven by a waveform generator (100 MSa/s sampling rate, 14 bits vertical resolution, frequency range 1 μHz~12 MHz), connected to a power amplifier (power amplification 50 dB). The focal region pressure reaches 1MPa (cavitation threshold~0.7 MPa [6]) when input voltage comes to 0.3 V.

2.3. High speed video camera system

A VRI Phantom V711 high speed video camera (2.156×10^5 fps@128×64 bit), together with a long-range microscope (magnification 0.11×~67.5×, working length 100 mm) were used to observe the behavior of cavitation cloud.

2.4. Laser systems

In order to induce stable cavitation clouds at prescribed locations, a Q-switched ruby laser (INNOLAS ltd. QSR laser, wavelength 694.3 nm, peak energy 1.5 J/pulse, pulse width 20 ns)was used for nuclei production in degassed NaCl solutions. An optical lens system was designed to focus the laser beam into the focal region of the HIFU transducer. The beam was first expanded through a concave lens, then converged through two convex lenses (figure 3).

Figure 3. Lens system.

2.5. Laser system

In order to induce stable cavitation clouds at prescribed locations, a Q-switched ruby laser (INNOLAS ltd. QSR, wavelength 694.3 nm, peak energy 1.5 J/pulse, pulse width 20 ns) was used for nuclei production in degassed NaCl solutions. An optical lens system was designed to focus the laser beam into the focal region of the HIFU transducer. The beam was first expanded through a concave lens, then converged through two convex lenses (figure 3), to have a large converge angle. A continuous wave (cw) He-Ne laser with a similar wavelengthwas utilized to precisely adjust the location of the nuclei.

2.6. Artificial stone samples

As stated above, two types of artificial stone samples, made from ceramic Begostone powders, and stainless steel, were fabricated, for the purposes of measuring volumetric loss and observing the behavior of cavitation clouds, respectively. As shown in reference [4], a curved wall can be expressed as a steady streamline as a result of the combined flows induced by two point sources with different strengths.

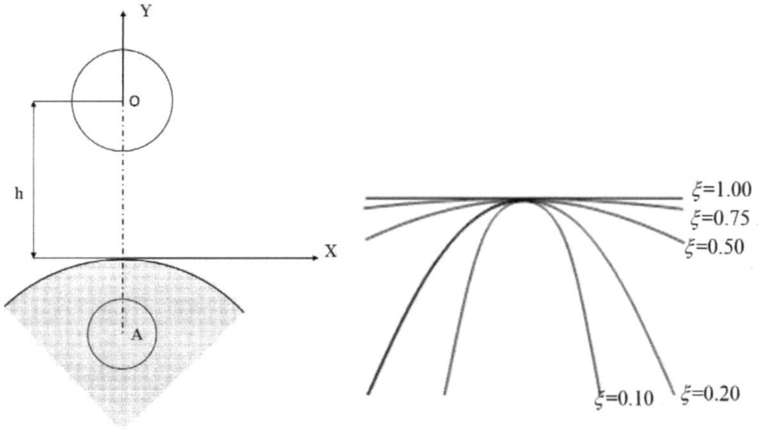

(a) (b)

Figure 4. Surface curvature.

The curved wall can be expressed as a stable streamline induced by two point sources. Cavitation cloud oscillates at $O(0, 0, h)$, and point source A is at $(0, 0, -\xi_h)$. The strength of point source B is ξ^2times that of A. The formula for the curve in the (r, z) plane is expressed as:

$$f(r,z)=\left[r^2+(h-z)^2\right]^{1/2}\left\{\xi^2(z+\xi h)+(1-\xi^2)\left[r^2+(z+\xi h)^2\right]^{1/2}\right\}$$
$$-(h-z)\left[r^2+(z+\xi h)^2\right]^{1/2}=0$$

In case of h=10, 5 different curves can be obtained when ξ=0.10, 0.20, 0.50, 0.75, 1.00, as shown in figure 4.

3. Results and Discussions

3.1. Calculus fragmentation observation

Figure 5 shows the high speed recording of the calculus fragmentation process, of the case ξ=0.2, HIFU inducer frequency 1.75 MHz, driving voltage 1.5 Vpp. Figure 6 shows the 3D scan of the erosion pattern during the fragmentation process, analyzed with Geomagic software package. A vertical bright streak in the middle of the view in figure 5 indicates the variation of refractive index caused by the strong acoustic field. The length of the horizontal black line in the figure is 1mm. It can be seen from these two figures that, in the 6 minutes of observation, two stages of the fragmentation can be distinguished. A circular erosion pattern occurs in the first stage, until the center of the surface is grinded flat. In the second stage, erosion mainly occurs in the center, forming a crater in the end.

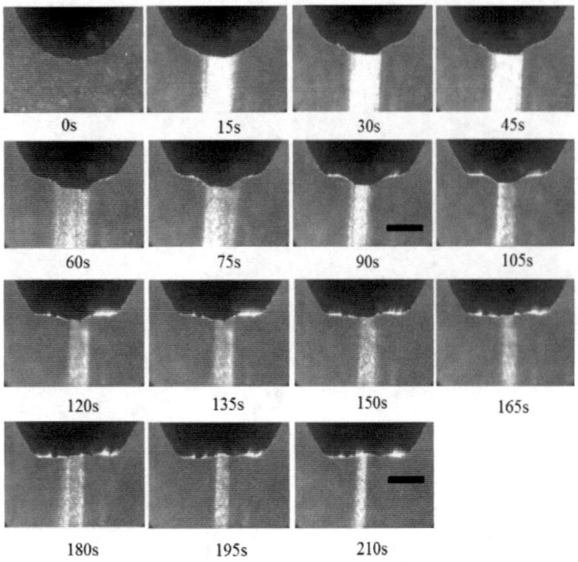

Figure 5. High speed photographic observation of calculus fragmentation process (ξ=0.2, HIFU inducer frequency 1.75 MHz, driving voltage 1.5 Vpp).

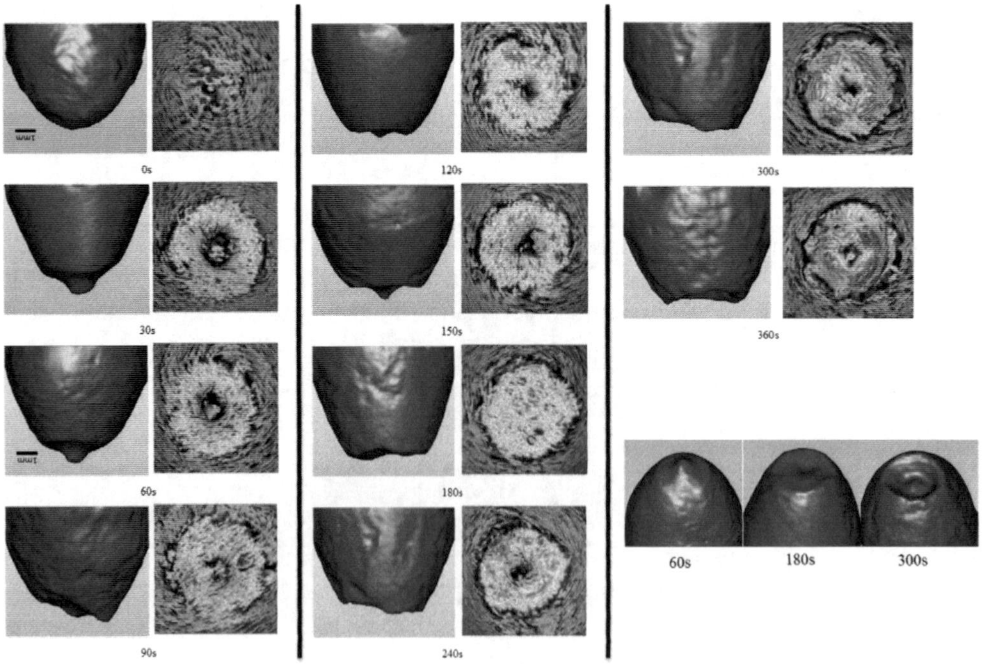

Figure 6. 3D scan of the erosion pattern.

The volumetric loss with different surface curvatures during the fragmentation process can be seen in figure 7. The overall rule of fragmentation efficiency is that, smaller the value of ξ is (the more convex the surface is), faster the samples are eroded. This trend is no longer obvious when ξ >0.5.

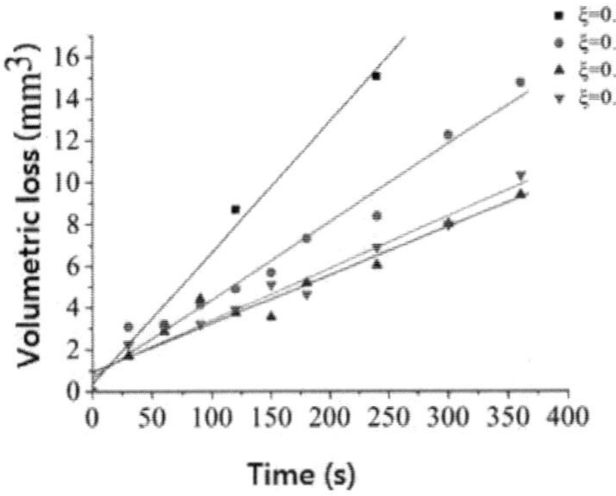

(a)

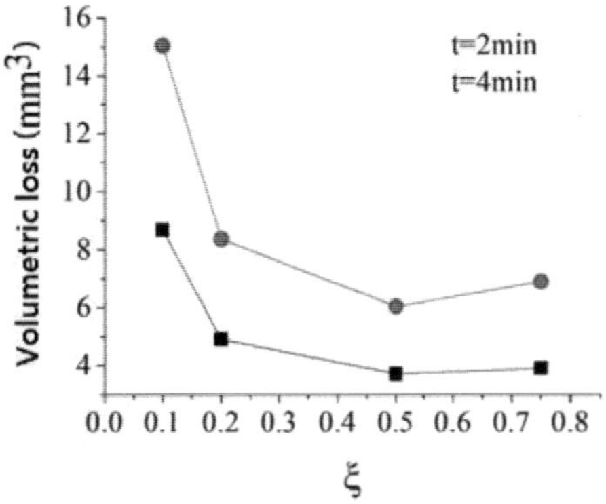

(b)

Figure 7. Volumetric loss with different surface curvatures.

3.2. High speed observation of cavitation cloud behaviour

The purpose of this experiment is to have a clearer observation on the behavior of cavitation cloud. Since minute particles in the liquid generated in previous fragmentation experiments on Begostone artificial stone samples will bring difficulties in clear visual observation of cavitation cloud, cavitation cloud was generated near stainless steel samples in NaCl solutions in this experiment. The schematic of this experiment is shown in figure 8. 20g/L NaCl solutions were used to reduce the content of dissolved gases. A converging laser beam with energy just below cavitation threshold was used to induce cavitation nuclei in the prescribed location. The HIFU transducer with 1.1 MHz frequency was driven with 0.5 Vpp voltage input to excite the nuclei. The behavior of the cavitation cloud was then observed by high speed photography.

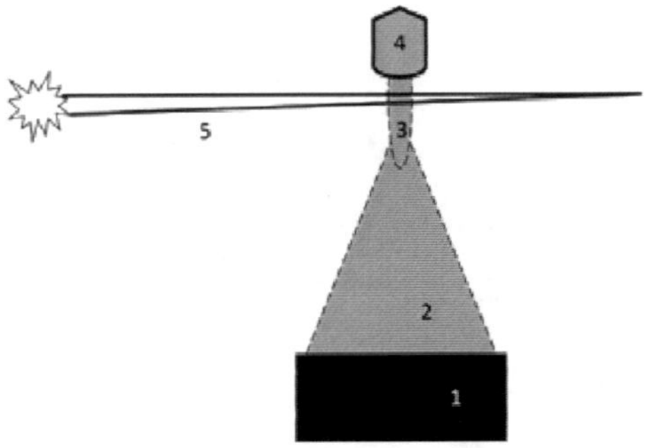

Figure 8. Schematic of high speed observation of cavitation cloud behavior: 1. HIFU transducer; 2. Converging acoustic field; 3. Focal region of the acoustic field; 4. Stainless steel sample; 5. Converging laser beam

Figure 9 shows the high speed observation results of ξ=0.2. The cloud grows from the nuclei induced by the laser beam, under the exposure to the acoustic field. It then moves towards the surface of the sample, due to the effect of acoustic streaming. A circular cavitation ring is then formed and oscillates on the surface, while the center of the surface is cavtation free.

Comparing with the erosion pattern described before, it is postulated that this form of cavitation ring is responsible for the ring-shaped damage in the first stage of fragmentation on the Begostone samples. More analysis shows that the diameter of the cavitation ring is related to the value of ξ. The bigger the value of ξ is, the larger the diameter. The moving speed of the cavitation cloud toward the surface is also related to the value of ξ. Smaller value of ξ corresponds to faster movement. The quantitative relationships are still under investigation.

3.3. Sonoluminescence tests

Sonoluminescence tests were carried out to study the cavitation intensity, and the behavior of acoustic streaming in the process. 0.01mol/L Luminol solutions and 0.1mol/L NaOH solutions were mixed for the purpose of observing cavitation sonoluminescence. A Nikon D7000 SLR camera (exposure time 10s) was used for the measurement of the light distribution. A Hamamatsu CR 105 photomultiplier (output resistance 82.5kΩ) was used for the measurement of the variation of light intensities.

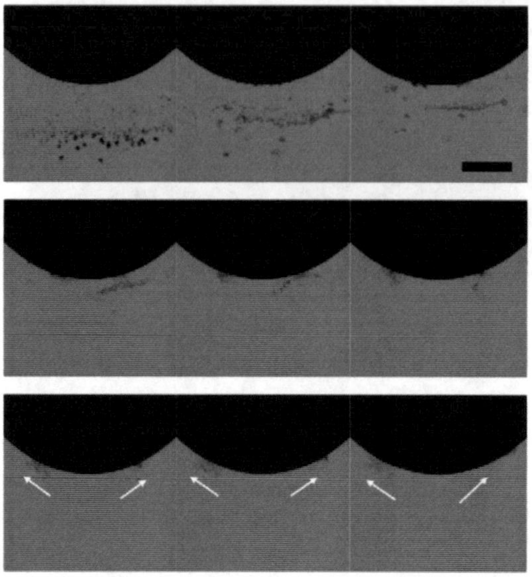

Figure 9. Oscillation of cavitation cloud(ξ=0.2, inter frame interval 0.2 ms, length of horizontal black line 1mm).

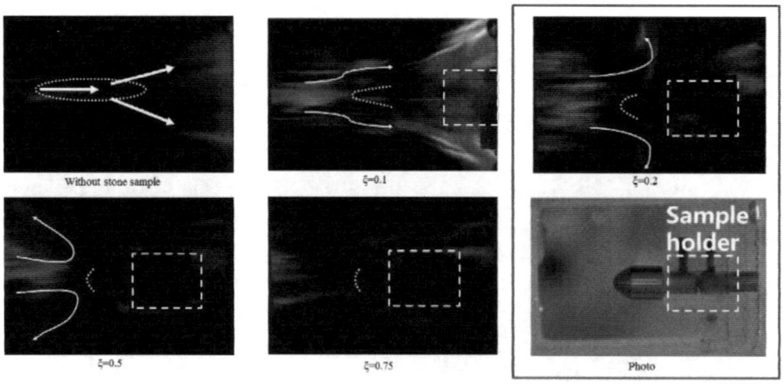

Figure 10. Cavitation sonoluminescence near surface with different curvatures.

Figure 10 shows the results of cavitation sonoluminescence near surface with different curvatures. Larger value of ξ corresponds to larger reverse flow, and lower light intensity.

Figure 11 shows the measured light intensity, which is an indication of cavitation intensity, during the process. It is observed that the larger value of ξ corresponds to lower light intensity, and slower acceleration rate of the light intensity. Changing in the value of ξ causes smaller variation in cavitation intensity, when the value of ξ is small.

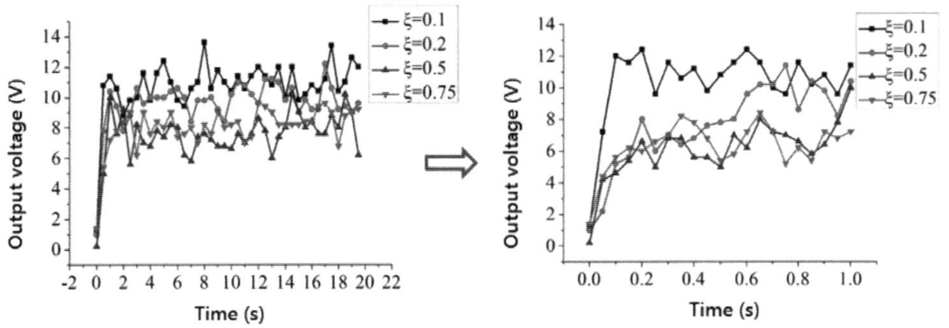

Figure 11. Light intensity vs. time.

4. Conclusions

A series of experiments was designed and carried out to investigate the influence of surface curvature on the behavior of HIFU induced cavitation cloud. Preliminary results show that:

1) The process of calculus fragmentation can be distinguished into two stages: a circular erosion pattern occurs in the first stage, until the center of the surface is grinded flat; in the second stage, erosion mainly occurs in the center, forming a crater in the end.

2) Smaller the value of ξ is (the more convex the surface is), faster the samples are eroded. This trend is no longer obvious when ξ>0.5.

3) The form of cavitation ring is responsible for the ring-shaped damage in the first stage of fragmentation on the Begostone samples. The diameter of the cavitation ring is related to the value of ξ.

4) Larger value of ξ corresponds to larger reverse flow, a lower cavitation intensity, and a slower acceleration rate of the cavitation intensity in the beginning of exposure to HIFU acoustic field.

References

[1] Benjamin T B and Ellis A T 1966 *Phil. Trans. R. Soc. Lond.* A **260** 221–240
[2] Crum L A 1975 *J. Acoust. Soc. Am.* **57** 1363–70
[3] Tomita Y and Shima A 1986 *J. Fluid Mech.* **169** 535–64
[4] Tomita Y, Robinson P B, Tong R P et al. 2002 *J. Fluid Mech.* **466** 259-283
[5] Gelet A, Chapelon J Y and Poissonnier L 2004 *Urology* **63** (4) 625-9
[6] Gerold B, Rachmilevitch I and Prentice P 2013 *New J. of Physics* **15** (3) 033044

IOP Conf. Series: Materials Science and Engineering **72** (2015) 062054 doi:10.1088/1757-899X/72/6/062054

Numerical simulation of the temperature effects on the performance of rotational supercavitating evaporator

Z Y Zheng[1], J P Cheng[1], F C Li[1, 3], M Zhang[1], Q Li[1] and V A Kulagin[2]
[1] School of Energy Science and Engineering, Harbin Institute of Technology, Harbin 150001, China
[2] Department of Heat Technology and Fluid Dynamics, Siberian Federal University, Krasnoyarsk 660041, Russia

E-mail: lifch@hit.edu.cn

Abstract. With the application of supercavitation effect, a novel device named Rotational Supercavitating Evaporator (RSCE) has been designed for desalination. In order to study the effect of temperature on the performance of RSCE and then direct the experimental study on RSCE for the next step, numerical simulations are conducted on the supercavitating flows in RSCE under different temperatures and rotational speeds. The results show that the rotational speed, resistance moment and mechanical energy consumed by the rotational cavitator under the critical state with the largest supercavity, decrease with the increase of temperature. And the area and volume of the supercavity increase exponentially with the increase of temperature under the same rotational speed.

1. Introduction

Fresh water deficit has been an increasingly critical global environmental issue, which promotes the development of desalination. The desalination methods can be classified into thermal method and membrane method in line with the different principles. And the current major large-scale industrial desalination methods include multi-stage flash, multiple effect distillation and reverse osmosis. However, the existence of solid walls in multi-stage flash and multiple effect distillation induces the generation of scaling and fouling which would decrease the heat transfer coefficient, and there must be need for rigorous pre-treatment in reverse osmosis. In order to overcome these shortcomings, a novel device named Rotational Supercavitating Evaporator (RSCE) was put forward and designed for desalination based on supercavitation effect [1-3].

Cavitation can be defined as the rupture of a liquid medium by the decrease in pressure under roughly constant temperature. It was originally considered as a harmful phenomenon that must be avoided because of its negative effects, such as wall erosion and the production of noise and vibrations. However, with the development of the researches on the mechanism of cavitation, the positive effects, including drag reduction and cavitating jet, have been applied to engineering practice. In this paper, large energy intensity and heat-mass transfer rate of the steam-water interface in the supercavitation, which is the most intense level of cavitation under low characteristic pressure or large characteristic velocity, are utilized to design RSCE for desalination. Most of the studies on the application of cavitation to desalination are conducted by Russian researchers, and the published relevant literatures

[3] Corresponding author.

are scarce. The only example is commercial cavitating desalination device "WATERFALL-1200" designed by the scientific and technical center "TJEROS-MIFI" in Russia [4]. Recently, Likhachev and Li [1-3] put forward RSCE and performed preliminary design and research.

The saturation vapor pressure increases with the increase of temperature, thus the rotational speed of RSCE at which the supercavity state is achieved can be lowered with increasing temperature, inducing the reduction of mechanical energy provided for the rotation of RSCE. Therefore, study on the effect of temperature on the performance of RSCE is further conducted in this paper to analyze the energy consumption of RSCE. For cavitating flows, the thermal effects play an important role. Numerical simulation by Kinzel et al. [5] indicated that the thermal effects were important only for specific applications that need high-frequency phenomena, while low-frequency measures needed for loading analysis appeared to be relatively insensitive to thermal effects. Goncalvès and Patella [6, 7] developed a compressible, multiphase and one-fluid solver to study and to predict thermodynamic effects in cavitating flows of cryogenics fluids. Numerical simulations were performed for a Venturi geometry with refrigerant R-114 and results were compared with experimental data.

In order to analyze the effects of the temperature on the performance of RSCE, numerical simulations are conducted in this paper on the supercavitating flows in RSCE under different rotational speeds and different temperatures of the water with different saturation pressures. The critical rotational speeds of RSCE at which the supercavity state are achieved under different temperatures are respectively obtained, as well as the change of the supercavity dimension with varying temperature under the same rotational speed. From the perspective of mechanical energy, the variation of the energy consumption of RSCE with temperature is further analyzed, giving the guidance to the experimental study on RSCE for the next step.

2. RSCE model

The core component of RSCE is a rotational cavitator made of two blades with wedge cross section and exit edge of alternative thickness, as shown in Figure 1. According to the size of the experimental device [3], a cylinder with the height of 100mm and the diameter of 430mm is selected as the computational domain. The diameters of the rotational cavitator d and the shaft d_0 are 400mm and 52mm, respectively.

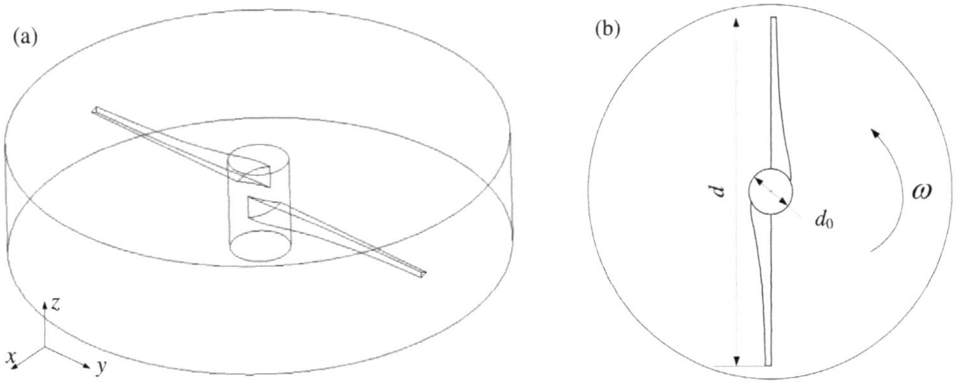

Figure 1. Schematic diagram of the rotational cavitator. (a) three-dimensional schematic diagram, (b) top view.

In order to obtain the blade shape of the rotational cavitator for the first step, the formulae in [3] are herein adopted for calculation. Note that the values of α at different radii in the original method didn't strictly equal to the wedge angles, i.e. 22.5 degree, at corresponding radii. Moreover, the deviation increases with the decrease of radius r. To address this issue, the improved method is proposed and shown in Figure 2, by which the blade shape of the rotational cavitator, shown in Table

1, is obtained through comparing the length of the arc with the length of the supercavity L which is acquired by the calculation with r_c instead of r.

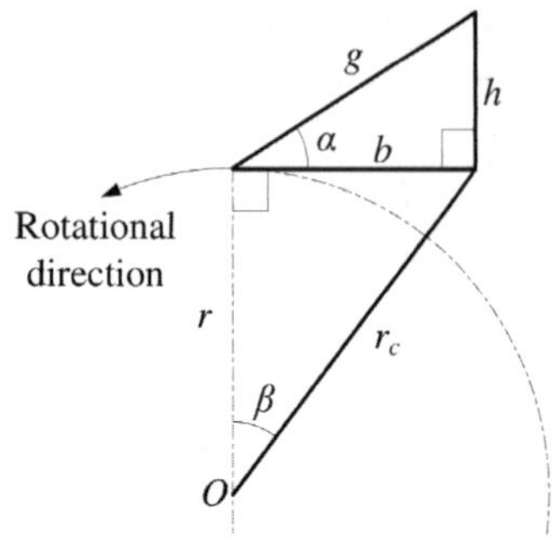

Figure 2. Schematic diagram of improved calculation method for the blade shape.

Table 1. Blade shape of the rotational cavitator.

r (mm)	*h* (mm)	r (mm)	*h* (mm)
30	9.42	120	4.06
40	8.64	130	3.78
50	7.80	140	3.53
60	7.01	150	3.32
70	6.31	160	3.13
80	5.70	170	2.96
90	5.19	180	2.80
100	4.76	190	2.66
110	4.38	200	2.54

3. Numerical method

3.1. Governing equations

Numerical simulations are conducted on the three-dimensional steady supercavitating flows in this paper. And supercavitating flow is a multiphase and turbulent flow. Mixture model based on the homogeneous equilibrium multiphase flow theory is selected as the multiphase flow model in the numerical simulation, in which the mixture of the gas and liquid phases is considered as homogeneous single-phase fluid. And the equations in this model are shown below.
Continuity equation of the mixture phase:

$$\nabla \cdot \left(\rho_m \vec{v}_m \right) = 0 \tag{1}$$

Momentum equation of the mixture phase:

$$\frac{\partial}{\partial t}\left(\rho_m \vec{v}_m \right) + \nabla \cdot \left(\rho_m \vec{v}_m \vec{v}_m \right) = -\nabla p + \nabla \cdot \left[\mu_m \left(\nabla \vec{v}_m + \nabla \vec{v}_m^T \right) \right] \tag{2}$$

Transportation equation of the volume fraction of the gas phase:

$$\nabla \cdot \left(\alpha_v \rho_v \vec{v}_m \right) = R_e - R_c \tag{3}$$

where ρ_m is the density of the mixture phase, $\vec{v}_m$ is the velocity, α_v is the volume fraction of the gas phase, ρ_v is the density of the gas phase, R_e and R_c are the production rate and condensation rate of the gas phase, respectively. In order to depict R_e and R_c, Schnerr-Sauer model [8] is adopted as the cavitation model. Turbulent flow in the numerical simulation is described by Realizable $k-\varepsilon$ model. And the Scalable wall function is adopted to solve the flow near the wall.

3.2. Mesh, boundary conditions and properties

The computational domain is meshed by the structured grids, and the meshes near the blades are arranged to be denser. Both the top and bottom of the computational domain are set to be pressure outlet with the pressure of 101325Pa. The cylindrical surface is set as fixed no-slip wall, and other boundaries are set to be no-slip wall with constant rotational speed. The thermodynamic properties of water and steam under different temperatures are set in line with the international standard for the thermodynamic properties of water and steam IAPWS95 [9].

3.3. Mesh independence verification

In order to guarantee the accuracy and economy of the calculations, the mesh independence is verified in advance. Five sets of mesh with 2084720, 3994000, 6042280, 8566000 and 9700120 grids are successively adopted for numerical simulations under the temperature of 30℃ with the rotational speed of 4000r/min. According to the results shown in Figure 3, the resistance moment that acts on the blades M_d and the volume of the supercavity V_c nearly remain constant for the meshes with the number of the grids above 8566000. Note that the profile of the supercavity is defined as the isosurface with the vapor volume fraction of 0.5. Therefore, the mesh with 8566000 grids is selected in the following numerical simulations.

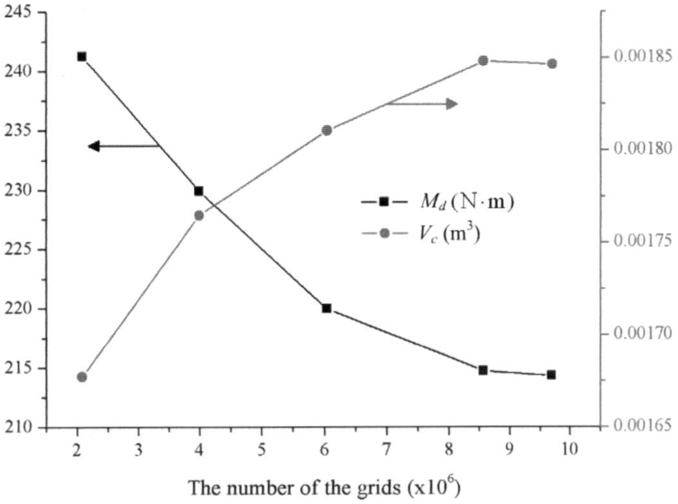

Figure 3. Verification of mesh independence.

4. Results and discussions

Numerical simulations are conducted on the supercavitating flows in RSCE under different rotational speeds and temperatures varied from 20℃ to 80℃ with increment of 10℃. The critical rotational speeds for different temperatures, at which the supercavitation states are achieved, are firstly obtained, as well as the resistance moments imposed on the blades, resulting in the acquisition of mechanical energy consumed by RSCE. And then the effects of temperature on the performance of RSCE are further studied by numerical simulations under the same rotational speed and different temperatures.

4.1. Results for critical rotational speed

Numerical simulations are performed on the supercavitating flows in RSCE under different temperatures and rotational speeds to obtain the critical rotational speed for each corresponding temperature, at which the largest supercavity is formed in RSCE. And above this critical value, the tail of the supercavity reaches to the next blade, causing the generation of asymmetric supercavities in the simulation. The profile of the supercavity at the critical rotational speed for 30℃ is shown in Figure 4. It can be seen that the top view of the supercavity profile is similar to symmetric sectors. Furthermore, re-entrant jet forms at the tail of the supercavities for smaller radii.

Figure 5 depicts the variations of critical rotational speed ω_c, resistance moment M_d and mechanical energy P_m with temperature T. And all decrease with the increase of temperature. As a result of the supercavities with the same dimensions generated by the rotational cavitator, cavitation numbers for different cases at corresponding critical rotational speeds under different temperatures should be the same. Therefore, the square of the critical rotational speed is proportional to the difference between the ambient pressure and saturation vapor pressure based on the definition of cavitation number, resulting in the decrease of critical rotational speed with the increase of the temperature for the rise of saturation vapor pressure. And then the resistance moment decreases with the reduction of rotational speed. Eventually, the mechanical energy consumed by rotational cavitator decreases because of the reduction of the rotational speed and resistance moment.

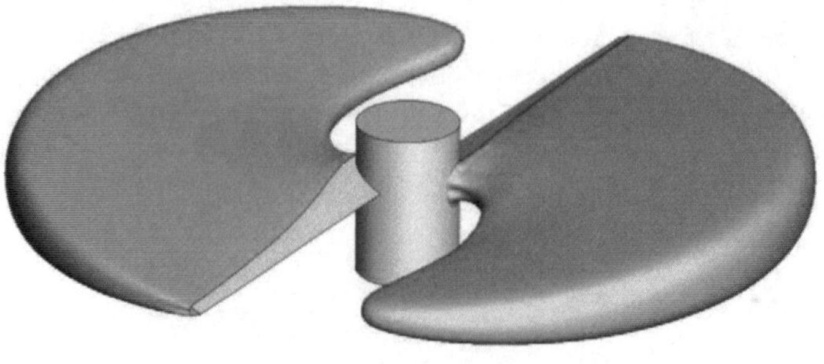

Figure 4. Profile of the supercavity.

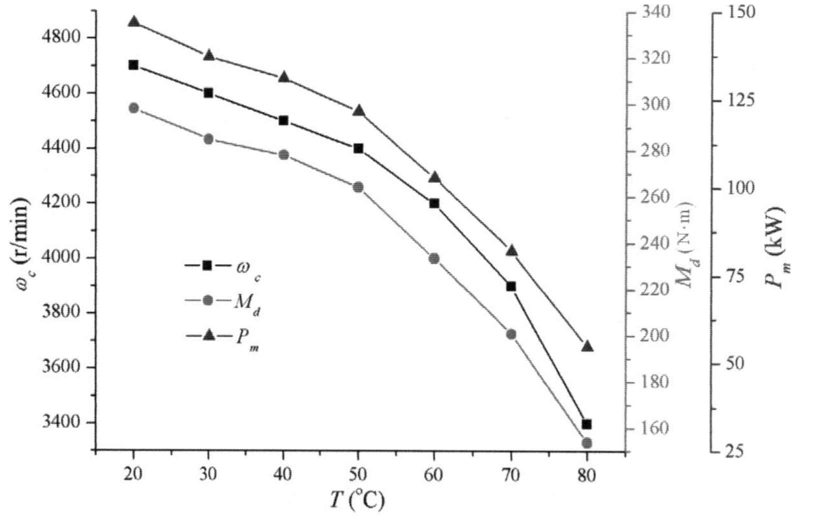

Figure 5. Variations of critical rotational speed, resistance moment and mechanical energy with temperature.

4.2. Results for the same rotational speed

In order to further study the effect of the temperature on the performance of RSCE, numerical simulations are conducted on the supercavitating flows in RSCE under different temperatures with the same rotational speed. The rotational speed of 3400r/min corresponding to critical rotational speed at 80℃ is selected for the numerical simulations in this section based on the results obtained above. The results, including the changes of resistance moment M_d, area A_c and volume V_c of supercavity with temperature, are shown in Figure 6. It can be seen that the resistance moments imposed on the rotational cavitator under different temperatures roughly keep unchanged for the constant rotational speed. Both A_c and V_c increase exponentially with the increase of temperature, resulting from the decrease of cavitation number induced by the increase of saturation vapor pressure.

5. Conclusions

Numerical simulations are successively conducted on the supercavitating flows in RSCE for the different temperatures under different rotational speeds and the same rotational speed to study the effect of temperature on the performance of RSCE. The supercavity profile and critical rotational speed for each temperature are obtained, as well as the variations of resistance moment, mechanical energy, area and volume of supercavity with temperature. The main conclusions are given below.

(1) The rotational speed, resistance moment and mechanical energy consumed by the rotational cavitator under the critical state, at which the largest supercavity is generated, decrease with the increase of temperature.

(2) The resistance moment is affected by the rotational speed rather than the temperature. The area and volume of the supercavity exponentially increase with the increase of temperature under the same rotational speed.

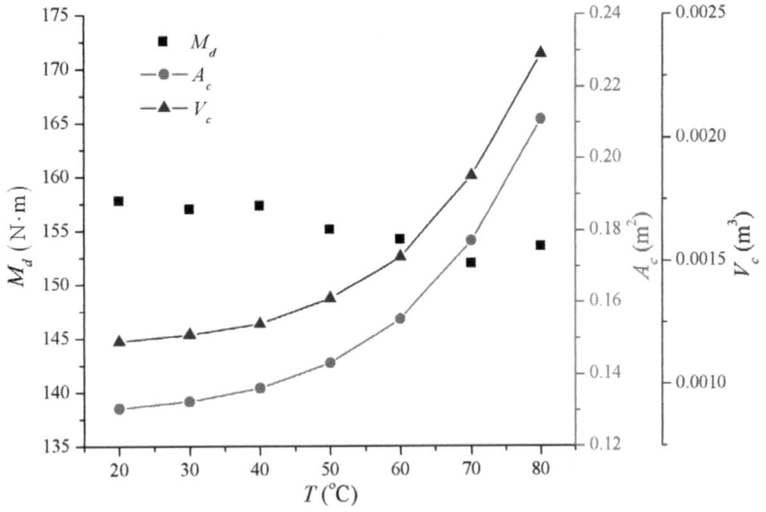

Figure 6. Variations of resistance moment, area and volume of supercavity with temperature at the same rotational speed.

Acknowledgments

This work is supported by National Natural Science Foundation of China (51276046). The authors would like to appreciate the valuable discussions with the members of Complex Flow and Heat Transfer Lab.

References

[1] Likhachev D S and Li F C 2013 *Desalination and Water Treatment* **51** 2836–49
[2] Likhachev D S and Li F C 2014 *Sci. China Phys. Mech. and Astronomy* **57**(3) 541–54
[3] Likhachev D S 2013 *Study on the hydrodynamic characteristics of rotational supercavitating evaporator* (Harbin: Harbin Institute of Technology PhD thesis)
[4] Rudenko B 2007 How to make up water resource *Journal of Science Life* **12** http://www.nkj.ru/archive/articles/12366/ (in Russian)
[5] Kinzel M P, Lindau J W and Kunz R F 2009 An examination of thermal modeling affects on the numerical prediction of large-scale cavitating fluid flows *Proc. the 7th International Symposium on Cavitation (Ann Arbor, Michigan, USA, 17–22 August 2009)* CAV2009–Paper No. 137
[6] Goncalvès E and Patella R F 2010 *Comput. Fluids* **39** 99–113
[7] Goncalvès E and Patella R F 2011 *Appl. Math. Comput.* **217** 5095–102
[8] Schnerr G H and Sauer J 2001 Physical and numerical modeling of unsteady cavitation dynamics *4th International Conference on Multiphase Flow (New Orleans, USA, 27 May-1 June 2001)*
[9] Wagner W and Pruß A 2002 *J. Phys. Chem. Ref. Data* **31** 387–535

IOP Conf. Series: Materials Science and Engineering **72** (2015) 062055 doi:10.1088/1757-899X/72/6/062055

Numerical simulation research on the optimization for blade shape of rotational supercavitating evaporator

Q Li[1], J P Cheng[1], Z Y Zheng[1], F C Li[1, 3] and V A Kulagin[2]
[1] School of Energy Science and Engineering, Harbin Institute of Technology, Harbin 150001, China
[2] Department of Heat Technology and Fluid Dynamics, Siberian Federal University, Krasnoyarsk 660041, Russia

E-mail: lifch@hit.edu.cn

Abstract. Rotational Supercavitating Evaporator (RSCE) has been proposed as a new technology for seawater desalination. However, it lacks systematic researches on the blade shape of RSCE. In this paper, numerical simulations were conducted on the supercavities formed behind two-dimensional (2D) wedge-shaped cavitators of the RSCE. The cavitating flows around the 2D wedge-shaped cavitators with several certain wedge angles varied from 30 to 150 degrees under different cavitation numbers were simulated, and the empirical formulae of supercavity dimensions about cavitation number at corresponding wedge angles were obtained.

1. Introduction

Water is the source of our lives and the material basis of the human beings' survival. There exists a severe shortage of fresh water around the world, which has led to development of seawater desalination methods. The most widely used technologies are thermal methods and membrane methods, which make up about 95% of worldwide fresh water production. However, the former is limited by the scale formation and thermal transfer coefficient, and the later is limited by the membrane fouling and the recovery factor. There is a technological bottleneck in the thermal methods to evaporate water into steam free from scale formation and possess a dramatic increase in thermal transfer coefficient. Water evaporates into steam on the surface of supercavity in the same way with boiling of water. Moreover, the heat transfer coefficient of evaporation from the surface of supercavity depends on the heat-flux density and there is no scale formation. So a device named Rotational Supercavitating Evaporator (RSCE) was put forward for desalination based on supercavitation effect. It also has a good application foreground in sewage disposal, such as nuclear waste water and polluted water containing heavy metals.

Cavitation can be defined as the breakdown of a liquid medium under very low pressures and applied to cases in which the liquid is either static or in motion. It was first discovered on the propeller in the warship of British Royal Navy in 1897. The study on cavitation primarily focused on the prevention of negative influence of cavitation phenomenon such as noise, vibration and surface

[3] Corresponding author.

erosion. Subsequently the great drag reduction effect of supercavitation on an underwater body was found, and it came an upsurge of research on the drag reduction of underwater vehicle based on supercavitation effect. 30 years ago, Machinski [1] introduced the stationary supercavitating cone evaporator, which can be connected to vacuum system for steam extraction for desalination. Recently Likhachev and Li [2-4] put forward RSCE and performed preliminary design and research.

In order to obtain systematic characteristics of cavity formed by the rotational supercavitating blades, numerical simulations are conducted on 2D wedge-shaped cavitators, which is a simplification of rotational supercavitating blades. The cavitating flows around the 2D wedge-shaped cavitator with certain dimension are simulated under different cavitation numbers to obtain the characteristics of cavity. Different wedge angles varied from 30 to 150 degrees of 2D wedge-shaped cavitators are selected to obtain the characteristics of cavity. As a conclusion, characteristics of planar symmetric cavity are analyzed and empirical formulae of supercavity dimensions as a function of cavitation number at corresponding wedge angles are obtained.

2. Numerical method

2.1. Governing equations

Numerical simulations are conducted on the two-dimensional steady supercavitating flows in this paper. In the numerical simulation, supercavitating flow is a multiphase and turbulent flow and mixture model based on the homogeneous equilibrium multiphase flow theory is selected as the multiphase flow model. The mixture of the gas and liquid phases is considered as homogeneous single-phase fluid and the equations in this model are shown below.

Continuity equation of the mixture phase:

$$\nabla\cdot\left(\rho_m\vec{v}_m\right)=0 \tag{1}$$

Momentum equation of the mixture phase:

$$\frac{\partial}{\partial t}\left(\rho_m\vec{v}_m\right)+\nabla\cdot\left(\rho_m\vec{v}_m\vec{v}_m\right)=-\nabla p+\nabla\cdot\left[\mu_m\left(\nabla\vec{v}_m+\nabla\vec{v}_m^T\right)\right] \tag{2}$$

Transportation equation of the vapour fraction of the gas phase:

$$\nabla\cdot\left(\alpha_v\rho_v\vec{v}_m\right)=R_e-R_c \tag{3}$$

where ρ_m is the density of the mixture phase, $\vec{v}_m$ is velocity, α_v is the vapour fraction of the gaseous phase, ρ_v is the density of the gaseous phase, R_e are R_c the production rate and condensation rate of the gaseous phase, respectively. In order to depict R_e and R_c, Schnerr-Sauer model [5] is adopted as the cavitation model.

For $p_v \geq p$,

$$R_e=\frac{\rho_v\rho_l}{\rho_m}\alpha_v(1-\alpha_v)\frac{3}{\Re_B}\sqrt{\frac{2}{3}\frac{(p_v-p)}{\rho_l}} \tag{4}$$

For $p_v \leq p$,

$$R_c=\frac{\rho_v\rho_l}{\rho_m}\alpha_v(1-\alpha_v)\frac{3}{\Re_B}\sqrt{\frac{2}{3}\frac{(p-p_v)}{\rho_l}} \tag{5}$$

where ρ_l is the density of the liquid phase, p_v is the saturation vapor pressure, $\Re_B$ is the radius of the bubble and defined as follows.

$$\Re_B = \left(\frac{\alpha_v}{1-\alpha_v} \frac{3}{4\pi} \frac{1}{n} \right)^{\frac{1}{3}} \tag{6}$$

Herein, n is the number of the microbubbles within the liquid of unit volume and set as 1×10^{13}.

Turbulent flow in the numerical simulation is described by Realizable $k-\varepsilon$ model and the Scalable wall function is adopted to solve the flow near the wall.

2.2. Boundary conditions and mesh

Because of the 2D wedge-shaped cavitator model is symmetric, half of the model can be used as the computational domain, which is shown in Figure 1. The half height of the base of the wedge-shaped cavitator is h_c=5mm and the wedge angles γ have several certain values ranged from 30 to 150 degrees. The computational domain is 5m in height and 20.5m in length, which is much bigger than the typical computational domain of an axisymmetric cavitator in case of that the cavity formed by a planar cavitator is much larger than that formed by an axisymmetric cavitator. The cavitator is placed 0.5m downstream from the inlet. Both the top and the inlet of the computational domain are set as velocity inlet with the velocity of certain values ranged from 25 to 60 m/s in axial direction. The axis is set as symmetry and the outlet is set as pressure-outlet with the pressure of 1 bar, while the other boundaries are set as fixed no-slip wall. The thermodynamic properties of water and steam are defined according to international standard for the thermodynamic properties of water and steam IAPWS95 [6] at 25℃. Density of water liquid and vapor are 997kg/m^3 and 0.023075kg/m^3, respectively. Viscosity of water liquid and vapor are 8.9011×10^{-4}Pa • s and 9.8669×10^{-6}Pa • s. The computational domain is meshed by the structured grids, and the meshes in the boundary layer near the blades surface are arranged to be denser, which are shown in Figure 2.

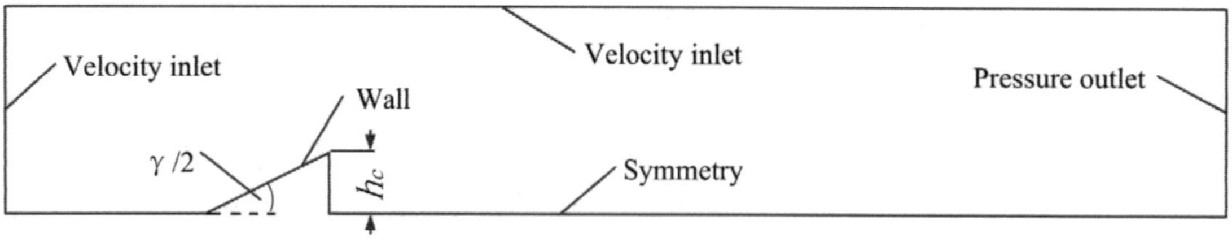

Figure 1. Schematic diagram of the computational domain.

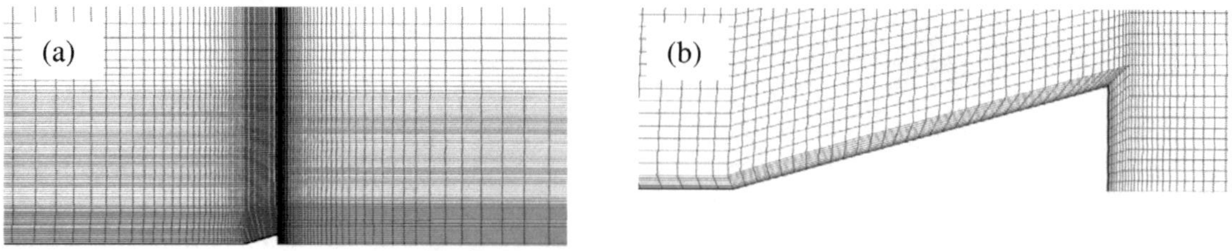

Figure 2. Schematic diagram of the mesh with: (a) enlarged part around cavitator, (b) meshes in the boundary layer.

3. Results and discussions

In order to obtain the characteristics of cavities under different cavitation numbers, the cavitating flows around each cavitator with certain wedge angles are simulated under different cavitation

IOP Conf. Series: Materials Science and Engineering **72** (2015) 062055 doi:10.1088/1757-899X/72/6/062055

numbers. The chosen wedge angles are 30, 45, 60, 90, 120 and 150 degrees. Cavitation number is defined as:

$$\sigma = \frac{p_0 - p_c}{\frac{1}{2}\rho v^2} \quad (7)$$

where p_0 is the ambient pressure, p_c is the pressure inside the cavity and equals to the saturation vapor pressure, ρ is the density of liquid, and v is the velocity. Different velocities, including 25, 30, 35, 40, 50, 60m/s, are selected to get different cavitation numbers.

3.1. Mesh independence verification

In order to guarantee the accuracy and economy of the calculations, the mesh independence is verified for the first step. Meshes with different grid number were generated with different grid sizes in the downstream of the cavitator for numerical simulation. Dimensionless parameters L/h_c and H/h_c are used to measure the shape of cavity, herein L is the length of cavity and H is the maximum thickness of cavity. The results of different meshes are shown in Figure 3. As the grid number up to about 0.45 million, the length of cavity remain unchanged. So the mesh of about 0.45 million grids is selected in the further research.

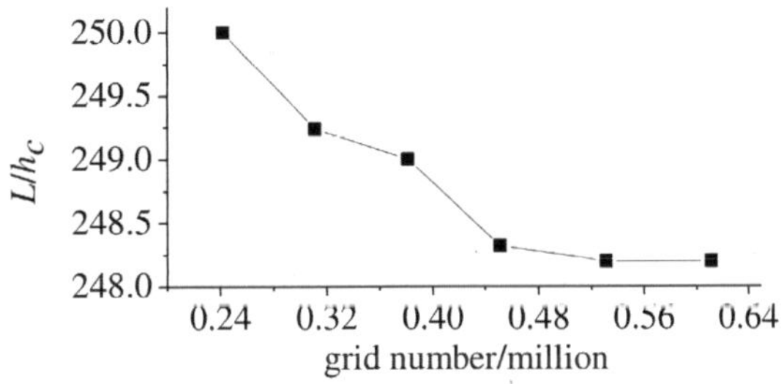

Figure 3. The results of mesh independence.

3.2. Results for different cavitation numbers

In order to obtain the dependences of cavity dimensions and drag coefficient on the cavitation number, the results for the cavitator with the wedge angle of 45 degrees are analyzed. In many cases, it is possible to model the experimental dependence of the cavity length, for low values of that parameter, by a power law:

$$L/h_c = A\sigma^{-n} \quad (8)$$

where the exponent n is found to equal to 2 if the body is located in an infinite medium [7]. The length and maximum thickness of the cavity formed by the cavitator with the wedge angle of 45 degrees are respectively fitted, as shown in Figures 4. The fitting functions are

$$L/h_c = 2.329\sigma^{-1.83} \quad (9)$$

$$H/h_c = 1.469\sigma^{-0.975} \quad (10)$$

The drag coefficient is fitted in parabola very well, as shown in Figures 5. The fitting function is:

$$C_x = 0.439 + 0.166\sigma + 0.743\sigma^2 \quad (11)$$

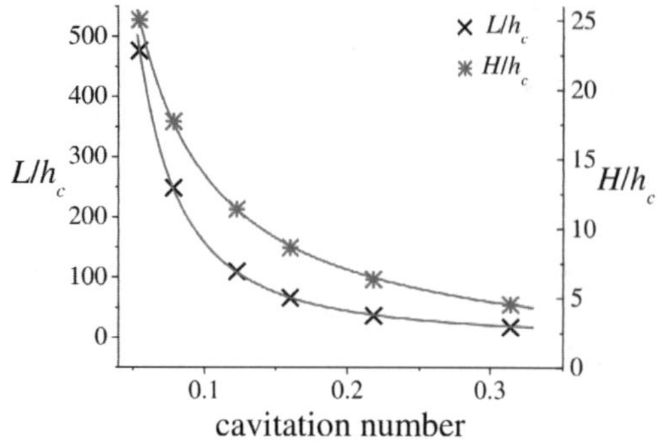

Figure 4. The fitting results of length and maximum thickness of cavity.

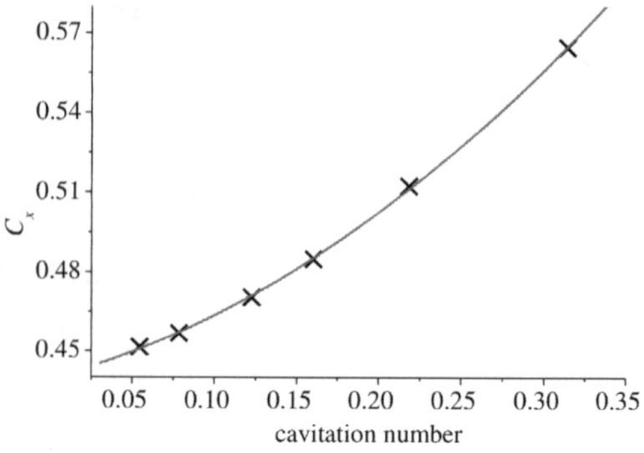

Figure 5. The fitting results of drag coefficient.

3.3. Results for different wedge angles

L/h_c and H/h_c of cavitator with different wedge angles are fitted respectively and the exponents of the fitting functions of L/h_c and H/h_c for different wedge angles are close to the values of -1.83 and -0.975. Thus the exponents of the fitting functions of L/h_c and H/h_c are fixed as -1.83 and -0.975 in the following research. The fitting results of L/h_c and H/h_c for the certain cavitation numbers under different wedge angles are shown in Figures 6 and 7. The cavity dimension increases rapidly as the wedge angle increases at a certain cavitation number, which means larger cavity can be formed by the cavitator with larger wedge angle under the same condition. The coefficient A of each fitting function are shown in Table 1, and it can be fitted in parabola as:

$$A_{(L/h_c)} = 0.496 + 2.602\gamma - 0.361\gamma^2 \tag{12}$$

$$A_{(H/h_c)} = 0.344 + 1.611\gamma - 0.241\gamma^2 \tag{13}$$

The drag coefficients of cavitator with different wedge angles are fitted in parabola:

$$C_x = a + b\sigma + c\sigma^2 \tag{14}$$

The fitting results are shown in Figures 8 and the coefficients, *a*, *b* and *c*, are shown Table 1. It can be seen that cavitators with larger wedge angle have larger drag coefficients. So more power is needed to obtain larger cavity. Finally, empirical formulae of supercavity dimensions as a function of cavitation number and wedge angles are obtained as:

$$L/h_c = (0.496 + 2.602\gamma - 0.361\gamma^2)\sigma^{-1.83} \tag{15}$$

$$H/h_c = (0.344 + 1.611\gamma - 0.241\gamma^2)\sigma^{-0.975} \tag{16}$$

Supercavity dimensions increase with the decrease of cavitation number in power law. What's more, the exponents of empirical formulae are constant for different wedge angles, which is consistent with the work of Franc and Michel [7] and the accuracy of the numerical simulation results is verified.

4. Conclusions

Numerical simulations are conducted on the cavitating flows around the 2D wedge-shaped cavitators with wedge angles of 30, 45, 60, 90, 120 and 150 degrees under different cavitation numbers, and the empirical formulae of supercavity dimensions are obtained.

Dimensions of supercavities formed by planar symmetric cavitators with different wedge angles are fitted in power law very well under different cavitation numbers, and the variation with wedge angles is parabolic. The empirical formulae will give the guidance to the optimization for blade shape of RSCE.

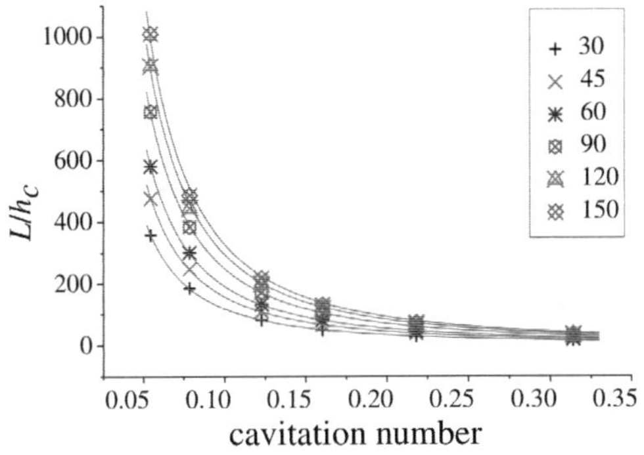

Figure 6. Fitting results for the length of cavities formed by cavitators with different wedge angles.

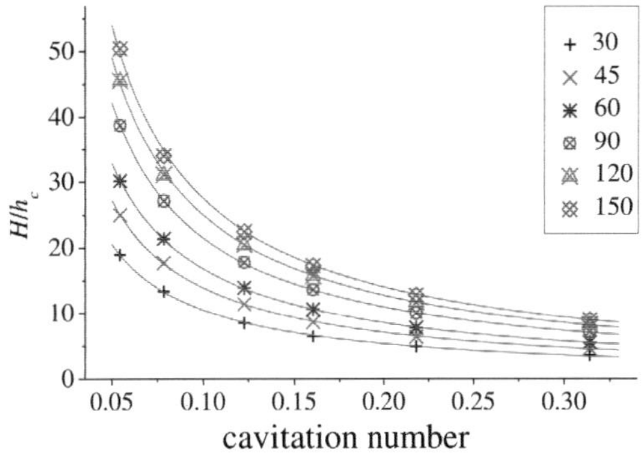

Figure 7. Fitting results for maximum thickness of cavities formed by cavitators of different wedge angles.

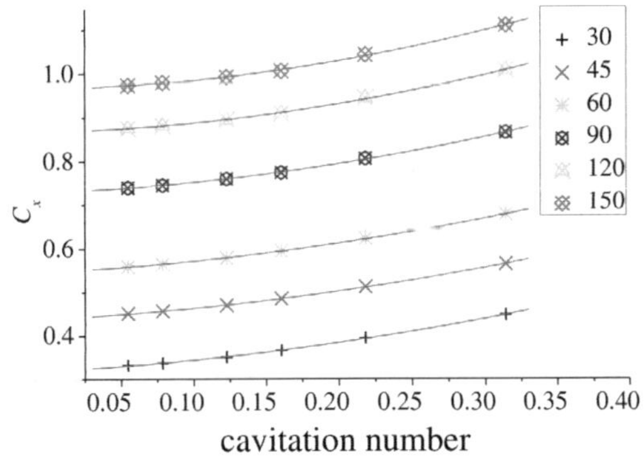

Figure 8. The fitting result of drag coefficients of cavitator with different wedge angles.

Table 1. Coefficients of each fitting function for L/h_c , H/h_c and C_x .

Angle(degree)	30	45	60	90	120	150
Coefficient A for L/h_c	1.748	2.329	2.835	3.684	4.364	4.843
Coefficient A for H/h_c	1.109	1.469	1.778	2.275	2.647	2.915
Coefficient a for C_x	0.320	0.439	0.547	0.730	0.868	0.967
Coefficient b for C_x	0.176	0.166	0.171	0.131	0.0951	0.0634
Coefficient c for C_x	0.741	0.743	0.789	0.966	1.129	1.273

Acknowledgments
This work is supported by National Natural Science Foundation of China (51276046). The authors would like to appreciate the valuable discussions with the members of Complex Flow and Heat Transfer Lab.

References

[1] Machinski A S. 1984 Hydrodymanics and thermal transfer characteristics of supercavitating evaporators for water desalination (Moscow: Kiev Order of Lenin Polytechnic Institute PhD thesis) 285 (in Russian)

[2] Likhachev D S and Li F C 2013 *Desalin. Water Treat.* **51** 2836–49

[3] Likhachev D S and Li F C 2013 *Sci. China Phys. Mech.* **56** 1–14

[4] Likhachev D S 2013 *Study on the hydrodynamic characteristics of rotational supercavitating evaporator* (Harbin: Harbin Institute of Technology PhD thesis)

[5] Schnerr G H and Sauer J 2001 Physical and numerical modeling of unsteady cavitation dynamics *4th International Conference on Multiphase Flow (New Orleans, USA, 27 May-1 June 2001)*

[6] Wagner W and Pruß A 2002 *J. Phys. Chem. Ref. Data* **31** 387–535

[7] Franc J P and Michel J M. 2006 *Fundamentals of Cavitation* (Berlin: Springer)

Numerical study of water entry supercavitating flow around a vertical circular cylinder influenced by turbulent drag-reducing additives

C X Jiang, J P Cheng and F C Li[1]
School of Energy Science and Engineering, Harbin Institute of Technology, Harbin 150001, China

E-mail: lifch@hit.edu.cn

Abstract. This paper attempts to introduce a numerical simulation procedure to simulate water-entry problems influenced by turbulent drag-reducing additives in a viscous incompressible medium. Firstly we performed a numerical investigation on water-entry supercavities in water and turbulent drag-reducing solution at the impact velocity of 28.4m/s to confirm the accuracy of the numerical method. Based on the verification, projectile entering water and turbulent drag-reducing solution at relatively high velocity of 142.7 m/s (phase transition is considered) is simulated. The cross viscosity equation was adopted to represent the shear-thinning characteristic of aqueous solution of drag-reducing additives. The configuration and dynamic characteristics of water entry supercavity, flow resistance were discussed respectively. It was obtained that the numerical simulation results are in consistence with experimental data. Numerical results show that the supercavity length in drag-reducing solution is larger than one in water and the velocity attenuates faster at high velocity than at low velocity; the influence of drag-reducing solution is more obvious at high impact velocity. Turbulent drag-reducing additives have the great potential for enhancement of supercavity.

1. Introduction

The water-entry phenomenon, intrinsically a transient process, is an important topic in naval hydrodynamic sector. Scientific research of water entry phenomenon initiated from the beginning of the 19th century [1, 2]. With the development of CFD techniques, numerical simulation has become a reliable and appropriate tool for the investigation of this complex unsteady nonlinear flow. Ref. [3-6] presented numerical solutions on water-entry problems.

In the research formerly, our work has obtained important discovery that at the same initial velocity: the size of supercavity in drag-reducing solution was larger than that in water and maintaining downstream distance of supercavity is longer in surfactant solution as well [7]. In order to have a deep insight into the mechanism of water-entry supercavitating flow influenced by turbulent drag-reducing activities, we first simulated projectile downward entering drag-reducing solution in moderate velocity and compared with the experimental data to confirm the accuracy of the numerical method employed. And then we make further efforts to numerically investigate water-entry supercavitating flow around a

[1] Corresponding author.

circular cylinder influenced by turbulent drag-reducing additives at high impact velocity and in which phase transition is considered.

2. Numerical simulation procedures

1.1. Governing equations
The volume of fluid (VOF) method is taken in this study to capture the interface between phases. The standard $k-\varepsilon$ model with standard wall functions was adopted to provide turbulence closure. Schnerr-Sauer cavitation model [8] is adopted to deal with phase transition.

1.2. Cross viscosity model
The The Cross viscosity model [9] was utilized herein to characterize the shear-thinning turbulent drag-reducing solution. The drag-reducing solution was characterized through modeling the dynamic viscosity by the Cross viscosity equation: $\mu=\mu_0[1+(\lambda\gamma)^m]^{-1}$. Where $\lambda=0.4\mu_s M[\mu]^2 c/R_0 T$, μ_0 is zero shear viscosity; λ is relaxation time; γ is shear rate; m is Cross power index and m=0.3 was utilized here; M is the mean mole mass of the solute in g/mol; μ_s is viscosity of solvent; c is the solution concentration in g/cm^3; R_0=8.31 J/mol•K; T is the absolute temperature in K; $[\mu]$ is intrinsic viscosity and $[\mu]=1.03\times10^{-2}M^{0.78}$ in ml/g. Together with the Cross equation, a decreased surface tension coefficient was used to characterize the drag-reducing solution. $\sigma_s=0.036$ N/m (half of that of water) was chosen for the solution [10].

1.3. Numerical method and conditions
The simulation model was then simplified to be 2-Ddimensional axially symmetric, as shown in Figure 1. Figure 1 (a) and (b) are the mesh systems for lower initial velocity and higher initial velocity, respectively. The User Defined Functions embedded in the FLUENT platform were used to control the movement of the projectile.

First, Numerical simulations were performed for water and surfactant solution ($\mu_0=0.01$ Pa·s, $\lambda=1$ s, m=0.3 and $\sigma_s=0.036$ N/m) flow cases at the initial velocity $V_0=28.4$ m/s, respectively. In these cases, only liquid phase and air phase are considered. Based on the first step, we moved on to the typically high impact velocity $V_0=142.7$ m/s in water and drag-reducing solution, and phase transition is added to this part.

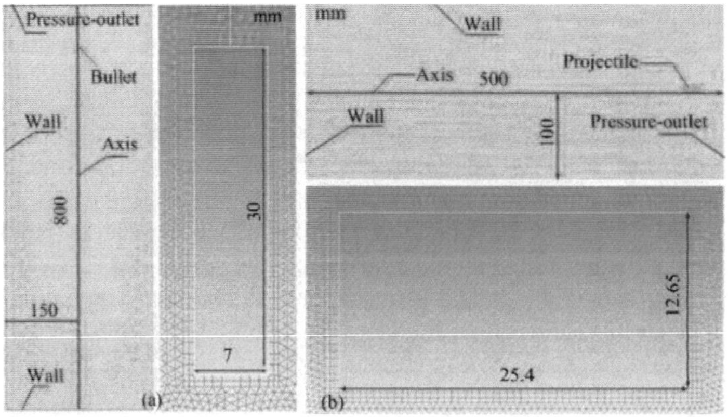

Figure 1. Physical model, computational region and grid system.

3. Results and discussion

IOP Conf. Series: Materials Science and Engineering **72** (2015) 062056 doi:10.1088/1757-899X/72/6/062056

3.1. Lower impact velocity of $V_0 = 28.4$ *m/s*

In the discussion of this section, the experimental data and analytical calculation used for the comparisons with the numerical simulation are all 1000 ppm in concentration from Ref. [7].

Figure 2 shows a comparison of the supercavity forming process of water entry between numerical simulation and experiments for the 1000 ppm CTAC solution case. The significant phenomenon is that, shortly after the impact of projectile on the liquid surface, upward moving jetting and a laterally expanding splash are formed above the surface, and the projectile is suddenly surrounded by the splash (seeing Figure 2 (a)). As shown in Figure 2 (b) for the "open-cavity phase": the flow separates from the projectile nose with the generation of the cavity; air rushes in from above the liquid surface to fill the cavity. In Figure 2 (c), it shows that the splash is forming a dome and closing over the entry point of the projectile to seal the cavity from the air above. As shown in Figure 2 (d), "surface closure" occurs and the cavity is then pushed down from the liquid surface. From this stage, the cavity has closed above the liquid surface and envelops the projectile, forming the supercavity (seeing Figure 2 (e)). In the process of water-entry, the shape of the cavity in solution is consistent well with the experimental observations. It demonstrates that the presently conducted numerical simulation procedure is suitable for simulating the supercavitating projectile in drag-reducing solutions.

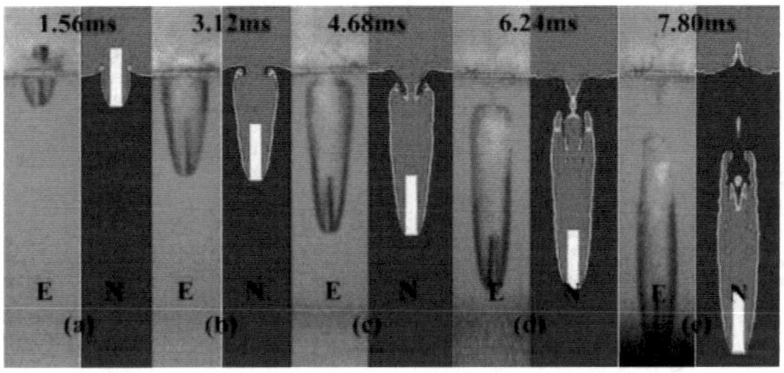

Figure 2. Comparisons of cavity configurations at water-entry between numerical simulation and experiments for the 1000 ppm CTAC solution case ("E" and "N" represent the experiments and numerical simulations, respectively).

The numerical simulation results for the variation of projectile velocity as compared with the analytical predictions are shown in Figure 3. It can be seen that, for water-entry supercavitation at initial velocity $V_0 = 28.4$ m/s, the drag obtained from the numerical simulation is comparable to that predicted using the analytical model, with the maximum discrepancy of 25%. In consideration of the impact process, because the projectile is launched at some distance above the water surface, when the projectile impacts the water surface, it will lose more energy than the analytical formulation. From the numerical simulation results, we can see that the velocity in solution decays slower than in water, which means that, at the same instant, the projectile in solution flies faster than in water.

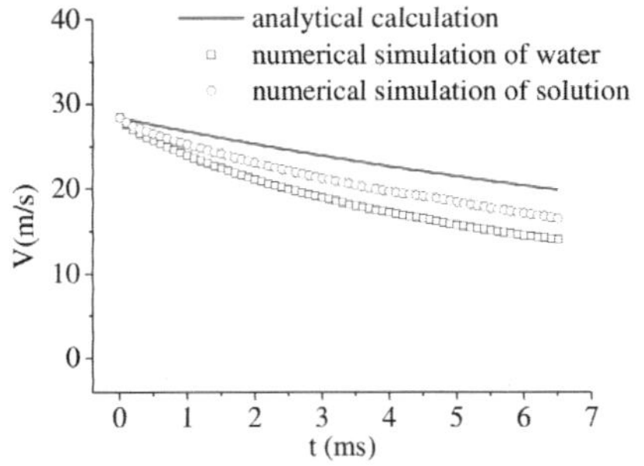

Figure 3. Comparison between analytical calculation and numerical simulation for velocity attenuation of the projectile with supercavitation.

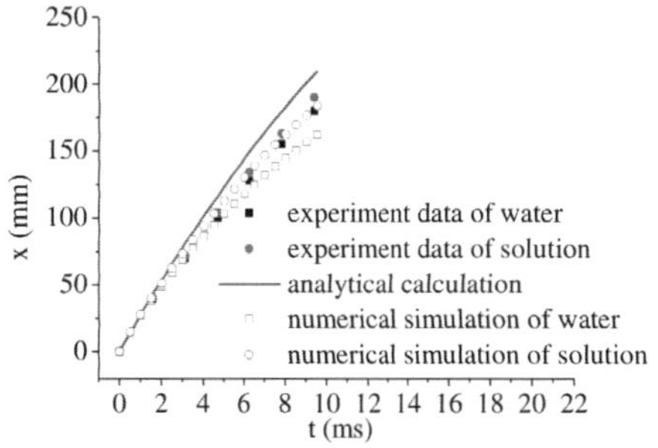

Figure 4. Comparison between analytical calculation and numerical simulation for velocity attenuation of the projectile with supercavitation.

The numerical simulation results compared to the predicted penetration distance based on the analytical formulation and the experiment data are shown in Figure 4. It is seen that the projectile in solution moves in longer distance than in water at the same instant. Actually, the projectile cannot move as far as the analytical calculation due to the loss of projectile kinetic energy at the water-entry process. From the results mentioned above, it is clear that the numerical simulation approach developed in the present study allows one to investigate the characteristics of water-entry induced supercavitation influenced by turbulent drag-reducing additives.

Figure 5 plots drag coefficients versus time for both water and drag-reducing solution cases. The drag coefficient C_d is defined below: $C_d = F / 0.5\rho V_p^2 A_0$, where F is the overall drag force. It can be seen that the drag coefficients rise sharply when the projectile impacts the liquid surface. The drag coefficient is obviously smaller in drag-reducing solutions than that in water. From these results, it can be conjectured that adding small amount of drag-reducing additives, projectile can move faster in the drag-reducing solution than in water provided the driving power is the same. It is expected that the actual drag coefficient would be greater than that predicted for a disk cavitator because there are drag forces on the projectile body.

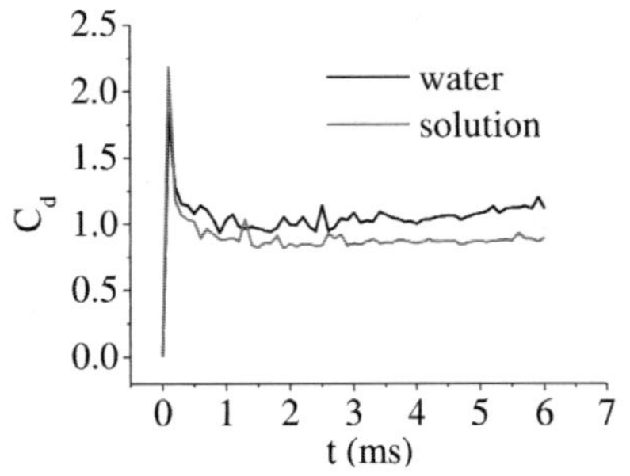

Figure 5. Drag coefficients for projectile in water and in drag-reducing solution varying with time by numerical simulation.

3.2. Higher impact velocity of $V_0 = 142.7$ m/s

In this part we discuss the unsteady behavior of water-entry supercavitating flows in water and influenced by turbulent drag-reducing additives at the impact velocity of $V_0 = 142.7$ m/s. The experimental data and analytical calculation used for the comparisons with the numerical simulation are all from Ref. [11].

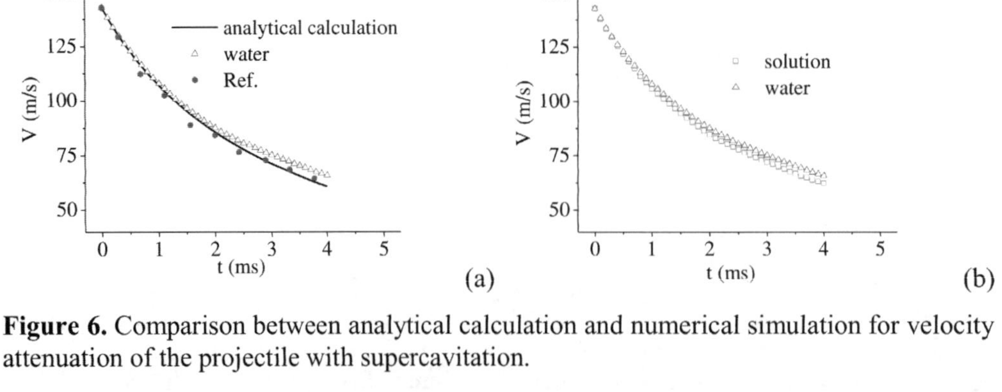

Figure 6. Comparison between analytical calculation and numerical simulation for velocity attenuation of the projectile with supercavitation.

The numerical simulation results for the variation of projectile velocity in water as compared with the analytical predictions and experimental data are shown in Figure 6 (a). It can be seen that, for water-entry supercavitation at initial velocity $V_0 = 142.7$ m/s, the velocity attenuation obtained from the numerical simulation is comparable to that predicted using the analytical model and experimental data. It can demonstrate the accuracy of the numerical method for high velocity water entry with phase transition. Figure 6 (b) gives the projectile velocity attenuation in water and drag-reducing solution. From the numerical simulation results, we can see that the velocity in solution decays slower than in water, which is the same as lower impact velocity. Again, it is proved that the projectile moves faster in drag-reducing solution than in water, no matter the impact velocity is high or not. Compared with Figure 3, it can be seen that the velocity attenuates faster at high velocity than at low velocity. Meanwhile, when t=4 ms, the velocity difference (ΔV) between water and solution at $V_0 = 28.4$ m/s is 2.7 m/s, while at $V_0 = 142.7$ m/s, $\Delta V = 3.3$ m/s. So it can be conclude that the influence of drag-reducing solution is more obvious at high impact velocity.

The numerical simulation results compared to the predicted penetration distance based on the analytical formulation and the experiment data are shown in Figure 7. It is seen that the projectile in solution moves in longer distance than in water at the same instant. From the results mentioned above, it is clear that the numerical simulation approach developed in the present study allows one to investigate the characteristics of water-entry induced supercavitation considering phase transition influenced by turbulent drag-reducing additives.

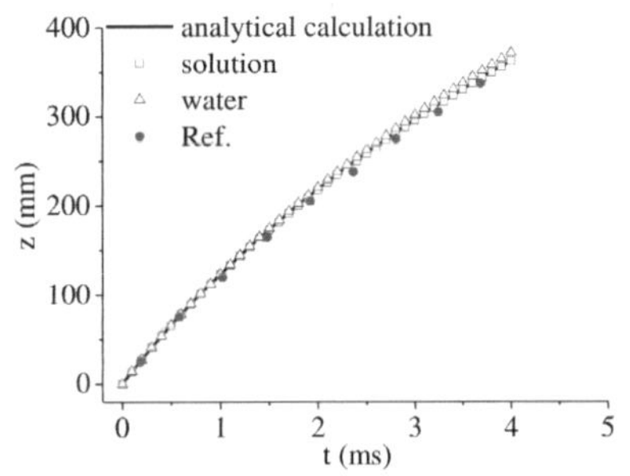

Figure 7. Comparison of experiments, analytical and numerical simulation for penetration distance.

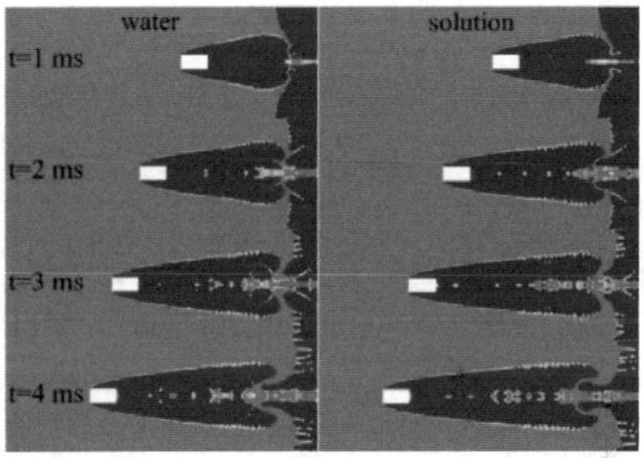

Figure 8. Comparisons of cavity configurations between water and drag-reducing solution.

Figure 8 shows a comparison of the supercavity forming process of the projectile entering water and drag-reducing solution. In consideration of the phase transition from water to vapor, we use the contour of the liquid phase to represent the cavity shape, in other words, the region under water where there is no water represents the cavity. At t=1 ms, the free surface is already closed in water, meanwhile owning to the surface closure a jet forms towards air; nevertheless, the cavity is still open to the air in drag-reducing solution, and the jet is towards the cavity. At t=2 ms, the reentrant jet forms in water and still, the intensity of the jet is weaker than that in solution and the aft part of the cavity is cupped in solution. Paying more attention to the process from t=2 ms to t=3 ms in solution, we'll find that the surface closure is different. The actual process is that the closure is disturbed after t=2 ms, and more air rushes into the cavity and then the cavity closed again, so the cavity becomes larger. From this process analyzed, the cavity does not close at one time, but first closes and then is disturbed and closes again. The closure process is periodic.

Figure 9 shows the components inside the cavity in water and in drag-reducing solution. It can be seen that the cavity is composed of vapor and air. The vapor is mainly distributed in the aft part and near the cavity boundary, and some other is behind the projectile. When the projectile impacts on the free surface, the air first rushes into the cavity, so most part of the cavity is immediately filled with air. With the movement of the projectile, the pressure inside the cavity decreases, due to the existence of the air, the vapor cannot form in the center of the cavity. As the result, the vapor transformed from water first generates in the cavity boundary in the aft part and a vapor film is formed in the front part of the cavity. And then the vapor region becomes larger. After the formation of the reentrant jet into the cavity, the low pressure transforms the liquid jet to the vapor; this is where the vapor behind the projectile comes from. Comparing the vapor region in water and in solution, it can be seen that the vapor region in solution is larger than that in water, which indicates that the drag-reducing solution enhances the formation of vapor.

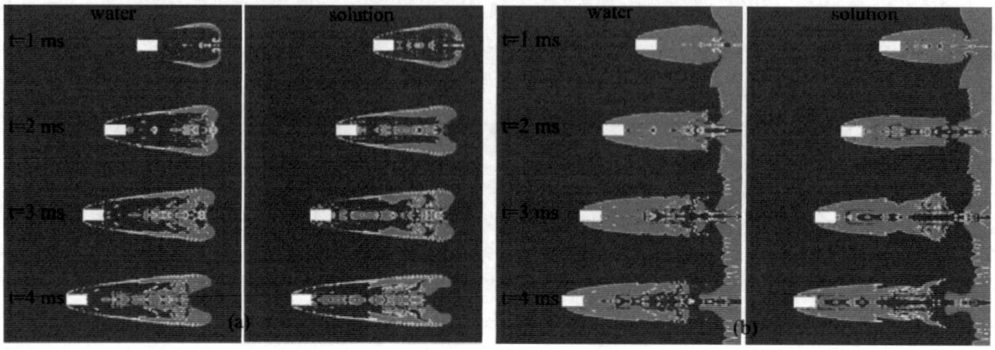

Figure 9. The components inside the cavity in water and in drag-reducing solution. (a) vapor component, (b) air component.

4. Conclusions

This paper introduced a numerical simulation procedure to simulate water-entry supercavitating flow influenced by drag-reducing solution at typically low velocity and high velocity. The configuration and dynamic characteristics of water entry supercavity, flow resistance were discussed respectively. It was obtained that the numerical simulation results are in consistence with experimental data; the numerical procedure proposed is believable. Detailed conclusions are as follows:

1) The projectile moves faster in solution than in water.

2) At higher impact velocity, the vapor is mainly distributed in the aft part and near the cavity boundary, and some other is behind the projectile. The drag-reducing solution enhances the formation of vapor. The actual cavity closure process is that the cavity does not close at one time, but first closes and then is disturbed and closes again. The closure process is periodic.

3) The velocity attenuates faster at high velocity than at low velocity. And it can be conclude that the influence of drag-reducing solution is more obvious at high impact velocity.

Acknowledgments

This work is supported by Foundation for Innovative Research Groups of the National Natural Science Foundation of China (51121004), National Natural Science Foundation of China (51276046), Specialized Research Fund for the Doctoral Program of Higher Education of China (No. 20112302110020).

References

[1] Worthington A M and Cole R S 1897 *Philos. Trans. R. Soc. London, Ser. A* **189** 137-48
[2] Worthington A M and Cole R S 1900 *Philos. Trans. R. Soc. London, Ser. A* **194** 175-99
[3] Panahi R 2012 *J. Theor. Appl. Mech.* **42** (2) 79-92
[4] Kim Y W, Kim Y, Liu Y M and Yue D 2007 *Proc. of 9th Int. Conf. on Numerical Ship Hydrodynamics (Michigan, USA, 5-8 Aug 2007)*
[5] Walhorn E, Kölke A, Hübner B and Dinkler D 2005 *Comput. Struct.* **83** 2100-11
[6] Yang Q and Qiu W 2012 *J. Offshore. Mech. Arct. Eng.* **134** 041101-1
[7] Jiang C-X and Li F-C 2014 *Adv. Mech. Eng.* **2014** 280643

IOP Conf. Series: Materials Science and Engineering **72** (2015) 062056 doi:10.1088/1757-899X/72/6/062056

[8] Schnerr G H and Sauer J 2001 *4th Int. Conf. on Multiphase Flow* New Orleans USA
[9] Bird R B, Curtiss C F, Armstrong R C and Hassager O 1987 *Dynamics of Polymeric Liquids, Fluid Mechanics* Volume 1 (John Wiley & Sons)
[10] Li F C, Dong Y, Kawaguchi Y and Oshima M 2008 *Exp. Therm. Fluid. Sci.* **33** 161-8
[11] Zhang W, Guo Z T, Xiao X K and Wang C 2011 *Explo. Shock. Wav.* **31**(6) 579-84

IOP Conf. Series: Materials Science and Engineering **72** (2015) 062057 doi:10.1088/1757-899X/72/6/062057

Numerical investigation on added mass and damping force coefficient of an underwater vehicle in cavitating flows

Z Y Wang, G Y Wang, C L Hu and Z Y Cui
Beijing Institute of Technology, Beijing 100081, China

Email:wangzhiying0965@126.com

Abstract. The objective of this paper is mainly to study the influence of cavitation on the added mass and damping force coefficient. Based on Reynolds averaged Navier-Stokes equations, the dynamic mesh is used to calculate the added mass, and the rotating coordinate frame method is applied to research on the damping force coefficient. In order to obtain fluid damping force coefficients, the movement pattern is set as a uniform circular motion. Then the additional force coefficient and pitch damping moment coefficient could be obtained using the method of least squares. The result shows that the method to calculate added mass is reliable by comparing with the analytical solution. With the cavitation number decreasing, the absolute value of the added mass of λ_{22} decreases and λ_{26} increases. What's more, both the absolute value of damping force and moment coefficient decrease substantially with the development of cavity when the cavitation number is larger than 0.45. However, with the cavitation number less than 0.45, the un-symmetric cavity is more prominent, the absolute value of damping force and moment coefficient increase slightly. This is probably caused by the strengthened pressure peak at the suction side induced by the re-entrant flow.

1. Introduction

Cavitation occurs in liquid flows if locally drops below the vapor pressure [1].This can be observed in hydraulic systems at high speed. In order to forecast the dynamic characteristic, it is important to research on the hydrodynamic force of a cavity underwater vehicle. The hydrodynamic force[2] of underwater vehicle design is usually classified into three parts, which are location force, inertial force and damping force. The damping force, such as pitch damping force and momentum, effects the trajectory stability and maneuverability dramatically[3][4].Therefore, the calculation of damping force is relatively very important during hydrodynamic design.

As the development of computational technique, the inertial force and damping force calculation using CFD method based on Reynolds averaged Navier-Stokes equations has been greatly improved. The dynamic mesh method[5][6][7] is used to calculate the added mass. A method to determine the added mass of running vehicle with cavitation is proposed based on calculating hydrodynamic forces of the vehicle swaying of rolling in numerical tunnel[8]. In hydrodynamics calculation, the unsteady method based on dynamic mesh has been applied to get the fluid damping force and moment coefficients[9][10].

In this paper, the dynamic mesh is used to research on the added mass and the rotating coordinate frame method has been applied to calculate damping force coefficient of an underwater vehicle in the cavitating flows.

2. Numerical Method

The commercial software CFD is used for modeling the hydrodynamics of a cavity underwater vehicle. The Navier-Stokes equations in their conservative form governing a Newtonian fluid without body forces and heat transfers in the Cartesian coordinates are presented below:

$$\frac{\partial \rho_{\mathrm{m}}}{\partial t}+\frac{\partial(\rho_{\mathrm{m}} u_j)}{\partial x_j}=0, \tag{1}$$

$$\frac{\partial(\rho_{\mathrm{m}} u_i)}{\partial t}+\frac{\partial(\rho_m u_i u_j)}{\partial x_j}=-\frac{\partial p}{\partial x_i}+\frac{\partial}{\partial x_j}\left[(\mu_{\mathrm{m}}+\mu_{\mathrm{T}})\left(\frac{\partial u_i}{\partial x_j}+\frac{\partial u_j}{\partial x_i}-\frac{2}{3}\frac{\partial u_k}{\partial x_k}\delta_{ij}\right)\right], \tag{2}$$

$$\frac{\partial \rho_1 \alpha_1}{\partial t}+\frac{\partial(\rho_1 \alpha_1 u_j)}{\partial x_j}=\dot{m}^{+}+\dot{m}^{-}, \tag{3}$$

$$\rho_{\mathrm{m}}=\rho_1 \alpha_1+\rho_{\mathrm{v}} \alpha_{\mathrm{v}}, \tag{4}$$

$$\mu_{\mathrm{m}}=\mu_1 \alpha_1+\mu_{\mathrm{v}} \alpha_{\mathrm{v}}, \tag{5}$$

where ρ_m is the mixture density, u is the velocity, p is the pressure, μ_m is the mixture laminar viscosity, μ_l and μ_t are respectively the liquid and vapor dynamic viscosities, and μ_t is the turbulent viscosity. The subscripts i, j, k are the axes directions. The source term $\dot{m}^{+}$ and the sink term $\dot{m}^{-}$ which is closured by the Kubota model [11] in this study represent the condensation and evaporation rates, respectively. Moreover, the standard k-ε turbulence model[12] is used here.

3. Numerical setup and description

During the numerical damping force calculation procedure, the added mass is ineluctably included and can not be separated effectively. In order to obtain damping force, added mass has to be calculated firstly to eliminate its effect. Thus the added mass calculation method based on dynamic mesh is introduced, and then the rotating coordinate frame method to calculate fluid damping force is presented.

3.1. Add mass calculation method

In this study, the underwater vehicle is simplified as a cylinder with a cone forehead.. The cone radius R=0.04m,length L=0.532m.Figure1 shows the computational domain and boundary condition.The mesh is shown in Figure2 . A no-slip boundary condition is imposed on the vehicle surface, and

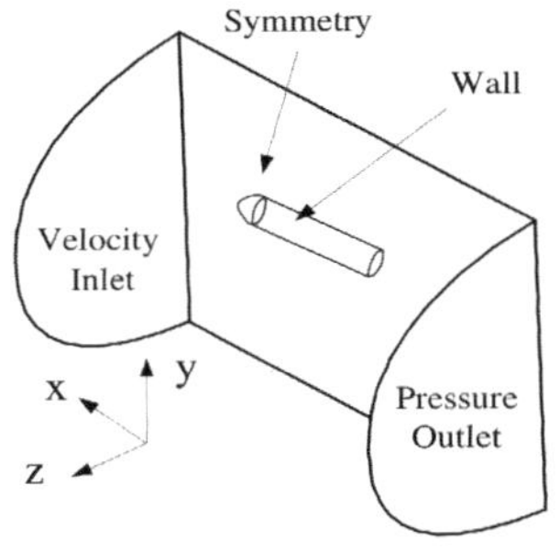

Figure 1. Computational domain and boundary conditions.

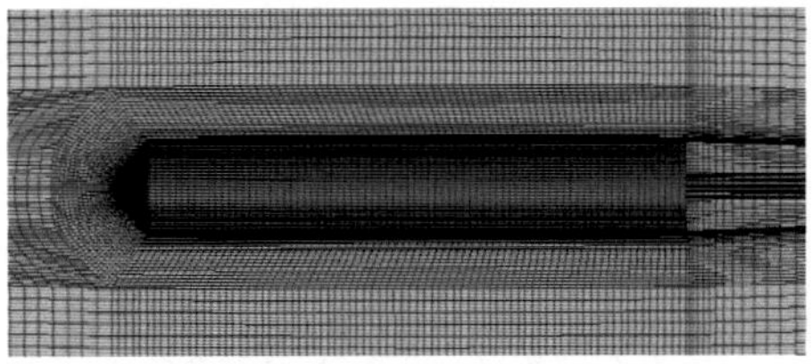

Figure 2. Mesh distributions

opening conditions are imposed on the side boundaries of the tunnel. The inlet velocity is set to be

V=4 m/s and the outlet pressure is set to vary according to the cavitation number, defined as $\sigma = (p_\infty - p_v)/(0.5\rho_l V^2)$ where p_∞ is the tunnel pressure.

The added mass is known to playing a significant role in defining the hydrodynamic forces acting on moving bodies due to the fact that the movement of the surrounding fluid requires an additional force over and above to accelerate the body itself. So the added mass can be obtained by transient motion with the dynamic mesh method. The movement pattern is shown in Figure3.

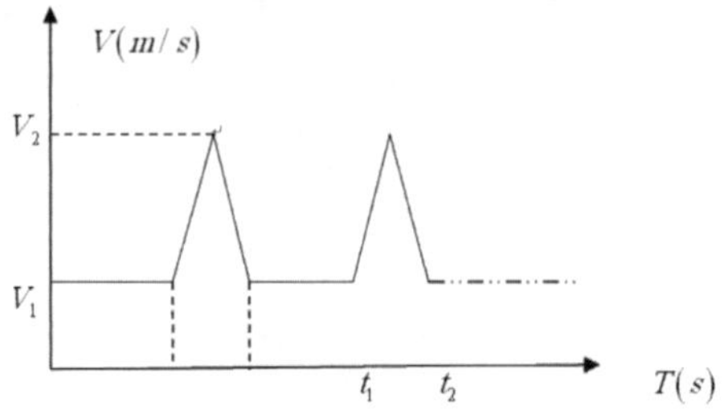

Figure 3. Curve of velocity

The added mass λ_{11},λ_{22}, λ_{26} can be obtain by the following equation.

$$\lambda_{11} = \frac{\Delta F_x}{a_x} = \frac{1}{2}\rho V^2 S\Delta Cd / a_x \tag{6}$$

$$\lambda_{22} = \frac{\Delta F_y}{a_y} = \frac{1}{2}\rho V^2 S\Delta Cl / a_y \tag{7}$$

$$\lambda_{26} = \frac{\Delta M_z}{a_y} = \frac{1}{2}\rho V^2 SL\Delta Cl / a_y \tag{8}$$

where ρ is the fluid density, S is the maximum cross-sectional area, a_i is the acceleration, Cd is the drag coefficient, Cl is the lift coefficient.

3.2. Fluid damping force calculation method

In order to obtain fluid damping force coefficients. The movement pattern is circumferential motion, and the rotating coordinate frame method is adopted to conduct steady numerical simulation.

Make the vehicle rotate uniform around the specified center, the distance between vehicle centroid and circular motion center is R, the rotating speed is defined as ω_{z1}, and the velocity at vehicle centroid is $V = R\omega_{z1}$. the hydrodynamic force is defined as follows.

$$qSC_{y1}(\alpha) + qSC_{y1}(\omega_{z1}) \approx qSC_{y1}^{\alpha}\alpha + qSC_{y1}^{\bar{\omega}_{z1}}\bar{\omega}_{z1} = N + \lambda_{11}V\omega_{z1}\cos\alpha \tag{9}$$

$$\begin{aligned} &qSlm_{z1}(\alpha) + qSlm_{z1}(\omega_{z1}) \approx qSlC_{m_{z1}}^{\alpha}\alpha + qSlC_{m_{z1}}^{\bar{\omega}_{z1}}\bar{\omega}_{z1} \\ &= M + \lambda_{26}V\omega_{z1}\cos\alpha + (\lambda_{22} - \lambda_{11})V^2\sin\alpha\cos\alpha \end{aligned} \tag{10}$$

The normal force N and pitch moment M could be obtained based on the numerical calculation of the vehicle with circumferential motion .Then make sure that the attack angle and the velocity is constant, just change the angular velocity. Then the damping force coefficient $C_{y1}^{\bar{\omega}_{z1}}$ and pitch damping moment coefficient $C_{m_{z1}}^{\bar{\omega}_{z1}}$ could be calculated by least squares method.

4. Results and discussion

IOP Conf. Series: Materials Science and Engineering **72** (2015) 062057 doi:10.1088/1757-899X/72/6/062057

4.1. Numerical result verification

4.1.1. Added mass calculation. To verify the calculation method, the calculation for the vehicle with the defined movement in the condition of without cavitation is conducted. The time evolution of the lift coefficient is shown in Figure 4, where the acceleration is a_y=40m/s^2.

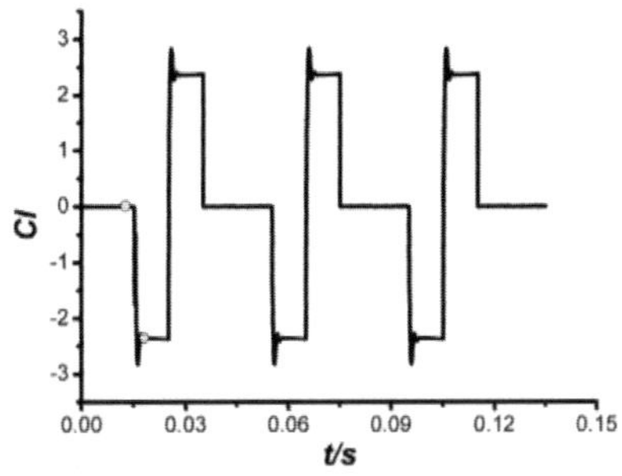

Figure 4. Curve of the lift coefficient

From the Eq(7) and (8), the added mass λ_{22} and λ_{26} could be obtained.And the λ_{11} can be obtained by the same method. Comparisons of the measured and predicted value are listed in Table 1. The numerical results showed a good agreement with the measured values.

Table 1. Comparisons of the calculation and predicted value

	λ_{22}	λ_{26}
predicted value	2.855227	-0.02105
calculation value	2.74857	-0.02095
error	3.74%	0.475%

4.1.2. Added mass in cavitation flow The reliability of the added mass calculation method is validated above. In this section, added mass in the different cavitation numbers will be calculated.

In order to present the variation of the added mass in different cavitation number, the corresponding graphs are shown in Figure 5.The trend of the curves show that the value of λ_{22}、λ_{26} decreases when the cavitation number decreases.

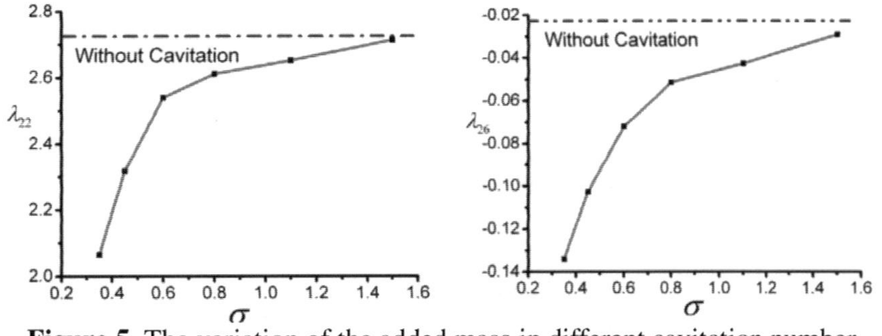

Figure 5. The variation of the added mass in different cavitation number

To further analyze the influence of cavitation on added mass, Figure 6 shows the pressure contours

IOP Conf. Series: Materials Science and Engineering **72** (2015) 062057 doi:10.1088/1757-899X/72/6/062057

of different cavitation numbers at two typical time (corresponding to the two moments in Figure 4). The cavity length increases with the cavitation number decreasing. From the pressure distributions, it is found that the difference between the up side and the down side in the longitudinal plane lies in the area without cavitation., as the saturated pressure is in the cavity. So the value of λ_{22} decreases when the cavitation number decreases. Figure 7 shows the corresponding pressure distributions on the both surfaces. In figure 7, the center of pressure moves toward the tail of the vehicle, when the cavitation number decreases. That causes the arm of force increases. So the absolute value of λ_{26} increases with the decrease of cavitation number.

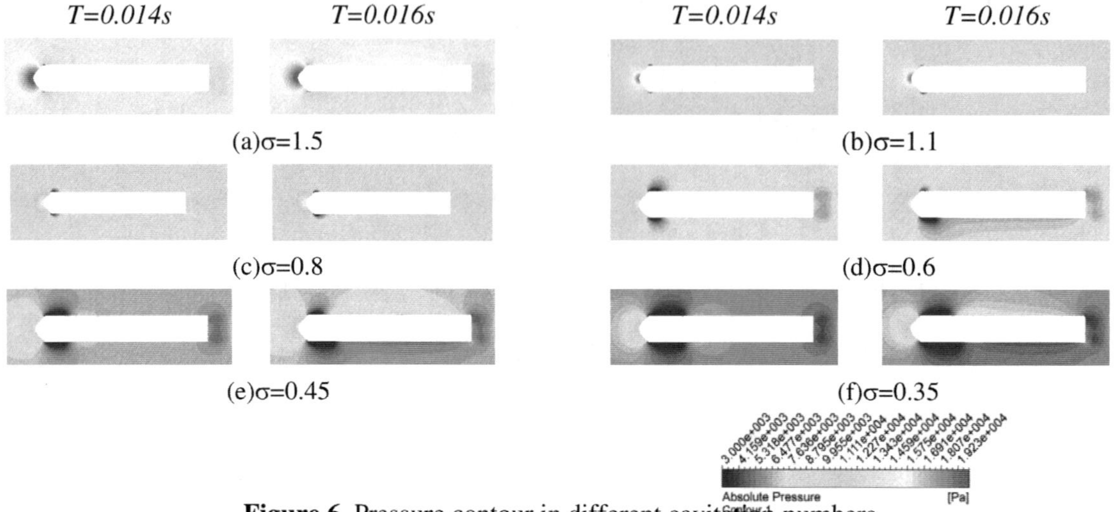

Figure 6. Pressure contour in different cavitation numbers

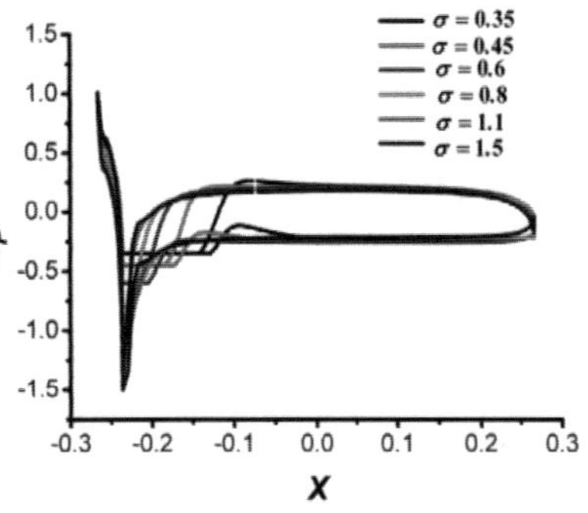

Figure 7. Curves of pressure coefficient in different cavitation numbers at the same time

4.2. The effect of cavitation on damping force coefficient

In the computation, the natural cavity number is decreased by the reduction of the operation pressure. Figure8 shows the cavity shapes in different cavitation regimes.

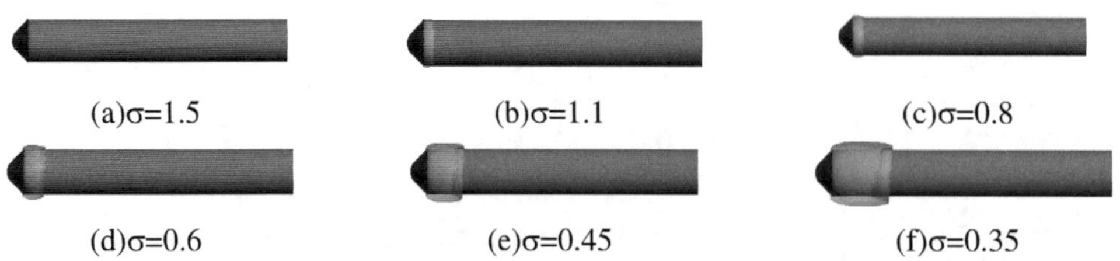

Figure 8. Cavity shape with different cavitation number

In order to obtain fluid damping force coefficients, the defined movement in the condition of without cavitation is conducted In this research, the attack angle α=0°.The corresponding normal force N and pitch moment M can be obtained at different angular velocity. From the Eq.(9) and(10),the coefficients can be calculation. Then the additional force coefficient and pitch damping moment coefficient could be obtained by using least squares method.

Figure 9 and 10 present the damping force and moment coefficient at different cavitation numbers, respectively. When the cavitation number decreases, the cavity coverage area increases, the increment of total normal force decreases. So the damping force coefficient decreases. While the size of cavity grows to a certain stage with the decrease of cavitation number, the cavity dissymmetry due to rotating motion becomes obvious. The cavity dissymmetry will cause larger normal force at closure zone. It could also be observed that the cavity dissymmetry become more obvious with the increment of angular velocity, and the corresponding normal force become larger and increases no-linear. When the cavitation number less than 0.45, the damping force coefficient increases with the decrement of cavitation number.

To be similar, as the cavitation number decreases, the increment of pitch moment decreases. With the decrease of cavitation number, the absolute value on pitch damping moment coefficient increases. When the cavitation number is less than 0.45, the influence of cavity dissymmetry is more obvious. With the increment of cavitation number, the absolute value of pitch damping moment coefficient become larger.

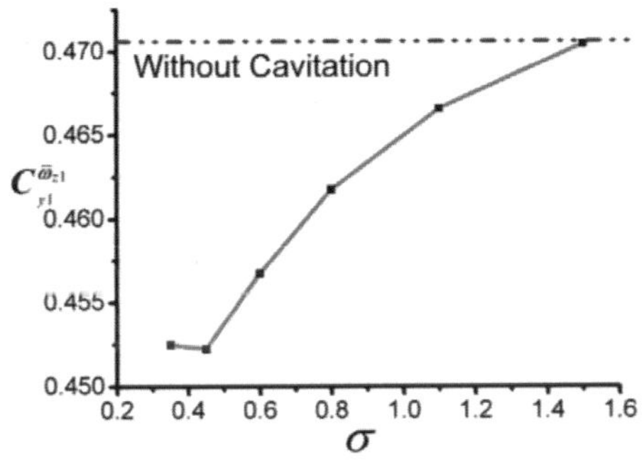

Figure 9. Fluid damping force with different cavitation number

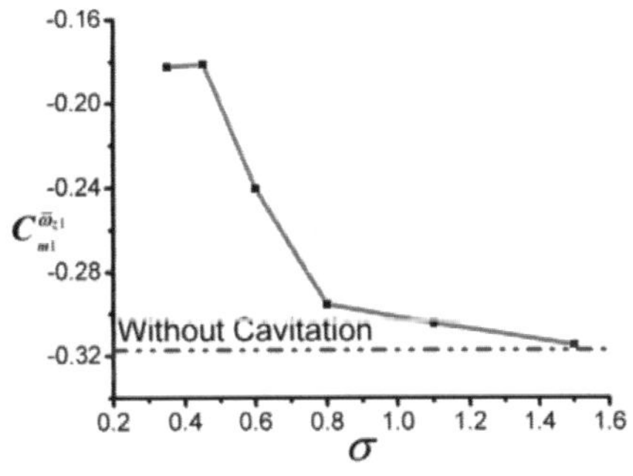

Figure 10. Pitch damping moment with different cavitation number

5. Conclusions

In this study, numerical simulation by CFD are applied to investigate the added mass and damping force coefficient of an underwater vehicle in cavitation flow. The main conclusions are below.

1) The numerical method of transient motion with the dynamic mesh method is used to research on the added mass of a vehicle with different cavitation number. The result shows that the method to calculate added mass is reliable by comparing with the analytical solution. With the cavitation number decreasing, the absolute value of the added mass of λ_{22} decreases and λ_{26} increases.

2) The numerical method of rotating coordinate frame method is used to realize the circumferential motion to research on damping force coefficient. It is shown that both the absolute value of damping force and moment coefficient decrease substantially with the development of cavity when the cavitation number is larger than 0.45. However, with the cavitation number of less than 0.45, the cavity dissymmetry is more obvious, the absolute value of damping force and moment coefficient increase slightly. This is probably caused by the strengthened pressure peak at the suction side which is induced by the re-entrant flow.

Acknowledgements

This work was supported by the National Natural Science Foundation of China (Grant Nos. 51306020) and Beijing Municipal Natural Science Foundation (Grant No. 3144043)

References

[1] Caupin F, Herbert E 2006 *Comptes Rendus Physique* **7** 1000-17
[2] Zhang Y W 1998 *Principle and Methods of Torpedo system Design* (Xi'an: Northwestern Polytechnical University Press) p 73
[3] Sen D 2000 *J. of Ship Research* **44** 186-96
[4] Jiang S J, Liu Y Q and Dang M L 2008 *J. of projectiles rockets* **28** 180-82
[5] Fu H P and Li J 2001 *J. of Harbin Engineering University* **32** 148-52
[6] Li J and Lu C J 2009 *The Chinese of Theoretical and Applied Mechanics.*(Zhengzhou)
[7] Ma Y and Shan X X 2007 *Computer Simulation* **24** 395-98
[8] H X and Lu C J 2009 *Chinese J. of Hydrodynamics* **24** 800-06
[9] He Z, Y R X and Hao P C 2010 *J. Marine. Sci. Appl* **9** 149-55
[10] Tyagi A and Sen D 2006 *Ocean Eng.* **33** 798-809
[11] Brennen C E 1995 *Cavitation and Bubble Dynamics.*(England:Oxford University Press)
[12] Launder B E and Spalding D B 1974 *Comput Methods Appt Mech Eng* **3** 269-89